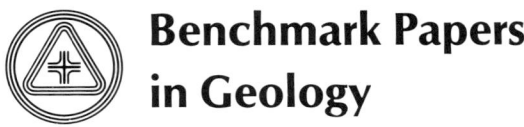
Benchmark Papers in Geology

Series Editor: Rhodes W. Fairbridge
Columbia University

A selection from the published volumes in this series

Volume
- 2 RIVER MORPHOLOGY / *Stanley A. Schumm*
- 16 GEOCHEMISTRY OF WATER / *Yasushi Kitano*
- 25 ENVIRONMENTAL GEOLOGY / *Frederick Betz, Jr.*
- 37 STATISTICAL ANALYSIS IN GEOLOGY / *John M. Cubitt and Stephen Henley*
- 39 BEACH PROCESSES AND COASTAL HYDRODYNAMICS / *John S. Fisher and Robert Dolan*
- 59 KARST GEOMORPHOLOGY / *M. M. Sweeting*
- 72 PHYSICAL HYDROGEOLOGY / *R. Allan Freeze and William Back*

A complete listing of volumes published in this series begins on p. 414.

**Benchmark Papers
in Geology / 73**

A BENCHMARK® Books Series

CHEMICAL HYDROGEOLOGY

Edited by
WILLIAM BACK
U.S. Geological Survey

and

R. ALLAN FREEZE
University of British Columbia

Hutchinson Ross Publishing Company

Stroudsburg, Pennsylvania

Copyright ©1983 by **Hutchinson Ross Publishing Company**
Benchmark Papers in Geology, Volume 73
Library of Congress Catalog Card Number: 82-11853
ISBN: 0-87933-440-1

All rights reserved. No part of this book covered by the copyrights hereon may be reproduced or transmitted in any form or by any means—graphic, electronic, or mechanical, including photocopying, recording, taping, or information storage and retrieval systems—without written permission of the publisher.

85 84 83   1 2 3 4 5
Manufactured in the United States of America.

LIBRARY OF CONGRESS CATALOGING IN PUBLICATION DATA
Main entry under title:
Chemical hydrogeology.
  (Benchmark papers in geology; 73)
  Includes bibliographical references and index.
  1. Hydrogeology—Addresses, essays, lectures.
2. Geochemistry—Addresses, essays, lectures. I. Back, William, 1925-   . II. Freeze, R. Allan. III. Series.
GB1004.C47 1983   551.49   82-11853
ISBN 0-87933-440-1

Distributed worldwide by Van Nostrand Reinhold Company Inc., 135 W. 50th Street, New York, NY 10020.

DISCARDED

LAMAR UNIVERSITY-BEAUMONT
0 00 01 0289975 4

501183

MARY AND JOHN GRAY
LIBRARY
LAMAR UNIVERSITY
BEAUMONT, TEXAS

CONTENTS

Series Editors' Foreword ... ix
Preface ... xi
Contents by Author ... xiii

Introduction ... 1

PART I: THE EVOLUTIONARY PERIOD

Editors' Comments on Papers 1 Through 9 ... 8

1. **PALMER, C.:** The Geochemical Interpretation of Water Analyses ... 12
   U.S. Geol. Survey Bull. **479:**5-9, 11-12, 31 (1911)

2. **ROGERS, G. S.:** Chemical Relations of the Oil-Field Waters in San Joaquin Valley, California ... 18
   U.S. Geol. Survey Bull. **653:**93-99 (1917)

3. **COLLINS, W. D.:** Graphic Representation of Water Analyses ... 25
   Indus. and Eng. Chemistry **15:**394 (1923)

4. **RENICK, B. C.:** Base Exchange in Ground Water by Silicates as Illustrated in Montana ... 26
   U.S. Geol. Survey Water-Supply Paper **520-D:**63-72 (1924)

5. **LANGELIER, W. F.:** The Analytical Control of Anti-Corrosion Water Treatment ... 36
   Am. Water Works Assoc. Jour. **28:**1500-1509, 1511-1514, 1520-1521 (1936)

6. **PIPER, A. M.:** A Graphic Procedure in the Geochemical Interpretation of Water-Analyses ... 50
   Am. Geophys. Union Trans. **25:**914-923 (1944)

7. **STIFF, H. A., Jr.:** The Interpretation of Chemical Water Analysis by Means of Patterns ... 60
   Jour. Petroleum Technology **3,** No. 10, Section 1, 15-16; Section 2, 3 (1951)

8. **FOSTER, M. D.:** The Origin of High Sodium Bicarbonate Waters in the Atlantic and Gulf Coastal Plains ... 63
   Geochim. et Cosmochim. Acta **1:**33-48 (1950)

9. **BACK, W.:** Origin of Hydrochemical Facies of Ground Water in the Atlantic Coastal Plain ... 79
   Internat. Geol. Cong., 21st, Copenhagen, 1960, Rept., Part I, pp. 87-95

PART II: OCCURRENCE AND GEOCHEMICAL SIGNIFICANCE OF SALTWATER

Editors' Comments on Papers 10 Through 16 ... 90

Contents

10 CARLSTON, C. W.: An Early American Statement of the Badon Ghyben-Herzberg Principle of Static Fresh-Water-Salt-Water Balance   98
Am. Jour. Sci. **261**:88-91 (1963)

11 KOHOUT, F. A.: Cyclic Flow of Salt Water in the Biscayne Aquifer of Southeastern Florida   102
Jour. Geophys. Research **65**:2133-2141 (1960)

12 BERRY, F. A. F., and B. B. HANSHAW: Geologic Field Evidence Suggesting Membrane Properties of Shales   111
Internat. Geol. Cong., 21st, Copenhagen, 1960, Abstracts, p. 209

13 WHITE, D. E.: Saline Waters of Sedimentary Rocks   112
Am. Assoc. Petroleum Geologists Mem. **4**:342-354, 363-366 (1965)

14 RUNNELLS, D. D.: Diagenesis, Chemical Sediments, and the Mixing of Natural Waters   129
Jour. Sed. Petrology **39**:1188-1201 (1969)

15 PLUMMER, L. N., H. L. VACHER, F. T. MACKENZIE, O. P. BRICKER, and L. S. LAND: Hydrogeochemistry of Bermuda: A Case History of Ground-Water Diagenesis of Biocalcarenites   143
Geol. Soc. America **87**:1301-1316 (1976)

16 HANOR, J. S.: Precipitation of Beachrock Cements: Mixing of Marine and Meteoric Waters vs. CO_2-Degassing   159
Jour. Sed. Petrology **48**:489-501 (1978)

PART III: THE EQUILIBRIUM APPROACH

Editors' Comments on Papers 17 Through 22   174

17 HEM, J. D.: Equilibrium Chemistry of Iron in Ground Water   179
Principles and Applications of Water Chemistry, S. D. Faust and J. V. Hunter, eds., John Wiley & Sons, New York, 1967, pp. 625-637, 642-643

18 HANSHAW, B. B., W. BACK, and M. RUBIN: Carbonate Equilibria and Radiocarbon Distribution Related to Groundwater Flow in the Floridan Limestone Aquifer, U.S.A.   193
Hydrology of Fractured Rocks, vol. 1, Proceedings of the Dubrovnik Symposium, 1965, International Association of Scientific Hydrology, 1965, pp. 601-614

19 THRAILKILL, J.: Chemical and Hydrologic Factors in the Excavation of Limestone Caves   207
Geol. Soc. America Bull. **79**:19-45 (1968)

20 LANGMUIR, D.: The Geochemistry of Some Carbonate Ground Waters in Central Pennsylvania   234
Geochim. et Cosmochim. Acta **35**:1023-1045 (1971)

21 BRICKER, O. P. and R. M. GARRELS: Mineralogic Factors in Natural Water Equilibria   257
Principles and Applications of Water Chemistry, S. D. Faust and J. V. Hunter, eds., John Wiley & Sons, New York, 1967, pp. 449-468

22 TRUESDELL, A. H. and B. F. JONES: WATEQ, A Computer Program for Calculating Chemical Equilibria of Natural Waters   277
U.S. Geol. Survey Jour. Research **2**:233-248 (1974)

PART IV: ISOTOPES IN GROUNDWATER

Editors' Comments on Papers 23, 24, and 25 294

23 PEARSON, F. J., Jr. and D. E. WHITE: Carbon 14 Ages and Flow Rates of Water in Carrizo Sand, Atascosa County, Texas
Water Resources Research **3**:251-261 (1967) 298

24 HITCHON, B. and I. FRIEDMAN: Geochemistry and Origin of Formation Waters in the Western Canada Sedimentary Basin—I. Stable Isotopes of Hydrogen and Oxygen
Geochim. et Cosmochim. Acta **33**:1321-1330, 1339-1343, 1346-1349 (1969) 309

25 DAVIS, G. H., C. K. LEE, E. BRADLEY, and B. R. PAYNE: Geohydrologic Interpretations of a Volcanic Island from Environmental Isotopes
Water Resources Research **6**:99-109 (1970) 328

PART V: HEAT AND MASS TRANSPORT

Editors' Comments on Papers 26 Through 29 340

26 CARTWRIGHT, K.: Tracing Shallow Groundwater Systems by Soil Temperatures
Water Resources Research **10**:847-855 (1974) 344

27 BREDEHOEFT, J. D. and G. F. PINDER: Mass Transport in Flowing Groundwater
Water Resources Research **9**:194-210 (1973) 353

28 SCHWARTZ, F. W. and P. A. DOMENICO: Simulation of Hydrochemical Patterns in Regional Groundwater Flow
Water Resources Research **9**:707-720 (1973) 370

29 ANDERSON, M. P.: Using Models to Simulate the Movement of Contaminants Through Groundwater Flow Systems
CRC Critical Rev. Environmental Control **9**:97-98, 139-146, 148-156 (1979) 384

Author Citation Index 397
Subject Index 407
About the Editors 413

SERIES EDITOR'S FOREWORD

The philosophy behind the Benchmark Papers in Geology is one of collection, sifting, and rediffusion. Scientific literature today is so vast, so dispersed, and, in the case of old papers, so inaccessible for readers not in the immediate neighborhood of major libraries that much valuable information has been ignored by default. It has become just so difficult, or so time consuming, to search out the key papers in any basic area of research that one can hardly blame a busy person for skimping on some of his or her "homework."

This series of volumes has been devised, therefore, as a practical solution to this critical problem. The geologist, perhaps even more than any other scientist, often suffers from twin difficulties—isolation from central library resources and immensely diffused sources of material. New colleges and industrial libraries simply cannot afford to purchase complete runs of all the world's earth science literature. Specialists simply cannot locate reprints or copies of all their principal reference materials. So it is that we are now making a concerted effort to gather into single volumes the critical materials needed to reconstruct the background of any and every major topic of our discipline.

We are interpreting "geology" in its broadest sense: the fundamental science of the planet Earth, its materials, its history, and its dynamics. Because of training in "earthy" materials, we also take in astrogeology, the corresponding aspect of the planetary sciences. Besides the classical core disciplines such as mineralogy, petrology, structure, geomorphology, paleontology, and stratigraphy, we embrace the newer fields of geophysics and geochemistry, applied also to oceanography, geochronology, and paleoecology. We recognize the work of the mining geologists, the petroleum geologists, the hydrologists, and the engineering and environmental geologists. Each specialist needs a working library. We are endeavoring to make the task of compiling such a library a little easier.

Each volume in the series contains an introduction prepared by a specialist (the volume editor)—a "state of the art" opening or a summary of the object and content of the volume. The articles, usually some twenty to fifty reproduced either in their entirety or in significant extracts, are selected in an attempt to cover the field, from the key papers of the last century to fairly recent work. Where the original works are in foreign languages, we

Series Editor's Foreword

have endeavored to locate or commission translations. Geologists, because of their global subject, are often acutely aware of the oneness of our world. The selections cannot therefore be restricted to any one country, and whenever possible an attempt is made to scan the world literature.

To each article, or group of kindred articles, some sort of "highlight commentary" is usually supplied by the volume editor. This commentary should serve to bring that article into historical perspective and to emphasize its particular role in the growth of the field. References, or citations, wherever possible, will be reproduced in their entirety—for by this means the observant reader can assess the background material available to that particular author, or, if desired, he or she too can double check the earlier sources.

A "benchmark," in surveyor's terminology, is an established point on the ground that is recorded on our maps. It is usually anything that is a vantage point, from a modest hill to a mountain peak. From the historical viewpoint, these benchmarks are the bricks of our scientific edifice.

RHODES W. FAIRBRIDGE

PREFACE

Chemical Hydrogeology and a companion volume on physical hydrogeology bring together much of the classic literature of groundwater and hydrogeology. We hope that these books provide a systematic guide to the literature for use in graduate courses and a historical perspective that will allow hydrogeologists to appreciate the heritage of our scientific roots. As scientists we must constantly look toward the future and explore new pathways; but even as we wander through unchartered wilderness we must often look over our shoulder to find the path back to the main stream, in order to achieve our goals of complete understanding. In addition, we hope that these classic papers will provide an introduction to hydrogeology for scientists in other parts of the multi-faceted discipline called geology.

Geochemistry represents the application of another major concept to hydrogeology similar in importance to the application of potential theory, flow-net analysis, analytical solutions, or numerical simulation. However, chemical hydrogeology has an ambivalent role in that it is a part of hydrogeology and yet certain aspects or manifestations provide it with a function of its own. It reaches out into divergent aspects of earth sciences that force us to include these diverse topics in the general field of hydrogeology. This expansion can be witnessed by noting the wide variety of nontraditional studies now undertaken by groundwater geologists, such as those pertaining to migration of ore-forming solutions, diagenesis of sediments, and geomorphic processes. Future guides to the classic literature will not be complete without inclusion of these ancillary activities.

In the Introduction, we outline our selection process and explain why we have chosen to emphasize certain aspects of the broad field of chemical hydrogeology and not to provide examples of many correlative topics.

Although the selection of papers in this volume is based on a seminar, "Geochemistry of Ground Water," taught at George Washington University, we solicited and received suggestions from many of our colleagues, particularly Mary Jo Baedecker, Ivan Barnes, John Cherry, Bruce Hanshaw, Brian Hitchon, Don Langmuir, Joe Poland, John Thrailkill, and Warren Wood. This book was greatly improved through the thoughtful review, comments, and suggested revisions of the manuscript given by Hilton Cooper, George Davis, Bruce Hanshaw, and Warren Wood. We are also most appreciative of the library

Preface

research and editorial assistance given by Laura Toran, U.S. Geological Survey. Joanne Taylor's cheerful cooperation in retyping the several versions has long made the preparation of this and other manuscripts a most pleasant task.

WILLIAM BACK
R. ALLAN FREEZE

CONTENTS BY AUTHOR

Anderson, M. P., 384
Back, W., 79, 193
Berry, F. A. F., 111
Bradley, E., 328
Bredehoeft, J. D., 353
Bricker, O. P., 143, 257
Carlston, C. W., 98
Cartwright, K., 344
Collins, W. D., 25
Davis, G. H., 328
Domenico, P. A., 370
Foster, M. D., 63
Friedman, I., 309
Garrels, R. M., 257
Hanor, J. S., 159
Hanshaw, B. B., 111, 193
Hem, J. D., 179
Hitchon, B., 309
Jones, B. F., 277
Kohout, F. A., 102
Land, L. S., 143
Langelier, W. F., 36
Langmuir, D., 234
Lee, C. K., 328
Mackenzie, F. T., 143
Palmer, C., 12
Payne, B. R., 328
Pearson, F. J., Jr., 298
Pinder, G. F., 353
Piper, A. M., 50
Plummer, L. N., 143
Renick, B. C., 26
Rogers, G. S., 18
Rubin, M., 193
Runnells, D. D., 129
Schwartz, F. W., 370
Stiff, H. A., Jr., 60
Thrailkill, J., 207
Truesdell, A. H., 277
Vacher, H. L., 143
White, D. E., 112
White, D. E., 298

Photograph of Mammoth Hot Springs on Gardiner's River in Yellowstone National Park, Wyoming, taken during the W. H. Jackson expedition, 1872. Springs had a significant influence on developing scientific interest in the chemistry of groundwater because of their spectacular mineral deposits, their reputed therapeutic benefits, and the extensive market for bottled spring water.

CHEMICAL HYDROGEOLOGY

INTRODUCTION

We have selected papers that demonstrate the historical development of the science of chemical hydrogeology. Some papers have had a revolutionary impact, some are benchmarks in the evolutionary sequence, and some are representative of substantial knowledge from other disciplines whose application to chemical hydrogeology has significant impact on our understanding. Most of the papers introduce, demonstrate, or develop new concepts. We did not always select papers where the concept was first formulated; often we selected the paper that made the particular concept an integral part of chemical hydrogeology.

The papers are grouped under five headings. The Evolutionary Period, Occurrence and Geochemical Significance of Salt Water, The Equilibrium Approach, Isotopes in Groundwater, and Heat and Mass Transport. Although the organization is by topic, within each topic the papers are listed more or less chronologically. We believe that these papers and discussions together with the regional studies in Part III of *Physical Hydrogeology* (Freeze and Back, 1983) represent the essence of chemical hydrogeology.

One topic of great significance to science and society not covered in sufficient detail is the chemical hydrogeology of contaminated systems. This omission is most unfortunate because chemical hydrogeology has a major role to play in disposal of radioactive and toxic wastes that will stimulate many research and applied investigations in the future. Case histories of contaminated groundwater systems and examples of studies on the application of organic geochemistry, the role of microbiology, and the occurrence and migration of metals and toxic elements are not included.

We have restricted our selection to North America studies despite numerous excellent papers available from the European literature, and we have included some post-1970 papers in this volume. The editors' comments mention other important papers, many of which are classics that could not be included. The Benchmark Papers along with the referenced papers, particularly the review articles, provide a structured entry to the scientific literature of chemical hydrogeology.

HISTORICAL PERSPECTIVE

The chemical aspects of hydrogeology developed quite differently from the physical aspects. Although well-construction techniques were practiced by many early civilizations, great demands on groundwater resources and the need for improved engineering techniques and well-drilling equipment did not develop until after Darcy's experiment in 1856. Therefore, by the time extensive groundwater exploration was necessary, Darcy's law provided the integrating concept and provided the rationale for exploration, well spacing, and pumping regimens.

Demands were significantly different for the chemical aspects in that people needed to know the chemical character of water—springs, lakes, and rivers—for diverse reasons. As each situation developed it was approached empirically with little theory available for extrapolation from previous experience. Some of the early chemical analyses were of water from mineral spas used as health resorts. Eventually, it became imperative that engineers know the chemical character of water available for both stationary and locomotive steam engines. The difficult and complex problems of boiler-scale formation and removal could not be economically resolved by engineering techniques of that time. It was therefore decidedly advantageous to avoid the use of water that would cause scale and corrosion problems in boilers. Later, knowledge of chemistry of water was required to select appropriate sources for municipal, agricultural, and industrial uses. This diversity of applications resulted in the multifaceted study of chemical hydrogeology.

As discussed in Part I the work of early chemists and engineers of the U.S. Geological Survey and the Reclamation Service was to provide descriptions and chemical characterizations of surface water and groundwater, economically justified by determining the fitness of the water for any intended use. The fundamental scientific need was to develop new analytical techniques and revise standard techniques giving the necessary precision for the desired descriptions.

These early studies of chemistry of water became part of hydrogeology when chemical data were incorporated into a hydrologic or geologic analysis of drainage basins such as that of Mendenhall et al. (1916). The coupling of chemical descriptions of water with local and regional flow systems within the geologic framework of aquifers demonstrated the existence of relationships that required explana-

Introduction

tions if the spatial distribution of chemical constituents within groundwater regimes were to be understood.

Investigation of the problem of saltwater encroachment into coastal aquifers stimulated development of both physical and chemical aspects of the science of hydrogeology and afforded early training for many future leading scientists. Whereas in the early years chemists and engineers had responsibility for studying chemistry of natural water, the saltwater problem brought the talents of geologists to bear on this topic. During the late 1920s, 1930s, and 1940s, the U.S. Geological Survey began many extensive and systematic studies of coastal areas. Examples of these works are given in Part II.

The evolutionary phase of chemical hydrogeology ended in the 1950s. Many of the important chemical reactions had been identified and shown to exert major controls on the chemical character of groundwater; such reactions as ion exchange, mineral solution, and sulfate reduction were well understood. Sanitary engineers understood the concept of mineral equilibria in the sense that some waters are corrosive and capable of dissolving minerals and plumbing systems, whereas other waters could precipitate minerals within the pipes. Also, biologists and limnologists had made great progress in understanding oxidation-reduction reactions as an environmental control. Zobell (1946) had written a significant paper discussing redox reactions in marine sediments, but this had not yet been applied to freshwater systems. Pourbaix (1949) had published his classic book on metal corrosion, but its significance was not realized by geologists for many years. Many of the necessary theoretical concepts were available for application to groundwater systems by the early 1950s. Papers 8 and 9 represent the end of the evolutionary period, a culmination of the work of many during the earlier decades. Never again would a paper be considered a significant contribution to chemical hydrogeology without emphasizing either mass balance, mass transfer, mineral equilibria, or kinetics.

During the 1950s little was published concerning the chemistry of groundwater. Three outstanding exceptions are Schoeller's (1955) classic work, which was published in French and was not immediately available in the United States; a series of papers by Chebotarev (1955) published in a prestigious journal but that perhaps caused more wonderment than enlightenment; and White's (1957a, 1957b) classic papers on deeper formation waters that stimulated much interest among the few groundwater geologists working in geochemistry.

Beginning in 1959, a renaissance in hydrogeochemistry led to the appearance of many significant publications. Two major texts

Introduction

were Garrels (1960) and Hem (1959); although Hem's was not marketed as a textbook, it is used extensively as one throughout the scientific world. Two significant papers on redox potential and natural environments were published by Sato (1960) and Baas Becking et al., (1960). Munnich and Vogel (1959) demonstrated the use of carbon-14 for groundwater studies; Craig (1961) published his classic paper on the isotopic variation of meteoric water; Back expressed the concept of hydrochemical facies (Paper 9) and provided the first application of chemical thermodynamics to groundwater (1961); Kohout (Paper 11) published his relevant paper on cyclic flow of saltwater in coastal aquifers; Carroll (1959) published a definitive review on ion-exchange characteristics of minerals; and Berry and Hanshaw (Paper 12) stimulated interest and investigations on geologic membranes in the development of anomalous head values and the origin of brines.

The revolutionary consequences of the equilibrium approach developed by R.M. Garrels, his associates, and students contrasts greatly to the evolutionary process. This period could be referred to as the "Garrelesian Revolution" because of the dominance of Garrels whose developments in aqueous geochemistry had such a profound impact on chemical hydrogeology. The vast majority of present-day mid-career aqueous geochemists either studied under him or worked directly with him, and most younger geochemists studied under his students. The law of mass action is for chemical hydrogeology what the Darcy equation is for physical hydrogeology. Chemical thermodynamics was first applied to low-temperature aqueous systems by Garrels and colleagues (references in Part III) in working on the exploration and genesis of uranium deposits in the Colorado Plateau. This fundamental approach led to the concept of mass balance and the use of equilibrium models and mass transfer concepts as exemplified by many of the papers in Part III.

These concepts are the basis for all geochemical studies of groundwater currently being undertaken. Mineral equilibrium studies have now been combined with isotopic analyses (Part IV) within the geologic and hydrologic framework to determine sources of water; the sources, fate, and behavior of chemical constituents; and controlling reactions and groundwater flow rates.

All these ideas and applications are contributing to our predictive capabilities by integration with groundwater flow models through the solute transport equation (Part V). The solute transport equation includes two mass-transport components—advection and hydrodynamic dispersion—and one mass-transfer component, or reactive term, which includes the effects of all chemical reactions. Groundwater geochemists evaluated the significance of the transfer term long before specific application of the solute transport equation to groundwater studies. Solute concentration is a function of mineral solubility, flowpath, and residence time. Essentially all investigations of groundwater chemistry are based to some extent on the transport equation, either in a formal mathematical manner or by intuitive consideration of the equation parameters. The transport equation indicates that movement of solutes in subsurface systems is controlled by three parameters: groundwater velocity, hydrodynamic dispersion, and the reaction term.

In personal communication, November 23, 1981, Warren Wood made the astute observation that he perceives the transport equation of chemical hydrogeology to be analogous to the Theis equation of physical hydrogeology and that much of the recent and current research is directed toward removing either the explicit or implicit assumptions required for evaluation of the mass-transfer term. Such assumptions included conservancy and nonreactivity of certain elements, mono-mineralogic aquifers, stoichiometric mineralogy, limited number of reactions, no radioactive decay, no sources or sinks for ions other than the reactions under consideration, and equilibrium conditions. These assumptions are gradually being removed as scientists gain experience, knowledge, and confidence to work on more complex systems. We expect that over the next several years groundwater geochemical studies will be oriented more directly toward evaluating the significance of the reaction term and determining the stoichiometry of the involved reactions. The papers in Part V show significance of heat transport in aquifer systems, show application of solute transport to groundwater studies, and review the current status of understanding solute transport.

REFERENCES

Baas Becking, L. G. N., I. R. Kaplan, and D. Moore, 1960, Limits of the Natural Environment in Terms of pH and Oxidation-Reduction Potentials, *Jour. Geology* **68**:243-284.

Back, W., 1961, Calcium Carbonate Saturation in Ground Water from Routine Analysis, *U.S. Geol. Survey Water-Supply Paper 1535-D*, pp. 1-14.

Carroll, D., 1959, Ion Exchange in Clays and Other Minerals, *Geol. Soc. America Bull.* **70**:749-780.

Chebotarev, I. I., 1955, Metamorphism of Natural Waters in the Crust of Weathering—Parts 1, 2, and 3, *Geochim. et Cosmochim. Acta* **8**:22-48; 137-170; 198-212.

Craig, H., 1961, Isotopic Variations in Meteoric Waters, *Science* **133**: 1702-1703. (Reprinted as Paper 8 in *Geochemistry of Water*, Kitano, ed., Benchmark Papers in Geology, vol. 16, Dowden, Hutchinson & Ross, Stroudsburg, Pa., 1975, p. 159.)

Introduction

Freeze, R. A., and W. Back, 1983, *Physical Hydrogeology*, Benchmark Papers in Geology, vol. 72, Hutchinson Ross Publishing Co., Stroudsburg, Pa., 448p.

Garrels, R. M., 1960, *Mineral Equilibria at Low Temperature and Pressure*, Harper & Bros., New York, 254p.

Hem, J. D., 1959, Study and Interpretation of the Chemical Characteristics of Natural Waters, *U.S. Geol. Survey Water-Supply Paper 1473*, 363p.

Mendenhall, W. C., R. B. Dole, and H. Stabler, 1916, Ground Water in San Joaquin Valley, Calif., *U.S. Geol. Survey Water-Supply Paper 398*, 310p.

Munnich, K. O., and J. C. Vogel, 1959, Altirsbestimmung von süsswasser-kalkablagerungen, *Naturwissenschaften* **46:**168–169.

Pourbaix, M. J. N., 1949, *Thermodynamics of Dilute Aqueous Solutions*, Edward Arnold and Co., London, 136p.

Sato, M., 1960, Oxidation of Sulfide Ore Bodies, Geochemical Environments in terms of Eh and pH, *Econ. Geology* **55:**928–961.

Schoeller, H., 1955, Géochimie des eaux souterraines, *Inst. Français Pétrole Rev.* **10**(3):181–213; **10**(4):219–246.

White, D. E., 1957a, Thermal Waters of Volcanic Origin, *Geol. Soc. America Bull.* **68:**1637–1658.

White, D. E., 1957b, Magmatic, Connate, and Metamorphic Waters, *Geol. Soc. America Bull.* **68:**1659–1682.

Zobell, C. E., 1946, Studies on the Redox Potential of Marine Sediments, *Am. Assoc. Petroleum Geologists Bull.* **30:**477–513.

Part I

THE EVOLUTIONARY PERIOD

Editors' Comments
on Papers 1 Through 9

1 **PALMER**
Excerpts from *The Geochemical Interpretation of Water Analyses*

2 **ROGERS**
Excerpt from *Chemical Relations of the Oil-Field Waters in San Joaquin Valley, California*

3 **COLLINS**
Graphic Representation of Water Analyses

4 **RENICK**
Excerpt from *Base Exchange in Ground Water by Silicates as Illustrated in Montana*

5 **LANGELIER**
Excerpts from *The Analytical Control of Anti-Corrosion Water Treatment*

6 **PIPER**
A Graphic Procedure in the Geochemical Interpretation of Water-Analyses

7 **STIFF**
The Interpretation of Chemical Water Analysis by Means of Patterns

8 **FOSTER**
The Origin of High Sodium Bicarbonate Waters in the Atlantic and Gulf Coastal Plains

9 **BACK**
Origin of Hydrochemical Facies of Ground Water in the Atlantic Coastal Plain

The systematic study of the chemistry of natural waters in North America has its roots in extensive surveys of rivers undertaken at the

Editors' Comments on Papers 1 Through 9

beginning of the twentieth century by the Reclamation Service, which had recently been reorganized from the U.S. Geological Survey (Dole, 1909; Stabler, 1911). As is obvious from reading Palmer (Paper 1), substantial accomplishments had been made by chemists in developing laboratory analytical techniques and by engineers in developing field-sampling procedures. Palmer's paper, in part, addresses the problem of characterization of the chemical components and clearly demonstrates that he understood that these constituents are not simply a dissolved load of rivers but rather a chemical system of balanced anions and cations. Much analytical experience had been gained previously by studying mineral and hot springs. Also, geologists studying formation of ore deposits were well aware of the need to study the chemical character of groundwater as Emmons and Harrington (1913) did to determine the processes involved in deposition by ascending hot meteoric waters. One of the first regional evaluations of the chemical suitability of water for irrigation was the classic study of San Joaquin Valley by Mendenhall et al. (1916).

In our present zeal to understand processes and reactions, it is commonly believed that early studies of the chemistry of water were concerned entirely with its classification and utilization. As gratifying as this chauvinistic perception may be, it is erroneous, as exemplified by Rogers (Paper 2) and Renick (Paper 4) whose early writings identified and articulated, respectively, the processes of sulfate reduction and ion exchange in natural water. Laboratory experiments by Thompson and Way (references in Paper 4) in the middle 1800s demonstrated that some properties of clay and other materials can change the composition of water passing through them. In the middle 1920s, Renick, while studying the Fort Union and Lance sequence of aquifers in Montana, recognized that this process was occurring naturally and that ion exchange exerted a major control on the chemical character of groundwater through the replacement of calcium by sodium.

In 1936, Langelier, professor of sanitary engineering at the University of California, published his report (Paper 5) that had a significant impact on groundwater chemistry twenty years later by providing a basis for calcite saturation studies.

The correlation and interpretation of the many variables associated with the chemical character of natural waters led to a need for suitable graphical techniques for data presentation. The three most commonly used graphs in North America were originally described by Collins (Paper 3), Piper (Paper 6), and Stiff (Paper 7). When Piper developed his trilinear diagram, he was unaware that a similar diagram had been prepared by Hill (1940) until after he had presented his paper; subsequent revision by Piper incorporated the references to

Hill's original diagram. This trilinear diagram is commonly referred to as the "Piper Diagram" because of the extensive use that Piper and his colleagues made of this diagram in understanding the problem of saltwater encroachment in southern California.

As mentioned in Part II, the saltwater encroachment problem provided an impetus for geologists to become aware of the need of their talents in studying the chemical character of groundwater. However, as late as 1950, the significant contributions were still being made by people whose formal training was in engineering and analytical chemistry as demonstrated by the papers by Hem (1950), Larson (1949), and Foster (Paper 8). At about the same time, Bond (1946) published a book that describes the regional chemical character of groundwater in South Africa, much appreciated by the few North American scientists trying to understand the relation of geology to the chemistry of water.

Foster's paper was a culmination of extensive work by her and others in the first half of the twentieth century. As early as 1933 (in Fiedler and Nye, 1933), she prepared a most advanced interpretation of the chemical character of water in the Roswell Basin of New Mexico. Paper 8 is an outstanding example of the scientific method in practice. She identified a significant scientific problem in the field—the occurrence of high bicarbonate waters in the coastal plain—developed a hypothesis that this was caused largely by ion-exchange, conceived a well-designed laboratory experiment to test the hypothesis, and then interpreted the field data in light of her experimental results.

Under the stimulation and guidance of Robert R. Bennett of the U.S. Geological Survey and building on Foster's work, Back (Paper 9; 1966) demonstrated that the study of groundwater chemistry need not be an exclusive purview of chemists but that groundwater geologists could make contributions by relating chemical data to groundwater in much the same manner as are other geologic and hydrologic data, such as heavy mineral analyses, lithologic logs, and hydraulic-head data. The concept of hydrochemical facies was developed to emphasize that chemical composition of water was not an unrelated entity but an integral part of hydrogeology. Detailed mapping of hydrochemical facies has been done in other areas, as for example, Seaber (1965), in his work on the Englishtown formation of New Jersey.

The current phase in demonstrating the role of ion-exchange is the sophisticated simulation approach used by Thorstenson et al. (1979). Building on these earlier ideas, they established by computer simulation the stoichiometric coefficients that conform with the chemical reactions which are due largely to ion-exchange and further demonstrated that in the presence of carbonates ion exchange can produce high pH values previously ascribed to only silicate dissolution reactions. Their choice of the Hell Creek (=Lance of Renick) Fox Hills aquifer system was a fortuitous coincidence that led them back to the formations in which Renick had originally identified the process; and so the evolutionary development has gone full circle—back to the region of its beginning.

REFERENCES

Back, W., 1966, Hydrochemical Facies and Ground-Water Flow Patterns in Northern Part of Atlantic Coastal Plain, *U.S. Geol. Survey Prof. Paper 498-A,* 42p.

Bond, G. W., 1946, A Geochemical Survey of the Underground Water Supplies of the Union of South Africa, *South Africa Geol. Survey Memoir 41,* 208p.

Dole, R. B., 1909, The Quality of Surface Waters of the U.S.: Part 1, Analyses of Waters East of the One Hundredth Meridian, *U.S. Geol. Survey Water-Supply Paper 236,* 123p.

Emmons, W. H., and G. L. Harrington, 1913, A Comparison of Waters of Mines and Hot Springs, *Econ. Geology* **8:**653–669.

Fiedler, Albert G., and S. Spencer Nye, 1933, Geology and Ground-Water Resources of the Roswell Artesian Basin, New Mexico, *U.S. Geol. Survey Water-Supply Paper 639,* 372p.

Hem, J. D., 1950, Geochemistry of Ground Water, *Econ. Geology* **45:**72–81.

Hill, R. A., 1940, Geochemical Patterns in Coachella Valley, *Am. Geophys. Union Trans.* **21:**46–53.

Larson, T. E., 1949, Geologic Correlations and Hydrologic Interpretation of Water Analyses, *Illinois Water Survey Circ. 27,* 8p.

Mendenhall, W. C., R. B. Dole, and Herman Stabler, 1916, Ground Water in San Joaquin Valley, Calif., *U.S. Geol. Survey Water-Supply Paper 398,* 310p.

Seaber, P. R., 1965, Variations in Chemical Character of Water in the Englishtown Formation, New Jersey, *U.S. Geol. Survey Prof. Paper 498-B,* 35p.

Stabler, H., 1911, Some Stream Waters of the Western United States with Chapters on Sediment Carried by the Rio Grande and the Industrial Application of Water Analyses, *U.S. Geol. Survey Water-Supply Paper 274,* 188p.

Thorstenson, D. C., D. W. Fisher, and M. G. Croft, 1979, The Geochemistry of the Fox Hills—Basal Hell Creek Aquifer in Southwestern North Dakota and Northwestern South Dakota, *Water Resources Research* **15:**1479–1498.

1

Reprinted from pages 5-9, 11-12, and 31 of *U.S. Geol. Survey Bull. 479*, 1911, 31p.

THE GEOCHEMICAL INTERPRETATION OF WATER ANALYSES.

By CHASE PALMER.

EXPRESSION OF CHEMICAL ANALYSES.

Terrestrial waters are essentially solutions of a few salts, and their chemical character, like that of solutions in general, depends on the nature and proportion of the substances they contain. The interpretation of the chemical character of a water from the results of analysis is necessarily uncertain and unsatisfactory if it is based merely on the amounts of the radicles determined. In analytical chemistry, as in other branches of the science, the chemist considers the inherent properties of the radicles of substances, and hence his statement of the results of a water analysis should be framed in accordance with the chemical nature and the proportional amounts of the radicles determined in a solution of mixed salts. There is no lack of information concerning the amounts of the various materials dissolved in natural waters and the mutual relations of their parts. What the chemist especially needs is a form of statement that will adequately express these relations and disclose the true proportions of the radicles.

The engineer has always recognized the importance of determining the properties of water without recourse to complete chemical analysis, and his attention is naturally directed to those properties which are objectionable. In 1841 Thomas Clark patented in England a process for removing the objectionable constituents of hard waters. The softening agent used by Clark was lime water, the action of which depends on a very simple principle. In contact with lime water the soluble calcium bicarbonate in hard water is changed to insoluble calcium carbonate and precipitated, the hardening constituent, calcium, being removed simultaneously from the hard water and from the softening agent.

The reaction may be expressed by the equation—

$$CaH_2(CO_3)_2 + Ca(OH)_2 = 2CaCO_3 + 2H_2O$$

This process of improving the quality of water at once acquired wide popularity. In response to many requests for information respecting his methods of examining waters, in 1847 Clark[1] addressed

---

[1] Clark, Thomas, On the examination of water for towns, for its hardness, and for the incrustation it deposits on boiling: Chemical Gazette, vol. 5, 1847, p. 100.

to friends a circular letter in which he states that his examination of waters involves two processes—one for ascertaining the hardness of water and one for ascertaining its alkalinity. The degrees of permanent hardness, temporary hardness, permanent alkalinity, and acidity are now capable of exact measurement by methods which do not involve the determination of the constituents of water. In the section on the properties of water definite limits, deducible from the results of a complete chemical analysis, will be set to the special properties. The limits there assigned conform to the measurements of hardness and alkalinity if made according to the exact method of Hehner as described by Sutton.[1] The acidity of water may be determined by direct measurement, made by neutralizing a known quantity of the acid water by a standard alkaline solution. Saltness caused by dissolved neutral salts is a general property of natural water. Since the alkalies and strong acids contribute largely to this property, it is essential that the proportional amounts of the alkalies and strong acids be separately determined. From the data thus obtained, the full value of the salinity—the saltness—of water may be determined.

It is to be observed that Clark fully recognized the propriety of looking to some of the properties of water for information concerning its fitness for domestic and industrial uses, and the benefit of his invention to modern civilization is beyond estimate. In problems involving the chemical action of water it is important to-day that the student consider all the properties conferred on water by all the substances dissolved in it, for in the totality of its properties lies the full power of water as a chemically active agent.

Two forms of stating the amounts of mineral materials dissolved in water have been widely used. These forms are typified by the following analysis of sea water:[2]

Composition of ocean water.

Amounts assigned to hypothetical combinations.		Amounts assigned to radicles.	
Combinations.	Milligrams per liter.	Radicles.	Milligrams per liter.
Sodium chloride (NaCl)	27,215	Sodium (Na)	10,710
Magnesium chloride ($MgCl_2$)	3,807	Potassium (K)	390
Magnesium sulphate ($MgSO_4$)	1,658	Calcium (Ca)	420
Calcium sulphate ($CaSO_4$)	1,260	Magnesium (Mg)	1,300
Potassium sulphate (K_2SO_4)	863	Sulphate (SO_4)	2,700
Magnesium bromide ($MgBr_2$)	76	Chloride (Cl)	19,350
Calcium carbonate ($CaCO_3$)	121	Bromide (Br)	60
		Carbonate (CO_3)	70
	35,000		35,000

[1] Volumetric analysis, 9th ed., p. 70.
[2] Mean of 77 analyses, by W. Dittmar, of sea water collected by the Challenger expedition: Challenger Report, Physics and chemistry, vol. 1, 1884, p. 203.

The older form, which represents the radicles as grouped together in arbitrary combinations, has by no means lost all adherents. It seems to be held in especial favor by the engineer because it gives the amount of dissolved material in terms which enable him to determine the corresponding amounts of substances necessary to fit a water for special industrial uses. The geologist, however, long ago realized that this form of expressing the chemical character of a water is inadequate to the exacting demands of research and has resorted to the form of statement in which the amounts of the radicles determined are given as independent units. In other words, he has practically abandoned a form of chemical expression and has adopted instead a statement of physical results. Chemical literature furnishes abundant evidence that the statement of water analyses in a form which does not recognize the *proportional reaction capacity* of the radicles fails to show the chemical character of the waters. Waters differing widely in character may be grouped together as similar if the classification is based on the preponderance of any radicle that may be considered as dominant in a solution of salts or on the apparent predominance of two or more radicles selected merely because they contribute largely to the weight of the mixture. Such classifications may be interesting from several points of view, but they are unreliable guides to the solution of geologic problems involving chemical processes. Furthermore, chemists, whose attention is fixed on the physical weights of the radicles, which are assumed to be free and independent, may easily fail to observe important facts concerning the chemical character of waters, especially facts relating to geology.

One advantage of the ionic form of stating water analyses is that it assigns weights directly to the chemically active parts of the dissolved substances instead of using those parts to build imaginary structures. The statement of the amounts of the radicles, however, indicates only the chemical composition of a water, not its character, for the *physical weight* of a radicle is no criterion of its *chemical value* in a system of dissolved salts such as exists in water. On the other hand, if the radicles are considered not as matter subject only to the law of gravitation, but rather as individuals acting together under the law of equivalent combining weights, contributing their proportional shares to the final balance of the system, the meaning of the results of a mineral analysis of water can be expressed clearly and precisely.

The reaction capacities of the radicles of the salts dissolved in water are the quotients obtained by dividing the weight of each radicle by its corresponding equivalent combining weight. The reaction capacity may be more logically determined by using for factors the reciprocals of the equivalent combining weights of the radicles, according to the practice of Herman Stabler, in interpreting the results of water analyses for industrial purposes.[1] Stabler defines "reaction coefficient" as the chemical reacting power of a unit weight of a radicle. The reaction coefficient of a radicle is the ratio of the reaction capacity of 1 part of that radicle to the reaction capacity of 8 parts of oxygen and is computed as follows:

	Atomic weights.	Equivalent weights.	Reaction coefficient.
Oxygen	16	$\frac{16}{2}=8$	
Hydrogen	1.008	$1.008=1.008$	$\frac{1}{1.008}=0.992$
Calcium	40.09	$\frac{40.09}{2}=20.045$	$\frac{1}{20.045}=.0499$
Magnesium	24.32	$\frac{24.32}{2}=12.16$	$\frac{1}{12.16}=.0822$

The other reaction coefficients are similarly obtained.

The product of the "reaction coefficient" by the amount of a radicle Stabler calls the "reacting value" of that amount of the radicle. These terms are peculiarly adapted to the chemical valuation of radicles determined in mixtures, and they will be adopted here in the chemical classification of waters. The following table shows the positive and negative radicles usually found in surface waters and their reaction coefficients:

Positive and negative radicles, with reaction coefficients.

Positive radicles.	Reaction coefficients.	Negative radicles.	Reaction coefficients.
Hydrogen (H)	0.992	Carbonate (CO_3)	0.0333
Ferrous iron (Fe)	.0358	Bicarbonate (HCO_3)	.0164
Aluminum (Al)	.1107	Sulphate (SO_4)	.0208
Calcium (Ca)	.0499	Chloride (Cl)	.0282
Magnesium (Mg)	.0822	Nitrate (NO_3)	.0161
Sodium (Na)	.0435		
Potassium (K)	.0256		

Stabler prefixes the letter r to the symbol of a radicle to designate the reacting value of the radicle, and the same symbolization will be used in this report.

Under the name "milligram equivalents" (that is, equivalents of milligrams of hydrogen) chemists have long used the reacting values of the radicles for two purposes—namely, to determine the accuracy of the analysis of a water and to obtain reliable factors to be used in the construction of hypothetical combinations. Stabler has shown that the reacting values may be put to a better use, for he has demonstrated mathematically that the analytical results can be

---

[1] Stabler, Herman, The mineral analysis of water for industrial purposes and its interpretation by the engineer: Eng. News, vol. 60, 1908, p. 356. Also, chapter on the industrial application of water analyses in Water-Supply Paper U. S. Geol. Survey No. 274, 1911, pp. 165–181.

interpreted far more satisfactorily directly from the reacting values than from their hypothetical combinations, and he has shown that the labor of calculating the amounts of remedial agents required to produce desired changes in the character of a water is thereby reduced to a minimum.

[*Editors' Note:* Material has been omitted at this point.]

PROPERTIES OF NATURAL WATERS.

Nearly all terrestrial waters have two general properties, salinity and alkalinity, on whose relative proportions their fundamental characters depend. Salinity is caused by salts that are not hydrolyzed; alkalinity is attributed to free alkaline bases produced by the hydrolytic action of water on solutions of bicarbonates and on solutions of salts of other weak acids.

All the positive radicles, including hydrogen, may participate in producing salinity; but of the negative radicles only those of the actively strong acids can perform a similar function. The principal strong acids in natural waters are represented by the sulphates, chlorides, and nitrates. Since salinity depends on the combined activity of equal values of both positive and negative radicles, and since its degree is limited only by the reacting values of the strong acids, the full value of salinity is obtained by multiplying the total value of the strong acid radicles by 2.

The full value of alkalinity and at the same time due recognition of the parent substances which are the source of alkalinity can be obtained by doubling the values of the bases in excess of the values of the strong acids.

The positive radicles determined in a water analysis, in accordance with their properties, fall naturally into three groups, as follows:

Group *a*. Alkalies (sodium, potassium, lithium). Their salts are readily soluble in water. They do not cause hardness.

Group *b*. Earths or alkaline earths. Calcium and magnesium are the chief representatives of this group. Many of their salts are sparingly soluble in water. They cause the property commonly known as hardness.

Group *c*. Hydrogen. Salts of hydrogen are acids and cause acidity in waters.

The groups of positive radicles are measured by the sum of the reacting values of their members, and in accordance with the prevalence of the reacting values of the groups of positive radicles in the system, five special properties are possible, namely:

1. Primary salinity (alkali salinity); that is, salinity not to exceed twice the sum of the reacting values of the radicles of the alkalies.

2. Secondary salinity (permanent hardness); that is, the excess (if any) of salinity over primary salinity, not to exceed twice the sum of the reacting values of the radicles of the alkaline earths group.

3. Tertiary salinity (acidity); that is, the excess (if any) of salinity over primary and secondary salinity.

4. Primary alkalinity (permanent alkalinity); that is, the excess (if any) of twice the sum of the reacting values of the alkalies over salinity.

5. Secondary alkalinity (temporary alkalinity); that is, the excess (if any) of twice the sum of the reacting values of the radicles of the alkaline earths group over secondary salinity.

In distinguishing the special properties, the values of radicles of the same sign are doubled. By this procedure the positive and negative radicles, which together induce the special properties, receive their full value. The use of the adjectives "primary" and "secondary" to qualify the general properties of the water solution associates naturally the alkalies with the oldest rock formations, of which the alkalies are the principal soluble decomposition products, and refers the alkaline earths to the more recent formations as their principal sources.

The character of natural waters with reference to the lithology of the region from which they are derived, to their solvent action on minerals with which they may come in contact, to sedimentary deposits that they are likely to form, to their effect on industrial processes, and to their chemical action in general can best be portrayed by a statement of as many of the five special properties above mentioned as may be found, expressed in percentages of their totality.

[*Editors' Note:* Material has been omitted at this point.]

CONCLUSION.

In this preliminary consideration of water analyses sufficient ground has been covered to justify the conclusion that natural water may be definitely characterized if the salts dissolved in it are recognized not as a load but as a chemical system of balanced values.

2

Reprinted from pages 93-99 of *U.S. Geol. Survey Bull. 653,* 1917, 119p.

CHEMICAL RELATIONS OF THE OIL-FIELD WATERS IN SAN JOAQUIN VALLEY, CALIFORNIA

G. Sherburne Rogers

[*Editors' Note:* In the original, material precedes this excerpt.]

CHEMICAL RELATIONS BETWEEN WATER AND THE HYDROCARBONS.

ALTERATION OF WATERS BY THE HYDROCARBONS.

NATURE OF ALTERATIONS.

It has been shown that the waters associated with the oil in the San Joaquin Valley oil fields are almost or quite sulphate-free, notwithstanding the fact that the shallower waters of the region are characterized by a large concentration of sulphate. An equally striking feature of many of the waters associated with the oil is the presence of alkali carbonate, which is lacking in the shallow waters on the west side of the valley. Between the sulphate and carbonate zones is a zone characterized by waters carrying hydrogen sulphide. Outside the oil fields sulphate and carbonate maintain the same mutual proportions to great depths and hydrogen sulphide waters are rare; the conclusion is therefore irresistible that a change in the composition of the oil-field waters has been caused directly or indirectly by constituents of the oil or gas. As sulphate is abundant in the shallower waters everywhere on the west side of the San Joaquin Valley, whereas sulphide is found only near the hydrocarbons, it is reasonable to suppose that the sulphide has been derived under special conditions through reduction of the sulphate. In regions where sulphates are rare or only locally distributed the alternative hypothesis—that sulphide has accompanied the oil from below and that sulphate is formed by its oxidation—may have to be considered, but as this condition does not prevail in the oil fields of San Joaquin Valley this hypothesis need not be discussed.

REDUCTION OF SULPHATE.

The observation that waters associated with oil contain no sulphate is by no means new, for it was pointed out in 1882 by Potilitzin[1] that the waters associated with oil in the Caucasian oil fields contain no sulphate, and this has been amply confirmed by later workers there

---

[1] Potilitzin, A., Zusammensetzung des die Naphta begleitenden und aus schlammvulkanen ausströmenden Wassers (abstract): Deutsche chem. Gesell. Ber., Band 15, p. 3099-b, 1882.

and in other fields. Höfer[1] mentions this widespread peculiarity of oil-field waters and presents a compilation of 27 analyses of sulphate-free water from various fields in Europe, Asia, and North America. However, neither Höfer nor, so far as the writer knows, any other writers on this subject have attempted to work out the chemical relations of the various types of water in any one field, but present merely isolated analyses of "oil-field water," the position of which with regard to the oil is generally not stated. The waters differ widely in chemical composition, some of them being concentrated brines very high in secondary salinity and others closely resembling the mixed (carbonate) type found in the San Joaquin Valley fields. The mixed type is less common but appears to be characteristic of the Russian and Galician fields. The nearly pure alkali carbonate water (reversed type) found in the Eastside Coalinga field has apparently not been found elsewhere. Some of the analyses show unusually large amounts of iodine and bromine, and several other rare elements have been reported, but the common characteristic of all the waters examined is the absence of sulphate. In some waters this is so complete that barium salts have been found in the solution.[2]

The earlier investigators apparently regarded these sulphate-free waters more as chemical curiosities than as normal and reasonable phenomena, but the absence of sulphate has since been attributed to the reducing action of the hydrocarbons. The reaction between sulphate and organic matter was suggested by Bischof[3] to explain the origin of certain sulphur deposits. It is supposed that the sulphate is reduced to sulphide, which passes off as hydrogen sulphide, and that an equivalent portion of the oil or gas is oxidized to carbon dioxide and carbonate. Höfer writes the reaction substantially as follows:

$$CaSO_4 + CH_4 = CaO + H_2S + CO_2 + H_2O$$

or

$$CaSO_4 + CH_4 = CaS + CO_2 + 2H_2O = CaCO_3 + H_2S + H_2O$$

These reactions, however, are hypothetic and are open to several objections. It has long been known that sulphate solutions are decomposed under some conditions in the presence of organic matter with the formation of hydrogen sulphide.[4] It was shown by Meyer[5] and more definitely by Plauchud,[6] however, that this decomposition

---

[1] Engler, C., and Höfer, H., Das Erdöl, Band 2, p. 28, 1909.
[2] Idem, p. 28.
[3] Bischof, G., Chemische und physikalische Geologie, 2, pp. 144–164, 1851.
[4] Lersch, B. M., Hydro-chemie, pp. 235–238, Berlin, 1864. Clarke, F. W., The data of geochemistry, 3d ed.: U. S. Geol. Survey Bull. 616, p. 111, 1916.
[5] Meyer, Lothar, Chemische Untersuchung der Thermen zu Landeck in der Grafschaft Glatz: Jour. prakt. Chemie, Band 91, pp. 5–6, 1864.
[6] Plauchud, E., Recherches sur la formation des eaux sulfureuses naturelles: Compt. Rend., vol. 84, p. 235, 1877; Sur la réduction des sulfates par les sulfuraires, et sur la formation des sulfures métalliques, naturels: Idem, vol. 95, p. 1363, 1882. Ètard, A., and Olivier, L., De la réduction des sulfates par les êtres vivants: Idem, vol. 95, p. 846, 1882.

is due not to the mere presence of dead organic matter but to the vital processes of microorganisms. Numerous observers have since studied these creatures, whose functions are diverse [1] and whose importance from the standpoint of geochemistry appears to be considerable. It has been found that certain bacteria have the function of reducing sulphate to sulphite or thiosulphate and that others reduce oxygenated sulphur compounds to hydrogen sulphide.[2] On the other hand, certain bacteria can exist only in solutions containing hydrogen sulphide, which they oxidize and secrete as sulphur. This sulphur is further oxidized in the course of metabolism to sulphate, but the excess of sulphur remaining in the organism after death may accumulate to form deposits of crystalline sulphur.[3] In general, the sulphide-producing bacteria are anaerobic, being able to exist in the absence of air, whereas those which secrete sulphur are probably aerobic. The hydrogen sulphide in many natural waters is thus doubtless derived from aqueous sulphate solutions by the action of bacteria.

The action of similar organisms in ocean water has also been studied. Van Delden,[4] in experimenting with a species that inhabits the estuaries on the coast of Holland, finds that these bacteria liberate in 27 days 843 milligrams of hydrogen sulphide per liter, which represents the reduction of 1,984 milligrams of sulphur trioxide. In this experiment he used sea water, to which was added a little potassium phosphate and organic matter; in another experiment with the same bacteria he used a prepared solution containing slightly more sodium chloride and more sulphate than sea water, and found that in 19 days 1,030 milligrams of hydrogen sulphide, equivalent to 2,424 milligrams of sulphur trioxide, were liberated. In the latter experiment the amount of sulphate reduced slightly exceeds that present in normal sea water. Van Delden notes also that the activity of this species increases with the concentration of sodium chloride up to 60,000 parts per million, but that the addition of more sodium chloride produces a marked diminution in their activity.

Hydrogen sulphide has been repeatedly observed in sea water and has been quantitatively determined by several observers. Lebedinzeff [5] finds that water from a depth of 8,290 feet in the

---

[1] Winogradsky, Sergius, Ueber Schwefelbakterien: Bot. Zeitung, Nos. 31 to 37, 1887.

[2] See, for example, Beyerinck, M. W., Ueber *Spirillum desulfuricans* als Ursache von Sulfatreduction: Centralbl. Bakteriologie, Band 1, Abt. 2, pp. 1–9, 49–59, 104–114, 1895. Also Saltet, R. H., Ueber Reduktion von Sulfaten in Brackwasser durch Bakterien: Idem, Band 6, Abt. 2, p. 648, 1900.

[3] For a summary see Stutzer, O., Die Wichtigsten Lagerstätten der Nicht Erze, Berlin, 1911; Phalen, W. C., The origin of sulphur deposits (translation from Stutzer's work): Econ. Geology, vol. 7, pp. 732–743, 1912.

[4] Van Delden, A., Beitrag zur Kenntnis der Sulfatreduktion durch Bakterien: Centralbl. Bakteriologie, Band 11, Abt. 2, pp. 92–94, 113–119, 1903.

[5] Lebedinzeff, A., Vorläufige Mitteilung über den chemischen Untersuchungen des Schwarzen und Asowischen Meeres in Sommer 1891: Soc. Naturalistes à Odessa Trav., vol. 16, fasc. 2, p. 149, 1891; abstract in Roy. Geog. Soc. Proc., new ser., vol. 14, p. 461, 1892.

Black Sea contains 6,550 parts per million of hydrogen sulphide, and Zelinsky [1] has identified in the bottom muds of the Black Sea several species of anaerobic bacteria that are very active in the formation of hydrogen sulphide. Murray and Irvine [2] report the formation of unstable sulphide in sea water associated with the blue muds on the floor of the ocean and by a series of analyses show that some of the sea water drained from the muds contains only 50 per cent as much sulphate as normal sea water. They note a concomitant increase in the alkalinity of the water, due principally to the formation of carbonate, and a slight loss of lime, due to the precipitation of calcium carbonate. This implies a decrease in secondary salinity and an increase in secondary alkalinity, limited by the solubility of the calcium carbonate; or, in other words, an approach to the oil-field brines along the lines explained above.

It is therefore well established that sulphate may be reduced by bacteria in the presence of organic matter, but the bearing of this process on the development of the composition of oil-field waters is conjectural. It may be assumed, if desired, that the connate water was completely altered by the action of bacteria shortly after being entrapped in the sediments, as suggested by the observations of Murray and Irvine, but the alteration of the meteoric water is more difficult to explain. There is no evidence to show that even anaerobic bacteria can continue to exist in the muds after they have been covered with a thousand feet or more of other sediments and elevated into land, and certainly some time must have elapsed after the elevation before meteoric water penetrated to the zone of alteration. In fact, the writer is inclined to believe that in some localities meteoric waters are percolating down to this zone and are being reduced at the present time. Hence, unless it be assumed that bacteria are present in the strata to depths of several thousand feet the formation of the sulphate-free waters can not be ascribed to bacterial action, except perhaps in part.

The belief that hydrocarbons can reduce sulphate at moderate temperatures in the absence of bacteria has been tacitly accepted for many years, and few attempts have been made to prove it in the laboratory. In the anhydrous condition gypsum ($CaSO_4$) is a very stable compound, and it has been found that a temperature of about 700° C. is required for its reduction, even with a fairly active reducing agent, such as carbon monoxide.[3] In solution, however,

---

[1] Zelinsky, N. [Sulphydric fermentation in the Black Sea]: Russ. Chem. Soc. Jour., vol. 25, pp. 298–303, 1894; abstract in Chem. Soc. Jour., vol. 66, pt. 2, p. 200, 1894. Andrussow, N., Physical exploration in the Black Sea: Roy. Geog. Soc. Geog. Jour., vol. 1, p. 49, 1893.

[2] Murray, John, and Irvine, Robert, On the chemical changes which take place in the composition of sea water associated with blue muds on the floor of the ocean: Roy. Soc. Edinburgh Trans., vol. 37, p. 481, 1892–93.

[3] Hofman, H. O., and Mostowitsch, W., The reduction of calcium sulphate by carbon monoxide and carbon, and the oxidation of calcium sulphide: Am. Inst. Min. Eng. Bull., pp. 913–939, 1910.

sulphate is more readily reduced. The experiments of the earlier workers are discredited by the fact that no precautions were taken to exclude bacteria, and certain more recent attempts proved unsuccessful, but recently Kharitschoff [1] has published a note on some simple experiments that were at least partly successful. He studied mixtures of equal volumes of 10 per cent sodium sulphate solution and kerosene or benzene under different conditions of temperature and pressure. Cadmium chloride was used to indicate the formation of sulphide. Three samples exposed to direct sunlight for six months at ordinary temperatures showed no sign of reduction. Other samples, sealed and heated for 420 hours on a water bath, under which conditions a pressure of not less than three atmospheres must have been developed, showed a very faint coloration due to the formation of a trace of sulphide. In still other samples left open and heated for 420 hours at 96° C. some sulphide was formed. A solution of magnesium sulphate mixed with kerosene and heated in the open for 420 hours underwent somewhat more reduction than the solution of sodium sulphate. Kharitschoff concludes from these experiments that the reduction of sulphate can be accomplished by hydrocarbons, but that high pressure and temperature during a long period of time are necessary to insure complete reduction.

If it be admitted that the reduction of sulphate is accomplished directly by the constituents of oil it must still be recognized that the reaction as generally written, involving methane, is improbable. Methane, being itself a decomposition product, is the most stable member of the paraffin series, which are the most inert of the hydrocarbons; and although methane becomes much more active at higher temperatures and pressures it seems that the reduction of a sulphate solution would be accomplished less readily by this hydrocarbon than by others. The different members of the hydrocarbon series probably react with sulphate solutions in different degree, but this phase of the subject has apparently not been investigated. The unsaturated chain compounds, such as the olefines, acetylenes, and terpenes, doubtless behave in different manner from the paraffins, the naphthenes, or the aromatic hydrocarbons, not only in the ease of reaction but in the stages involved. In the reactions between some substances hydrolysis is probably important, and in those between other substances the action of oxidizing agents may enter. It is quite possible that certain constituents of the oil other than true hydrocarbons are active in the reduction of sulphate solutions, although for the sake of brevity the term hydrocarbon is used in this report to include all oil constituents. In any event, the reaction as written by Höfer (p. 94) can be considered only a condensed repre-

---

[1] Kharitschoff, K. V., The waters in petroleum wells: Petroleum Rev., vol. 29, p. 368, 1913.

sentation of the type of change that takes place, the intermediate stages in the decomposition of the hydrocarbons on the one hand and of the sulphate on the other being as yet unknown.

Just as sulphate by reduction yields sulphide, so sulphide under other conditions may oxidize to sulphate. Whether alkaline sulphide is the first product of reduction or not, the final product is hydrogen sulphide, and this gas is readily oxidized to produce free sulphur, probably in accordance with the equation

$$2H_2S + O_2 = 2H_2O + 2S$$

Thus, free sulphur has been found in a number of marine muds, where it is doubtless formed by the oxidation of hydrogen sulphide derived from the sulphate in the sea water.[1] Under more strongly oxidizing conditions, or in the presence of certain bacteria, the sulphur becomes thiosulphate, sulphite, and finally sulphate. The complete reversion of hydrogen sulphide to sulphate is probably not widespread in deeply buried strata, but the change to free sulphur, which may take place even on the floor of the ocean, must be taken into account. It may also be noted that the oxidation of hydrogen sulphide to sulphur or sulphate results in the evolution of much heat,[2] and if the earth temperatures in oil regions are higher than elsewhere, as suggested by Koenigsberger and Mühlberg,[3] some of the excess may be contributed by this reaction.

FORMATION OF CARBONATE.

In the few published accounts of oil-field waters special stress is laid on the absence of sulphate as discussed in the preceding section, but no attempt has been made, so far as the writer can learn, to work out the relations of the various types of water or to explain the significance of the carbonate. If the reduction of the sulphate is to be ascribed to the action of hydrocarbons, however, the formation of carbonate is a necessary concomitant, and the presence of unusual amounts of carbonate in oil-field waters may be explained by this reaction.

The proportion of carbonate formed during the reduction of a definite amount of sulphate is not known and can not be determined until the stages involved in the reaction have been critically studied. Murray and Irvine report that the increase in the alkalinity of sea water associated with bottom muds is proportional to its loss in sul-

---

[1] Buchanan, J. Y., On the occurrence of sulphur in marine muds and nodules, and its bearing on their mode of formation: Roy. Soc. Edinburgh Proc., vol. 18, p. 17, 1890–91; Clarke, F. W., The data of geochemistry, 3d ed.: U. S. Geol. Survey Bull. 616, p. 514, 1916.

[2] Becker, G. F., Geology of the quicksilver deposits of the Pacific slope: U. S. Geol. Survey Mon. 13, p. 254, 1888.

[3] Koenigsberger, J., and Mühlberg, M., Über Messungen der geothermischen Tiefenstufe: Neues Jahrb., Beilage Band 31, pp. 107–157, 1911.

phate, but this observation merely indicates that the two changes are the result of the same process and does not throw much light on the proportions involved. In the transition from normal to altered waters in the oil fields the increase in carbonate is roughly proportional to the decrease in sulphate, but the loss by precipitation of alkaline-earth carbonates prevents the deduction of exact figures. The assumption made on page 88 that for the value of sulphate removed from the water an equivalent value of carbonate is introduced, is perhaps the best that can be made at the present time and is fairly adequate if the water alone is considered. As a matter of fact, however, the waters in the zone of alteration contain sufficient half-bound carbon dioxide to allow the formation of bicarbonate almost exclusively, and in addition many of these waters contain considerable free carbon dioxide. The amount of free and half-bound carbon dioxide in the zone of alteration seems disproportionately large in relation to the amount of hydrogen sulphide, even if the ready oxidation of hydrogen sulphide is taken into account.

The apparent disparity between the total amount of carbon dioxide formed and the amount of sulphate removed may be due to the fact that all the carbon dioxide is not derived from the oxidation of hydrocarbons. Several reactions are known by which carbonate may be derived from inorganic sources. Hilgard [1] finds that a solution containing free carbon dioxide in the presence of sodium sulphate dissolves calcium carbonate and forms sodium bicarbonate and a precipitate of gypsum. This reaction would partly explain the disappearance of sulphate and the formation of carbonate, but it would not account for the formation of hydrogen sulphide. If it is assumed, however, that the hydrogen sulphide is derived through the reduction of sulphate, the presence of free carbon dioxide may be explained by the following reaction, first investigated by Béchamp:[2]

$$CaCO_3 + 2H_2S = Ca(SH)_2 + H_2O + CO_2.$$

Under other conditions hydrogen sulphide may unite with calcium carbonate to form calcium sulphate and sulphur.[3] It is evident, therefore, that the disappearance of sulphate and the formation of carbonate may be the net result of several reactions. As the strata in the oil fields of the San Joaquin Valley do not contain much calcium carbonate, the reactions just discussed have probably not entered largely into the development of the chemical character of the waters, but the possibility that they have played some part should be duly considered.

---

[1] Hilgard, E. W., The geologic efficacy of alkali carbonate solution: Am. Jour. Sci., 4th ser., vol. 2, pp. 100-107, 1896.
[2] Béchamp, A., Recherches sur l'état du soufre dans les eaux minérales sulfurées: Annales chimie et phys., 4th ser., vol. 16, p. 234, 1869.
[3] Spezia, G., Sull' origine del solfo nei giacimenti solfiferi della Sicilia, Torino, 1892.

[*Editors' Note:* Material has been omitted at this point.]

Graphic Representation of Water Analyses[1]

By W. D. Collins

U. S. GEOLOGICAL SURVEY, DEPARTMENT OF THE INTERIOR, WASHINGTON, D. C.

In connection with the study of waters in their relations to one another and to the geologic formations from which they come, the graphic representation of analyses has been found exceedingly helpful. The method used in the U. S. Geological Survey is like others that have been published in that it represents analyses by plotting areas proportional to the milligram equivalents of the radicals that take part in the equilibrium of the system. The advantage of the method is in the order of arrangement and in the use of certain color combinations.

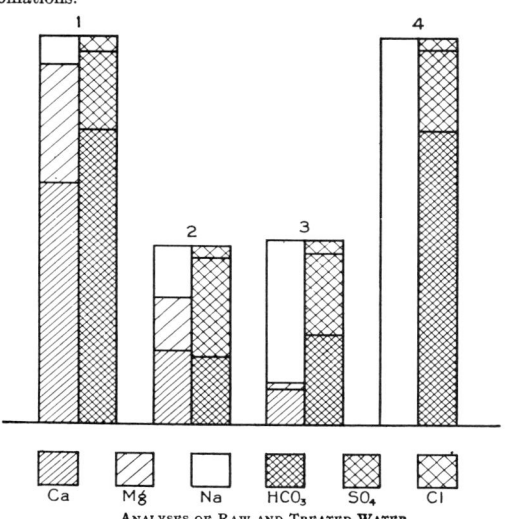

ANALYSES OF RAW AND TREATED WATER

The accompanying figure shows in graphic form the four analyses given in the table. Printing in the text necessitates

[1] Published by permission of the Director, U. S. Geological Survey.

the use of patterns which are harder to make and less useful than colors for showing the quantities of the different radicals. In regular practice colored crayons or pencils are used as follows:

Calcium	Red	Chloride	Light green
Magnesium	Orange	Sulfate	Light blue
Sodium	Yellow	Bicarbonate	Violet or purple

Potassium is included with sodium, nitrate with chloride, and carbonate with bicarbonate. If it is desired to represent separately the quantities of these radicals, black lines are drawn across the areas in which they are included so as to show potassium and nitrate at the top and carbonate at the bottom of the diagram.

The inevitable error in balance of an analysis may be indicated by inequality of length of the basic and acid columns, but it is generally better to distribute the error between the calcium and bicarbonate so as to make the columns of the same height. If silica is shown, it is represented by a black area extending across the top of the two columns. Suspended matter in an unfiltered water is shown by a brown or sepia area, frequently several times the combined width of the acid and basic columns. The areas for silica and suspended matter are each calculated on the basis of a combining weight of 30.

ANALYSES OF RAW AND TREATED WATER

	1[a]	2[b]	3[c]	4[d]
Parts per million				
Ca	65	19	8.4	Trace
Mg	20	9	0.9	Trace
Na	8	16	45	120
HCO_3	240	53	70	240
SO_4	54	64	54	54
Cl	6	6	6	6
Milligram equivalents				
Ca	3.24	0.95	0.42	....
Mg	1.64	0.74	0.07	....
Na	0.34	0.69	1.94	5.22
HCO_3	3.93	0.87	1.14	3.93
SO_4	1.12	1.33	1.12	1.12
Cl	0.17	0.17	0.17	0.17

[a] Raw water.
[b] Water treated with alum, lime, and soda ash for a public supply.
[c] Water treated hot with lime and soda ash to reduce the hardness to 25 p. p. m.
[d] Water softened by an exchange silicate.

Reprinted from pages 63-72 of *U.S. Geol. Survey Water-Supply Paper 520-D,* 1924, pp. 53-72

BASE EXCHANGE IN GROUND WATER BY SILICATES AS ILLUSTRATED IN MONTANA

B. Coleman Renick

[*Editors' Note:* In the original, material precedes this excerpt.]

BASE-EXCHANGE SILICATES.

On the basis of some experiments carried on in 1845 Thompson [19] pointed out that certain soils possess the power of decomposing and retaining the salts of ammonia. He stated that he had not studied the reaction sufficiently to account for the manner in which it was accomplished. About the same time Way,[20] an English investigator, observed that this power of soils is not confined to ammonium salts but that the bases of different alkaline salts may be separated from solution and retained by ordinary soils. Summarizing his previous work in a later paper [21] he made the following pertinent statements:

But further, this power of the soil was found not to extend to the whole salt of ammonia or potash, but only to the alkali itself. If, for instance, sulphate of ammonia were the compound used in the experiments, the ammonia would be removed from solution, but the filtered liquid would contain sulphuric acid in abundance—not in the free or unconfined form, but united to lime; instead of sulphate of ammonia we would find after the experiment sulphate of lime in solution; and this result was obtained whatever the acid of the salt experimented on might be. * * * It was satisfactorily proved that the quantity of lime acquired by solution corresponded exactly to that of ammonia removed from it; the action was therefore a true chemical decomposi-

---

[19] Thompson, H. S., On the absorbent power of soils: Roy. Agr. Soc. Jour., vol. 11, pp. 68–74, 1850.
[20] Way, J. T., The power of soils to absorb manure: Roy Agr. Soc. Jour., vol. 11, pp. 313–379, 1850.
[21] Way, J. T., Roy. Agr. Soc. Jour., vol. 13, pp. 123–143, 1852.

tion. * * * It was found that the process of filtration was by no means necessary; by the mere mixing of an alkaline solution with a proper quantity of soil, as by shaking them together in a bottle and allowing the soil to subside, the same result was obtained; the action, therefore, was in no way referable to any physical law brought into operation by the process of filtration.

Again it was found that the combination between soil and alkaline substance was rapid, if not instantaneous. * * * It was shown that the power to absorb alkaline substances did not exist in sand; that the organic matter of the soil had nothing to do with it; that the addition of carbonate of lime to a soil did not increase its absorptive power for these salts; and indeed that a soil in which carbonate of lime did not occur might still possess in a high degree the power of removing ammonia or potash from solution, and it was evident that the active ingredient in all these cases was clay. * * * The stiffest and most tenacious clays taken from considerable depths, which had never since their deposition been exposed to atmospheric influences, and which also were absolutely free from organic matter, or carbonate of lime, possessed to the fullest extent the absorptive property. By these experiments the subject was so far narrowed that the origin of the power in question had been traced to the clay existing in all soils. * * * It soon became evident that the idea of the clay as a whole being the cause of the absorptive property was inconsistent with all the ascertained laws of chemical combination. * * *

I was, indeed, convinced at a very early period of this inquiry that the absorptive property was due to a small quantity of some definite chemical compound existing in the clay and possibly not constituting more than 4 or 5 per cent of its whole weight. I had hoped that, although I might not be able to separate this substance from clay—for of that there was little prospect—it might yet be possible to form it artificially from other sources at the disposal of the chemist, and by producing a compound, or compounds, having the same properties as those shown to be possessed by clay to prove their identity with the active principles of clay itself and thus indirectly establish its real nature.

After eliminating lime (CaO), lime carbonate, sulphate, nitrate, and other simple salts of lime from consideration as the material that gave to soils this reactive power, Way concluded that it must be due to some silicate. After preparing a simple lime silicate, which he found did not possess this property, he turned his attention to the preparation of double silicates of alumina with the alkalies and alkaline earths. He observed that these silicates possessed the property of exchanging their bases and concluded that the base-exchange material in soil was similar in chemical composition but that it was not feldspar or other undecomposed minerals from granitic rocks. Way furthermore brought out the important point that these silicates contain water of combination and that if it is driven off by strong heating the base-exchange property is destroyed, the silicate being no longer reactive.

From the foregoing description of the work of Way it is evident that much was known about base-exchange silicates at a relatively early date. It has long been known that the mineral zeolites are capable of readily exchanging their bases, and after Way's discovery of base-exchange silicates in soils many agriculturists came to regard the reactive material as zeolites. This base-exchange material in soils is referred to as the zeolitic portion of soils even to the present time, though ordinary soils almost certainly do not contain zeolites.

Sullivan,[22] in addition to summarizing the work of earlier investigators on base exchange in natural and artificial silicates, made a noteworthy contribution to the subject. He effected numerous base-exchange reactions between natural silicates and the base of numerous salts in solution. Some of his conclusions follow:

> The fact of prime significance geologically seems to be that by a process of simple chemical exchange the metal may be removed from solution and fixed in a solid state and thus concentrated, by contact with even the most stable of the silicates. The changes under consideration involve the action of the alkali or alkaline-earth salt of a weak acid (silicic or alumino-silicic) and are thus analogous to the more familiar behavior of sodium carbonate with solutions of salts of the metals. Owing to hydrolysis the precipitates caused by sodium carbonate tend to split up into the acid and base (carbonic acid and metal oxide or hydroxide), and the weaker the base the more marked is this action. The precipitate from solutions of salts of strong bases, such as calcium chloride, is the normal carbonate; a weaker base, such as nickel, is precipitated as basic carbonate or a mixture of the normal carbonate with hydroxide or oxide; while the very weak bases, as iron in ferric salts, are precipitated as hydroxide or oxide containing little or no carbonate, and the corresponding quantity of carbon dioxide is set free.[23]

Sullivan discussed the mechanics of the reaction and by way of summary said:

> The natural silicates precipitate the metals from solutions of salts, while at the same time the bases of the silicates are dissolved in quantities nearly equivalent to the precipitated metals. The bases most commonly replacing the metals in these processes are potassium, sodium, magnesium, and calcium. Where exact equivalence is wanting, it is attributable either to solubility of the mineral in pure water or to the precipitation of basic salts.
>
> The specific materials on which work was done are albite, amphibole, augite, biotite, enstatite, garnet, clay gouge, kaolin, microcline, muscovite, olivine, orthoclase, prehnite, shale, talc, tourmaline, and vesuvianite, with cupric sulphate solution; and orthoclase with salts of sodium, potassium, magnesium, calcium, strontium, barium, manganese, iron, nickel, copper, zinc, silver, gold, and lead. Experiments were also made on the action of kaolin on solutions of salts of zinc and iron, and of glass, fluorite, and pyrite on cupric sulphate and of carbonic and sulphuric acids on orthoclase.[24]

Sullivan's conclusions differ somewhat from those previously arrived at by Lemberg,[25] who says:

> In addition to these [the zeolites] I have experimented with the various feldspars, hornblende, cordierite, serpentine, and scapolite, but up to the

---

[22] Sullivan, E. C., Interaction of minerals and water solutions: U. S. Geol. Survey Bull. 312, 1907. Also an earlier shorter paper, The chemistry of ore deposition—precipitation of copper by natural silicates: Econ. Geology, vol. 1, pp. 67–73, 1905.
[23] Sullivan, E. C., op. cit. (Bull. 312), pp. 61–62.
[24] Sullivan, E. C., op. cit., p. 64.
[25] Lemberg, J., Deutsch. geol. Gesell. Zeitschr., vol. 22, p. 335, 1870; vol. 24, p. 187, 1872; vol. 28, p. 591, 1876. Cited by E. C. Sullivan, U. S. Geol. Survey Bull. 312, p. 23, 1907.

present only in the case of hornblende could an exchange of substance be proved with certainty.

This conflict in statement may be due to a difference in viewpoint—that is, Lemberg might not have regarded an exchange as taking place unless it was fairly complete, while Sullivan regarded an experiment as highly successful and demonstrative of base exchange if only a very small part of the silicate was exchanged.

Although these base-exchange silicates had been prepared and most of their properties understood for over half a century, Gans, by publishing two papers [26] in 1905 and 1906, aroused considerable interest in them. He apparently was the first one to conceive of the idea of utilizing artificially prepared sodium base-exchange silicates for softening water by allowing hard water containing salts of calcium and magnesium to flow over it, the calcium and magnesium being removed by exchange with the sodium to give a mixed calcium, magnesium, and sodium base-exchange silicate and soft water. This artificial water softener, which he considered must contain the essential molecules soda, alumina, silica, and water of combination, he named permutite and patented.

When the sodium capable of exchange has been exhausted the base-exchange silicate is regenerated by passing a strong solution of a sodium salt, preferably chloride, through it. Regarding the ability of these double silicates to be regenerated by reversing the reaction Way [27] erred when he wrote, "Of course, the reverse of this action can not occur."

In 1907 Feldoff [28] reported on the success of this method of softening water for use in boilers, and furthermore showed that permutite could be used to remove iron and manganese quantitatively from drinking water. Gedroiz [29] has since stated that this exchange may take place between the base of any metallic salt and a base-exchange silicate. As pointed out in a recent thesis by Baker,[30] artificial softeners have been patented which substitute the oxides of zinc, tin, lead, titanium, zirconium, chromium, and iron for alumina, and boric acid has been used in place of silica.

Since 1907 the use of so-called artificial zeolites has become of considerable economic importance in municipal and industrial water-

---

[26] Gans, Robert, Zeolites and similar compounds, their constitution and their importance for technology and agriculture: Preuss. geol. Landesanstalt Berlin Jahrb., Band 26, Heft 2, pp. 179–211, 1905; The constitution of zeolites, processes of obtaining and technical importance: Idem, Band 27, Heft 1, p. 63, 1906.

[27] Way, J. T., Power of soils to absorb manure: Roy. Agr. Soc. Jour., vol. 13, p. 132, 1852.

[28] Feldoff, A., Natural and artificial zeolites (permutite) and their technical application: Centralbl. Zuckerindustrie, vol. 15, pp. 1307–1310, 1907; Chem. Abstracts, vol. 1, p. 2755, 1907.

[29] Gedroiz, K. K., Colloidal chemistry as related to soil science: Russia Bur. Agr. and Soil Sci. Communication 8, p. 25, 1912 (U. S. Dept. Agr. translation, p. 18). Recent papers of Dr. Gedroiz, of Petrograd, published between 1912 and 1923, have been translated into English by Dr. S. A. Waksman and mimeographed by the United States Department of Agriculture in order that they may be available to American investigators.

[30] Baker, G. C., Water softening by base exchange: Am. Waterworks Assoc. Jour., vol. 11, pp. 128–149, 1924.

softening plants. Many investigators in Europe and America have studied the permutite reaction, but in spite of much detailed work, different investigators have arrived at different conclusions with regard to the mechanics of the reaction, some contending that this exchange is ionic, and others that it is a phenomenon of adsorption.[31] Rideal,[32] in referring to the reaction of carbonate of lime with natural mineral zeolites, said, " The reaction suggests another way by which nearly pure alkali may originate in nature." Numerous papers by Gedroiz are especially noteworthy, as they contain not only much valuable original work but also discussions of the papers of earlier workers. The mechanics of the reaction has also been discussed by numerous investigators, including Wiegner,[33] Don,[34] Raumann, Marz, Biesenberger, and Spengel,[35] Rothmund and Kornfeld,[36] and Raumann and Junk.[37]

Gans, like his predecessors, assumed that the soil contains mineral zeolites. Since Gans's papers appeared many writers on this subject have referred to these complex aluminum silicates capable of base exchange found in the soil as " zeolites." As ordinary soil does not contain mineral zeolites, it seems that the usurping of a definite mineralogic term to describe any complex hydrated aluminum silicate capable of base exchange gives an erroneous concept. For such material the term " base-exchange silicates " instead of zeolite seems appropriate.

Gedroiz[38] concludes that this exchange is ionic and states that

> The zeolitic (and humic) part of every soil contains a well-defined quantity of zeolitic cations. These cations can be replaced by any cation or mixture of any cations. The replacement takes place as a result of the reaction of mutual exchange of the cations between the zeolitic (and humic) part of the soil and the solution of the salt or acid taken. As a reaction of double exchange, the replacement takes place in equivalent concentrations.

Besides artificial base-exchange silicates, many of which utilize kaolin, quartz, and feldspar, there are certain natural minerals and rock materials that will soften water and have been used for that purpose. These natural materials are rendered more efficient by

---

[31] Gedroiz prefers the use of the word absorption. Some writers include both phenomena under the general term sorption.

[32] Rideal, S., Origin of carbonate of soda in natural waters and mineral deposits: Chem. World, vol. 1, p. 16, 1912.

[33] Wiegner, George, The exchange of bases in cultivated soil: Jour. Landw., vol. 60, pp. 197–222, 1912: Chem. Abstracts, vol. 6, pp. 2477, 3304, 1912.

[34] Don, J., The use of permutit and polarit in water purification: Glasgow Kolloid Zeitschr., vol. 15, pp. 132–134, 1914: Chem. Abstracts, vol. 9, p. 676, 1915.

[35] Raumann, E., Marz, S., Biesenberger, K., and Spengel, A., The exchange of bases of silicates—Exchange of alkalies and ammonium by hydrous aluminum-alkali silicates permutites): Zeitschr. anorg. allgem. Chemie, vol. 95, pp. 115–128, 1916; Soc. Chem. Industry Jour., vol. 35, p. 1129, 1916: Chem. Abstracts, vol. 11, p. 2174, 1917.

[36] Rothmund, V., and Kornfeld, G., Basic exchange in permutit: Zeitschr. anorg. allgem. Chemie, vol. 103, pp. 129–163, 1918; Chem. Abstracts, vol. 13, p. 2823, 1919.

[37] Raumann, E., and Junk, H., Basic exchange in silicates—III: Zeitschr. anorg. allgem. Chemie, vol. 114, pp. 90–104, 1920; Chem. Abstracts, vol. 15, p. 2592, 1921.

[38] Gedroiz, K. K., op. cit., p. 26 (U. S. Dept. Agr. translation, p. 19).

various treatments, many of which are relatively simple. The most useful of these minerals are greensand or glauconite, bentonite, and clay or kaolin. Bentonite, though containing more than one mineral, always consists chiefly of leverrierite, or one of the group of micaceous clay minerals which includes the mineral commonly known as leverrierite. This group possesses the property of readily exchanging its bases and is considerably more reactive than the ordinary natural clays. The bentonite from Ardmore, S. Dak., consists mostly of leverrierite but contains a considerable quantity of disseminated carbonate (probably calcite) and a few grains of biotite and muscovite. This leverrierite-bearing material from Ardmore after being treated to render it more reactive and to prevent it from swelling (a characteristic property of leverrierite) is used as a commercial water softener.

ORIGIN OF THE SOFT WATERS.

It is believed that the difference in the composition of water in deep and shallow wells in this area of Lance and Fort Union rocks can be explained as the result of natural softening. This exchange of the calcium and magnesium in the water for sodium can be accounted for by the minerals of the leverrierite group, which exchange their bases easily and which are plentiful in these formations. The hypothetical reactions might be written as follows:

(1) $Na(or\ K)$ base-exchange silicate $+ Ca(HCO_3)_2$ (or $Mg(HCO_3)_2$) $=$ Ca (or Mg) base-exchange silicate $+ 2NaHCO_3$ (or $2KHCO_3$).

(2) Na (or K) base-exchange silicate $+ CaSO_4$ (or $MgSO_4$) $= Ca$ (or Mg) base-exchange silicate $+ Na_2SO_4$ (or K_2SO_4).

Analyses of minerals of the leverrierite group.[a]

	1	2	3	3a	4	5	6	7	8
SiO_2	47.28	47.84	47.56	789	47.95	49.90	50.55	49.4	48.43
Al_2O_3	20.27	20.88	20.57	256	32.67	37.02	19.15	45.1	41.63
Fe_2O_3	8.68	8.48	8.58		.23	3.65			
MnO		.24	.24				4.40		
CaO	2.75	2.52	2.52	89	.41	Tr.	.63		2.13
MgO	.70	.91	.80		.46	.30			2.13
Na_2O	.97	1.58	1.28		2.47				
K_2O	Tr.	Tr.			.24	1.13			
H_2O+	} 19.72	6.65	6.65		7.03	8.65	} 24.05	5.6	7.70
H_2O+		10.95	12.01		8.56	(?)		(?)	(?)
	100.37	100.05	100.21		99.36	100.65	98.78	100.1	99.89
$Al_2O_3 : SiO_2$			1:2.76		1:2.34	1:2.11	1:3.95	1:1.86	1:1.94

[a] Larsen, E. S., and Wherry, E. T., Leverrierite from Colorado: Washington Acad. Sci. Jour., vol. 7, No. 8, p. 213, 1917.

1 and 2. Material from Beidell, Colo. New analyses by E. T. Wherry.
3. Average of 1 and 2.
3a. Molecular proportions of 3.
4. Average of two analyses of "rectorite," Garland County, Ark. Brackett, R. N., and Williams, J. F., Am. Jour. Sci., 3d ser., vol. 42, p. 16, 1891.
5. Leverrierite, Rochelle, France. Termier, P., Soc. min. Bull., vol. 22, p. 29, 1899. Analysis made on material dried at 110°–130°. Older analyses show 13.21 and 18.0 per cent of total water.
6. "Montmorillonite, var. delanouite," Millac, France. Quoted from Lacroix, A., Minéralogie de la France.
7. "Batchelorite," Tasmania. H_2O stated as "combined H_2O." Gregory, J. W., Australian Inst. Min. Eng. Trans., vol. 10, p. 187, 1905.
8. Kryptotile. Quoted from Dana. The original article reports H_2O without a statement as to whether it represents total water or water above 100°.

These analyses of minerals of the leverrierite group show that there is considerable variation in the quantity of bases in different species and thus suggest easy base exchange among the leverrierite minerals.

Although it is believed that leverrierite is the principal mineral that brings about this natural softening of the water in the area here considered, it is recognized that other hydrated aluminum silicates, such as kaolin, feldspars, and mica, are also capable of exchanging wholly or in part their sodium and potassium for other bases.

An inspection of Plate V shows that the calcium and magnesium have been essentially removed by exchange for sodium by the time the water reaches a depth of 125 feet and in some localities before it reaches 80 feet. The diagrams might convey the impression that the calcium and magnesium are more readily exchanged in the Fort Union rocks than in those of the Lance formation. This may or may not be the case, for this apparent difference may be accounted for by the fact that no analyses of samples of water from wells in the Lance formation between the depths of 30 and 100 feet are at hand.

It is not intended to convey the idea that the hard near-surface ground water is softened by direct downward percolation through a given number of feet of these leverrierite-bearing strata, because most of the deeper soft water has moved laterally through many feet and even miles, and it is impossible to say just how necessary or important this lateral movement may be. But owing to the facts that these reactions between dissolved salts and base-exchange silicates are rapid and that the exchange has been accomplished in all waters beyond a given depth, the conclusion seems justified that ground water will have its calcium and magnesium essentially removed by percolating through relatively few feet of rock containing leverrierite.

It is apparent from Plate V that there are notable differences in the acid radicles in the Lance and Fort Union waters. The most striking feature is that some of the waters contain sulphate and others do not. It is not within the scope of this paper to discuss the acid radicles in these waters. The cause of the elimination of the sulphate, which, on the whole, is less abundant in the waters from the deeper wells, is considered in another paper.[39] Bicarbonate and carbonate are usually the most abundant acid radicles, especially in the waters from the deeper wells.

The noteworthy fact that the mineral content of these waters from the Lance and Fort Union formations does not appreciably

---

[39] Renick, B. C., Some geochemical relations of ground water and associated natural gas in the Lance formation, Montana (to be published in Jour. Geology).

increase with increasing depth indicates that the amount of dissolved salts is determined relatively near the surface, and that any subsequent change that the basic radicles suffer is in the nature of an exchange. This exchange is probably not of the nature of an absorption phenomenon but is more likely an ionic exchange, as contended by Gedroiz.[40]

The possibility of the removal of calcium and magnesium by means of a reaction between calcium bicarbonate and magnesium sulphate dissolved in the ground water, with the consequent deposition of gypsum ($CaSO_4.2H_2O$) and nesquehonite ($MgCO_3.3H_2O$), has been considered, but because of the relatively slight concentrations existing in these waters such a reaction seems improbable. If calcium and magnesium had been removed in this way, the total solids in the deeper waters would be less (calcium and magnesium having been removed), and, as pointed out on page 62, there is no essential difference in the amount of total solids in the shallow and in the deep waters.

The dissolved salts in the upper hard waters are derived from the soluble materials resulting from the decomposition of minerals in the sedimentary beds and also to some extent from soluble salts deposited in the interstices between mineral grains. The distance through which it is necessary for these hard waters to percolate before they are softened depends upon the quantity of leverrierite and related mineral species in the rocks, and these minerals, although distributed throughout the Lance and Fort Union section in this area, are more abundant in some places than in others. It is probable that the distance may also depend somewhat on the character of the material—that is, whether it is rock in place or alluvium—because aggregates of leverrierite swell and go to pieces when wet, and this would happen when the Lance and Fort Union beds are converted into soil and alluvium. The leverrierite might thereby lose its effectiveness by being disintegrated and decomposed when the Lance and Fort Union beds are converted into alluvium, while that in the Lance and Fort Union beds would be prevented from being disintegrated when wet by the containing walls formed by adjacent mineral grains.

The rate of erosion would, no doubt, be another factor influencing the depth necessary for softening. Where the land was being rapidly degraded it would probably not be necessary for the water to pass through as great a distance as in a region where degradation was relatively slow, for in the region of rapid erosion the base-exchange material would be removed at a rate somewhat proportional to the rate at which its property to exchange alkali for alkaline-earth bases was exhausted, whereas in the region of rela-

---

[40] Gedroiz, K. K., op. cit., p. 26 (U. S. Dept. Agr. translation, p. 19).

tively slow erosion there would accumulate a considerable thickness of rock débris whose base-exchange material had exhausted its property to exchange alkali for alkaline-earth bases, and under these conditions unsoftened water would be encountered at a somewhat greater depth. In this connection it is interesting to note that Way (see p. 64) long ago found that the deeper clays which had not been exposed to weathering were the most effective in producing base exchanges.

Other factors that affect the depth requisite for softening include the structure of the rocks, which would influence the rate of lateral flow underground, and the texture and porosity of the strata, which would influence the rate of downward and lateral percolation.

The discussion of the origin of the soft waters given in this paper is based entirely on data obtained from the Lance (Tertiary?) and Fort Union (Tertiary) formations. It seems very probable that this exchange of bases has also taken place in the underlying Upper Cretaceous beds in this region, but complete data to establish this point are not yet available.

Soft sodium bicarbonate or carbonate and sodium sulphate waters at depths comparable to those of the waters described above are known to occur at many places in the United States, but their origin and their depth relations have not been explained. It is likely that they originated in much the same way as the soft waters in the Lance and Fort Union formations in central Montana and that the natural softening was effected by some mineral having base-exchange properties, not necessarily of the leverrierite group but very probably closely related to it.

SUMMARY.

Studies of ground water in an area of Lance (Tertiary?) and Fort Union (Tertiary) formations in east-central Montana, in the Great Plains province, show that near the surface the water is relatively high in calcium and magnesium, which, with increasing depth, are exchanged for sodium (and potassium?), the result being a natural softening. The minerals of the leverrierite group, which are plentiful though disseminated in these formations and are believed to be derived from the decomposition in place of the glassy constituents of rock fragments, are considered the principal agents in effecting this exchange of bases, though the exchange may be aided by such minerals as kaolinite, feldspar, and mica, which are also present in these rocks. This exchange of bases is accomplished by the time a depth of 125 feet or less is reached. There is no tendency for the water to acquire more dissolved material with increasing depth. The amount of total dissolved solids is therefore determined relatively near the surface.

In this paper the discussion of base exchange by silicates is confined to reactions involving the alkalies and alkaline earths, but similar exchange reactions take place between the most resistant mineral silicates and the bases of salt solutions of the heavy metals. Perhaps the data presented will be of assistance in estimating the number of feet through which it is necessary for ground water carrying salts of the heavy metals in solution to percolate under natural conditions in order to deposit the bases of the metals by exchange with the bases of silicates.

THE ANALYTICAL CONTROL OF ANTI-CORROSION WATER TREATMENT

By W. F. Langelier

(Associate Professor of Sanitary Engineering, University of California, Berkeley, Calif.)

It is the purpose of this paper to discuss certain chemical relationships involved in the action of natural oxygen containing waters on the interior of iron or galvanized pipe and to attempt to place upon a more rational basis the analytical control of preventive treatment. It is hoped that the data presented will be subjected to critical examination by water technologists in various parts of the country, in order that they may ascertain the degree of correlation which exists between theory and laboratory experiment as presented herein and their observations in practice.

It is a well established fact that the composition of water is only indirectly a factor in corrosion. Any water, regardless of its composition, after a certain period of contact with a clean iron surface will cause corrosion of the metal. If, however, certain of the products of the corrosive action are held at the boundary surface, the corrosion rate is reduced and, if the conditions are favorable, further corrosion will be practically eliminated. The problem therefore resolves itself into one of obtaining a suitable protective coating. Users of pipe recognize this and usually specify pipe which has been coated by the manufacturer. Although these coatings are generally useful and necessary, they frequently lack permanence under conditions of use, and often the need of a self-forming or self-healing coating or film, the formation and permanence of which is assured by the composition of the water, is indicated.

In natural oxygen containing water, calcium carbonate is the salt which is most useful in forming or, together with rust,[1] assisting in the formation of a self-healing or natural protective coating. In

---

[1] A rust coating not containing $CaCO_3$ is too porous and is not sufficiently dense and continuous to offer protection, and on the contrary may even accelerate corrosion in the form of pitting.

order that this salt may deposit on the pipe interior, it is necessary that the product of the concentration of the calcium ions and the carbonate ions present in the water *at a given point in the system* shall exceed a certain value known as the "solubility product constant," or in more modern chemical terminology, the "activity product," of the salt in question. .If the water is deficient in either of these ions so that the activity product is not equaled, not only will a carbonate film not form, but also, any existing film will be dissolved. Many surface waters are of this type, and these in general are the ones known to be most corrosive. In such waters, a sufficient increase in either or both of these constituents is all that is needed to attain saturation. Any base added to water will convert existing bicarbonate ions into carbonate ions; but the addition of lime has the advantage of increasing both the calcium and the carbonate ions simultaneously. Because lime is relatively cheap and readily available, it is the chemical which is most often used for this purpose.

It is important to note that as the film is laid down in the pipe, the separate calcium and carbonate ions in the water may decrease to the point where the activity product is no longer exceeded, in which case the water is incapable of furnishing any further protection. Applied to a given distribution system, this would mean that the pipes nearest the source would receive the greatest protection. This tendency, however, is offset by the fact that the deposition from a supersaturated solution is not instantaneous but requires a certain induction period, which for low degrees of supersaturation and at low temperatures may be as great as several hours. Also, there is reason to believe that under conditions of operation there is a leveling process whereby material deposited at one point may be picked up by solution and deposited at a point farther out in the system. This could be brought about by irregularities in treatment and by temperature changes.

Where corrosion of water pipes is not a problem, difficulties due to excessive incrustation with a carbonate scale are often encountered. In these cases, the calcium ion and carbonate ion concentrations exceed too greatly the activity product. This is likely to occur where the water has been softened by the lime-soda process. Here the calcium ion concentration has been decreased by the treatment, but this has been more than offset by an increase in the carbonate ion concentration. A remedy for this condition is recarbonation with CO_2 gas. This is merely a chemical device to convert some of the carbonate ions into bicarbonate ions, in which form they do not participate in the activity product equation.

A survey of the water-supply literature has failed to reveal an adequate and practical formulation of the chemical mechanism of the carbonate film theory. We owe a great deal, however, to the German chemist Tillmans (1) for his work in this field. It was he who in 1912 proposed the carbonate saturation theory of pipe protection. In several papers on this subject he has elaborated his earlier views, but his formulation of carbonate saturation is in terms of free carbon dioxide, a constituent which in the small quantity present near the saturation point for soft waters is not directly determinable in the laboratory. Moreover, his equation does not take into consideration the total ion concentration of the water, an item which may be of major importance.

In the United States, Baylis (2) has been the leader in the application of this theory to practice, and he has written several excellent papers on the subject. He has published a graph, based upon his experiences at Baltimore and elsewhere, of carbonate saturation in terms of pH and total alkalinity. He states, however, that his equilibrium curve must be modified for different waters. Since it fails to take into consideration the calcium content of the water, it of course could not have a general application. Oceanographers have formulated the mechanism of buffer action in sea waters, and bio-chemists have considered the precipitation of calcium salts in tissues. Physical chemists, notably Johnson (3), have studied the chemistry of carbonate solutions and have supplied the several constants needed in the solution of problems in applied chemistry.

In the Sanitary Engineering Laboratory of the University of California this problem has been considered and an attempt has been made to formulate calcium carbonate saturation in natural waters in terms of pH, calcium, alkalinity, total salinity, and temperature. By the use of certain limiting assumptions of conditions not encountered in practice, a final equation has been obtained which is extremely simple to use and which is believed to be entirely rational in its development.

DERIVATION OF THE GENERAL EQUATION

The derivation of the equation for pH_s, which is the pH at which a water of given calcium content and alkalinity is in equilibrium (neither over nor under-saturated with calcium carbonate) is based upon three well known mass law equations and one stoichiometric equation, as follows:

$$(Ca^{++}) \times (CO_3^{--}) = K_s' \tag{1}$$

$$\frac{(H^+) \times (CO_3^{--})}{(HCO_3^-)} = K_2' \tag{2}$$

$$(H^+) \times (OH^-) = K_w \tag{3}$$

$$(Alk) + (H^+) = 2(CO_3^{--}) + (HCO_3^-) + (OH^-)* \tag{4}$$

In these and all subsequent equations, the chemical symbols imply concentrations of the respective ions. Also, all concentrations are molal, with the exception of (Alk) which is an equivalent concentration, i.e. titratable equivalents of base per liter. The two constants K_s' and K_2' are apparent constants, applicable only in solutions of the same total mineral content and at a given temperature, but which for any given condition can be computed from the corresponding known thermodynamic constants, commonly written without the primes. Equation 4 defines what is meant by the term "Total Alkalinity to Methyl Orange," as used in Standard Methods of Water Analysis. It represents the equivalent concentration of titratable base, and not the hydroxyl ion concentration which is sometimes called the "true alkalinity." The other terms in this equation have the same significance as in the mass law equations. The validity of this equation as applied to natural waters assumes only that the salts of weak acids other than carbonic acid are absent.

Proceeding with the derivation,

from (2):

$$(CO_3^{--}) = \frac{(K_2') \times (HCO_3^-)}{(H^+)} \tag{2a}$$

and from (4):

$$(CO_3^{--}) = \frac{(Alk) - (HCO_3^-)}{2}$$

* Within the range of natural waters (pH 4.5 to 10.3) the values of (H$^+$) and (OH$^-$) are relatively so small that they may be neglected in this equation without error.

therefore

$$(HCO_3^-)\left(\frac{K_2'}{(H^+)} + \frac{1}{2}\right) = \frac{(Alk)}{2}$$

and

$$(HCO_3^-) = \frac{(Alk)}{1 + \frac{2K_2'}{(H^+)}} \qquad (5)$$

By substituting this value of (HCO_3^-) in (2a), we obtain

$$(CO_3^{--})^* = \frac{K_2'}{(H^+)} \times \frac{(Alk)}{1 + \frac{2K_2'}{(H^+)}} \qquad (6)$$

Now, by substituting this value of (CO_3^{--}), we obtain from (1)

$$(Ca^{++}) \times \frac{K_2'}{(H_s^+)} \times \frac{(Alk)}{1 + \frac{2K_2'}{(H_s^+)}} = K_s' \qquad (7)$$

Since it is the practice to represent (H^+) in terms of pH, which is a symbol for $\log \frac{1}{(H^+)}$, it is convenient to rearrange (7) in terms of logarithms using this same convention, whence we obtain for pH_s, the pH at saturation:

$$pH_s = (pK_2' - pK_s') + pCa + pAlk + \log\left[1 + \frac{2K_2'}{(H_s^+)}\right] \qquad (8)$$

A more general equation, not limited to waters in the range specified above, is obtained by correcting the alkalinity term for hydrogen and hydroxyl ions, as follows:

$$pH_s = (pK_2' - pK_s') + pCa$$
$$+ p\left[Alk + (H^+) - \frac{K_w}{(H^+)}\right] + \log\left[1 + \frac{2K_2'}{(H_s^+)}\right] \qquad (9)$$

* It should be stated here that the method of determining the concentration of carbonate, as given in Standard Methods by means of a two indicator titration is (within a certain range of pH) too inaccurate for our present purpose. Equations 5, 6, and similarly derived equations for H_2CO_3 and OH could well be included in Standard Methods to supplement the present discussion of Alkalinity.

The correction to (Alk) as well as the term, $\log\left[1 + \frac{2K_2'}{(H_s^+)}\right]$, is ordinarily small, and in all cases where pH_s falls between approximately 6.5 and 9.5, both can be omitted, and we have for this range:

$$pH_s = (pK_2' - pK_s') + pCa + pAlk \qquad (10)$$

In the event that pH_s as obtained by (10) is greater than 9.5, the corresponding value of (H_s^+) may be substituted in the last term of (8) and a closer approximation to the correct pH_s will be obtained by that equation. In order to facilitate the use of equations 8 or 9, applicable when pH_s is greater than approximately 9.5, table 1 has

TABLE 1

Values of $\log\left[1 + \frac{2K_2'}{(H^+_s)}\right]$ *as a Function of* $(pK_2' - pH_s)$

$(pK_2' - pH_s)$	0.0	0.1	0.2	0.3	0.4	0.5	0.6	0.7	0.8	0.9	1.0	1.2	1.3
$\log\left[1 + \frac{2K_2'}{(H^+_s)}\right]$	0.48	0.41	0.35	0.30	0.25	0.21	0.18	0.15	0.12	0.10	0.08	0.06	0.05

been included. This table gives values of $\log\left(1 + \frac{2K_2'}{(H_s^+)}\right)$ for different values of $(pK_2' - pH_s)$. The correction of the (Alk) for caustic ions need not be made unless pH_s exceeds 10.5. This condition is encountered so rarely that a table to simplify its computation has not been included.

The algebraic difference between the actual pH of a sample of water and its computed pH_s is the logarithm of its degree of carbonate saturation. We have called this the "Calcium Carbonate Saturation Index." Actually this index is the logarithm of the ratio of the hydrogen ion concentration which the sample must have if saturated (without change in composition) to its actual hydrogen ion concentration. If the index is zero, the sample is in equilibrium. A plus sign before the index indicates over-saturation and a tendency to crystalize, or to lay down a protective coating of $CaCO_3$ in the pipe, and a minus sign indicates under-saturation, or a tendency to dissolve an existing carbonate coating. Expressed mathematically:

Saturation Index = $\text{pH}_\text{actual} - \text{pH}_\text{saturation} = \log \dfrac{1}{(H^+)} - \log \dfrac{1}{(H_s^+)}$ $= \log \dfrac{(H_s^+)}{(H^+)}$. The significance of this Index and its applications to pipe corrosion problems will be discussed later.

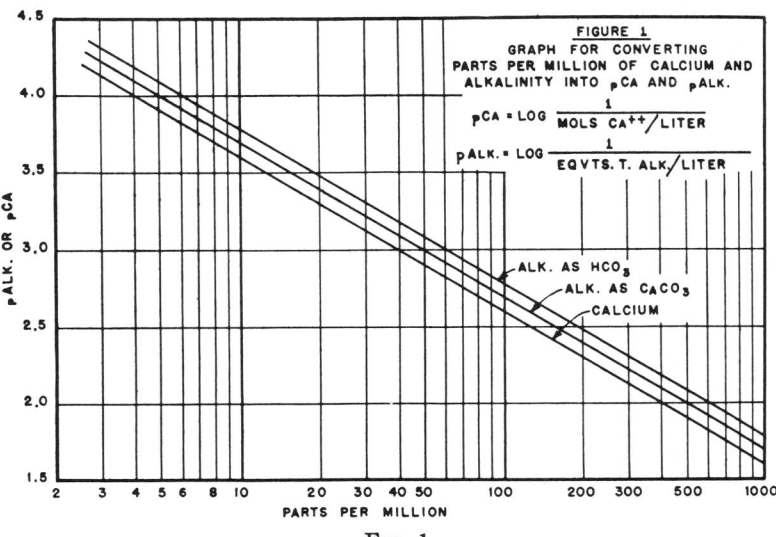

Fig. 1

PROVISION FOR TOTAL SALINITY OR IONIC STRENGTH

In the proposed equation for pH_s, the effects of variable salinity upon saturation are provided for in the term, $(pK_2' - pK_s)$. For any given water the exact values of these constants can be computed from the true thermodynamic constants K_2 and K_s (the values of which are known) by multiplying each with a certain activity coefficient, in accordance with the principles of the "activity" concept of the theory of electrolytic dissociation. This concept, which has only recently come into general acceptance, assumes complete dissociation of all the salts in a dilute solution, but requires that all concentration terms appearing in a mass law equation shall be converted into activity terms by combining each with an activity coefficient. These coefficients are related to the total salt content of the solution and to the valences of the ions present. A convenient method of expressing the combined effect of salinity and valence of the ions present is in terms of "ionic strength" represented by the symbol, μ. The ionic strength of a solution is defined as one half the sum of the molality of each ion in solution multiplied by the square of its valence, or if c is the molal concentration of the ion and v is the valence of that ion, the ionic strength of the solution μ equals $1/2[c_1v_1^2 + c_2v_2^2 + \cdots]$. An example in the computation of the ionic strength of the water of the Mississippi River at New Orleans, from a mineral analysis of same, follows:

MONOVALENT IONS, P.P.M.	MOLAL CONCENTRATION	DIVALENT IONS, P.P.M.	MOLAL CONCENTRATION
$Na^+ = 18$	0.00078	$Mg^{++} = 10$	0.0004
$Cl^+ = 28$	0.00079	$Ca^{++} = 39$	0.0010
$HCO_3^- = 116$	0.00190	$SO_4^{--} = 42$	0.0004
$\Sigma(c_1)$	0.00347	$\Sigma(c_2)$	0.0018
$\Sigma(c_1v_1^2)$	0.00347	$\Sigma(c_2v_2^2)$	0.0072

Ionic strength $\mu = 1/2(0.00347 + 0.0072) = 0.0054$

It will be noted that for this water, which has a total dissolved solids content of 220 p.p.m., the sum of the equivalent concentrations of the monovalent ions is about equal to the sum of the equivalent concentrations of the divalent ions, and it is believed that his is reasonably typical of other natural waters with respect to the relative proportions of monovalent and divalent ions. In this example, 40 parts per million of total solids is equal to 0.001 unit of ionic strength. Computation of the ionic strength of several typical waters indicates that for the purpose under consideration this value may be used as a general relationship applicable to most waters.

With this means of computing the ionic strength of a sample of water from its total dissolved solids, we are now able to compute the activity coefficient f for any given ion by the equation, $-\log f = 0.5v^2\sqrt{\mu}$. This is the Brönsted-La Mer modification of the Debye-Hückel equation. In this equation, which has been found to hold for ionic strengths up to 0.02, v is the valence of any ion having the activity coefficient f.

Consider now the application of the individual ion activity coefficients obtained in this way to the computation of the value pK_s' from pK_s, the true thermodynamic constant which varies only with

TABLE 2

Values of pK_2' and pK_s' at 25°C. for various ionic strengths and of the difference $(pK_2' - pK_s')$ for various temperatures

IONIC STRENGTH	TOTAL DIS-SOLVED SOLIDS	25°C.			$(pK_2' - pK_s')$							
		pK_2'	pK_s'	$pK_2'-pK_s'$	0°C.	10°C.	20°C.	50°C.	60°C.	70°C.	80°C.	90°C.
.0000	0	10.26	8.32	1.94	2.20	2.09	1.99	1.73	1.65	1.58	1.51	1.44
.0005	20	10.26	8.23	2.03	2.29	2.18	2.08	1.82	1.74	1.67	1.60	1.53
.001	40	10.26	8.19	2.07	2.33	2.22	2.12	1.86	1.78	1.71	1.64	1.57
.002	80	10.25	8.14	2.11	2.37	2.26	2.16	1.90	1.82	1.75	1.68	1.61
.003	120	10.25	8.10	2.15	2.41	2.30	2.20	1.94	1.86	1.79	1.72	1.65
.004	160	10.24	8.07	2.17	2.43	2.32	2.22	1.96	1.88	1.81	1.74	1.67
.005	200	10.24	8.04	2.20	2.46	2.35	2.25	1.99	1.91	1.84	1.77	1.70
.006	240	10.24	8.01	2.23	2.49	2.38	2.28	2.03	1.94	1.87	1.80	1.73
.007	280	10.23	7.98	2.25	2.51	2.40	2.30	2.05	1.96	1.89	1.82	1.75
.008	320	10.23	7.96	2.27	2.53	2.42	2.32	2.07	1.98	1.91	1.84	1.77
.009	360	10.22	7.94	2.28	2.54	2.43	2.33	2.08	1.99	1.92	1.85	1.78
.010	400	10.22	7.92	2.30	2.56	2.45	2.35	2.10	2.01	1.94	1.87	1.80
.011	440	10.22	7.90	2.32	2.58	2.47	2.37	2.12	2.03	1.96	1.89	1.82
.012	480	10.21	7.88	2.33	2.59	2.49	2.39	2.13	2.04	1.97	1.90	1.83
.013	520	10.21	7.86	2.35	2.61	2.50	2.40	2.15	2.06	1.99	1.92	1.85
.014	560	10.20	7.85	2.36	2.62	2.51	2.41	2.16	2.07	2.00	1.93	1.86
.015	600	10.20	7.83	2.37	2.63	2.52	2.42	2.17	2.08	2.01	1.94	1.87
.016	640	10.20	7.81	2.39	2.65	2.54	2.44	2.19	2.10	2.03	1.96	1.89
.017	680	10.19	7.80	2.40	2.66	2.55	2.45	2.20	2.11	2.04	1.97	1.90
.018	720	10.19	7.78	2.41	2.67	2.56	2.46	2.21	2.12	2.05	1.98	1.91
.019	760	10.18	7.77	2.41	2.67	2.57	2.47	2.21	2.12	2.05	1.98	1.91
.020	800	10.18	7.76	2.42	2.68	2.58	2.48	2.22	2.13	2.06	1.99	1.92

temperature. In accordance with the activity concept, we may write for the activity product of calcium carbonate

$$f_{Ca} \cdot (Ca^{++}) \times f_{CO_3} \cdot (CO_3^{--}) = K_s$$

Also, equation (1)

$$(Ca^{++}) \times (CO_3^{--}) = K_s'$$

therefore

$$K_s' = \frac{K_s}{f_{Ca} \times f_{CO_3}}$$

Since the calcium and carbonate ions have the same valence f_{Ca} will be equal to f_{CO_3}, and therefore

$$pK'_s = pK_s - 2 \log f_{Ca} = pK_s - 4\sqrt{\mu} \qquad (11)$$

The values of pK'_s for the various ionic strengths (or salinities) given in table 2 were computed by this method from Frear and Johnston's (4) value of K_s which is 4.8×10^{-9} at 25°C.

The values of pK'_2 given in this table are from MacInnes and Belcher (5).

[*Editors' Note*: Material has been omitted at this point.]

SALINITY AND TEMPERATURE EFFECTS IN PIPE PROTECTION

It has been shown that in natural fresh waters an increase in salinity causes an increase in the solubility of calcium carbonate such that a water of 800 p.p.m. total solids requires about threefold more of the salt at saturation than does distilled water. This increase in solubility is very appreciable in brackish waters and brines and is undoubtedly one of the principle reasons why such waters, unable to form a protective carbonate film on pipe interiors and exteriors, are so very corrosive. It is known that calcium carbonate is approximately 500-fold more soluble in sea than in fresh water. Attention is called to the fact that equation (11), with the salinity correction as given, holds only for waters containing not more than 800 parts per million total solids. Other correction factors are available for more saline waters.

In practice it is commonly observed that the tendency of water to corrode or incrust pipe is greatest at elevated temperatures. Actually, an increase in temperature exerts two effects: (1) a shift of the Index in the direction of higher saturation, and (2) an increase in the speed of the reaction in either direction, as a result of decreasing viscosity. Each of these effects is important and will be discussed further.

(1) The effect of temperature on the 25°C. saturation index. The values of pH, pK_2' and pK_s', each of which is used in computing the Saturation Index, vary with temperature, and although ordinarily for any given water it is sufficient to know the value of the Index at 25°C., it is desirable to have a conversion factor for estimating the Index for any other temperature encountered in practice. Considering first the temperature coefficients of pH and pK_2, it is unfortunate that these are not known throughout the complete temperature range of 0°C. to 100°C. However, it is known that for at least the lower part of this range, pH and pK_2' decrease at approximately the same rate, i.e. about 0.01 units per degree C. Assuming this to be correct for the entire range, and since it is with their difference that we are concerned, the 25°C. values may be used in the equation for any temperature. This is an advantage because pH measurements are most conveniently made at room temperature. Thus it is neces-

TABLE 4

Effect of temperature on solubility of calcium carbonate

	t, °C.								
	0	10	20	25	50	60	70	80	90
r	1.8	1.4	1.1	1.0	0.62	0.51	0.44	0.37	0.32
pK_s'	8.06	8.17	8.27	8.37	8.53	8.61	8.60	8.75	8.82

sary to consider only the temperature correction for pK_s'. This correction is given by Frear and Johnston as $\log r = (830/T) - 2.78$. In this equation r is the ratio of solubility of $CaCO_3$ at t°C. to that at 25°C., the values of which, together with the values of the corresponding pK_s', are given in table 4.

It should be noted here that the effect of increasing temperature from 25°C. to 90°C. is to decrease the solubility of calcium carbonate[2] threefold. The significance, of course, is that heating of water will

[2] In the literature of water supply, there seems to be much confusion as to the correct solubility of $CaCO_3$ and also the effect of temperature upon its solubility. The lower values (13 p.p.m.) refer to pure water in which there is no free CO_2, and the higher values (35 to 50 p.p.m.) are for waters in equilibrium with the CO_2 of the air. As to the direction of change with temperature, several water technologists state that the solubility of $CaCO_3$ increases with temperature rise. The work of Johnston and of other authorities indicates that this is incorrect.

cause a change in the Index toward saturation. For example, a water which has an Index of −0.3 at 25°C. will have an Index of +0.2 at 90°C.

(2) Effect of temperature on reaction velocity. Temperature is the most important factor in controlling reaction velocity. In general, for reactions of this kind, involving a solid and a liquid, a rise in temperature of 10°C. increases the speed of the reaction about twofold. The importance of this is noted when it is realized that a temperature rise from 25°C. to 75°C. increases the speed of the reaction about 32-fold. The higher the temperature, the more vigorous the activity in either direction.

SIGNIFICANCE OF THE SATURATION INDEX

From the foregoing discussion it might be assumed that in practice a zero Saturation Index would be ideal. This is correct, provided that the Index is zero at those points in the system where the water is most active. If such activity occurs principally in hot water systems, obviously the high-temperature Index should be zero; on the other hand, if cold-water mains or services require protection, the low-temperature Index should be zero. Since most corrosion troubles occur in hot-water systems, it would seem that an Index of, say, −0.4 at 20°C. (which is equivalent to 0.0 at 70°C.) might prove best for most supplies.

It should be emphasized that the Saturation Index is an indication of directional tendency and of driving force but that it is in no way a measure of capacity. The capacity to coat or corrode the pipe will depend upon the property of the water to resist change in the value of its Index following attack. Obviously, a high concentration of both calcium and carbonate ions will permit of a more extensive deposit of calcium carbonate than will a low concentration, yet both waters might well have the same Index. This buffer property, or ability to resist change, will be least in soft waters; more accurately, those with the higher values of pH_s. The upper limit for this value in practice is not known, but it would seem to be somewhere in the neighborhood of 9.6. In waters of this type, the total alkalinity may need building up and for this purpose limestone contact, possibly followed with lime dosage, would seem to be the best treatment.

Moreover, the Saturation Index does not indicate the quantity of chemical required to establish equilibrium. In general the lower the value of pH$_s$ the greater will be the dosage. It can be computed, but is best determined by treating several portions of the water with the chemical which is to be used and computing the Index of each from its analysis. This method would seem to possess considerable merit over the time-consuming and not two dependable marble test. It makes use of standard analytical procedures and yields more complete information.

[*Editors' Note:* Material has been omitted at this point.]

SUMMARY AND CONCLUSIONS

An equation is derived for pH$_s$, the pH at which a given water is in equilibrium with solid calcium carbonate. This equation, in its simplest form and applicable within the pH range 7.0 to 9.5, is: pH$_s$ = (pK$_2'$ − pK$_s'$) + pCa^{++} + pAlk. The two latter terms are negative logarithms of the molal and equivalent concentrations of calcium and titratable base, respectively. pK$_2'$ and pK$_s'$ are the negative logarithms of the second dissociation constant for carbonic acid and the activity product of CaCO$_3$, respectively. The difference (pK$_2'$ − pK$_s'$) varies with ionic strength (salinity) and temperature. Its value for a soft water at 20°C. is 2.1 and at 80°C. is 1.6. A table of values for varying salinities and temperatures is given. Experiments with a wide variety of raw and treated waters confirm the validity of the equation at 25°C. The difference (pH$_{actual}$ − pH$_s$) is called the Saturation Index. If the Index is zero, the water is in equilibrium at that temperature. A positive Index indicates oversaturation and a tendency to crystallize or to lay down a protective coating of CaCO$_3$ in the pipe, and a negative Index indicates undersaturation, or a tendency to dissolve an existing carbonate coating. A rise of temperature increases the algebraic value of the Index, and for a given water very materially increases the degree of activity.

The average Index for the several largest unadjusted municipal water supplies in the United States was found to be −1.58 equivalent to 2.63 percent of saturation. In those cities where pH adjustment is practised, either for corrosion or incrustation control, the tendency is to undertreat.

The effects of various forms of water conditioning, i.e. zeolite softening, lime-soda softening, coagulation with alum, contact with limestone, dosing with caustic soda or lime, etc., upon the Saturation Index, are discussed and examples from practice are given.

The Index is shown to correlate with the published results of a six-year study of interior pipe corrosion by the New York City Water Department.

The Index furnishes a new method which can be used in the laboratory control of anti-corrosion water treatment.

LITERATURE CITED

(1) TILLMANS, J., "Die Chemische Untersuchung von Wasser und Abwasser," 2nd Ed., Wilhelm Knapp Halle (Saale) 1932.
(2) BAYLIS, JOHN R., Jour. Amer. W. W. Assoc., 27, 220–234 (1935).
(3) JOHNSTON, JOHN, Jour. Amer. Chem. Soc., 38, 947–983 (1916).
(4) FREAR, G. L., AND JOHNSTON, J., Jour. Amer. Chem. Soc., 51, 2082–2092 (1929).
(5) MACINNES, D. A., AND BELCHER, D., Jour. Amer. Chem. Soc., 55, 2630 (1933).
(6) THRESH, JOHN C., BEALE, J. F., AND SICKLING, E. V., "The Examination of Waters and Water Supplies," 4th Ed., p. 647 (1933).
(7) VIESOHN, VOM WASSER, Jahrbuch IV, G. M. B. H., Berlin (1930).
(8) HALE, FRANK E., Jour. Amer. W. W. Asso., 26, 1315–1347 (1934); also 27, 1199–1224 (1935).

6

Copyright ©1944 by the American Geophysical Union
Reprinted from *Am. Geophys. Union Trans.* **25**:914-923 (1944)

A GRAPHIC PROCEDURE IN THE GEOCHEMICAL INTERPRETATION OF WATER-ANALYSES

Arthur M. Piper

General considerations--This paper outlines certain fundamental principles in a graphic procedure which appears to be an effective tool in segregating analytical data for critical study with respect to sources of the dissolved constituents in waters, modifications in the character of a water as it passes through an area, and related geochemical problems. The procedure is based on a multiple-trilinear diagram (Fig. 1) whose form has been evolved gradually and independently by the writer during the past several years through trial and modification of less comprehensive antecedent forms. Neither the diagram nor the procedure here described is a panacea for the easy solution of all geochemical problems. Many problems of interpretation can be answered only by intensive study of critical analytical data by other methods.

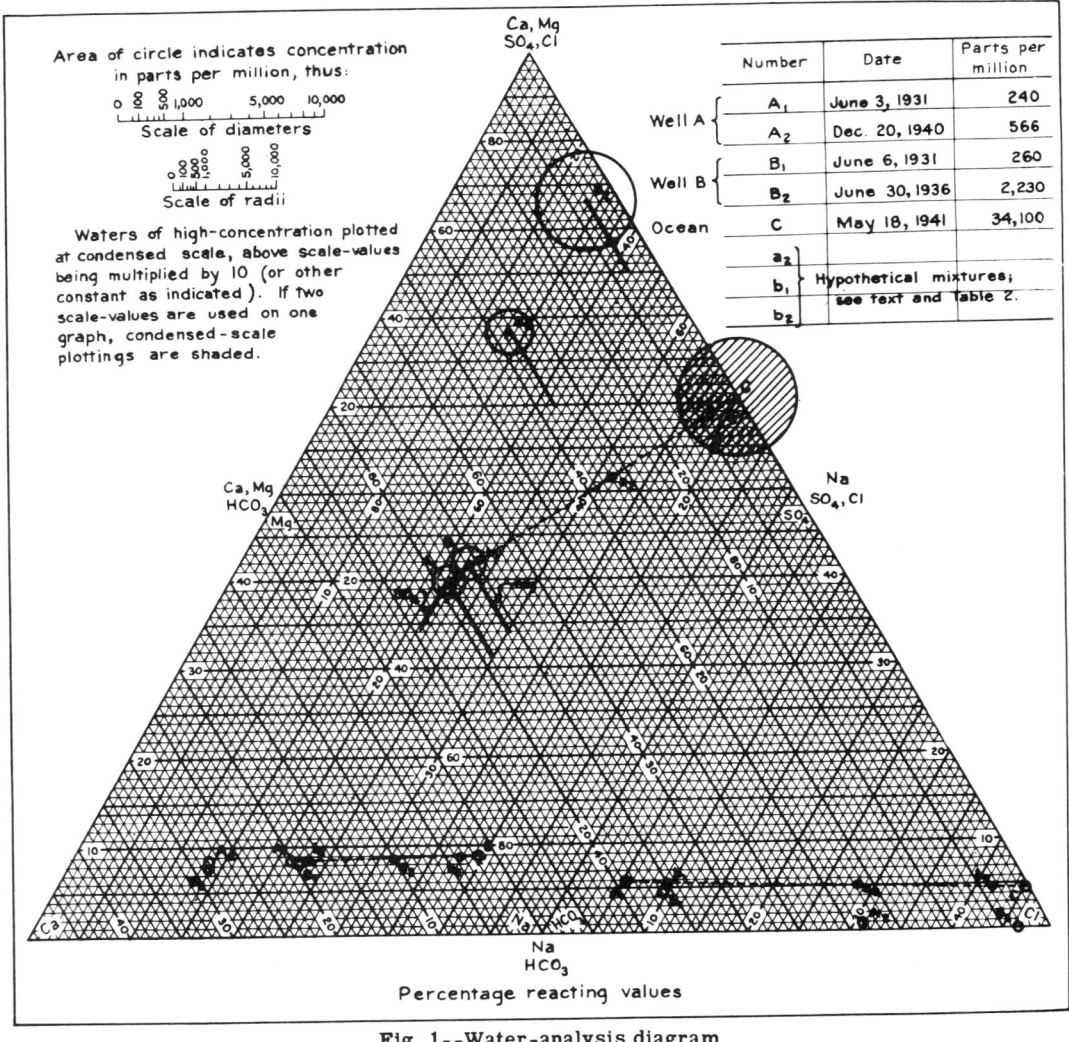

Fig. 1--Water-analysis diagram

Interpretation of Water-Analyses

In certain respects the diagram is analogous to, but in other respects differs fundamentally from, the "geochemical chart" described recently by HILL [see 1 of "References" at end of paper] and from the "water classification diagram" described by LANGELIER and LUDWIG [2]. As presented in Figure 1, the diagram is the mirror image of a prototype which in 1942 was circulated among colleagues in the United States Geological Survey and other coworkers in hydrology; thus, it conforms substantially to LANGELIER'S adaptation [3] of HILL'S diagram and to the conventional practice of arranging diagrams of water-analyses with cations shown to the left of the anions. Some details in Figure 1 embody adaptations which serve constructive criticisms of the prototype by colleagues and coworkers; for these criticisms the writer is grateful.

Most natural waters contain relatively few dissolved constituents, with cations (metals or bases) and anions (acid radicles) in chemical equilibrium with one another; commonly the waters contain some silicon, iron, and aluminum but these constituents are usually assumed to occur in the colloid state as oxides and not to be in chemical equilibrium with the ionized constituents. Ordinarily the most abundant cation constituents are two "alkaline earths", calcium (Ca) and magnesium (Mg), and also one "alkali", sodium (Na). Potassium (K) also occurs commonly, but ordinarily is much less abundant than sodium. Still other cation-constituents occur in appreciable quantities in highly concentrated natural waters and in some waters of unusual composition. For the graphic methods treated in this paper all these less abundant constituents are summed with the major three constituents to which they are respectively related in chemical properties, as indicated by the two ranks of entry in Table 1. The most common anion-constituents are one "weak acid" bicarbonate (HCO_3); also two "strong acids", sulphate (SO_4) and chloride (Cl). Less common anion-constituents are listed in Table 1; for plotting, these are summed with the major three anion to which they are respectively related. Thus, for much of the graphic methods here described, a natural water is treated substantially as though it contained only three cation-constituents and three anion-constituents.

Table 1 -- **Common and minor constituents of natural waters**

Cations	Reciprocal of combining weight	Anions	Reciprocal of combining weight
Alkaline earths		Weak acids	
Calcium (Ca^{++})	0.04990	Bicarbonate (HCO_3^-)	0.01639
Barium (Ba^{++})	0.01456	Carbonate ($CO_3^=$)	0.03333
Strontium (Sr^{++})	0.02282	Tetraborate ($B_4O_7^=$)	0.01288
Magnesium (Mg^{++})	0.08224	Orthophosphate ($PO_4^\equiv$)	0.03157
Alkalies		Strong acids	
Sodium (Na^+)	0.04348	Sulphate ($SO_4^=$)	0.02082
Potassium (K^+)	0.02558	Chloride (Cl^-)	0.02820
Caesium (Cs^+)	0.00752	Iodide (I^-)	0.00788
Rubidium (Rb^+)	0.01170	Bromide (Br^-)	0.01251
Lithium (Li^+)	0.14409	Fluoride (F^-)	0.05263
Ammonium (NH_4^+)	0.05543	Nitrate (NO_3^-)	0.01613
		Nitrite (NO_2^-)	0.02174

Notes: Of the second-rank constituents only potassium, carbonate, fluoride, and nitrate are commonly determined in a "complete" analysis. Reciprocals of combining weights are based on the international atomic weights of 1938.

In substantially all natural waters the cations are in chemical equilibrium with the anions. Accordingly, if the concentrations of the several dissolved constituents are measured in terms of percentage of reacting value--that is, according to their "equivalents per million" expressed as a percentage of the sum of the equivalents for all the constituents [the concentration of any constituent in equivalents per million (milligram equivalents per kilogram) is computed by multiplying its concentration in parts per million by the reciprocal of its combining weight]--the subtotals of the cations and anions are necessarily each 50 per cent of the whole.

Thus, to the extent that a natural water can be treated in terms of three cation-variables and three anion-variables, as has been outlined, and because the subtotals of its cations and anions are

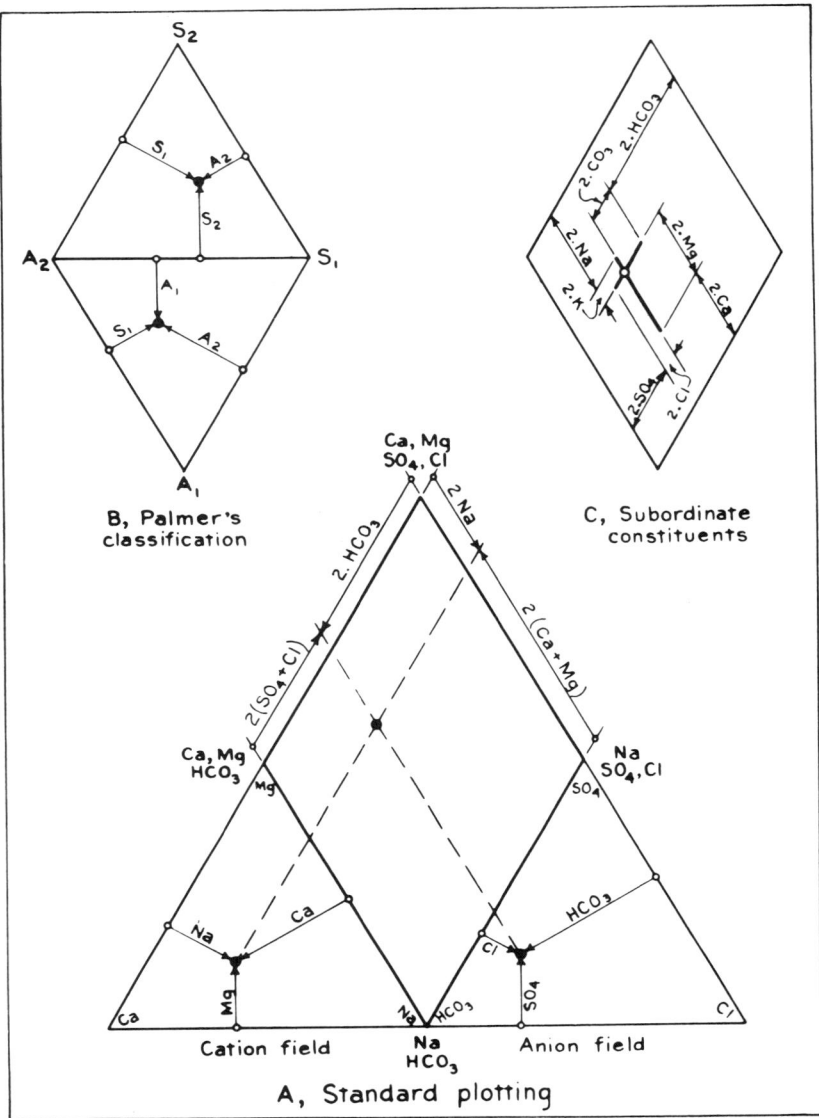

Fig. 2--Plotting key for water-analysis diagram (In diagram B, A_1 indicates primary alkalinity or "carbonate alkali", A_2, secondary alkalinity or "carbonate hardness", S_1, primary salinity or "noncarbonate alkali", S_2, secondary salinity or "non-carbonate hardness")

each 50 per cent of the total reacting value, the essential chemical character of the water can be indicated graphically by single-point plotting on trilinear coordinates. This is the basis of the diagram herein described.

Methods of plotting

The diagram herein described combines three distinct fields for plotting--two triangular fields at the lower left and lower right, respectively, with percentage scales reading in 50 parts; also an intervening diamond-shaped field with scales reading in 100 parts (see Figs. 1 and 2-A). In the triangular field at the lower left, the percentage reacting values of the three cation-groups (Ca, Mg, Na) are plotted as a single point according to conventional trilinear coordinates. The three anion-groups (HCO_3, SO_4, Cl) are plotted likewise in the triangular field at the lower right. Thus, two points on the diagram--one in each of the two triangular fields--indicate the relative concentrations of the several dissolved constituents of a natural water.

The central diamond-shaped field is used to show the over-all chemical character of the water by a third single-point plotting, which is at the intersection of rays projected from the plottings of cations and anions as indicated on Figure 2-A. Using the scales of Figure 1, the position of this plotting indicates the relative composition of a water in terms of the cation-anion pairs that correspond to the four vertices of the field. This central-field plotting can also be taken directly from the analytical data according to the vectors shown along the outer margins of the field on Figure 2-A. For such plotting only one cation-variable and one anion-variable need be used--either alkaline earths or alkalies with either weak acids or strong acids; the two percentage reacting values selected from the analytical data are doubled to suit the numerical scales of Figure 1.

The three trilinear plottings just described will show the essential chemical character of a water according to the relative concentration of its constituents, but not according to the absolute concentrations. Because the absolute concentrations commonly are decisive in many problems of interpretation, it is convenient to indicate the plotting in the central field by a circle whose area is proportional to the absolute concentration of the water. Figure 1 shows such plottings for several dissimilar waters.

The diamond-shaped field of the writer's diagram is essentially a mirror image of LANGELIER'S diagram, sheared 30° to transform the latter from Cartesian to trilinear coordinates. Also, plottings in that field can be made or interpreted according to PALMER'S classification [4], as explained in the following paragraphs. This scheme of classification has many advantages but has not found universal favor, possibly because it implies certain specific combinations of dissolved constituents, which are hypothetical rather than real.

The classification by PALMER designates the alkaline cations (Na, K) as the "primary" constituents, the alkaline-earth cations (Ca, Mg) as the "secondary" constituents, the strong-acid anions (SO_4, Cl, NO_3) as the "saline" constituents, and the weak-acid anions (CO_3, HCO_3) as the "alkaline" constituents. It ascribes "primary salinity" to a water to the extent that the alkalies of that water are balanced by strong acids and "secondary alkalinity" to the extent that the alkaline earths are balanced by weak acids. Further, it ascribes "primary alkalinity" to the water to the extent that alkalies exceed strong acids and are balanced by weak acids, or "secondary salinity" to the extent that alkaline earths exceed weak acids and are balanced by strong acids. Because the latter two properties are mutually exclusive, a water can not possess both. Thus, the chemical character of most natural waters can be expressed by PALMER'S classification in terms of three hypothetical properties; in terms of percentage reacting value, the three must sum up to unity (analytical errors adjusted), of course. Accordingly, chemical character can be plotted as a single point with respect to trilinear coordinates.

A very few natural waters contain free acid in substantial quantity--that is, hydrogen is present as a cation; this cation PALMER designates as "tertiary". The chemical character of such waters can not be fully represented on the diagram.

If on the water-analysis diagram the two rays projected from the plottings of cations and anions intersect in the lower triangular half of the diamond-shaped field, the water has primary alkalinity (A_1), secondary alkalinity (A_2), and primary salinity (S_1) (see Fig. 2-B). Conversely, if the rays intersect in the upper triangular half of the field, the water has secondary salinity (S_2) rather than primary alkalinity.

Obviously, the plottings in the diamond-shaped field do not bring out critical ratios between certain constituents, such as the ratios between sodium and potassium, calcium and magnesium, carbonate and bicarbonate, or sulphate and chloride. When pertinent, these ratios can be indicated graphically by vectors that indicate four of the common eight constituents, as shown by Figure 2-C. In scaling these vectors, the percentage reacting values of the analytical data are doubled to suit the numerical scales of Figure 1.

With respect to the sources of dissolved constituents, or to progressive changes in chemical character within a particular area, many problems involve waters which differ only slightly in character or in which significant differences are masked by some common but preponderant constituent. Under such circumstances, the small differences in character can be emphasized by using the full diagram to represent proportionate subdivisions of the standard three plotting fields,

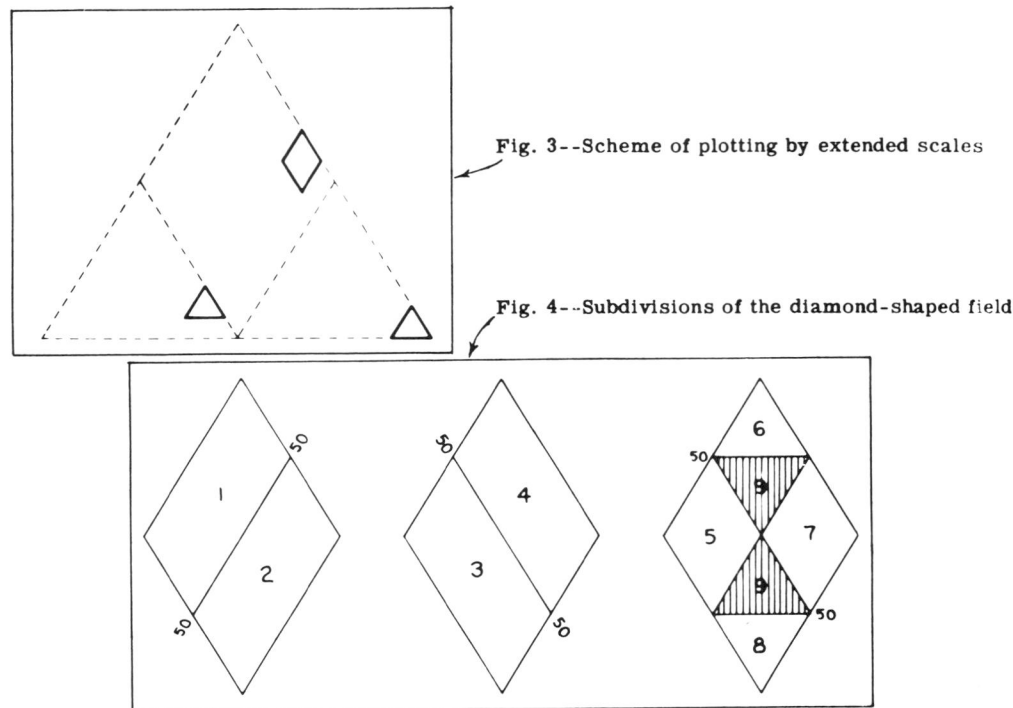

Fig. 3--Scheme of plotting by extended scales

Fig. 4--Subdivisions of the diamond-shaped field

and plotting the constituents with extended scales. For example, among brines similar to ocean water the minor constituents can be differentiated by plotting at five times the standard scale and using the full diagram to represent the small areas outlined in Figure 3. If desired, minor or accessory constituents can be so emphasized by vectors in the diamond-shaped field, as described in the preceding paragraph.

Differentiation of water-types

Certain distinct types can be quickly discriminated by their plottings in certain subareas of the diamond-shaped field, as indicated by Figure 4 and the following explanation: Area 1, alkaline earths exceed alkalies; Area 2, alkalies exceed alkaline earths; Area 3, weak acids exceed strong acids; Area 4, strong acids exceed weak acids; Area 5, secondary alkalinity ("carbonate hardness") exceeds 50 per cent--that is, chemical properties of the water are dominated by alkaline earths and weak acids; Area 6, secondary salinity ("non-carbonate hardness") exceeds 50 per cent; Area 7, primary salinity ("non-carbonate alkali") exceeds 50 per cent--that is, chemical properties are dominated by alkalies and strong acids--ocean water and many brines plot in this area, near its right-hand vertex; Area 8, primary alkalinity ("carbonate alkali") exceeds 50 per cent--here plot the waters which are inordinately soft in proportion to their content of dissolved solids; Area 9, no one of the cation-anion pairs in PALMER'S classification exceeds 50 per cent.

These subareas might serve as a basis for numerical or other symbols to designate specific classes, types, and subtypes of water. Symbols for this purpose have been introduced by PALMER [4] and HILL [1]; however the writer feels that inflexible classifications of this sort tend to confuse by over-emphasizing differences in composition that may not be significant to the problem under consideration.

To serve most needs for classifying waters by types the writer proposes--in lieu of symbols such as those introduced by PALMER, HILL, and others--to designate a water by a binomial symbol written in the form of a decimal fraction, whose two terms are (1) the percentage of hardness-causing constituents among the bases and (2) the percentage of bicarbonate (and carbonate, if present) among the acids. For example, the symbol 64.80 would indicate a water in which the hardness-causing constituents (Ca + Mg) amount to 64 per cent of all the bases, in terms of reacting values (equivalents); also in which the weak acids (CO_3 + HCO_3) amount to 80 per cent of all the acids, in like terms. Numerically, the first term is twice the percentage reacting value of calcium and magnesium from analytical data in which the percentage base is the sum of all dissolved constituents, both bases and acids. The first term can be read directly from the numerical scales on the diamond-shaped field of the diagram, in accord with Figure 2-A. Likewise, the second term of the symbol is twice the percentage reacting value of carbonate and bicarbonate, or is scaled directly from the diagram. This form of symbol has the distinct advantage of indicating the general character of a water specifically, without the disadvantage of implying that two waters have distinctly different characters merely because their analyses plot on either side of a boundary betwee arbitrary subdivisions of any water-analysis diagram. For treatment in a text, waters can be grouped according to limiting values for the two terms of the symbol here proposed, and those limits can be varied at will to suit the discussion of the problem at hand.

This decimal-fraction symbol indicates numerous characteristics of a water simply but specifically. Thus, the more common type of natural water contains chiefly calcium, magnesium, and bicarbonate; its symbol approaches 100.100 as a limit. For its fairly common opposite, the alkalicarbonate water, the symbol approaches 0.100 as a limit. The first term of the symbol indicates relative hardness in percentage of total equivalents. If the second term exceeds the first, all the

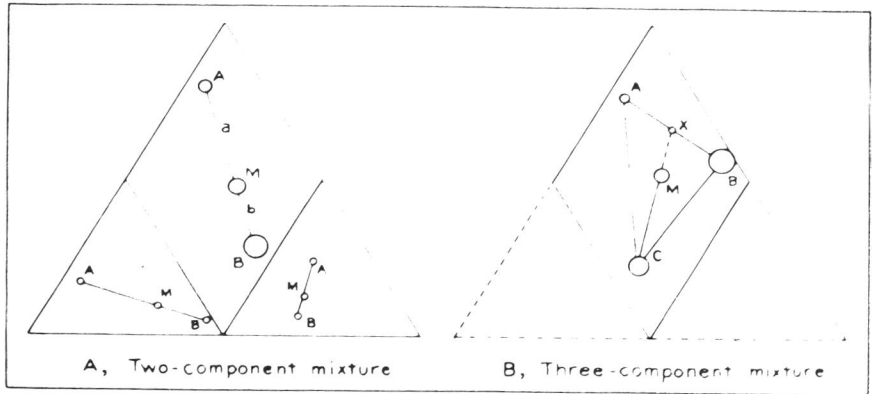

Fig. 5--Preliminary confirmation of mixtures

hardness is carbonate or "temporary" hardness. However, if the second term is smaller, some of the hardness is non-carbonate or "permanent" and the relative amount of non-carbonate hardness is indicated by the numerical difference between the two terms. The first term of the symbol is the percentage complement of the "per cent sodium" introduced by SCOFIELD [5] to measure the effect of a water on the physical properties of a soil when applied for irrigation. Thus, if this term is greater than about 50 the physical condition of the soil is not likely to be impaired seriously, but if the term is less than about 40 such impairment may result.

Mixtures of waters

Many hydrologic problems involve apparent mixtures of natural waters, which the investigator seeks to confirm or disprove. The solution of such problems is facilitated by use of the diagram as described beyond; this use has been anticipated in the initial paper by HILL and in the paper by LANGELIER and LUDWIG, which have been cited.

Mixtures of two waters in all proportions, if all products remain in solution, plot in the three fields on the respective straight lines that join the points representing the respective chemical characters of the two waters mixed. Thus, in Figure 5-A the straight lines AB will include the plottings of every possible mixture of two waters whose chemical characters are represented by points A and B, respectively. Point M represents a possible mixture in one particular proportion.

There is an obvious application of this procedure in demonstrating a cause for deterioration of water quality in a coastal area--whether or not due to simple invasion by ocean water. If so, chemical analyses of the progressively deteriorated water must, within reasonable limits of error, plot on a set of three vectors directed toward the composition of ocean water. If the analyses do not so plot, simple admixture of sea-water is not a valid and adequate explanation of the deterioration. For example, on Figure 1 there are plotted the chemical characters of five natural waters from a long shore-area in whose ground-waters, at certain places, the content of chloride and certain other constituents has changed very substantially in recent years. The five waters include: Two from well A, in 1931 and 1940, respectively; two from adjacent well B, in 1931 and 1936, respectively; and one from the ocean a few miles away. From the plottings on Figure 1 it is obvious that this deterioration in ground-water quality could not have been caused by a simple intermingling of the fresh ground-water with ocean water.

To demonstrate conclusively that a certain water is a quantitative mixture of two other waters--neither diluted, concentrated, nor chemically modified after the mixing--one graphic criterion and one graphic-algebraic criterion must be satisfied. First, by the graphic criterion, in all three fields of the diagram the apparent mixture must plot on straight lines between the plottings of its two inferred components. Also, the area-concentration plottings in the central field must conform to the principle that the concentration of a mixture is necessarily greater than the least, but less than the greatest, of the several concentrations of its components. This graphic criterion alone is not decisive because it involves only percentage reacting values and does not involve absolute concentrations. Neither is any other simple graphic construction on the diagram decisive. The second and decisive criterion requires satisfaction of the following equations:

With reference to Figure 5-A let: V_a = proportionate volume in mixture M of water having composition A; V_b = proportionate volume of water having composition B; E_a = concentration of water A, in equivalents; E_b = concentration of water B, in equivalents; E_m = concentration of the mixture, in equivalents; a = intercept between the plottings of A and M, measured in any of the three fields of the diagram and at any convenient scale; and b = intercept between the plottings of B and M. Then it follows and can be shown that

$$(a/b) = (V_b \cdot E_b / V_a \cdot E_a) \quad \dots \dots \dots \dots \dots \dots \dots \dots \dots \dots \dots (1)$$

$$(V_a / V_b) = (b \cdot E_b / a \cdot E_a) \quad \dots \dots \dots \dots \dots \dots \dots \dots \dots \dots \dots (2)$$

$$E_m = [E_a \cdot E_b (a+b)] / (a \cdot E_a + b \cdot E_b) \quad \dots \dots \dots \dots \dots \dots \dots \dots (3)$$

$$V_a = b \cdot E_b / (a \cdot E_a + b \cdot E_b) \text{ and } V_b = a \cdot E_a / (a \cdot E_a + b \cdot E_b) \quad \dots \dots \dots \dots (4)$$

Further, with respect to any particular chemical constituent, let: C_a = concentration in component water A, in equivalents or parts per million as desired; C_b = concentration in component B; and C_m = concentration in the mixture. Then

$$C_m = (C_a \cdot V_a) + (C_b \cdot V_b) \quad \dots \dots \dots \dots \dots \dots \dots \dots \dots \dots \dots (5)$$

Decisive proof of a quantitative mixture is accomplished when, for the water of intermediate composition, agreement is shown between analytical data and corresponding values computed from equation (3) for total concentration and from equations (4) and (5) for concentration of individual chemical constituents. Equation (3) is numerically equivalent to a corresponding equation derived by LANGELIER and LUDWIG [2, pp. 350-351], but is expressed in a form that facilitates the necessary computations.

As has been pointed out by HILL [1 (1940), pp. 48-49], the chemical character of a mixture of waters will plot at the center of gravity of the plottings of the respective components, each having been weighted according to its concentration and its proportionate volume in the mixture. Based on this principle, graphic-algebraic criteria for decisive proof of quantitative mixtures are perhaps feasible for a three-component system but become involved for systems with more than three components. Problems involving three or more components are likely to be infrequent; accordingly, it seems most practicable to solve them by an adaptation of the two-component criteria given above. Thus, on Figure 5-B let M represent a water presumed to be a mixture of waters A, B, and C. To prove or disprove a quantitative mixture project a straight line on the diagram through C and M to intersect AB at point X, which represents the chemical composition of a hypothetical mixture of waters A and B in the same proportions that these waters would enter a mixture of composition M. From preceding equations (3), (4), and (5) determine the composition and concentration of water X. Then, treat water M as a presumed mixture of waters X and C. As desired, this procedure is readily adaptable to mixtures of more than three components.

Application to geochemical problems

The greatest utility of the diagram herein described probably is in "screening" a large number of water-analyses for critical study with respect to sources of the dissolved constituents, modifications in chemical character as a water passes through an area, and related geochemical problems.

Changes in the chemical character of a natural water by solution of progressively increasing amounts of some particular mineral must plot on a set of straight-line vectors directed in each of the three fields toward the point representing the chemical composition of the mineral. This case is analogous to mixing one water with another whose concentration is infinitely great.

A natural water may be concentrated progressively by evaporation until it becomes saturated with respect to certain constituents, which then separate out in the solid phase. Until a saturation-point is reached, the chemical character of the concentrating water is represented by a single fixed point in each of the three fields of the diagram. If the material separating in the solid phase is of constant composition then, as evaporation continues, the character of the water will be traced on the respective fields by straight-line vectors directed away from the points that represent the composition of the separating solid. If the solid phase is a simple compound of one cation and one anion, the vector in the central field will be directed away from one of the apexes of that field, each of which represents a particular simple salt or group of salts. In the two triangular fields, the vector will be directed away from the respective apexes that represent the cation and anion composing the solid phase. For example, on Figure 6-A point A represents the composition of a hypothetical water that is saturated with respect to calcium sulphate ($CaSO_4$). As evaporation then

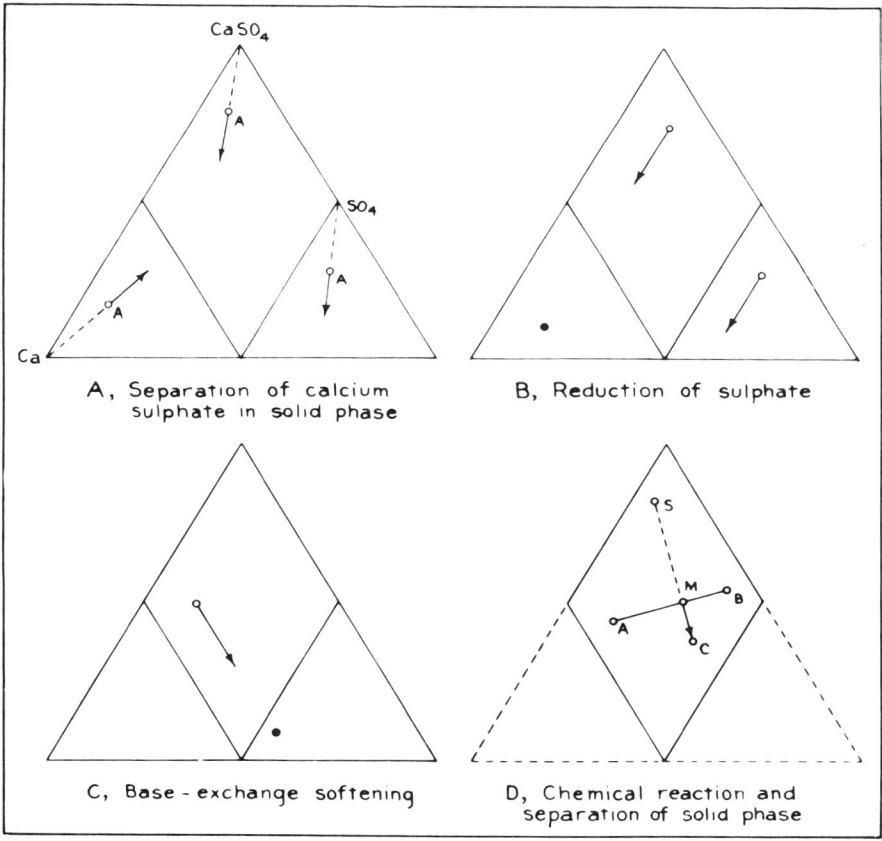

Fig. 6--Vectors characteristic of certain geochemical processes

continues and calcium sulphate forms in solid phase, the changing composition of the liquid phase will trace the three vectors shown. If a separating solid phase is a mixture of salts in a constant proportion, a straight line will be traced in the central field trending away from the point that corresponds to the constant composition of the mixture; straight lines may or may not be traced in the two triangular fields.

Certain changes in the chemical character of a water are caused by chemical reactions which in effect substitute one cation or one anion for another, molecule for molecule. Thus, reduction of sulphate [6] is equivalent to substitution of bicarbonate (HCO_3) for an equivalent amount of sulphate (SO_4); natural softening by reacting with base-exchange minerals [7], to substitution of sodium and potassium (Na, K) for calcium and magnesium (Ca, Mg). These two chemical changes are traced on the diagram by straight-line vectors parallel to the bases of the central field as shown by Figures 6-B and 6-C, respectively.

If two waters (or a water and a mineral) react chemically when brought together and some product or products of the reaction form in solid phase, the chemical character of the products remaining in solution will not plot on the straight line joining the points that represent the two reacting waters. Rather, it will plot on the extension of the straight line drawn from (1) the point that represents the composition of the solid phase to (2) the point that indicates the proportionate volumes and compositions of the two reacting waters. Thus, in Figure 6-D suppose that waters A and B are brought together in a proportion indicated by point M on the straight line AB, and that a solid phase of composition S results; then, the soluble products will plot on the extension of the straight line SM, as at point C. If the precipitate is a compound of one base and one acid, this relation between points A, B, M, S, and C will apply likewise in the two triangular fields (not shown) and point S will fall on one apex of each field. If the two waters are brought together in various proportions and the product S is appreciably soluble, the composition of the liquid phase will traverse line AB up to the point of saturation with respect to product S and then, beyond the point of saturation, will deflect away from point S.

The chemical composition of a natural water may undergo complex changes by an interplay of several or numerous processes. The causes are commonly obscure. However, when comprehensive chemical data are available the diagram herein described can assist greatly in a preliminary discrimination of causes, by application of the principles just described. Doubtless other useful principles will be developed as this diagram and similar diagrams are more widely used.

Preliminary analysis of a typical problem

The plottings on Figure 1, to which reference has been made, are typical of a problem in the saline contamination of fresh ground-water in a longshore area. Table 2 gives the corresponding numerical data.

From their analytical data one could infer that waters B_1, A_2, and B_2 represent progressive stages in the contamination of water A_1 by some unknown high-chloride source. Because these are longshore ground-waters from wells only about 200 feet deep, the ocean is an obvious potential source of a high-chloride contaminant. However, Figure 1 shows conclusively that the contaminated ground-waters are not simply a mixture of ocean water with uncontaminated water A_1, because their plottings do not conform to the graphic criterion for a simple mixture as already developed. Specifically: (1) In the cation-triangle, waters A_2 and B_2 do not fall between the plottings of A_1 and C, although all the plottings are in substantial alignment; (2) in the anion-triangle, B_1, A_2, and B_2 all plot below the line A_1C; and (3) in the central diamond-shaped field B_1 plots very slightly above, but A_2 and B_2 plot far above the line A_1C. If the analysis of water B_2 had not been available, these discordances would not have been obvious in the analytical data. Neither would the analysis of water B_1 have shown clearly that it represented the incipient stage of contamination.

With reference to the corresponding two hypothetical mixtures, waters A_2 and B_2 contrast sharply in two respects: (1) Their content of calcium is much greater and that of sodium is much less; in percentage reacting value the excess of calcium is substantially equal to the deficiency of sodium, as though the hypothetical mixtures had been hardened by an ion-for-ion exchange of bases with the water-bearing material (see Fig. 6-C). (2) Their content of sulphate is substantially deficient, as would be expected if sulphate had been reduced to bicarbonate (see Fig. 6-B). These two contrasts also appear to apply in a small measure to water B_1, thus tending to confirm the inference that this water represents an incipient stage in a common process of contamination.

Table 2--Principal chemical constituents of certain longshore ground-waters and of ocean water

Constituent	A_1	B_1	b_1	A_2	a_2	B_2	b_2	C
Parts per million								
Calcium (Ca)	39	40	39	102	42	466	65	393
Magnesium (Mg)	10	10	11	19	22	77	98	1,228
Sodium (Na) / Potassium (K)	47	52	56	54 / 3.6	152	255	808	10,220[a] / 353
Carbonate (CO_3)	....	....	....	0	....	....	....	0
Bicarbonate (HCO_3)	204	207	204	203	203	166	199	139
Sulphate (SO_4)	24	21	26	6.7	49	0	207	2,560
Chloride (Cl)	16	32	32	199	199	1,346	1,346	18,360
Percentage reacting values (adjusted)								
Calcium (Ca)	20.2	19.6	18.4	28.0	10.0	28.6	3.5	1.7
Magnesium (Mg)	8.5	8.1	8.6	8.6	8.6	7.8	8.7	8.8
Sodium and potassium (Na + K)	21.3	22.3	23.0	13.4	31.4	13.6	37.8	39.5
Totals	50.0	50.0	50.0	50.0	50.0	50.0	50.0	50.0
Bicarbonate (HCO_3)	38.9	35.9	34.9	18.3	16.7	3.3	3.6	0.2
Sulphate (SO_4)	5.8	4.6	5.7	0.8	5.1	0	4.7	4.6
Chloride (Cl)	5.3	9.5	9.4	30.9	28.2	46.7	41.7	45.2
Totals	50.0	50.0	50.0	50.0	50.0	50.0	50.0	50.0

[a]Calculated.

Notes: A_1 and A_2 indicate water from Well A on June 3, 1931, and December 20, 1940, respectively. B_1 and B_2 indicate water from Well B on June 6, 1931, and June 30, 1936, respectively. C indicates water from the ocean a few miles from wells A and B. a_2, b_1, and b_2 indicate hypothetical mixtures of waters A_1 and C in such proportions that their chloride contents are equal to those of A_2, B_1, and B_2, respectively.

Obviously the data here presented are not adequate fully to define this water-quality problem, in part because the analyses of waters A_1, B_1, and B_2 are approximate only. However, as an elementary example of procedure they are especially effective because they afford a striking comparison but involve only water-quality and time as principal variables. A complete solution of the problem here suggested involves data so voluminous that it is not feasible to introduce them.

References

[1] R. A. HILL, Geochemical patterns in Coachella Valley, Trans. Amer. Geophys. Union, Part I, pp. 46-49, 1940; also Salts in irrigation waters, Trans. Amer. Soc. Civ. Eng., v. 107, pp. 1,478-1,493, 1942.
[2] W. F. LANGELIER and H. F. LUDWIG, Graphical methods for indicating the mineral character of natural waters, J. Amer. W. W. Assn., v. 34, pp. 335-352, 1942.
[3] Op. cit., Fig. 5.
[4] CHASE PALMER, The geochemical interpretation of water analyses, U. S. Geol. Surv., Bull. 479, 31 pp., 1911.
[5] C. S. SCOFIELD, South Coastal Basin investigation, quality of irrigation waters, California Div. Water Resources, Bull. 40, pp. 22-23, 1933.
[6] ROGER REVELLE, Criteria for recognition of sea-water in ground-waters, Trans. Amer. Geophys. Union, Part III, pp. 595-596, 1941.
[7] B. C. RENICK, Base exchange in ground-water by silicates as illustrated in Montana, U. S. Geol. Surv., W.-S. Paper 520, pp. 53-72, 1925.

U. S. Geological Survey,
    Portland 9, Oregon

Copyright ©1951 by the Society of Petroleum Engineers of AIME
Reprinted from *Jour. Petroleum Technology* **3**, No. 10, Section 1, 15–16; Section 2, 3 (1951)

THE INTERPRETATION OF CHEMICAL WATER ANALYSIS BY MEANS OF PATTERNS

HENRY A. STIFF, JR., MEMBER AIME, ATLANTIC REFINING CO., DALLAS, TEX.

ABSTRACT

The classification and correlation of water analysis data presents many problems which can be solved by graphic methods. The pattern system, a new type of graphic procedure described in this communication, is believed to have several advantages over older methods. Examples of the application of the pattern system to the solution of problems encountered in petroleum production are given.

INTRODUCTION

Several graphic methods for presenting analytical water data have been developed and are now in use.[1,2,3,4] It is believed, however, that a recently developed type of graph called the "pattern" offers several advantages over other methods. This system presents a better picture of the total salt concentration than is usual in such graphs. The effect of dilution or concentration has been reduced to a minimum, and at the same time distinction between various types of water has been improved. The system is extremely versatile, yet so simple it can be plotted on ordinary graph paper and adapted to almost any type of filing system.

DISCUSSION

The essential feature of the pattern system is the graph shown in Fig. 1. Horizontal lines extending right and left from a vertical line at zero form the graph. Positive ions are plotted to the left while negative ions are plotted to the right. The figure immediately beneath each ion gives the scale. Most oil field waters can be plotted on a scale where 100 milliequivalents of sodium and chloride and 10 milliequivalents of each of the other ions are represented by one scale unit. For highly concentrated brines, 1,000 milliequivalents of sodium and chloride and 100 milliequivalents of each of the other ions are represented by one unit. For convenience sodium, potassium, lithium, etc., ordinarily determined as the difference between the positive and negative ions, are referred to as sodium.

It will be noted that the chemical unit of "milliequivalents per liter" is employed. If the results of the analysis are in parts per million, they can be readily converted by dividing by the equivalent weight in milligrams or multiplying by its reciprocal. Appropriate conversion factors can be found in standard chemical handbooks.[5]

When the points have been properly placed on the graph, they are connected by lines as shown in Fig. 2, thus forming a closed "pattern." These patterns present a variety of shapes

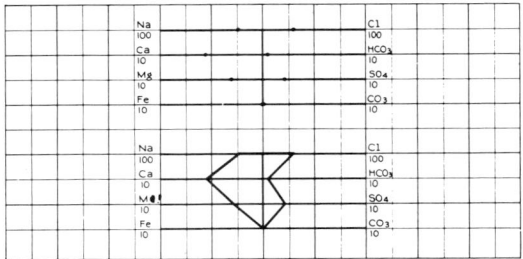

FIG. 2 — METHOD OF CONSTRUCTING A PATTERN.

and sizes, each easily recognized and remembered and each characteristic of a certain water.

Fig. 3 shows some common water patterns, including those of fresh water, sea water, chemical solutions and oil field brines. The straight line pattern of fresh water shown in the first diagram results from the use of a scale ordinarily employed for oil field brines. Any system of scales thought suitable to a particular operation may be used. Single scales, in which all ions are represented by an equal number of units, are used in plotting fresh waters, while multiple scales, in which all ions are not represented by the same number of units, find application in work with oil field waters. Various types of multiple scales can be used to advantage in water treatment work, corrosion control, and injection water studies.

One of the distinctive features of the system is the tendency of the pattern to maintain its characteristic shape as the sample becomes dilute. In this way the total salt concentration as well as the chemical composition of the water is shown by the pattern.

The availability of various scales makes it possible to select one which emphasizes the differences and similarities of the waters being studied, thereby making direct comparison and correlation between these waters possible.

Another very valuable feature of this system is its extreme simplicity. The pattern can be constructed by anyone on

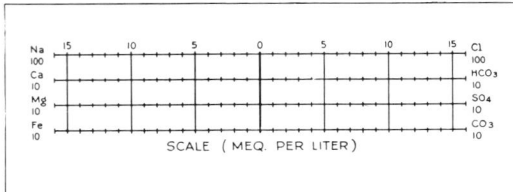

FIG. 1 — ESSENTIAL FEATURE OF THE PATTERN ANALYSIS SYSTEM.

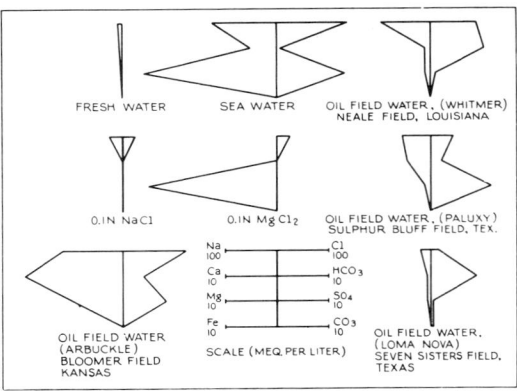

FIG. 3 — COMMON WATER PATTERNS.

Interpretation of Chemical Water Analysis

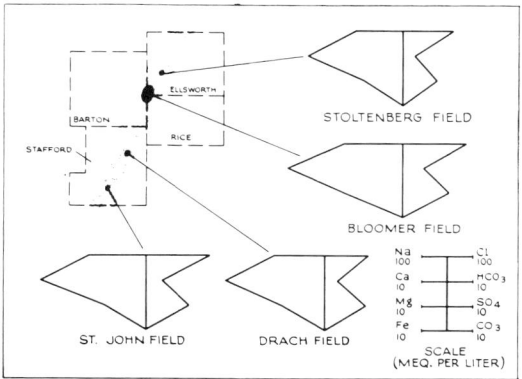

FIG. 4 — COURSE OF ARBUCKLE FORMATION THROUGH KANSAS SHOWN BY WATER PATTERNS.

ordinary graph paper. The use of a printed card containing a graph similar to that shown in Fig. 1, and in addition blank spaces for pertinent well data, is very convenient for this work. By this means a large and very useful library of patterns can be built up in a small space.

APPLICATION

The applications of the pattern system are many and varied. It is believed that this system will facilitate the solution of almost any problem in which water analysis is a factor. The following examples illustrate its application to various problems encountered in petroleum production.

Correlation of Producing Formations

Since the pattern tends to maintain its shape upon concentration or dilution, a formation may be expected to yield water of a characteristic pattern. A study of the water patterns can, in many cases, be utilized to identify different producing strata and correlate them in a given locality. In Kansas, for example, the Arbuckle group of the Ordovician period can be traced from Ellsworth County down through Barton County and into Stafford County by means of water patterns. Fig. 4 shows patterns from Stoltenberg Field (Ellsworth County), Bloomer Field (Barton County), and Drach and St. John's Fields (Stafford County). The characteristic pattern of the Arbuckle formation can be easily followed.

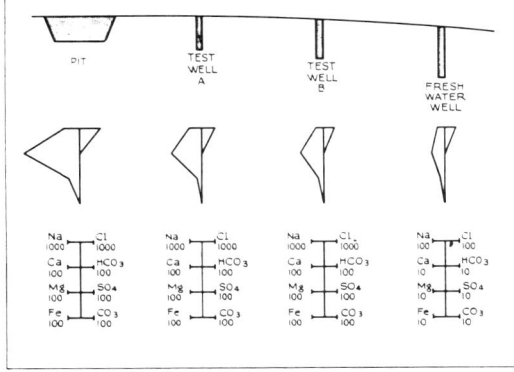

FIG. 5 — PATTERN ANALYSIS USED AS A TRACER.

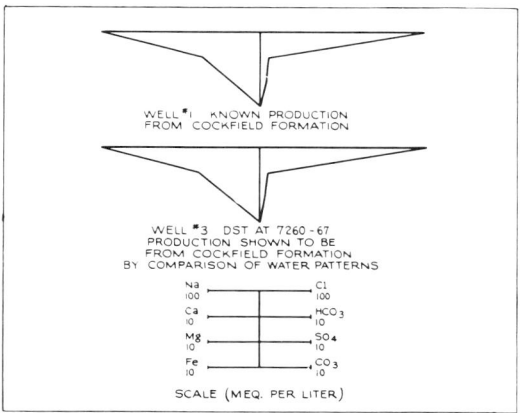

FIG. 6 — PATTERN ANALYSIS USED IN DRILL STEM TESTING.

Tracer Problems

It is often possible to trace the passage of water through a sand by a study of the patterns obtained from properly spaced test holes along its path. In the case presented in Fig. 5, seepage of salt water from a disposal pit into a fresh water well was suspected. Shallow test holes were drilled at A and B and allowed to fill by seepage. Samples were taken, the water analyzed, and the patterns plotted. The appearance of the disposal pit pattern, somewhat reduced in size, definitely established seepage in the direction of the well. When a sample from the fresh water well plotted on a magnified scale gave the same pattern as the pit, contamination became evident.

Drill Stem Testing

Because of the uncertainty regarding contamination of water samples by drilling mud encountered in drill stem testing, attempts to establish the formation from which such samples originate are often unsuccessful. In some cases, however, this information is extremely helpful. Because of the ease with which one pattern can be compared with another, the pattern system is especially useful for this purpose. Fig. 6 shows the

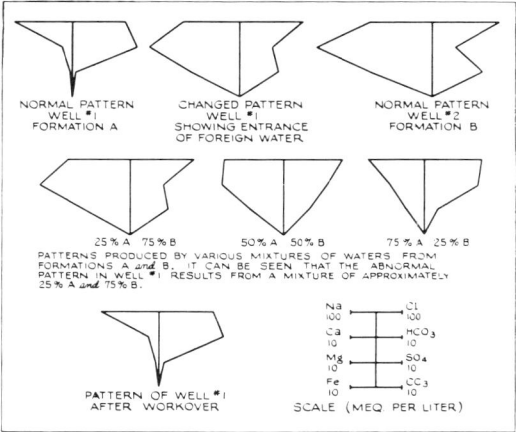

FIG. 7 — DETECTION OF FOREIGN WATER AND DETERMINATION OF ITS SOURCE.

results of a case where pattern analysis was successfully utilized to determine the formation from which a drill stem test sample originated.

Detection of a Foreign Water and the Determination of Its Source

Oil wells are completed so as to yield a minimum of water, but a break in the cement or a leak in the casing may allow extraneous water to enter. Water yielded by a well in the normal course of operation is usually bottom water, that is, water from the producing formation, while water entering through a leak or cement break may come from some other strata. If sufficient information is available on the composition of the waters in the area, the source of a foreign water may be discovered, and its exclusion brought about with a minimum of trouble and expense.

Fig. 7 presents a case in which water patterns were used to determine the source of water entering a well. Well No. 1 had been producing water of a consistent pattern from formation A since its completion, when suddenly the water pattern changed. Another well in the field was producing from formation B which was located about 150 ft above A. This formation had, of course, been cemented off during completion of Well No. 1. The patterns of various mixtures of A and B were calculated and plotted as shown. It then became evident that the new pattern resulted from a mixture of approximately 25 per cent water from formation A and 75 per cent water from formation B. It was therefore concluded that foreign water was entering from formation B either through a casing leak or through a break in the cement. Accordingly, the casing was pressure tested and since no leak was found, a cement job was undertaken. At the conclusion of this operation the well was again put on production and at the end of several months the pattern was again similar to that of formation A.

CONCLUSIONS

The pattern analysis system presented in this paper offers a simple, practical means of characterizing, comparing, and correlating ground waters. It is particularly useful in facilitating the solution of many petroleum production problems, but can be used to advantage in the study of any question in which water analysis is a factor.

ACKNOWLEDGMENT

The author wishes to thank The Atlantic Refining Co. for permission to publish this paper, and to acknowledge the advice and encouragement given him by J. H. Sullivan and C. M. Pounders.

Credit for the drawings is extended to Gene Nigh.

REFERENCES

1. Tickell, E. G.: *Report of the California State Oil and Gas Supervisor* (1921) **6**, (9), 5.
2. Reistle, C. E.: *U. S. Bureau of Mines Technical Paper 404* (1927).
3. Parker, J. S., and Southwell, C. A. P.: *Journal of Institution of Petroleum Technologists*, London (1929) **15**, 138-182.

Corps, E. V.: *Proc. World Petr. Congress 1* (1933) 338.

Lange, A. L.: *Handbook of Chemistry*, Handbook Publishers, Inc., Sandusky, O. (1946) 757.

The origin of high sodium bicarbonate waters in the Atlantic and Gulf Coastal Plains*

MARGARET D. FOSTER

U.S. Geological Survey, Washington, D.C.

(*Received* 7 *April 1950*)

ABSTRACT

Some sodium bicarbonate waters at depth in the Atlantic and Gulf Coastal Plains have the same bicarbonate content as the shallower calcium bicarbonate waters in the same formation and appear to be the result of replacement of calcium by sodium through the action of base-exchange minerals. Others, however, contain several hundred parts per million more of bicarbonate than any of the calcium bicarbonate waters and much more bicarbonate than can be attributed to solution of calcium carbonate through the action of carbon dioxide derived from the air and soil.

As the waters in the Potomac group (Cretaceous) are all low in sulphate and as the environmental conditions under which the sediments of the Potomac group were deposited do not indicate that large amounts of sulphate are available for solution, it does not seem probable that carbon dioxide generated by chemical or biochemical breakdown of sulphate is responsible for the high sodium bicarbonate waters in this area.

Sulphate as a source of oxygen is not necessary for the generation of carbon dioxide by carbonaceous material. Oxygen is an important constituent of carbonaceous material and carbon dioxide is a characteristic decomposition product of such material—as, for example, peat and lignite.

Experimental work showed that distilled water, calcium bicarbonate water, and sodium bicarbonate water, after contact with lignite, calcium carbonate, and permutite (a base-exchange material), had all increased greatly in sodium bicarbonate content and had become similar in chemical character and in mineral content to high sodium bicarbonate waters found in the Coastal Plain. The tests indicated that carbonaceous material can act as a source of carbon dioxide, which, when dissolved in water, enables it to take into solution more calcium carbonate. If base-exchange materials are also present to replace calcium with sodium, a still greater amount of bicarbonate can be held in solution. The presence of carbonaceous material, together with calcium carbonate and base-exchange minerals in a formation is, therefore, sufficient to account for the occurrence in it of high sodium bicarbonate waters.

INTRODUCTION

In shallow waters in the water-bearing sands of the Atlantic and Gulf Coastal Plains the dissolved mineral matter is usually made up predominantly of calcium and bicarbonate, but in many of the deeper waters in the same formations the dissolved mineral matter is predominantly sodium and bicarbonate [8], [9]. Such a difference in character between the shallower and deeper waters has been attributed to exchange of calcium in solution in the waters for sodium of base-exchange minerals in the rock materials [14], [15]. This base-exchange reaction involves no change in the content of bicarbonate or other acidic constituents—sulphate, chloride or nitrate—present in solution, and many of the sodium bicarbonate waters are similar in content of bicarbonate—and other acidic constituents—to the shallower calcium bicarbonate waters in the same formation. Others, however, contain several hundred parts per million more of bicarbonate than the shallower calcium bicarbonate waters from which they are presumably derived. The origin of the excessive bicarbonate content of such waters is the subject of the present study.

---

* *Published by permission of the Director, U.S. Geological Survey.*

Ground Water Conditions in the Atlantic and Gulf Coastal Plains

The Atlantic and Gulf Coastal Plains are underlain by sedimentary deposits of gravel, sand, clay, shale, limestone, chalk, and marl, ranging in age from Lower Cretaceous to Quaternary. The boundary between the Coastal Plain sediments and the hard crystalline rocks of the Piedmont province is called the "Fall Line." This is in reality a broad zone, marked by falls and rapids, within which there is an abrupt change in the gradient of streams as they flow from the area underlain by crystalline rocks to that underlain by unconsolidated sediments. Along the Fall Line the crystalline rocks are at or near the surface but coastward they lie at progressively greater depths and the Coastal Plain sediments progressively thicken. In general the sedimentary strata dip gently seaward, successively older formations cropping out farther and farther inland thence dipping beneath younger beds and passing to greater and greater depths in the direction of the coast. Meteoric water enters the formations along their landward margins, where they are at or near the surface, and moves down the dip of the permeable beds toward the coast. Stratigraphically and structurally conditions are favourable for the accumulation of artesian pressure. The artesian aquifers are the water-bearing gravels, sands, and limestones and the confining members are the clays, shales, and other strata of low permeability.

Carbon Dioxide—Bicarbonate Relations in the Ground Waters

The principal soluble material in the sediments of the Atlantic and Gulf Coastal Plains is calcium carbonate. The clastic constituents of the deposits, being the weathered residues of older rocks, have already been more or less altered and leached of their soluble constituents. Solution of calcium carbonate is, therefore, the primary action when meteoric waters containing in solution carbon dioxide from the air and soil move down through the permeable beds.

Calcium carbonate is relatively insoluble in pure water: 1 litre of distilled water dissolves only 14 mg at 25° C [10]. In the presence of carbon dioxide, however, its solubility is increased to a marked degree owing to the formation of the more soluble bicarbonate. The amount of calcium bicarbonate that can be held in solution depends on the carbon dioxide content of the water, and this in turn depends on the partial pressure of carbon dioxide in the atmosphere over the water. The most complete summary and theoretical analysis of the solubility of calcium carbonate in water is found in the papers of Johnston [11], [12]. Johnston presents the complete solubility curve of calcium carbonate, showing the concentration of calcium in solution at equilibrium in the system CaO—H_2O—CO_2. As Johnston's graph is the logarithm of the concentration of the calcium ion plotted against the logarithm of the partial pressure of carbon dioxide, and as only a small section of his curve applies to the solubility of calcium carbonate in natural waters, Adams and Swimmerton [1] redrew his graph, showing only the relationship of calcium carbonate dissolved to partial pressure of carbon dioxide. This curve is reproduced in Figure 1. From the shape of the graph it is indicated that at low pressures a small increase in the partial pressure of CO_2 corresponds to a relatively large increase in the solubility of calcium carbonate.

In their paper Adams and Swimmerton discuss the source of the carbon dioxide which enables ground waters to take into solution the amounts of calcium bicarbonate found in them. Under normal conditions the partial pressure of carbon dioxide in the air is 0·0003 atm. At this partial pressure only 63 parts per million of calcium carbonate (equivalent to 76·8 parts per million of bicarbonate) can be dissolved at 16° C. It is apparent, therefore, that meteoric ground waters must obtain carbon dioxide from some other source to enable them to dissolve the concentrations of calcium bicarbonate found in them. The most obvious source of this increased concentration of carbon dioxide is the soil-air, which is characterized by a much higher partial pressure of carbon dioxide than atmospheric air. The biochemical oxidative decomposition of organic matter and the respiration of plant roots result in the liberation in the soil of large amounts of carbon dioxide. ROBINSON [16] states that the air of grassland soils may contain as much as 1·5% (0·015 atm) of carbon dioxide and arable soils up to 0·5% (0·005 atm). BOUSSINGAULT and LEWY [2] found that air from a soil poor in humus and not manured for a long time contained at least 25 times as much carbon dioxide as atmospheric air; air from humus rich soil contained 90 times and that from recently manured soil 250 times that of the atmosphere. Although the composition of soil-air varies from time to time and from place to place with difference in the energy of microbiological

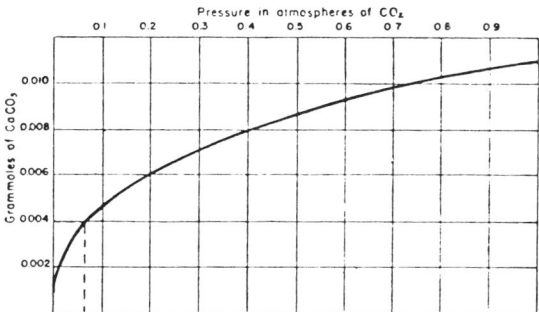

Fig. 1—Solubility of $CaCO_3$ at various pressures of CO_2, temperature = 16° C. Reproduced from ADAMS and SWIMMERTON.

Table 1—*Analyses of waters from limestone and marl in Florida* [4]

Constituent	Parts per million							
	1	2	3	4	5	6	7	8
Silica (SiO_2)	17	13	33	12	25	—	28	35
Iron (Fe)	1·5	·12	·16	·38	·12	·60	·09	Tr
Calcium (Ca)	66	54	83	137	48	52	60	98
Magnesium (Mg)	11	27	14	3·0	20	29	11	20
Sodium (Na)	14	5·5	11	28	11	20	16	47
Potassium (K)	1·8							
Bicarbonate (HCO_3)	233	288	325	423	223	295	253	337
Sulphate (SO_4)	6·0	6·5	3·0	2·0	16	4	1·7	7·6
Chloride (Cl)	26	9·0	10	42	9·0	34	14	94
Nitrate (NO_3)	Tr	Tr	Tr	Tr	Tr	—	Tr	Tr
Dissolved Solids (T.S.)	264	265	319	472	248	360	258	494
Total Hardness as $CaCO_3$ (T.H.)	210	246	265	355	202	249	195	327
Depth of well (in ft)	710	250	200	85	400	250	630	181

1. Panama City, Bay County. Municipal Supply.
2. River Junction, Gadsden County. Well of W. L. Shepard.
3. Whitehouse, Duval County. Well of Chas. F. Cox.
4. St. Augustine, St. Johns County. Municipal Well.
5. Lake City, Columbia County. Municipal Supply.
6. Arcadia, De Soto County. Courthouse Well.
7. Plant City, Hillsborough County. Municipal Well.
8. Seabreeze, Volusia County. Municipal Well.

decomposition, temperature, and other factors, it is evident that the carbon dioxide content of soil-air is much greater than that of atmospheric air. Meteoric waters moving through this carbon dioxide rich zone dissolve carbon dioxide and, thus enriched, are able to dissolve greater amounts of calcium carbonate than if they had been in contact only with normal atmospheric air.

The bicarbonate content (as calcium bicarbonate) of waters from calcareous rocks like limestone and marl, in which there is an abundance of calcium carbonate available for solution, indicates that the carbon dioxide in waters derived from the air and soil is seldom high enough to permit the waters to take into solution more than 450 parts per million of bicarbonate as calcium bicarbonate. This amount of bicarbonate in solution corresponds to a partial pressure of carbon dioxide of about 0·05 atm, 170 times the partial pressure of carbon dioxide in the atmosphere. Most waters from limestones and marls, however, contain less than 300 parts per million of bicarbonate, which corresponds to a carbon dioxide partial pressure of less than 0·03 atm. Analyses of typical limestone and marl waters from the Coastal Plain in Florida [4] are given in Table 1.

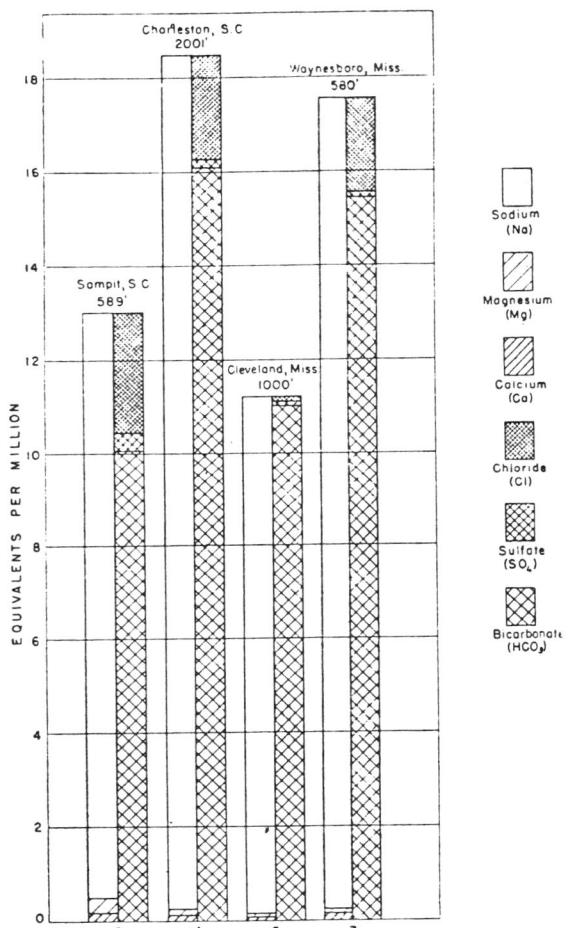

Fig. 2—High sodium bicarbonate waters in the Coastal Plain.

The (calcium) bicarbonate content of waters in the sand and gravel beds of the Coastal Plain, like that of waters from limestones, does not ordinarily exceed 450 parts per million. In deposits that are relatively calcareous, the capacity of the waters to take calcium (and magnesium) carbonate into solution through the action of carbon dioxide is exhausted at shallow depths, but in formations that contain relatively little calcareous material the waters may have to travel to some depths before the carbon dioxide in them is fully utilized in solution of calcium carbonate. However, regardless of the calcium carbonate content of the materials through which they pass, the carbon dioxide in the waters derived from the air and soil is seldom high enough to permit the waters to take into solution more than 450 parts per million of bicarbonate as calcium bicarbonate. In most waters the carbon dioxide concentration thus derived is sufficient to permit the waters to take into solution only up to 250 parts per million of bicarbonate.

In some of the sodium bicarbonate waters, however, the bicarbonate content is more than 800 parts per million and a few have been analysed that contained more than 1200 parts per million. Analyses of high sodium bicarbonate waters from the Coastal Plain in South Carolina [5] and Mississippi [20] are shown in

Table 2—*Analyses of high sodium bicarbonate waters in the Coastal Plain* [5, 20]

Constituent	South Carolina				Mississippi				
	1	2	3	4	5	6	7	8	9
				Parts	per million				
Silica (SiO_2)	—	—	—	32	18	21	25	14	25
Iron (Fe)	0.11	0.15	0.05	1.0	7.0	1.1	1.4	0.30	0.50
Calcium (Ca)	4.0	4.0	5.6	3	1.1	3.0	2.9	2.4	12
Magnesium (Mg)	4.1	3.9	4.4	0.4	0.5	1.4	0.7	3.2	12
Sodium (Na)	281	288	277	421	254	290	396	571	651
Potassium (K)				4.4	3.7	6.0	6.0	4.0	
Carbonate (CO_3)	26	36	34	54	0	0	0	38	79
Bicarbonate (HCO_3)	647	540	535	872	672	782	942	1355	1552
Sulphate (SO_4)	2.0	18	21	7.2	1.2	2.0	1.6	9.0	29
Chloride (Cl)	43	91	82	92	5.0	4.7	72	54	12
Nitrate (NO_3)	.25	.35	1.2	—	—	—	—	.30	.50
Dissolved Solids (T.S.)	727	743	797	1051	626	714	970	1368	1591
Total Hardness as $CaCO_3$ (T.H.)	27	26	32	9	5	13	10	19	79
Depth of well (ft)	365	589½	800	2001	1000	422	580	1375	480?

    1. Well of E. Stalvey at Stalvey, Horry County, S.C.—flowing well.
    2. Well of B. D. Bourne Estate at Sampit, Georgetown County, S.C.—flowing.
    3. Well of Gas & Electric Co. at Summerville, Dorchester County, S.C.
    4. Well of Charleston Consolidated Railway and Lighting Co. at Charleston, Charleston County, S.C.—flowing well.
    5. Well of town of Cleveland, Bolivar County, Miss.—flowing well from Claiborne group.
    6. Well of Mr. Weems at Subuta, Clarke County, Miss.—flowing well from Sparta sand of Claiborne group.
    7. Well of town of Waynesboro, Webster County, Miss.
    8. Well of Southern Lumber Co. at Jackson, Hinds County, Miss.—flowing well from Wilcox formation.
    9. Well of City of Greenville, Washington County, Miss.—from Wilcox formation.

Table 2. Four of the analyses are represented graphically in Figure 2. In some of these waters, sodium and bicarbonate make up 90% or more of the dissolved mineral matter; calcium and magnesium are generally not present in them in excess of 5 parts per million. Sulphate and chloride are usually very low also, except near the coast, where the ground waters may be high in chloride due to contamination by sea water.

ALTERATION IN CHARACTER OF WATERS WITH DEPTH IN FORMATION

The relations between sodium waters of high bicarbonate content and calcium and sodium waters of moderate bicarbonate content in the Potomac group (Lower and Upper Cretaceous) in Virginia were discussed in a paper by CEDERSTROM [3]. The

Potomac group, which is nonmarine in origin, consists of alternating sands, clays, and sandy clays. The sands of the group are recharged from rain water which enters these strata where they are at or near the surface along the Fall Zone. Down dip of the recharge area, the water in the sands is under artesian pressure and rises in wells which penetrate into the formation. The areal distribution of the different

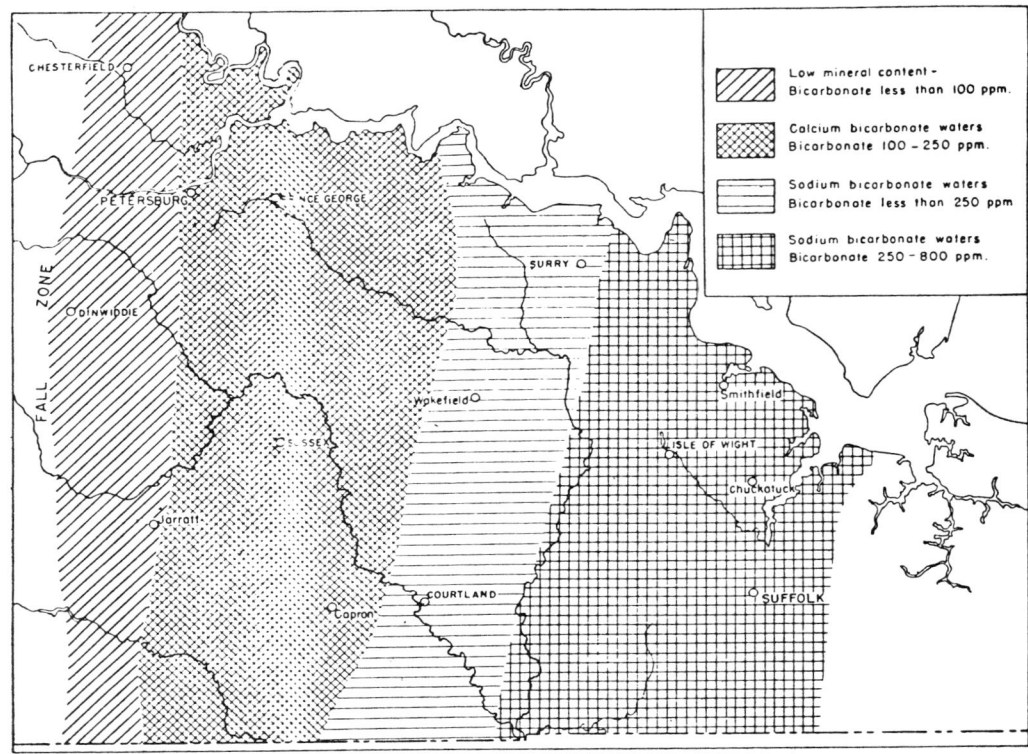

Fig. 3—Map of southeastern Virginia showing geographic distribution of the different types of water in the Potomac group.

types of water found in the sands of the Potomac group is shown in Figure 3. Along the Fall Zone, where meteoric waters enter the formation, the sands of the Potomac group yield waters that are soft and low in bicarbonate and in total mineral content. But as the waters move eastward down the dip of the permeable beds, they increase in calcium bicarbonate and in total mineral content until they contain approximately 170 to 250 parts per million of bicarbonate as calcium and magnesium bicarbonate. Then they seem to undergo a rather abrupt change in character—and become sodium bicarbonate waters. A few waters of intermediate character indicate that the change in character is due to a decrease in calcium and magnesium with an equivalent increase in sodium. During the alteration in character and for some distance to the eastward the bicarbonate content remains relatively constant. Still farther eastward from the zone of softening, however, the waters in the sands of the Potomac group again increase in bicarbonate content and in sodium content. In the vicinity of Suffolk, water from the Potomac group contains 550 to 650 parts per million of bicarbonate; and one sample of water, from a well south of Suffolk, had 775 parts per million of bicarbonate. Graphic representations of analyses showing these changes with depth in the character of waters from the Potomac group are shown in Figure 4. To the eastward of Suffolk the waters in the Potomac group contain increasing amounts of chloride due to contamination by marine water.

Determinations of free carbon dioxide on a few samples of water from shallow wells along the Fall Zone that draw from the Potomac group indicate that the waters in this part of the area contain considerable free carbon dioxide. For example, one sample which contained 60 parts per million of bicarbonate had 45 parts per million of free carbon dioxide. Others had from 20 to 35 parts per million. On the other hand, moderately hard calcium bicarbonate waters contain very little free carbon dioxide —most of those examined for carbon dioxide had less than 2 parts per million. The amount of calcium bicarbonate in the hard waters is about that to be expected if the free carbon dioxide in the shallow waters were completely utilized in solution of calcium carbonate. For example, if the 45 parts per million of free carbon dioxide found in the Fall Zone water mentioned above were entirely utilized in solution of calcium carbonate, the bicarbonate thus formed plus the 60 parts of bicarbonate already in the water would total 171 parts per million.

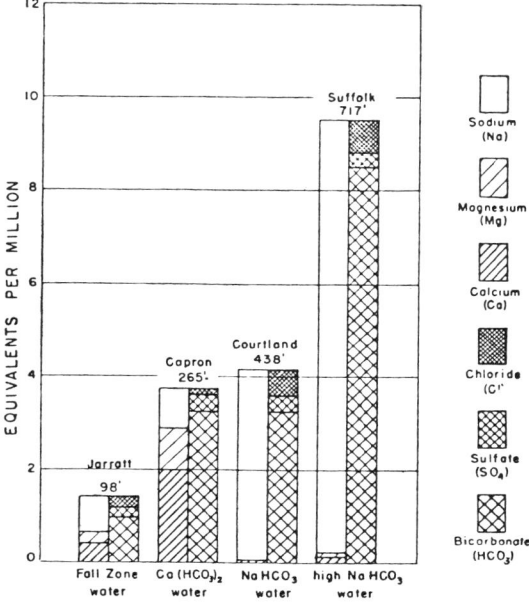

Fig. 4—Types of waters in southeastern Virginia Coastal Plain.

This amount of bicarbonate approximates that found in about 50% of the calcium bicarbonate waters from the Potomac group that have been analysed.

Although the bicarbonate content of the high sodium bicarbonate waters in the Potomac group is not as great as that of many deep sodium bicarbonate waters in other parts of the Coastal Plain, as illustrated by the analyses in Table 2, it is greatly in excess of the amount that can be attributed to the solvent action of carbon dioxide derived from the air and soil in shallow waters in the Potomac sediments along the Fall Zone. If the bicarbonate in these waters is presumed to have been taken into solution as calcium bicarbonate and the calcium subsequently exchanged for sodium, a carbon dioxide partial pressure of 0·10 to 0·20 atm would have been required, compared with a carbon dioxide partial pressure of about 0·01 atm for the bicarbonate in solution in the calcium bicarbonate waters. For concentrations of 1200 parts per million and more found in some Coastal Plain waters, carbon dioxide pressures of 0·7 atm or more would have been required for solution of calcium carbonate.

Theories as to Source of Additional Carbon Dioxide

Reduction of sulphate by carbonaceous material as a source of carbon dioxide.

CEDERSTROM [3] attributes the increase in bicarbonate in the Virginia waters to chemical or biochemical breakdown of sulphate by carbonaceous material, with liberation of carbon dioxide and subsequent solution of calcium carbonate and replacement of calcium by sodium. It is true that the high sodium bicarbonate waters in southeastern Virginia are low in sulphate—most of them contain less than 16 parts per million. However, analyses indicate that few waters in the formation, from the waters of low mineral content along the Fall Zone to the high sodium bicarbonate waters in the vicinity of Suffolk, contain more than 16 parts per million of sulphate. Thus the sulphate content of the high sodium bicarbonate waters is comparable to that of other waters in the formation. The continuous generation of small amounts of carbon dioxide postulated by CEDERSTROM would require continuous reduction of sulphate and this, in turn, would require continuous solution of sulphate to maintain the sulphate content of these waters on a par with that of waters to the west and updip in the formation. To generate sufficient carbon dioxide to effect an increase in bicarbonate of 350 parts per million—the average increase in bicarbonate of the waters from the zone of softening to the vicinity of Suffolk—would require solution of 200 mg of sulphate per litre of water. As the waters in the formation are all low in sulphate and as the environmental conditions under which the sediments of the Potomac group were deposited do not indicate that large amounts of sulphate are available for solution it does not seem probable that carbon dioxide generated by chemical or biochemical breakdown of sulphate is responsible for the high sodium bicarbonate waters in this area.

Generation of carbon dioxide by carbonaceous material without intervention of sulphate

Sulphate as a source of oxygen is not necessary for the generation of carbon dioxide from carbonaceous material. Oxygen, as well as carbon, is an important constituent of such material and carbon dioxide is a characteristic product of vegetal alteration. Organic deposits as they are found buried in the earth's strata have practically everywhere undergone both chemical and physical changes. Loaded beneath hundreds or thousands of feet of sediments, brought downward into zones of greater temperature, and perhaps subjected to shearing stresses, the carbonaceous material undergoes changes which are referred to as dynamochemical in contradistinction to the primary biochemical action. The outstanding result of the dynamochemical process is progressive elimination of the volatile matter from organic sediments [21]. The alteration consists primarily of internal molecular readjustment which causes the detachment and elimination of simple volatile substances from the great unstable molecules that characterize vegetal debris. These emanations consist almost entirely of the simplest compounds of carbon, hydrogen, and oxygen—carbon dioxide, methane, and water, and in smaller proportions carbon monoxide and nitrogen, and sometimes traces of hydrogen sulphide and higher hydrocarbons. From peat and lignite the predominant gases evolved are the oxides, water and carbon dioxide; in later stages of alteration methane predominates [13]. Thus without the intervention of sulphate, alteration of carbonaceous material could provide carbon dioxide to underground waters filling the interstitial spaces of surrounding rock materials.

Many sediments in the Atlantic and Gulf Coastal Plains were deposited under environmental conditions favourable for the accumulation of plant debris. For example, the character of the Potomac (Lower and Upper Cretaceous) sediments indicates that they were laid down in estuaries or along shore in comparatively shallow waters that were fresh or, at most, brackish, and the clays of the formation are more or less charged with vegetable remains, either silicified or in the condition of lignite [7]. Leaves and wood were found in Potomac sediments in a deep well in Norfolk [6] and lignite was found in a well at Claremont, Va. The sands, clays, and lignites which predominantly compose the typical beds of the Black Creek (Upper Cretaceous) formation in North Carolina were deposited in shallow sea water near shore, or perhaps in part in shallow bays and estuaries. The clays, as a rule, are dark to black in colour on account of contained carbonaceous matter; and thin seams of finely comminuted, lignitized vegetable particles are common in the formation. Pieces of lignite, ranging in size from small particles to twigs, branches, and even large trunks of trees, occur scattered irregularly through the formation [19]. A few thin seams of lignite and carbonaceous clay have been noted in the Black Creek formation in South Carolina [5].

In Texas the strata of the Wilcox (Eocene) group comprise a heterogeneous series, several hundred feet thick, of sandy lignitiferous littoral clays, cross-bedded sands, compact, noncalcareous lacustrine or lagoonal clays, lignite lentils, and stratified deltaic silts [17]. The Wilcox in Georgia [18] and Mississippi [20] also contains lignitic sand or clay and in the latter State, interbedded lignite layers.

The calcium bicarbonate waters and the sodium bicarbonate waters of comparable bicarbonate content in the Coastal Plain as a rule are colourless, but many of the high sodium bicarbonate waters are noticeably coloured. Deep waters in the Yazoo Delta region in Mississippi, which have a bicarbonate content (as sodium bicarbonate) of 800 to 1500 parts per million, have a yellow to dark brown colour, similar to the colour of swamp waters, suggesting that they may have been in contact with carbonaceous material.

It is suggested, therefore, that carbon dioxide generated by alteration of carbonaceous material in the sediments is a determining factor in the occurrence of the high sodium bicarbonate waters in certain formations in the Coastal Plain. Waters in formations containing carbonaceous material acquire additional carbon dioxide, which permits them to dissolve more calcium carbonate, the calcium thus taken into solution being then replaced by sodium through the action of base-exchange minerals —the end result of these reactions being an increase in the sodium bicarbonate content of the waters.

Experimental

Lignite, as a representative of carbonaceous material, was used in the following series of tests to determine its effect, in different environments, on the mineral content of different types of waters.

1a—500 ml of distilled water on 5 gm of lignite.
1b—500 ml of distilled water on 5 gm of lignite and 1 gm of calcium carbonate.
1c—500 ml of distilled water on 5 gm of lignite, 1 gm of calcium carbonate, and 10 gm of permutite.
2a—500 ml of sodium bicarbonate water on 5 gm of lignite.
2b—500 ml of sodium bicarbonate water on 5 gm of lignite and 1 gm of calcium carbonate.
2c—500 ml of sodium bicarbonate water on 5 gm of lignite, 1 gm of calcium carbonate, and 10 gm of permutite.

3a—500 ml of calcium bicarbonate water on 5 gm of lignite.
3b—500 ml of calcium bicarbonate water on 5 gm of lignite, and 1 gm of calcium carbonate.
3c—500 ml of calcium bicarbonate water on 5 gm of lignite, 1 gm of calcium carbonate, and 10 gm of permutite.

The lignite used in these tests, which was from Hoyt, Texas, was kindly furnished by Miss TAISIA STADNICHENKO of the Geological Survey. The calcium carbonate was MALLINCKRODT'S C. P. grade.

The waters and materials, in 500 ml rubber stoppered ERLENMEYER flasks, were mixed and shaken twice daily. After 14 days the waters were decanted through a filter paper and analysed. Analyses of the waters before and after the tests are given in Table 3 and are represented graphically in Figure 5.

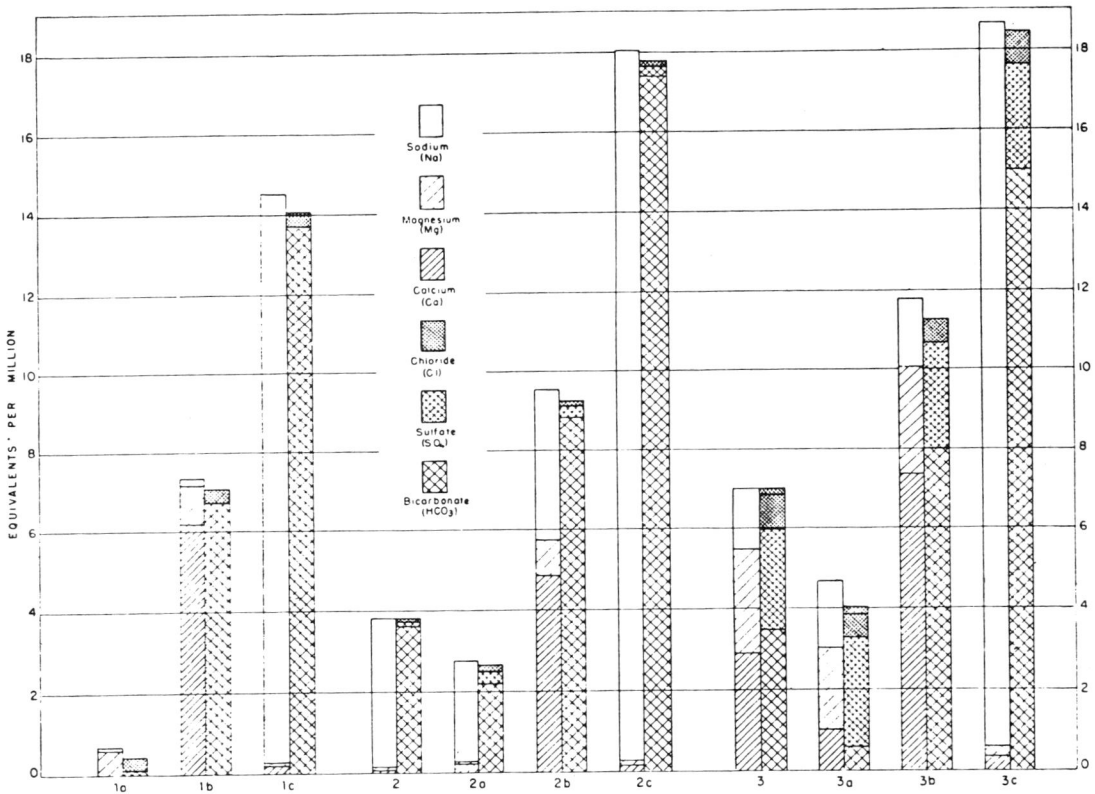

Fig. 5—Change in mineral content of (1) distilled water, (2) sodium bicarbonate water, and (3) calcium-bicarbonate water, after treatment with (*a*) lignite, (*b*) lignite and calcium carbonate and (*c*) lignite, calcium carbonate and permutite.

In contact with lignite alone, distilled water gained a little mineral matter, chiefly calcium and sulphate, but the sodium bicarbonate and the calcium bicarbonate waters both lost mineral matter, the sodium bicarbonate water losing 87 parts per

Table 3—Analyses showing change in mineral content of distilled water, NaHCO₃ and Ca(HCO₃)₂ waters on treatment with lignite, lignite and CaCO₃, and lignite, CaCO₃ and permutite

	Distilled water			NaHCO₃ water				Ca(HCO₃)₂ water			
	Before	After lignite	After lignite and CaCO₃	After lignite CaCO₃ and permutite	Before	After lignite	After lignite and CaCO₃	After lignite CaCO₃ and permutite	Before	After lignite	After lignite and CaCO₃	After lignite CaCO₃ and permutite
					Parts per million							
Total Solids at 180°C	—	60	391	843	211	172	485	1007	414	322	667	1103
Loss on ignition of dried residue	—	34	36	40	11	22	21	30	45	63	50	43
SiO₂	—	1·6	1·0	10	10	8·8	8·1	13	20	19	18	18
Ca	—	7·5	124	4·0	1·2	2·7	97	3·8	58	20	148	7·4
Mg	—	—	12	1·0	·6	·9	11	1·0	31	25	33	2·7
Na	—	1·1	3·5	319	84	58	86	403	35	36	37	408
K	—	1·1	1·6	9·6	1·6	1·8	3·7	10	2·4	2·4	2·7	11
CO₃	—	—	0	24	12	0	0	18	6·9	0	0	0
HCO₃	—	Acid*	410	786	197	134	540	1024	200	36	488	912
SO₄	—	14	17	16	3·3	14	13	13	117	129	128	128
Cl	—	0	0	1	4·0	5·0	4·0	4·0	25	21	22	28
NO₃	—	0	0	0	·8	—	—	—	11	8·3	0	0
pH	6·2	4·5	7·0	8·7	8·7	6·3	7·0	8·3	8·1	5·7	6·9	8·3
Colour	0	25	25	45	0	45	25	45	0	25	25	40

* 6·0 ppm HCO₃ after aeration.

million of bicarbonate as sodium bicarbonate and the calcium bicarbonate water losing 178 parts per million of bicarbonate as calcium and magnesium bicarbonate. This loss in mineral content is probably the result of adsorption by the lignite. The greater loss of calcium and magnesium bicarbonate from the calcium water than of sodium bicarbonate from the sodium water is in accordance with the general rule that bivalent ions are adsorbed more strongly than univalent ions. The greater loss of calcium than of magnesium from the calcium bicarbonate water, although the two were originally present in equivalent amounts, also follows the general rule that calcium is more strongly adsorbed than magnesium.

In contact with lignite and calcium carbonate, all the waters gained *calcium* bicarbonate—the increase in bicarbonate being 410, 319, and 274 parts per million for the distilled water, the sodium bicarbonate water, and the calcium bicarbonate water, respectively. In control tests to determine the amount of calcium carbonate taken into solution by these waters in the absence of lignite, the distilled water and the sodium bicarbonate water gained only 48 and 40 parts per million of calcium bicarbonate, respectively. This increase may be attributed to calcium carbonate dissolved through the action of the carbon dioxide in the waters under the partial pressure of the carbon dioxide of the air above the waters. The calcium bicarbonate water lost 38 parts per million of bicarbonate—this may be attributed to loss of carbon dioxide from the water under the low partial pressure of carbon dioxide in the air above the water. Only a very small amount of the calcium carbonate taken into solution when the waters were in contact with lignite and calcium carbonate can, therefore, be attributed to the action of free carbon dioxide in solution in the waters under the normal partial pressure in the air of carbon dioxide. The great increase in calcium bicarbonate in the calcium bicarbonate water in contact with lignite and calcium carbonate as compared with the loss of calcium bicarbonate from the calcium bicarbonate water when in contact with calcium carbonate alone indicates greatly increased partial pressure of carbon dioxide in the atmosphere above the water when lignite was present.

There was not the loss in sodium content from the sodium bicarbonate water after contact with lignite and calcium carbonate that there was after contact with lignite alone. Presumably, as calcium is more strongly adsorbed than sodium, some of the calcium taken into solution when calcium carbonate is present with lignite is preferentially adsorbed and the sodium content remains unchanged.

In contact with lignite, calcium carbonate, and permutite all the waters increased greatly in sodium bicarbonate content—the increase in terms of bicarbonate being 786 parts, 803 parts, and 698 parts per million for the distilled water, the sodium bicarbonate water, and the calcium bicarbonate water, respectively. The resulting waters are similar in chemical composition and mineral content to many high sodium bicarbonate waters from deep wells in the Coastal Plain.

During the tests the waters, which were colourless at the start, acquired a yellowish colour, presumably from the lignite. This colour and the fact that the sum of the equivalent values of the basic constituents, calcium, magnesium, sodium, and potassium, in the waters after the tests was in excess of the sum of the equivalent values of the acidic constituents, bicarbonate, sulphate, chloride, and nitrate, suggest that the excess basic constituents are paired with some undetermined organic acid radical.

In a second series of tests 20 gm of powdered lignite was added to 4 litres of a sodium bicarbonate water and the bicarbonate content and hardness of the water were determined at 24-hr intervals. When successive determinations showed no change, 500 ml of the water was withdrawn for analysis. Two grams of calcium carbonate was then added to the main portion of water and the bicarbonate and hardness were determined at 24-hr intervals until they were again constant. After withdrawal of a second portion of the water for analysis, 10 gm of permutite was added to the main portion. During the first 6 hrs after addition of the permutite, the bicarbonate and hardness were determined at 72-min intervals. Thereafter the determinations were made at 24-hr intervals until successive determinations showed no change in the bicarbonate content

Table 4—*Change in bicarbonate* (HCO_3) *content and total hardness* (TH) *of a sodium bicarbonate water after addition of lignite, calcium carbonate, and permutite*

Day	$NaHCO_3$ water		After addition of lignite		After addition of calcium carbonate		After addition of permutite	
	HCO_3	TH as $CaCO_3$	HCO_3	TH as $CaCO_3$	HCO_3	TH as $CaCO_3$	HCO_3	TH as $CaCO_3$
				Parts per million				
Start	310	30						
1			240	21	376	102	544	57
2			214	12	416	126	676	30
3			200	6	438	144	—	—
4			—	—	—	—	806	21
5			200	0	453	156	856	21
6					464	162	892	21
7					468	165	916	18
8							932	15
9							948	15
10							—	—
11							966	15
12							978	15
13							984	15
14							993	15
15							1001	15
21							1010	15

or hardness. The water was then filtered and analysed. Throughout the period of the test the flask was shaken at half-hour intervals during the working day.

The values obtained for bicarbonate and hardness during the test are shown in Table 4 and are plotted against time (in days) in Figure 6. Analyses of the water at the end of each step in the test are given in Table 5 and are represented graphically in Figure 6 at the point on the curve where the respective samples were taken for analysis.

Although in this test the materials, lignite, calcium carbonate, and permutite, were added successively after reaction with the materials already added had ceased, the mineral content of the water at the end of each step was similar to that of the waters obtained in the first series of tests (2a, 2b, and 2c) when a sodium bicarbonate water was in contact with the respective combinations of materials from the start.

In both sets of tests the bicarbonate in solution when base-exchange minerals (permutite) were present was considerably greater than when they were not present. The amount of calcium bicarbonate that could be held in solution reached a saturation point and no more calcium carbonate could be dissolved. This is clearly shown the second step of the second test. When, however, calcium is replaced by sodium from base-exchange minerals, much more bicarbonate can be held in solution—as shown in the third step of the second test.

Table 5—Analyses showing changes in mineral content of sodium bicarbonate water after contact first with lignite, then with lignite and calcium carbonate, and finally with lignite, calcium carbonate, and permutite

Constituent	At start	After lignite	After lignite and $CaCO_3$	After lignite, $CaCO_3$ and permutite
		Parts per million		
Total Solids	289	233	447	972
Loss on ignition	5·2	11	29	29
SiO_2	2·6	2·4	3·0	14
Ca	7·6	—	51	4·1
Mg	2·7	—	9·7	1·2
Na	107	88	111	374
K	4·0	2·4	2·1	16
CO_3	6·9	·0	0	0
HCO_3	282	200	470	1014
SO_4	13	18	19	22
Cl	4·0	4·0	4·0	4·0
NO_3	1·2	—	—	—
TH	30	0	165	15
pH	8·7	7·9	7·5	8·8
Colour	0	35	35	70

Conclusions

The tests support the hypothesis that carbonaceous material may act as a source of carbon dioxide which, when adsorbed by water, enables the water to dissolve more calcium carbonate. If base-exchange materials are also present to replace calcium with sodium, a still greater amount of bicarbonate can be held in solution and high sodium bicarbonate waters like those in the Coastal Plain result.

If a water-bearing formation lacks any one of the three materials—calcium carbonate, base-exchange minerals, or carbonaceous matter—it is not likely that waters of high sodium bicarbonate content will be found in it. If the formation lacks calcium carbonate, even the carbon dioxide derived from the soil and air cannot be utilized and the water remains low in dissolved mineral matter, as the sand and gravel beds contain very little soluble material except calcium carbonate. Any base-exchange minerals and carbonaceous material in such a formation have no opportunity to act. If the formation lacks base-exchange minerals but contains calcium carbonate and carbonaceous materials, the waters of the formation may be expected to be hard, calcium bicarbonate waters, the amount of calcium bicarbonate in solution depending not only on carbon dioxide derived from the air and soil but also on carbon dioxide evolved by carbonaceous material in the formation. In such a formation, therefore, a calcium bicarbonate water may contain calcium bicarbonate in excess of that equivalent to the amount of calcium carbonate that can be dissolved through the action of carbon dioxide derived from the air and soil.

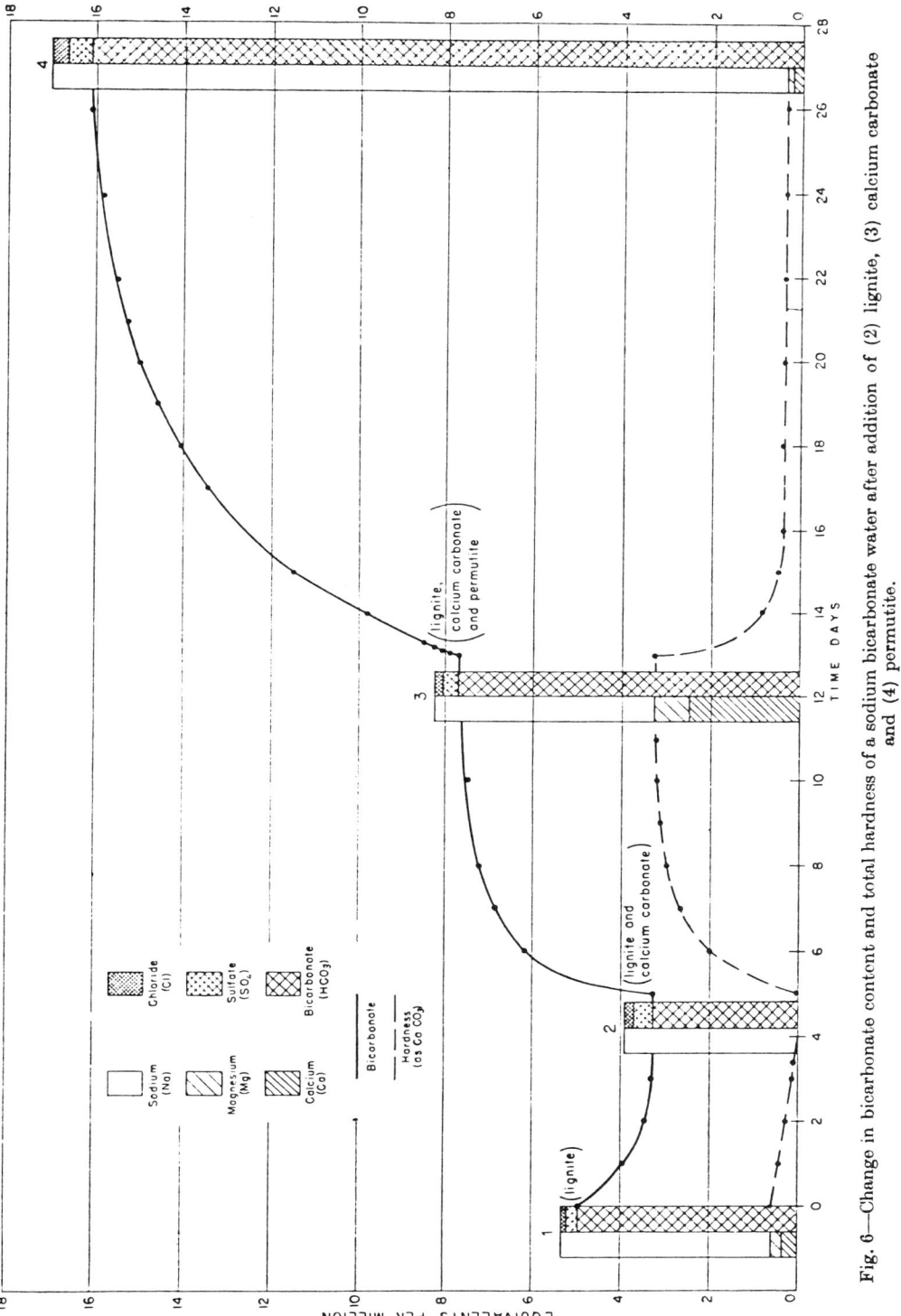

Fig. 6.—Change in bicarbonate content and total hardness of a sodium bicarbonate water after addition of (2) lignite, (3) calcium carbonate and (4) permutite.

If a formation contains calcium carbonate and base-exchange minerals but no carbonaceous material, it is to be expected that the shallower waters will be calcium bicarbonate waters, altering with depth to sodium bicarbonate waters and both types of waters having an amount of bicarbonate in solution equivalent to the carbon dioxide content of the waters entering the formation. The depth at which the alteration in character begins to take place differs in different formations but alteration does not usually occur at depths less than 100 ft and in many formations does not occur at depths less than several hundred feet. It is not to be presumed from this fact that base-exchange minerals are necessarily lacking in the shallower materials, but more probably that any base-exchange minerals in them have been exhausted of their capacity to exchange sodium for calcium.

Only in a formation containing all three materials—calcium carbonate, base-exchange minerals, and carbonaceous materials—may waters of high sodium bicarbonate content be expected and these, usually, only at some depth in the formation. Conversely, the occurrence of such waters in a formation may be taken as indicative of the presence of these three materials.

It seems probable that carbon dioxide evolved by carbonaceous material would permeate water in the interstices of the rock materials in its environment and that in order to obtain carbon dioxide it would not be necessary for the waters to come in direct contact with the carbonaceous material. Nor would it be necessary that the carbonaceous material be present in definite, recognizable deposits. Disseminated organic debris in quantities not readily detected without chemical or microscopic aid could also contribute carbon dioxide to water.

References

[1] ADAMS, C. S. and SWIMMERTON, A. C.; Solubility of limestone. Amer. Geophys. Union Trans. 1937 18 504–507. [2] BOUSSINGAULT, J. B. and LEWY, B.; Sur la composition de l'air confine dans la terre vegetale. Ann. Chim. Phys. 1853 37 5–50. [3] CEDERSTROM, D. J.; Genesis of ground waters in the Coastal Plain of Virginia. Econ. Geol. 1946 41 No. 3 218–245. [4] COLLINS, W. D. and HOWARD, C. S.; Chemical character of waters of Florida. U.S. Geol. Survey Water-Supply Paper 596–G, 1928. [5] COOKE, C. W.; Geology of the Coastal Plain of South Carolina. U.S. Geol. Survey Bull. 867, 1936. [6] DARTON, N. H.; U.S. Geol. Survey Geol. Atlas, Norfolk, Va., N.C., Folio No. 80, 1902. [7] FONTAINE, W. M.; The Potomac formation in Virginia. U.S. Geol. Survey Bull. 1896 145 147. [8] FOSTER, M. D.; The chemical character of the ground waters of the South Atlantic Coastal Plain. Washington Acad. Sci. J. 1937 27 405–412. [9] FOSTER, M. D.; Ground waters of the Houston-Galveston area. Eng. and Ind. Chemistry, Ind. Ed., 1939 31 1031–1035. [10] HODGEMAN, C. D.; Handbook of chemistry and physics, 29th ed., p. 359. Cleveland, Ohio: Chemical Rubber Publishing Co. 1945. [11] JOHNSTON, J.; The solubility-product constant of calcium and magnesium carbonates. Amer. Chem. Soc. J. 1915 37 2001. [12] JOHNSTON, J. and WILLIAMSON, E. D.; The complete solubility curve of calcium carbonate. Amer. Chem. Soc. J. 1916 38 975–983. [13] LEWIS, J. V.; The evolution of mineral coals, Part 1. Econ. Geol. 1934 29 1–38. [14] RENICK, B. C.; Base-exchange in ground water by silicates as illustrated in Montana. U.S. Geol. Survey Water-Supply Paper 520–D, 1924 [15] RIFFENBURG, H. B.; Chemical Character of ground waters of the northern Great Plains. U.S. Geol. Survey Water-Supply Paper 560–B, 1926. [16] ROBINSON, G. W.; Soils, their origin, constitution and classification. p. 32. London: Thos. Murby and Company 1932. [17] SELLARDS, E. H., ADKINS, W. S. and PLUMMER, F. B.; Geology of Texas. Univ. Texas Bull. 1932 3232 573. [18] STEPHENSON, L. W. and VEATCH, J. O.; Underground waters of the Coastal Plain of Georgia. U.S. Geol. Survey Water-Supply Paper 341, p. 71, 1915. [19] STEPHENSON, L. W.; The Cretaceous formations of North Carolina. North Carolina Geol. and Econ. Survey, 1923 5 7–11. [20] STEPHENSON, L. W., LOGAN, W. N. and WARING, G. A.; The ground water resources of Mississippi. U.S. Geol. Survey Water-Supply Paper 576, 1928. [21] TWENHOFEL, W. H.; Treatise on sedimentation, p. 415. Baltimore: Williams and Wilkins Co. 1935.

ORIGIN OF HYDROCHEMICAL FACIES
OF GROUND WATER
IN THE ATLANTIC COASTAL PLAIN*

By WILLIAM BACK
U.S.A.

ABSTRACT

The application of the concept of facies to the chemical aspects of ground water shows that the kinds of ions in solution and their concentration result from chemical processes responding to the lithology and the hydrologic flow pattern of a particular region. The Atlantic Coastal Plain was selected as a field model in which to study the portion of the geochemical cycle of elements that is controlled by the circulation of ground water. Significant characteristics of hydrochemical facies can be illustrated by methods similar to those used in lithofacies studies—trilinear diagrams that show the types of facies present in any area or formation; panel diagrams that show the overall facies distribution; and maps showing isopleths of chemical constituents within certain formations. Within the Coastal Plain sediments the calcium magnesium facies occurs in areas of high head (areas of recharge); the sodium facies occurs in downgradient areas of lower head. Mapping of these facies demonstrates that the outcrop area of Cretaceous and Eocene sediments in southern Maryland is the discharge area for ground water, rather than a recharge area as is more normally the role of the outcrop of artesian aquifers.

INTRODUCTION

THE part of the geochemical cycle of the elements that probably has been studied the least is the portion in which the circulation of ground water controls the concentration and distribution of chemical constituents within particular environments of the earth's crust. Chemical analyses and interpretive papers pertaining to the hydrosphere are related almost entirely to the fields of oceanography, potamology, and limnology. Ground water has been largely ignored in the geochemical cycle. Although the amount of water stored in, and circulating through, the sedimentary formations is a small percentage of the total water of the earth, it is this water with its contained chemical constituents that is largely responsible for the chemical nature and quantity of the dissolved load carried to the oceans by streams. This paper is concerned with this part of the geochemical cycle and partially summarizes a study by the U.S. Geological Survey, with the purpose of clarifying the relations between the chemical character of ground water and the geology and hydrology of the region in which the water occurs. The Atlantic Coastal Plain was selected as a suitable field model in which to study these relationships.

About 3,000 chemical analyses of ground water and several hundred logs of geologic formations were studied. Stratigraphic classification of well logs were obtained from published ground water reports resulting from investigations of Geological Survey in cooperation with state agencies. Figure 1 and 2 were prepared by

* Publication authorized by the Director, U.S. Geological Survey.

projecting the geologic logs of the wells closest to the lines of profiles for the fence diagrams, to establish the geologic framework for presentation of the hydrochemical facies. Nearly 300 chemical analyses, judged to be typical of a particular area and formation by group plotting on trilinear diagrams similar to that used for figure 3, were used to map the facies distribution. Selected chemical analyses and the locations of wells from which samples were collected are not given in this paper but will be included in the final report of the study.

The part of the Coastal Plain discussed in this paper includes parts of New Jersey, Delaware, Maryland, and Virginia. The Coastal Plain is generally flat to gently rolling and ranges from sea level to about 300 feet above. The highest areas are in the northern part of New Jersey and in southern Maryland south of Washington, D. C. The eastern shores of Maryland and Virginia and the State of Delaware are on the lowest part of the Coastal Plain, where the elevation is generally 50 feet or less.

SUMMARY OF STRATIGRAPHY

As shown on the fence diagrams (figs. 1 and 2) the Coastal Plain consists of a wedge-shaped mass of unconsolidated and semiconsolidated sediments ranging in age from Cretaceous to Recent. Underlying the Coastal Plain sediments is a complex of crystalline rocks of pre-Cretaceous age.

In the southern part of this area the lower part of the Cretaceous sediments consist of interbedded clay and sand, which are of fresh water origin. Overlying these sediments are beds of marine clay and sand. In New Jersey the Cretaceous sediments are essentially all marine sand and clay. In these diagrams (figs. 1–2) the Paleocene sediments are grouped with the Eocene in all the area except Virginia, where the Mattaponi formation contains deposits of both Paleocene and Upper Cretaceous. The Eocene sediments of marine sand, clay and marl which are commonly glauconitic.

As shown in the fence diagrams, the Miocene sediments attain a thickness of more than 1,000 feet along the coast. These sediments are marine and consist primarily of gray and blue sand, clayey sand, and clay. In New Jersey the equivalents of these sediments are subdivided into the Miocene Kirkwood formation and the overlying Miocene(?) Cohansey sand. The Pliocene(?) sediments of the Coastal Plain generally cap the higher hills and consist of slightly cemented silty sand and gravel. The Pleistocene sediments, which are primarily of fluvial origin and are at a lower altitude than the Pliocene(?) sediments, occur over much of the Coastal Plain.

HYDROCHEMICAL FACIES

The notion of hydrochemical facies is used in this paper to denote the diagnostic chemical aspect of water solutions occurring in hydrologic systems. They reflect the response of chemical processes in the lithologic framework and the pattern of water flow in it.

The following terms are used to designate particular cation facies: the calcium magnesium facies occurs where calcium and magnesium ions comprise 90 percent or more of the total cations (on the basis of equivalents per million, not parts per million); the calcium sodium facies exists where the calcium and magnesium content exceedss 50 percent but is less than 90 percent of the total cations; and the sodium

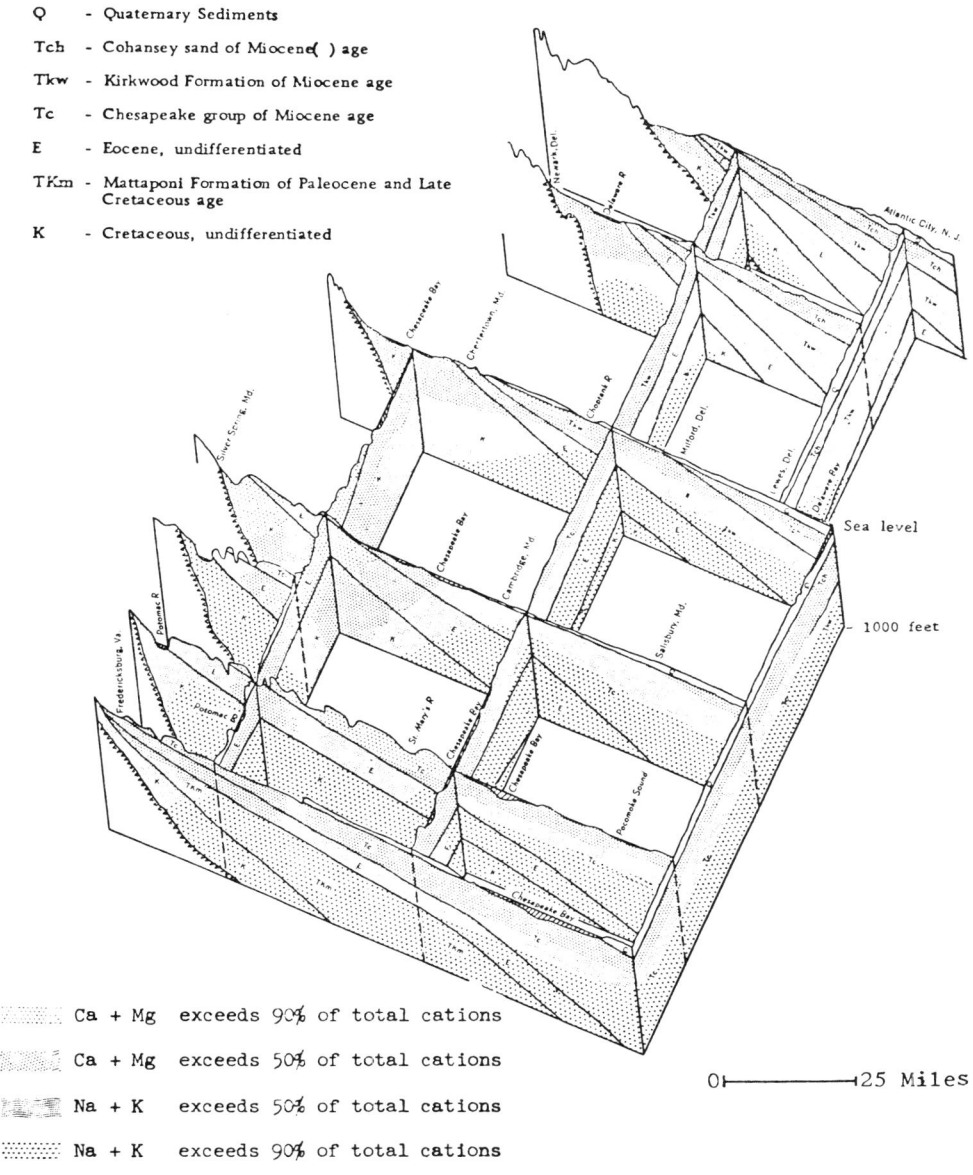

FIG. 1.—*Fence diagram showing the distribution of the cation hydrochemical facies in part of the Atlantic Coastal Plain.*

and potassium content exceeds 10 percent but is less than 50 percent; the sodium calcium facies designates the chemical character of water in which the content of the sodium and potassium ions exceeds 50 percent but is less than 90 percent of the total cations; the sodium facies occurs where the sodium and potassium ions exceed 90 percent and the calcium and magnesium are equal to less than 10 percent. Although

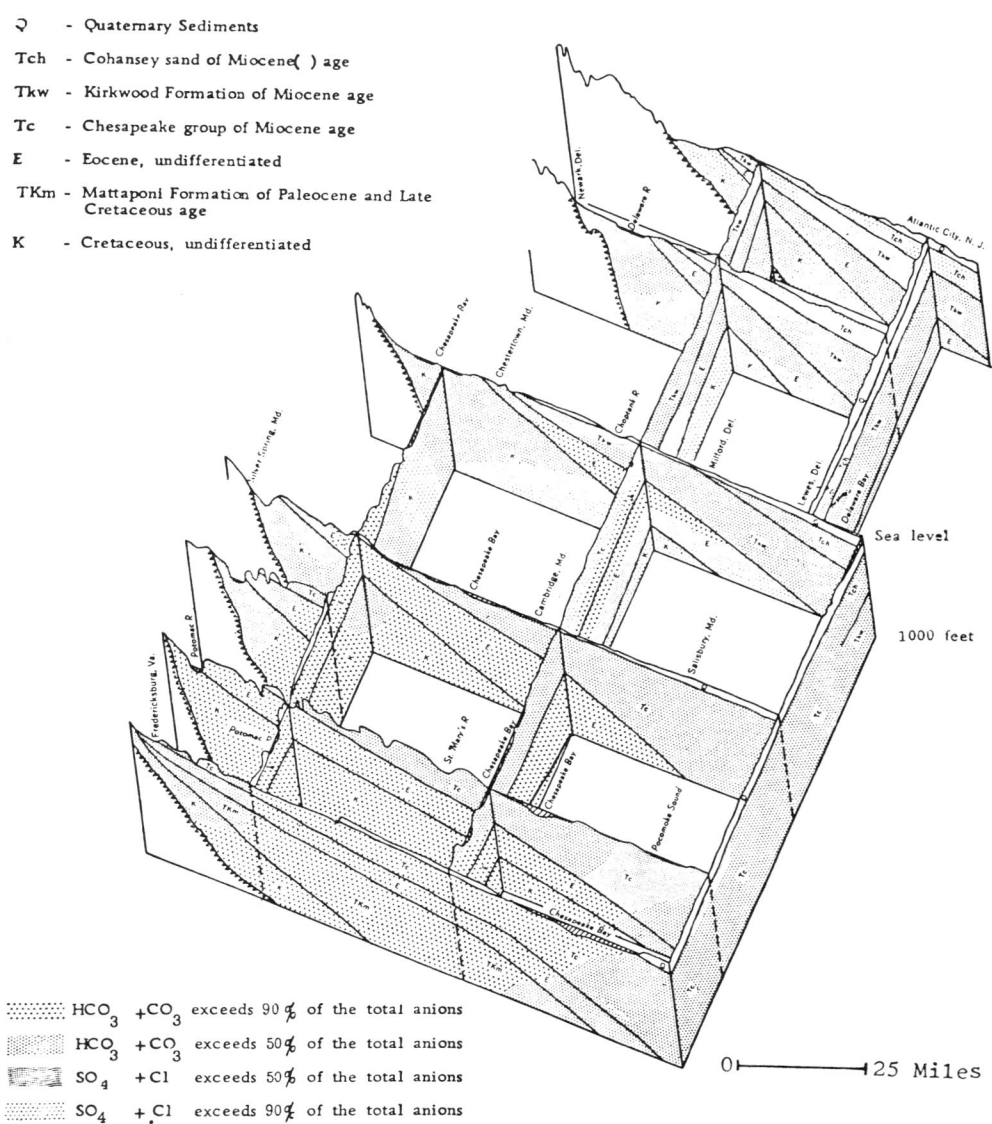

FIG. 2.—*Fence diagram showing the distribution of the anion hydrochemical facies in part of the Atlantic Coastal Plain.*

the potassium ion occurs in far lower concentration than the sodium ion, it is included in the facies classification because often it is not separated from the sodium ion in the chemical analysis.

Essentially the same limits are used in figure 2 for the designation of the facies; that is, the bicarbonate facies occurs where the bicarbonate plus carbonate exceeds 90 percent of the total anions; the bicarbonate chloride sulfate facies indicates that bicarbonate is less than 90 percent but exceeds 50 percent, and the content of the other anions exceeds 10 percent and is less than 50 percent of the total anions; the chloride sulfate bicarbonate facies occurs where the chloride and sulfate content exceeds 50 percent and is less than 90 percent of the total anions. In most areas of the chloride sulfate facies the chloride ion occurs in a concentration greater than the sulfate ion, and this terminology can be simplified to "chloride facies".

The distribution of cation facies in figure 1 shows that the calcium magnesium facies is generally predominant near the surface and the sodium potassium facies at greater depth. The sodium potassium facies is less common in New Jersey than it is farther south in Maryland and Virginia.

In the Eocene and Cretaceous sediments the full range of cation facies is represented. Near the surface, within these formations the calcium magnesium facies is common and at progressively lower elevations and, therefore, lower head, the facies changes from calcium through calcium sodium and sodium calcium to the complete sodium facies. The occurrence of these facies within one formation or group of formations in which the mineralogy remains essentially constant indicates that the flow characteristics of the aquifer systems control the distribution of the facies. For example the sodium facies is developed by the process of ion exchange in which the clay minerals and the glauconitic sand of the Eocene and Cretaceous formations act as natural water softeners. As the water enters the recharge areas and percolates downward the calcium magnesium bicarbonate facies develops. As this water continues to move through the Eocene and Cretaceous formations, the ion-exchange materials remove the calcium and magnesium ions from the water and replace them with sodium ions.

In the calcium magnesium facies the ground water is not saturated with calcium carbonate. The water has a low concentration of dissolved solids and a low pH. Therefore, if more calcareous material were available, or if the water had a longer residence time in these areas, more calcium carbonate would go into solution.

The anion facies distribution, shown on figure 2, indicates that the bicarbonate facies and the bicarbonate chloride sulfate facies are predominant. The bicarbonate ions originate from the solution of calcium carbonate by ground water made acid by dissolving carbon dioxide gas from the atmosphere and soil.

The chloride sulfate facies and the chloride sulfate bicarbonate facies have several modes of origin in various areas of the Coastal Plain. The major cause, is the occurrence of sodium chloride water in the deep Cretaceous sediments. In the Miocene sediments of New Jersey and Maryland in which these facies occur, the water solution is so dilute (generally less than 2 equivalents per million of total ions) that the development of any particular facies is the sensitive response of minor changes in the lithologic material and a small addition of calcareous material would change the facies to the bicarbonate type. In those areas along the western margin of the Coastal Plain where these facies occur in the Eocene and Cretaceous sediments, the origin is due to lack of calcareous material and the presence of iron sulfide minerals in which the sulfide is oxidized to sulfate.

The distribution of the concentration of dissolved solids shows that the waters of New Jersey and Delaware have generally a lower content than those in Maryland and Virginia. The Miocene formations in New Jersey are composed of cleaner sand and have fewer clay beds than do the equivalent formations farther south. Therefore, it may be that less soluble material is available in the near-surface formations of New

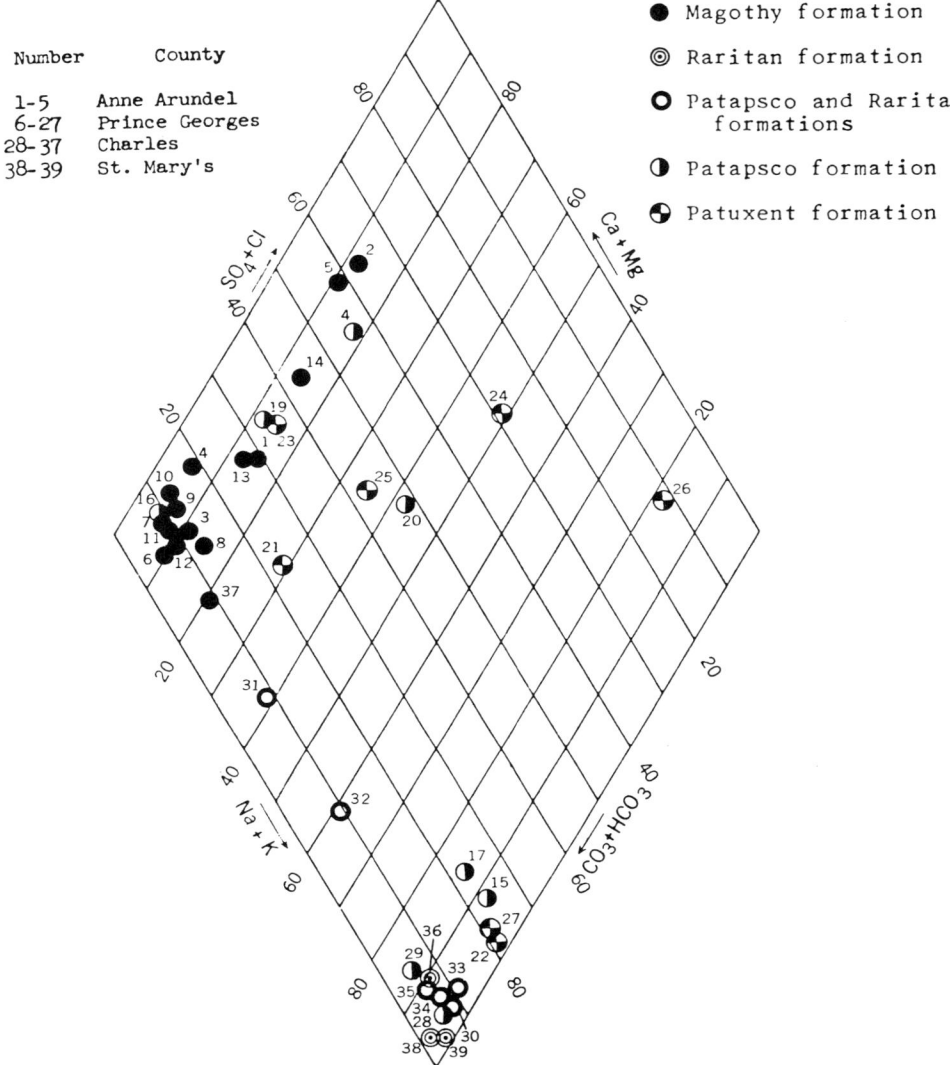

FIG. 3.—*Water analysis diagram showing the hydrochemical facies in the Cretaceous formations in southern Maryland.*

Jersey than is available in Maryland and Virginia. Another possible explanation for the low dissolved-solids concentration of ground water in New Jersey is that the bedrock is much shallower in that area and the water does not have as long a flow path as it does farther south. It would appear that there is less upward leakage into the Miocene beds in New Jersey than in Maryland. Another area of discharge for Cretaceous formations in New Jersey is along the Delaware River under conditions similar to those described for Maryland. In Maryland and Virginia discharge from

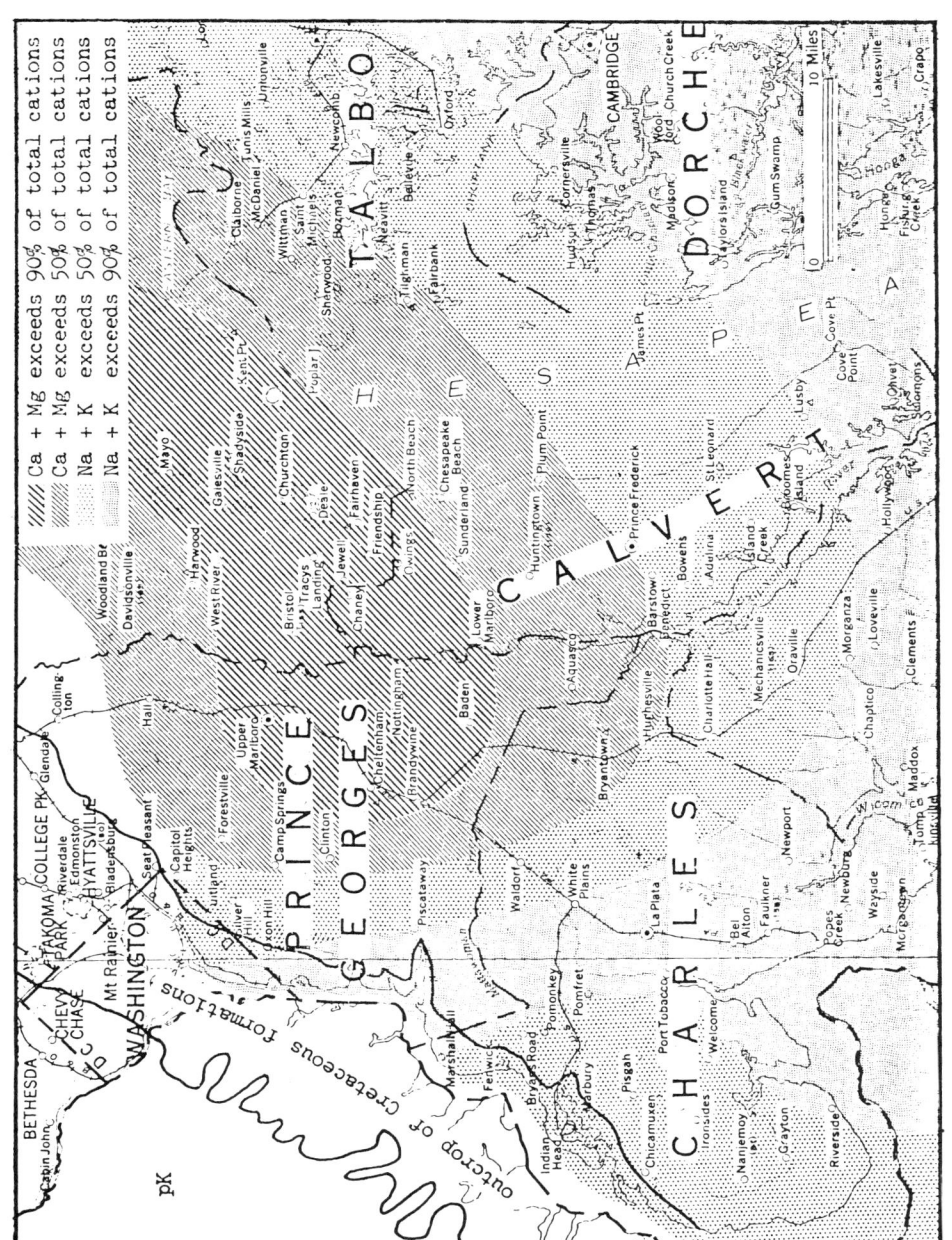

FIG. 4.—Map of southern Maryland showing the distribution of the cation facies in the Cretaceous formations.

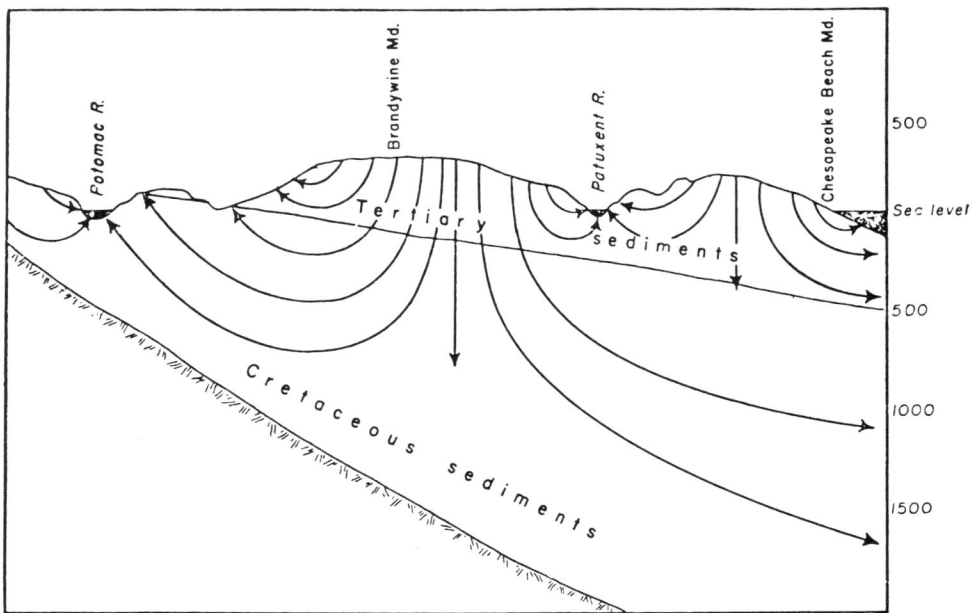

Fig. 5.—*Diagrammatic cross-section through southern Maryland showing the lines of ground-water flow.*

the Cretaceous sediments occurs by upward leakage that causes a concentration of dissolved solids higher in the shallow formations than would be the case if such leakage did not occur.

Figure 3 shows the characteristic hydrochemical facies of the Cretaceous formations in southern Maryland. The cation facies show the complete range of distribution from the calcium to the sodium facies. Most of these samples represent the bicarbonate facies. The sulfate and chloride facies occur in the more dilute waters. Although these analyses are identified by the formation in which the well is completed, the similarity of the lithology of the formations permits the development of the same facies within each formation. Therefore, the grouping of the analyses in the high calcium and high sodium areas reflects the changes caused by the movement of the water rather than by change in type of geologic materials.

The cation distribution shown in figure 4 results from several different processes. For instance, the sodium facies can occur in areas having no source of calcium carbonate material. In a recharge area that has been functioning for a long time the calcium carbonate material may be leached and the predominate cation may be sodium or potassium derived from the marine clays or weathered feldspars. However, in such an area the dissolved solids will be very low, and it should be possible to identify the recharge areas as those showing the lowest dissolved-solids concentration. On the other hand, an area where the high sodium facies with a high dissolved-solids concentration is likely to be the discharge area, because the amount of dissolved solids in the water would reflect the length of residence time in the aquifers.

The calcium facies shown in figure 4 includes the topographic high southeast of

Washington. This suggests that the topographic high is a recharge area and that part of the water flows westward, up the dip, and part of it flows southward and southeastward, generally down the dip. This hypothesis is further supported in that the water of low dissolved solids content (100 to 150 ppm) occurs in the calcium and calcium sodium facies. The dissolved solid content of the water in the sodium facies ranges from about 200 to 350 ppm.

It follows that in part of the area along the Potomac River the outcrop of the Cretaceous sediments is a discharge area. Therefore, the outcrop area of artesian aquifers need not be an area of recharge but may function also as an area of discharge. Figure 5 is a diagrammatic cross section showing the direction of groundwater flow as interpreted from the mapping of hydrochemical facies.

[*Manuscript received September* 1, 1959]

Part II
OCCURRENCE AND GEOCHEMICAL SIGNIFICANCE OF SALTWATER

Editors' Comments
on Papers 10 Through 16

10 CARLSTON
   An Early American Statement of the Badon Ghyben-Herzberg Principle of Static Fresh-Water-Salt-Water Balance

11 KOHOUT
   Cyclic Flow of Salt Water in the Biscayne Aquifer of Southeastern Florida

12 BERRY and HANSHAW
   Geologic Field Evidence Suggesting Membrane Properties of Shales

13 WHITE
   Excerpts from *Saline Waters of Sedimentary Rocks*

14 RUNNELLS
   Diagenesis, Chemical Sediments, and the Mixing of Natural Waters

15 PLUMMER et al.
   Hydrogeochemistry of Bermuda: A Case History of Ground-Water Diagenesis of Biocalcarenites

16 HANOR
   Precipitation of Beachrock Cements: Mixing of Marine and Meteoric Waters vs. CO_2-Degassing

   The potential problem of saltwater encroachment was recognized in the United States as soon as the first water well was drilled in 1824. As pointed out by Carlston (Paper 10), a description and explanation of the relationship between freshwater and saltwater in coastal areas was published by an American, Du Commun, in 1828, more than half a century before the work of Badon Ghyben and Herzberg who are credited with the first enunciation of this principle.
   As early as 1854 the problem was recognized on Long Island, and saltwater contamination became a threat elsewhere when additional wells were drilled in coastal areas. One hundred years after drilling

the first well, the problem had become so widespread with so much public concern that O. E. Meinzer had Brown (1925) do a study of the coastal groundwater. In what became a classic paper he summarized the American experience and reviewed much of the European work. By that time, many of the principles required for management of coastal aquifers were well understood. Examples of these principles include the upward coning of saltwater from depth in response to water levels lowered by pumping (Pennick, 1904) and recognition that it was permissible to have water levels below sea level and that the "actual danger point comes when cone of depression in the water table caused by pumping reaches the sea, not when the water level at the well falls below sea level" (Brown, 1925, p. 39).

The study of saltwater encroachment made significant contributions to the development of the science of hydrogeology. Areas in which investigations produced classical studies are the Atlantic Coastal Plain (Sanford, 1910); New Jersey (Thompson, 1928); Long Island (Jacob, 1943/1944, 1944/1945; Perlmutter et al., 1959); North Carolina (Harris, 1967); Brunswick, Georgia (Counts and Donsky, 1963; Wait, 1965; Hanshaw et al., 1965; Paper 27); Florida (Brown and Parker 1945; Parker et al., 1955; Cooper, 1963; Vernon, 1961); California and Texas (Piper et al., 1953; Poland et al., 1959; Winslow et al., 1957), where land-surface subsidence is associated with saltwater encroachment; and Hawaii (Wentworth et al., 1955; Stearns and MacDonald, 1942). For example, significant advances were made in understanding the chemistry of water by Piper et al. (1953) and Poland et al. (1959) during a study of saltwater contamination of Southern California in which they demonstrated the role of ion exchange for both water softening and water hardening within aquifers. They also showed the importance of sulfate reduction and suggested techniques for differentiating brines from seawater.

The static conditions assumed by both Badon Ghyben and Herzberg are not compatible with field conditions. As discussed by Carlston (Paper 10), Hubbert (1940, p. 872) first refined their equation by taking into account movement of freshwater. The next significant modification of this principle was by Cooper (1959) who demonstrated that circulation of seawater in coastal aquifers is a necessary consequence of the zone of dispersion or "mixing zone" that exists between freshwater and seawater. The work of Kohout (Paper 11) is a classic field application and demonstration of the cyclic flow in the zone of dispersion as described by Cooper (1959). Understanding the movement of freshwater and saltwater at the interface has continued to be an area of active research (Cooper et al., 1964; Bear and Dagan, 1964; Pinder and Cooper, 1970; Shamir and Dagan, 1971; Larson et al., 1977; Mercer et al., 1980).

Saltwater also occurs in the form of brines and formation waters not associated with encroachment of ocean water. These waters have not yet received the hydrogeologic study that is warranted. We need far more studies like those exemplified by Berry and Hanshaw (Paper 12), White (Paper 13), Hitchon (Paper 24), and Graf et al. (1966).

In the 1930s, 1940s, and the early 1950s, several papers were published that postulated a filtration mechanism to account for subsurface concentration of brines. During the middle 1950s and 1960s, this theoretical work was tested in laboratories by experiments in which salt solutions were forced at high pressure through compacted clays. As had been hypothesized earlier, these studies demonstrated that clays were capable of acting much the same as osmotic membranes. Application of the results of this theoretical and experimental work to field studies and particularly to topics of interest to hydrogeology was undertaken in the late 1950s when Berry and Hanshaw, along with their colleagues at Petroleum Research Corporation under the leadership of Gilman Hill, demonstrated the importance of osmotic pressure and salt filtration effects in generation of brines that caused anomalous head values in deeper basins (Hanshaw and Hill, 1969). Hanshaw (1962) also investigated the membrane properties of clays and provided the theory necessary to explain quantitatively the results of his experiments. Coplen and Hanshaw (1973) and Hanshaw and Coplen (1973) discussed the results of experiments regarding ultrafiltration by compacted clay membranes and showed that not only were dissolved solids excluded from passage through membranes but that the passage of water also affected the isotopic composition of the water molecules themselves. Hitchon and Friedman (Paper 24) discussed the isotopic and chemical composition of formation water in the Alberta basin and suggested that ultrafiltration was responsible for some of the observed variations. Graf et al. (1965) suggested that the highly concentrated calcium chloride brines of the Michigan basin result from ultrafiltration by shale membranes. Billings et al. (1969) also postulated that the geochemistry and origin of formation water in sedimentary basins in Washington and Canada are partially the results of membrane concentration of more dilute formation water.

White (Paper 13) has long been the leader in studying chemistry of groundwater from the point of view of origin of brines and ore deposits. Much work remains to be done on this, and groundwater geologists have yet to make the contributions to understanding the genesis of ores of which they are capable. An earlier geologist, Lindgren (1903), developed an outstanding reputation based, in part, on his awareness of the role of groundwater in ore deposition. His early work as a groundwater geologist took place in Hawaii with special concern about the relationship of saltwater to freshwater.

Hitchon and his colleagues (see: references in Paper 24), have produced a number of classic papers during their study of the formation waters in the basins of western Canada. The only other deep basin to receive such extensive investigation from the hydrogeologic point of view is the Illinois basin, the subject of several classic papers (Clayton et al., 1966; Graf et al., 1965, 1966).

Carpenter (1978) summarized the chemical reactions and discussed the origin and chemical evolution of brines in sedimentary basins. Our understanding of the origin of saline water has been augmented by papers such as those of Mackenzie and Garrels (1966); and Mackenzie et al. (1967). Jones (1966), whose work on saline lakes demonstrated that silicate hydrolysis, an important weathering reaction, is buffered by CO_2 from the atmosphere and thereby prevents pH from rising above about 10. The Eugster and Jones paper (1979) is a most systematic and detailed presentation of the reactions that produce saline water. Wood (1976) provided an application of ion filtration to explain the distribution of dissolved solids in shallow freshwater aquifers.

The chemical character of groundwater has become a topic of interest to geochemists and carbonate petrologists who study diagenesis of sediments. Much of the work on this subject has been done in the zone of dispersion between the freshwater and saltwater. Runnels (Paper 14) was among the earliest to recognize the significance of this zone. Using the concepts and principles discussed in Part III, he demonstrated the geochemical reactivity of the zone of dispersion. He also reviewed much of the earlier literature and provided the geochemical explanation for the cause of the chemical reactions that occur in this highly reactive zone. Because of the continuous inflow of seawater, the mixing zone has been hypothesized as a suitable environment for dolomitization by Hanshaw et al. (1971) and Land (1973). This idea has been further developed by Badiozamani (1973) who named it the Dorag (Persian word for *mixed blood*) Model and by Folk and Land (1975) whose work strengthens the validity of Hanshaw's hypothesis by their excellent geochemical interpretation of the mineralogic reactions.

Plummer et al. (Paper 15) applied many of the geochemical principles to understand the chemistry of groundwater and its diagenetic effects on the carbonate sediments of Bermuda. More recent studies indicative of the large amount of research now being done on chemistry of mixing zones are (1) Back et al. (1979) in which the importance of the mixing zone and discharge of groundwater in formation of geomorphic features on the east coast of the Yucatan are investigated; (2) Knauth (1979) who proposed that many nodular cherts in limestone have formed in the mixing zone where the dissolution of biogenic opal produces water highly supersaturated with respect to quartz; and (3) the work of Hanor (Paper 16) exemplifies the work of carbonate petrologists and evaluates the relative significance of the reactions in the mixing zone emphasizing the importance of the loss CO_2 gas in the deposition of beachrock.

REFERENCES

Back, W., B. B. Hanshaw, T. Pyle, L. N. Plummer, and A. Weidie, 1979, Geochemical Significance of Ground-Water Discharge and Carbonate Solution to the Formation of Caleta Xel Ha, Quintana Roo, Mexico, *Water Resources Research* **15:**1521-1535.

Badiozamani, K., 1973, The Dorag Dolomitization model—Application to the Middle Ordovician of Wisconsin, *Jour. Sed. Petrology* **43:**965-984.

Bear, J., and G. Dagan, 1964, Moving Interface in Coastal Aquifers, *Am. Soc. Civil Engineers Proc., Jour. Hydraulics Div.* **90(HY4):**103-216.

Billings, G. K., B. Hitchon and D. R. Shaw, 1969, Geochemistry and Origin of Formation Waters in the Western Canada Sedimentary Basin. 2. Alkali Metals, *Chem. Geology* **4:**211-223.

Brown, J. S., 1925, A Study of Coastal Ground Water with Special Reference to Connecticut, *U.S. Geol. Survey Water-Supply Paper 537,* 101p.

Brown, R. H., and G. Parker, 1945, Salt Water Encroachment in Limestone at Silver Bluff, Miami, Florida, *Econ. Geology* **40:**235-262.

Carpenter, A., 1978, Origin and Chemical Evolution of Brines in Sedimentary Basins, *Oklahoma Geol. Survey Circ. 79,* pp. 60-77.

Clayton, R. N., I. Friedman, D. L. Graf, T. K. Mayeda, W. F. Meents, and N. F. Shimp, 1966, The Origin of Saline Formation Waters—I. Isotopic composition, *Jour. Geophys. Research* **71:**3869-3882. (Reprinted as Paper 11 in *Geochemistry of Water,* Y. Kitano, ed., Benchmark Papers in Geology, vol. 16, Dowden, Hutchinson & Ross, Stroudsburg, Pa., pp. 208-221.)

Cooper, H. H., Jr., 1959, A Hypothesis Concerning the Dynamic Balance of Fresh Water and Salt Water in a Coastal Aquifer, *Jour. Geophys. Research* **64:**461-467.

Cooper, H. H., Jr., 1963, Type Curves for Nonsteady Radial Flow in an Infinite Leaky Artesian Aquifer, *U.S. Geol. Survey Water-Supply Paper 1545C,* pp. 48-55.

Cooper, H. H., Jr., F. A. Kohout, H. R. Henry, and R. E. Glover, 1964, Sea Water in Coastal Aquifers, *U.S. Geol. Survey Water-Supply Paper 1613-C,* 84p.

Coplen, T. B., and B. B. Hanshaw, 1973, Ultrafiltration by a Compacted Clay Membrane—I. Oxygen and Hydrogen Isotopic Fractionation, *Geochim. et Cosmochim. Acta* **37:**2295-2310.

Counts, H. B., and E. Donsky, 1963, Salt-water Encroachment, Geology, and Ground-Water Resources of Savannah Area, Georgia and South Carolina, *U.S. Geol. Survey Water-Supply Paper 1611,* 100p.

Eugster, H. P., and B. F. Jones, 1979, Behavior of Major Solutes during Closed-Basin Brine Evolution, *Am. Jour. Sci.* **279:**609-631.

Folk, R. L., and L. S. Land, 1975, Mg/Ca Ratio and Salinity: Two Controls over Crystallization of Dolomite, *Am. Assoc. Petroleum Geologists Bull.* **59:**60-68.

Graf, D. L., I. Friedman, and W. F. Meents, 1965, The Origin of Saline Formation Waters. II. Isotopic Fractionation by Shale Micropore Systems, *Illinois Geol. Survey Circ. 393,* 32p.

Graf. D. L., W. F. Meents, I. Friedman, and N. F. Shimp, 1966, The Origin of Saline Formation Waters. III. Calcium Chloride Waters, *Illinois Geol. Survey Circ. 397,* 60p.

Hanshaw, B. B., 1962, Membrane Properties of Compacted Clays, Unpub. Ph.D. thesis, Harvard University, Cambridge, Mass.

Hanshaw, B. B., and T. B. Coplen, 1973, Ultrafiltration by a Compacted Clay Membrane—II. Sodium Ion Exclusion at Various Ionic Strengths, *Geochim. et Cosmochim. Acta* **37:**2311-2327.

Hanshaw, B. B., and G. Hill, 1969, Geochemistry and Hydrodynamics of the Paradox Basin Region, Utah, Colorado, and New Mexico, *Chem. Geology* **4:**263-294.

Hanshaw, B. B., W. Back, M. Rubin, and R. Wait, 1965, Relation of Carbon-14 Concentrations to Saline Water Contamination of Coastal Aquifers, *Water Resources Research* **1:**109-114.

Hanshaw, B. B., W. Back, and R. G. Deike, 1971, A Geochemical Hypothesis for Dolomitization by Ground Water, *Econ. Geology* **66:**710-724.

Harris, W. H., 1967, Stratification of Fresh and Salt Water on Barrier Islands as a Result of Differences in Sediment Permeability, *Water Resources Research* **3:**89-97.

Hubbert, M. K., 1940, Theory of Ground-Water Motion, *Jour. Geology* **48:**785-944.

Jacob, C. E., 1944, Correlation of Ground-Water Levels and Precipitation on Long Island, New York, Part I. Theory, *Am. Geophys. Union 24th Ann. Meeting, 1943, Section of Hydrology, Reports and Papers,* National Research Council, Washington, D. C., pp. 564-572. (Reprinted with Part II in *New York Water Power and Control Commission Bulletin* **GW-14:**563-573, 929-939.)

Jacob, C. E., 1945, Correlation of Ground-Water Levels and Precipitation on Long Island, New York, Part II. Correlation of Data, *Am. Geophys. Union 25th Ann. Meeting, 1944, Section of Hydrology, Papers,* National Research Council, Washington, D.C., pp. 928-939. (Reprinted with Part I in *New York Water Power and Control Commission Bulletin* **GW-14:**563-573, 929-939.)

Jones, B. F., 1966, Geochemical Evolution of Closed Basin Waters in the Western Great Basin, *Second Symposium on Salt,* J. I. Rau, ed., vol. 2, Northern Ohio Geological Society, Cleveland, Ohio, pp. 181-200.

Knauth, L. P., 1979, A Model for the Origin of Chert in Limestone, *Geology* **7:**274-277.

Land, L. S., 1973, Contemporaneous Dolomitization of Middle Pleistocene Reefs by Meteoric Water, *North Jamaica Bull. Marine Sci.* **23:**64-92.

Larson, S. P., S. S. Papadopulos, H. H. Cooper, Jr., and W. L. Burnhan, 1977, Simulation of Wastewater Injection into a Coastal Aquifer System near Kahului, Maui, Hawaii, in *Hydraulics in the Coastal Zone,* Proc. 25th Annual Hydr. Division Spec. Conf., Am. Soc. Civil Engineers, pp. 107-116.

Lindgren, W., 1903, The Water Resources of Molokai, *U.S. Geol. Survey Water-Supply Paper 77,* 62p.

Mackenzie, F. T., and R. M. Garrels, 1966, Chemical Mass Balance between Rivers and Oceans, *Am. Jour. Sci.* **264:**507–525. (Reprinted as Paper 6 in *Geochemistry of Water,* Y. Kitano, ed., Benchmark Papers in Geology, vol. 16, Dowden, Hutchinson & Ross, Stroudsburg, Pa., pp. 120–138 and as Paper 5 *Sea Water,* J. Drever, ed., Benchmark Papers in Geology, vol. 45, Dowden, Hutchinson & Ross, Stroudsburg, Pa., pp, 97–115.)

Mackenzie, F. T., R. M. Garrels, O. P. Bricker, and F. Bickley, 1967, Silica in Sea Water: Control by Silica Minerals, *Science* **155:**1404–1405. (Reprinted as Paper 6 in *Sea Water,* J. Drever, ed., Benchmark Papers in Geology, vol. 45, Dowden, Hutchinson & Ross, Stroudsburg, Pa., pp. 116–117.)

Mercer, J. W., S. P. Larson, and C. R. Faust, 1980, Simulation of Saltwater Interface Motion, *Ground Water* **18:**374–385.

Parker, G. G., G. E. Ferguson, S. K. Love, and others, 1955, Water Resources of Southeastern Florida with Special Reference to the Geology and Ground Water of the Miami Area, *U.S. Geol. Survey Water-Supply Paper 1255,* 965p.

Pennick, J. M. K., 1904, Investigations for Ground-Water Supplies, *Am. Soc. Civil Engineers Trans., **54,***Part D, pp. 169–181.

Perlmutter, N. M., J. J. Geraghty, and J. E. Upson, 1959, The Relation between Fresh and Salty Ground Water in Southern Nassau and Southeastern Queens Counties, Long Island, New York, *Econ. Geology* **54:**416–435.

Pinder, G. F., and H. H. Cooper, Jr., 1970, A Numerical Technique for Calculating the Transient Position of the Saltwater Front, *Water Resources Research* **6:**875–882.

Piper, A. M., A. A. Garrett and others, 1953, Native and Contaminated Water in the Long Beach—Santa Ana Area, California, *U.S. Geol. Survey Water-Supply Paper 1136,* 320p.

Poland, J. F., A. A. Garrett, and A. Sinnott, 1959, Geology, Hydrology, and Chemical Character of Ground Waters in the Torrance—Santa Monica Area, California, *U.S. Geol. Survey Water-Supply Paper 1461,* 425p.

Sanford, S., 1910, Saline Artesian Waters of the Atlantic Coastal Plain, *U.S. Geol. Survey Water-Supply Paper 258,* pp. 75–86.

Shamir, V., and G. Dagan, 1971, Motion of the Seawater Interface in Coastal Aquifer: A Numerical Solution, *Water Resources Research* **7:**644–657.

Stearns, H. T., and G. MacDonald, 1942, Geology and Ground-Water Resources of the Island of Maui, Hawaii, *Hawaii Div. Hydrology Bull. 7,* 344p.

Thompson, D. G., 1928, Ground-Water Supplies of the Atlantic City Region, *New Jersey Dept. Conserv. and Development Bull. 30,* 138p.

Vernon, R. O., 1961, The Geology and Hydrology Associated with a Zone of High Permeability (Boulder Zone) in Florida, *Soc. Mining Engineers, Preprint 69-AG-12,* 24p.

Wait, R. L., 1965, Geology and Occurrence of Fresh and Brackish Ground Water in Glynn County, Georgia, *U.S. Geol. Survey Water-Supply Paper 1613-E,* 94p.

Wentworth, C. K., A. C. Mason, and D. A. Davis, 1955, Salt-Water Encroachment as Induced by Sea-Level Excavation on Angaur Island, *Econ. Geology* **50:**669–680.

Winslow, A. G., W. W. Doyle, and L. A. Wood, 1957, Salt Water and its Relation to Fresh Ground Water in Harris County, Texas, *U.S. Geol. Survey Water-Supply Paper 1360-F,* pp. 375–405.

Wood, W., 1976, A Hypothesis of Ion Filtration in a Potable-Water Aquifer System, *Ground Water* **14:**233–244.

AN EARLY AMERICAN STATEMENT OF THE BADON GHYBEN-HERZBERG PRINCIPLE OF STATIC FRESH-WATER-SALT-WATER BALANCE*

C. W. CARLSTON
U. S. Geological Survey, Washington, D. C.

ABSTRACT. In 1828 the American Journal of Science published a paper by Dr. Joseph Du Commun, a West Point Military Academy teacher of French, which outlined qualitatively and quantitatively the principle of hydrostatic balance of fresh water and sea water in coastal regions (1828). Du Commun's statement antedated by 61 years Badon Ghyben's (1889) statement of this principle and Herzberg's (1901) statement by 73 years. Badon Ghyben and Herzberg have hitherto been credited for the first enunciation of the principle.

In recent years Hubbert (1940) has shown that the fresh-water-salt-water balance is hydrodynamic rather than hydrostatic because a flow system must exist in the fresh water. Wentworth's (1948) demonstration of a zone of diffusion on either side of the fresh-water-salt-water interface has led Cooper (1959) to the hypothesis that a flow system also exists in the salt water near the interface. Cooper's hypothesis and his further deductions have been confirmed by field observations in the Miami area (Kohout, 1960). Glover (1959) and Henry (1959, 1960) have also contributed to this field of inquiry.

INTRODUCTION

The principle of static equilibrium between fresh water and salt water in coastal regions is widely known as the Badon Ghyben-Herzberg principle. The principle was applied to hydrologic problems of seacoasts by W. Badon Ghyben, a Dutch captain of engineers, in 1889. Because there was little interest in coastal ground water at that time his paper on the subject received little attention. Alexander Herzberg's studies of the fresh-water-salt-water balance on the island of Norderney was published in 1901 at a time when coastal water supplies in the Netherlands were of considerable interest to hydrologists. Herzberg apparently had no knowledge of Badon Ghyben's work and arrived at the same conclusions independently. Herzberg's paper was widely acclaimed and it was not until later that the priority of Badon Ghyben was established. The names of both men were then credited for the principle.

The principle states:

$$z = \frac{\rho_f}{\rho_s - \rho_f} h$$

in which z is the depth below sealevel to a point on the fresh-water-salt-water interface vertically below where h is measured, ρ_f is the density of fresh water and is assumed to be 1 g/cm^3, ρ_s is the density of sea water, and h is the head of fresh water above sealevel datum.

PRIORITY OF DU COMMUN

The writer has discovered that credit for the first enunciation of this principle should be given to an American of French birth whose statement in the American Journal of Science antedated the Badon Ghyben paper by 61 years and the Herzberg paper by 73 years. Dr. Joseph Du Commun,[1] a teacher

---

* Publication authorized by the Director, United States Geological Survey.

[1] According to the records of the West Point Military Academy, Dr. Du Commun was born in France, was appointed to the Academy from New York on March 1, 1818 and served as a civilian Second Teacher of French until his resignation on August 31, 1831.

of French at West Point Military Academy, clearly outlined qualitatively and quantitatively the principle of coastal balance of fresh and sea water in 1828 in the American Journal of Science (Du Commun, 1828).

In May, 1824, Levi Disbrow, the first American water-well driller, bored his first water well at a distillery in New Brunswick, New Jersey (Carlston, 1943, p. 123). By 1827, this well, or a later drilled well, was the subject of discussion in two newspapers of the day, the "National Gazette" and the "Harmony Gazette". Questions were asked as to why this fresh-water well had a water level that rose from eight to fourteen feet above the Raritan River and why the yield of the well varied in exact phase with the rise and fall of the tide.

To explain these phenomena, Du Commun, applying the basic physical principles developed by Archimedes and Pascal, used the analogy of a U-tube. This tube, ACB, is shown in figure 1, as copied from Du Commun's illustration. Du Commun's explanation of the ebb and flow of the well followed these logical steps:

1. Water poured into this tube will fill it in part to the same height (a and b).

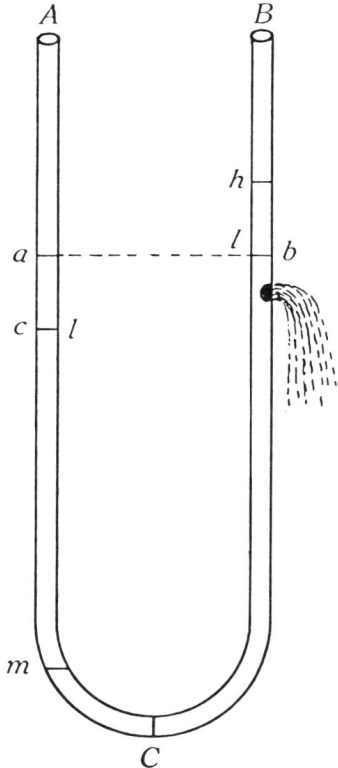

Fig. 1. Du Commun's illustration of density balance of liquids.

2. An inch of mercury at *m* will support, in branch *B*, 13 inches of rain water.

3. A liquid heavier than fresh water, such as salt water in branch *A*, will support, in branch *B*, a column of fresh water rising to a higher level than that of the salt water in inverse ratio to their densities. The height of the fresh-water column is represented by level *b*, and the height of the salt water column is represented by level *c*.

4. Branch *B* can represent fresh water in the rocks under the well, and Branch *A*, the salt water of the ocean lying beneath the fresh water.

Du Commun then states: "If so, it explains why the fresh water in boring by the sea shore is raised and flows above the level of the sea water . . . " (p. 174). He then proceeds to reason in this fashion:

> "Let us suppose that a hole has been opened in the branch *B*, a little below *lb* the level of the water at ebb; the water will then flow with a velocity that may be represented by 1, but at high tide the water might be supported at the height *h*, if the opening in the tube did not permit it to flow out, and it then must flow with the same velocity as if pressed under a column of fluid of that elevation. The quantity of water so running may be as 3, 4, 5, & c, according to the height of the tide; and finally, it must continually and exactly follow its oscillations." (p. 175).

Du Commun described the fresh-water-salt-water balance in the following words:

> "If we calculate the particular case here given, we shall find, the density of fresh water being represented by 1000, that of sea water by 1029 (Dr. Murray), the difference of the levels being 15 feet, we shall find, I say that the depths at which they join underground must be five hundred feet."

Although Du Commun did not give the formula for his computation, he obviously used a form equivalent to the Badon Ghyben-Herzberg equation:

$$z = \frac{1}{1.029-1.000} \; 15 = 517 \text{ feet below sealevel, or rounded off to 500 feet.}$$

Du Commun conceived of ground water as "subterranean streams" and thought that this principle explained springs on high ground and at the top of mountains (p. 175). Such misconceptions are expectable in the context of knowledge of ground-water hydrology in 1828 and should not detract from his principal hydrologic contribution.

Discovery of the very clear priority of Du Commun over both Badon Ghyben and Herzberg adds one more problem in nomenclature. For example, a Lt. Colonel J. Drabbe was the senior author of Badon Ghyben's paper. It has been assumed that Drabbe did not participate in writing the paper, but became senior author by virtue of his military rank. For many years it was thought that Badon was the given name of Ghyben, so that only his last name was used in connection with the principle. Now it is known that Badon Ghyben is a compound surname and that this full name should be used, although habitual usage of only the last name has resulted in strong resistance to adding the name of Badon to Ghyben. Finally, it was thought for many years that Herzberg's first name was Baurat, whereas this was his title, translatable roughly as "State Architect".

RECENT ADVANCES

Since the time of Du Commun, Badon Ghyben, and Herzberg very considerable advances have been made in the formulation of a fresh-water-salt-water interface model in coastal aquifers. Hubbert (1940, p. 872) has stated that the fresh-water-salt-water equilibrium is not hydrostatic but is instead a dynamic equilibrium between flowing fresh water and static salt water. He pointed out that because the fresh-water-salt-water balance is hydrodynamic rather than hydrostatic the Badon Ghyben-Herzberg principle yields results that are more or less in error, depending on the potential gradient. Recently, solutions for the positions of the interface under selected boundary conditions and hydrodynamic balance have been derived by Glover (1959) and Henry (1959).

The rise and fall of the interface in response to tidal fluctuations and variations in recharge creates a zone of diffusion between the fresh water and salt water (Wentworth, 1948). Cooper (1959) has reasoned that, as a consequence of this zone of diffusion, sea water circulates from the floor of the sea to the zone of diffusion and back to the sea, and that this flow tends to decrease the distance from the shore line to the fresh-water-salt-water front. This circulation and the seaward displacement of the front predicted by Cooper have been confirmed by field observations in the Miami area (Kohout, 1960) and are in accord with analytical developments for selected boundary models of the problem (Henry, 1960).

Thus, the principal advances beyond Du Commun, Badon Ghyben, and Herzberg have been Hubbert's demonstration of a hydrodynamic balance between flowing fresh water and static salt water and Cooper's demonstration that the hydrodynamic balance is between flowing fresh water and flowing salt water.

REFERENCES

Badon Ghyben, W., 1889, Nota in verband met de voorgenomen put boring nabij Amsterdam: Koninkl. Inst. Ing. Tijdschr., 1888-89 [The Hague], p. 21.
Carlston, C. W., 1943, Notes on the early history of water well drilling in the United States: Econ. Geology, v. 38, p. 119-136.
Cooper, H. H., 1959, A hypothesis concerning the dynamic balance of fresh water and salt water in a coastal aquifer: Jour. Geophys. Research, v. 64, p. 461-467.
Du Commun, Joseph, 1828, On the cause of fresh water springs, fountains, & c.: Am. Jour. Sci., 1st ser., v. 14, p. 174-176.
Glover, R. E., 1959, The pattern of fresh-water flow in a coastal aquifer: Jour. Geophys. Research, v. 64, p. 457-459.
Henry, H. R., 1959, Salt intrusions into coastal aquifers: Jour. Geophys. Research, v. 64, p. 1911-1919.
─────── 1960, Salt intrusions into coastal aquifers: Internat. Assoc. Sci. Hydrology, Comm. Subterranean Waters Pub. 52, p. 478-487.
Herzberg, Alexander, 1901, Die Wasserversorgung einiger Nordseebader: Gasbeleucht. u. Wasserversorg. Jahrb. Jahrg. 44 [Munich].
Hubbert, M. King, 1940, The theory of ground water motion: Jour. Geology, v. 48, p. 785-944.
Kohout, F. A., 1960, Flow pattern of fresh and salt water in the Biscayne aquifer of the Miami area, Florida: Internat. Assoc. Sci. Hydrology, Comm. Subterranean Waters Pub. 52, p. 440-448.
Wentworth, C. K., 1948, The growth of the Ghyben-Herzberg transition zone under a rinsing hypothesis: Am. Geophys. Union Trans., v. 29, no. 1, p. 97-98.

Cyclic Flow of Salt Water in the Biscayne Aquifer of Southeastern Florida

F. A. Kohout

Abstract. Observations over a period of nearly 20 years confirm the fact that the salt-water front in the Biscayne aquifer along the coast of the Miami area, Florida, is dynamically stable at a position seaward of that computed according to the Ghyben-Herzberg principle. During periods of heavy recharge the fresh-water head is high enough to cause the fresh water, the salt water, and the zone of diffusion between them to move seaward. In addition to this bodily movement of the system, there is a seaward flow of diluted salt water in the zone of diffusion. When the fresh-water head is low, salt water in the lower part of the aquifer intrudes inland, but some of the diluted sea water in the zone of diffusion continues to flow seaward. Cross sections showing equipotential lines in terms of equivalent fresh-water head show that the sea water flows inland, becoming progressively diluted with fresh water, to a line along which there is no horizontal component of flow, after which it moves upward and returns to the sea. The cyclic flow acts as a deterrent to the encroachment of sea water because of return to the sea of a part of the inland flow.

Introduction. The basic premise of the Ghyben-Herzberg principle is that the position of the interface between fresh water and salt water in a coastal aquifer will be governed by a hydrostatic equilibrium between fresh water and the more dense sea water. *Hubbert* [1940, pp. 924–926] showed, however, that because fresh water was known to flow seaward, the position of the interface would be governed by a dynamic equilibrium between flowing fresh water and static salt water. This concept is shown in Figure 1 where the depth to a point on the interface (z) would be equal to the head of fresh water (h) with reference to sea level at the point on the interface multiplied by the ratio of the density of fresh water (ρ_f) to the difference between the densities of sea water (ρ_s) and fresh water. Observations over a period of nearly 20 years indicate that the salt front in the Biscayne aquifer of the Miami, Florida, area is dynamically stabilized seaward of the theoretical position given by either concept (Fig. 2). Recent studies indicate that the lack of agreement results from the fact that two assumptions inherent in the above developments are not fulfilled in the Biscayne aquifer. These assumptions are (1) that a sharp interface exists between fresh water and salt water in an aquifer and (2) that the salt water in the aquifer is static.

It is the intent of this paper to illustrate by field observations that the salt water is not static but flows in a cycle from the floor of the sea into the zone of diffusion and back to the sea and that the cycle acts to lessen the extent to which the salt water occupies the aquifer. The hypothesis of cyclic flow has been expressed by *Cooper* [1959].

Geologic and hydrologic characteristics. The Biscayne aquifer consists of solution-riddled limestone and calcareous sandstone. It is a water-table aquifer and extends from land surface to an average depth of 100 feet below msl (mean sea level). In general, coefficients of permeability are in the range 50,000 to 70,000 gpd/sq ft [*Parker*, 1951, p. 824].

The zone of diffusion. In the Biscayne aquifer the zone of diffusion is a zone of substantial

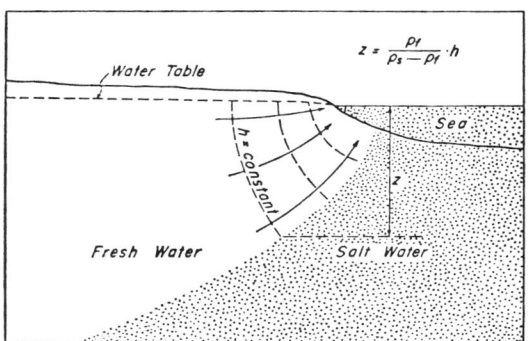

Fig. 1. Balance between fresh water and salt water in a coastal aquifer with the salt water static.

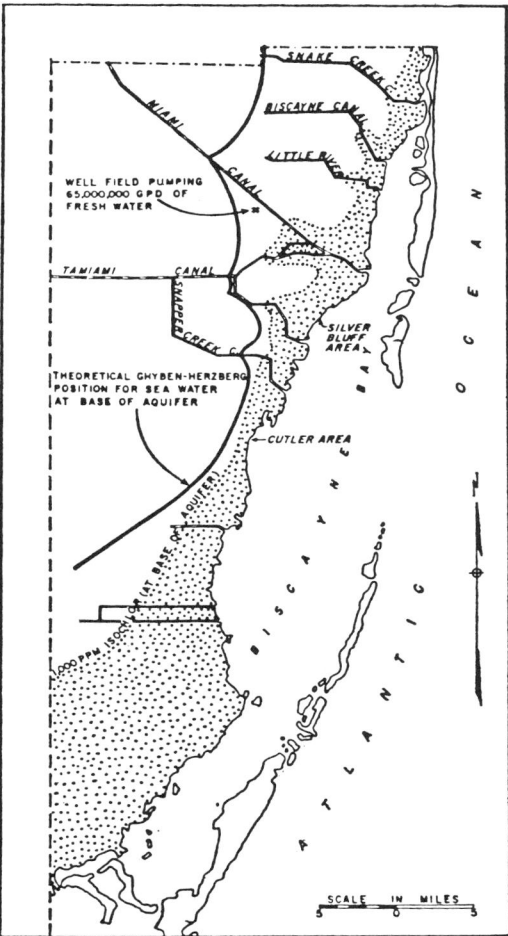

Fig. 2. Map of Dade County comparing the theoretical Ghyben-Herzberg position with the actual field position of salt water at the base of the Biscayne aquifer.

thickness in which there is a gradation of salt content from that of fresh water, 16 ppm (parts per million) chloride, to that of sea water, about 19,000 ppm chloride. Figures 3 and 4 are cross sections of the zone of diffusion in the Silver Bluff and Cutler areas; the locations of these areas are shown in Figure 2.

The distance from the bay to the inland toe of the wedge of salty water is more than 12,000 feet in the Silver Bluff area, but only about 1600 feet in the Cutler area. The toe of the wedge is blunt-nosed in each area. This convex-upward configuration of the salt front at the inland extremity is anomalous to the concave-upward parabola that would be present if assumptions of a sharp interface and static salt water were fulfilled in the field.

Seaward flow of salty water. The fluctuations of chloride content in well G 519A, which is 400 feet from Biscayne Bay in the Silver Bluff area, are shown in Figure 5 (also see Fig. 3 for position of the open-hole part of well G 519A in the zone of diffusion). The rapid decrease in chloride content at the three sampling depths during October 1953 resulted from a large increase in fresh-water head following heavy rainfall in early October. Salt water was rapidly expelled from the aquifer and the zone of steep concentration gradient (just below well G 519A in Fig. 3) was depressed downward and seaward.

A ground-water velocity test, with fluorescein dye used as a tracer, was performed at the site of well G 519A on January 4, 1954. The results of the test indicated that water containing 1500 to 2000 ppm of chloride (open-hole part of well G 519A, Fig. 3) was flowing seaward at a rate greater than 70 feet per day. This was considered quite significant because, obviously, a large quantity of salt water was flowing toward the ocean.

A rough calculation of the quantities of ground-water and salt outflow through a vertical section of the aquifer at the shoreline is pertinent. If the base of the seaward-flowing fresh-water section in the Silver Bluff area is assumed to be at the 5000 ppm isochlor (Fig. 3), the thickness of the flow section at well G 519A is about 35 feet. If the average velocity through this thickness is 70 feet per day and the effective porosity of the limestone is 0.2, the discharge of water through a 1-foot-wide strip of the fresh-water flow section is 490 ft^3/day.

From Figure 3, the average chloride content of the water discharging at the shoreline is estimated to be about 1900 ppm. Ten units of 1900-ppm water are closely equivalent in salt content to one unit of 19,000-ppm water; therefore, the equivalent of 49 ft^3/day of ocean water must be incorporated into the fresh-water flow section in order that 490 ft^3/day of 1900-ppm water be discharged from each 1-foot-wide strip of the aquifer. This rough calculation of the seaward movement of salt indicates that the ocean water being integrated into the fresh-water flow section

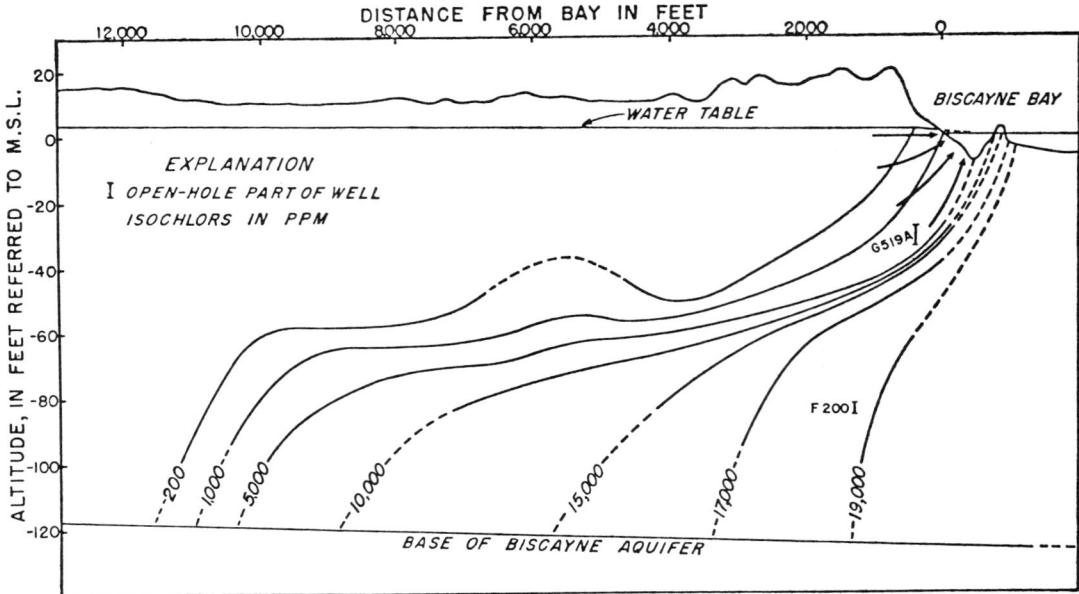

Fig. 3. Cross section through the Silver Bluff area showing the zone of diffusion, November 2, 1954.

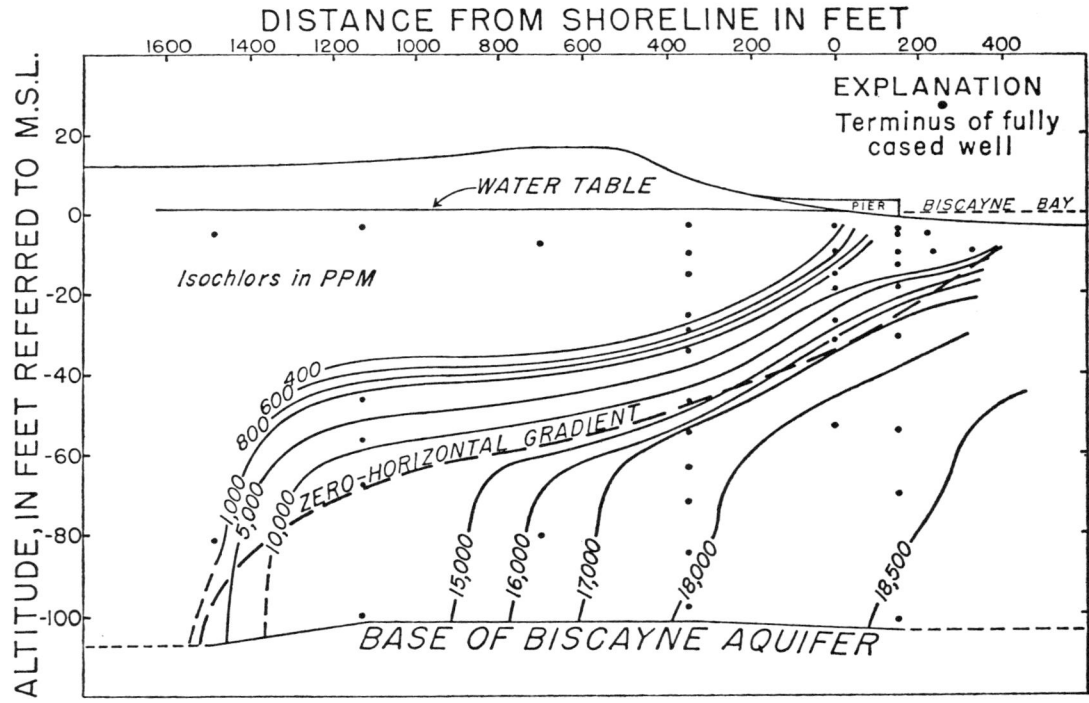

Fig. 4. Cross section through the Cutler area showing the position of the zero-horizontal gradient line within the zone of diffusion (traced from Fig. 10), September 18, 1958.

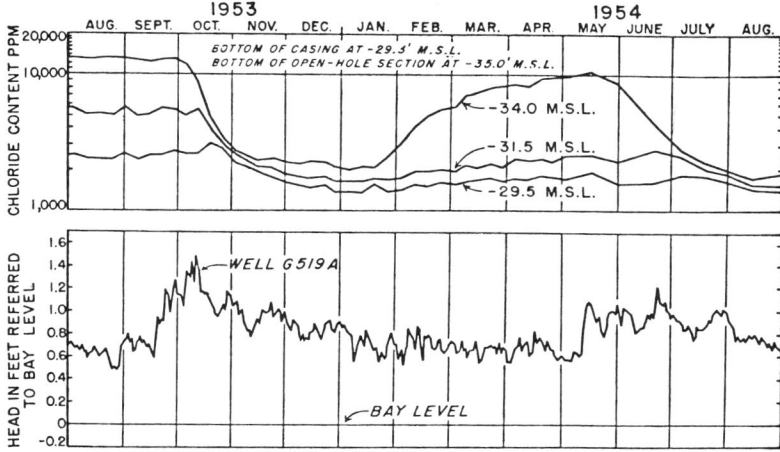

Fig. 5. Graph showing fluctuations of chloride content and water level in well G 519A in the Silver Bluff area.

may amount to 10 per cent (or more) of the total seaward flow of water.

Dispersion. As observations show that the diffused zone remains essentially unchanged although large quantities of salt are flushed back to the sea, it must be concluded that a mechanism much stronger than molecular diffusion is acting to recreate the zone of diffusion.

The growth of a zone of diffusion under a rinsing hypothesis has been described by *Wentworth* [1948, pp. 97–98]. More recent studies, by *Day* [1956], *Rifai, Kaufman, and Todd* [1956], *Kaufman and Orlob* [1956], *Orlob and Radhakrishna* [1958], and *Eriksson* [1958], have shown that variations of fluid velocity across the pores of a permeable medium will combine with molecular diffusion to cause a rapid intermingling of fluids of different concentration. This intermingling process is referred to as dispersion and consists of two separate mechanisms: convection, the mechanical transfer of one fluid into the region of another, and molecular diffusion [*Bosworth*, 1949, p. 465].

For an assumed set of conditions in a hypo-

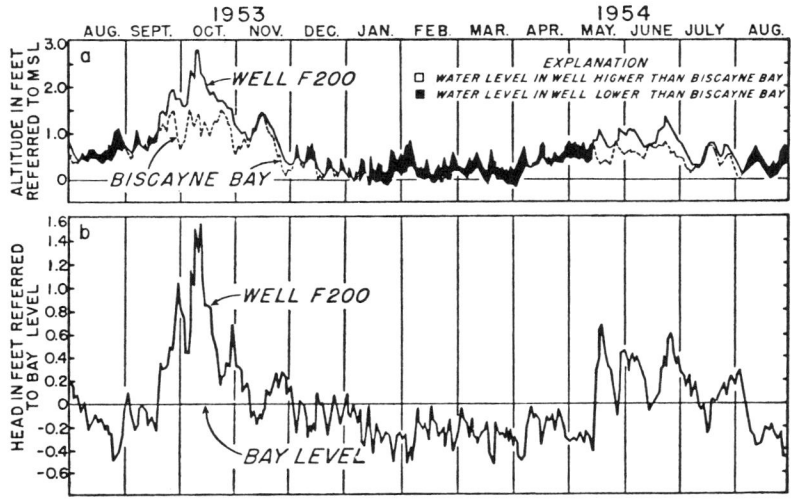

Fig. 6. Hydrographs of daily-average water level in well F 200 and Biscayne Bay.

thetical aquifer of sand under the influence of tidal action, *Cooper* [1959] calculated the coefficient of dispersion to be about 100 times greater than the coefficient of molecular diffusion. Also, he suggested that the coefficient of dispersion in an aquifer of nonuniform permeability may be considerably greater than that for a homogeneous aquifer. In the suggested mechanism, elements of salt water under tidal stimulus move greater horizontal distances in permeable beds than in adjacent less-permeable beds, and the salt-water projections thus formed will be integrated by the upward, cross-bed flow of fresh water.

Movement of ground water caused by the tide has both horizontal and vertical components. Clearly, a mechanism that permits a very rapid transportation and dispersion of salt is available.

Hydraulic gradient in the salt-water zone. Evidently the dispersion must occur at a rate large enough to maintain the zone of diffusion while a large quantity of salt water discharges seaward. To maintain this equilibrium, some means of transporting the salts from the floor of the sea through the aquifer and into the zone of diffusion must be available. However, in the regions below and seaward of the zone of diffusion (Figs. 3 and 4), the concentration gradient is too small for appreciable transportation of salt by dispersion. Therefore, the salts must be transported by hydraulic flow with an accompanying loss of head. In Figure 6, the daily-average water levels in well F 200 are compared with the daily-average water levels of Biscayne Bay. The chloride content ranged from 18,300 to 18,800 ppm during the period shown. The water level of the well, which closely represents the head of ocean water in the aquifer, is higher than the surface of the bay during heavy rainfall periods, as is shown in the unshaded intervals in Figure 6a, and lower than the bay during dry periods, as is shown in the shaded intervals. Clearly, the negative heads reflect the head losses of salt water as it flows landward through the aquifer. In Figure 6b, the daily-average head of the bay has been algebraically subtracted

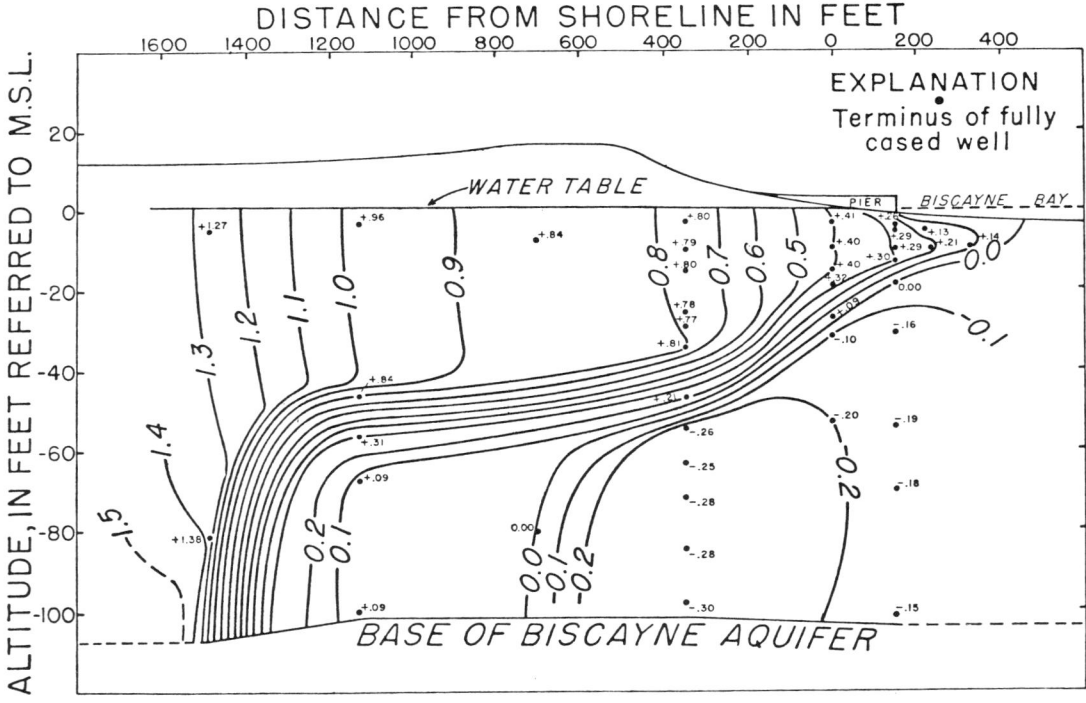

Fig. 7. Cross section through the Cutler area showing equipotential lines in terms of the environmental water, average head, September 18, 1958.

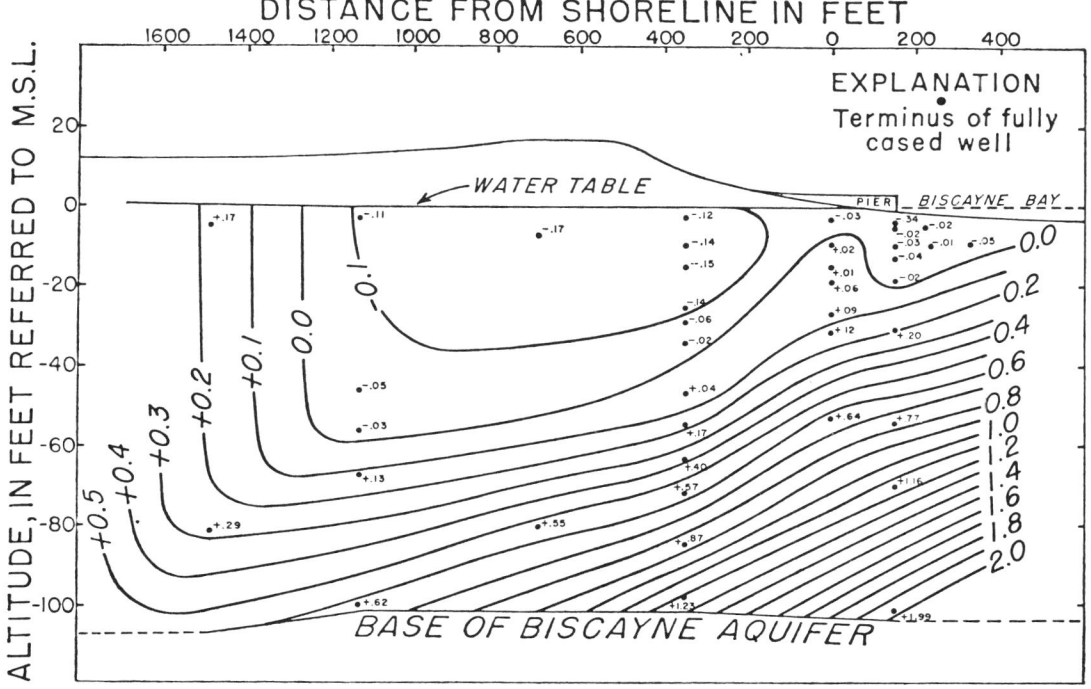

Fig. 8. Cross section through the Cutler area showing equipotential lines in terms of equivalent freshwater head at 1335 EST (bay high tide), September 18, 1958.

from the daily-average head of well F 200, so that the head in the well each day is referred to a bay head of zero for that day. When the hydrograph is positive, salt water flows seaward, and when the hydrograph is negative, salt water flows inland. The hydrograph of well G 519A (Fig. 5), which is constructed in a manner similar to that of well F 200, is never negative. This indicates that the movement of water is always seaward at that point in the aquifer. Therefore, during the intrusion part of the salt-water flow cycle, there must be a line somewhere between the termini of these two wells (within the zone of diffusion, Fig. 3) where the water has the same head as the ocean and where the water is flowing neither inland nor seaward. The water along this line cannot be stagnant, however, because continuity requirements would then be violated. We may conclude, therefore, that salt water along this line must be flowing vertically upward along a return path to the ocean. Thus, a continuous cyclic flow of salt water—inland from the floor of the sea to the zone of diffusion, and then seaward through the fresh-water flow section—has been postulated.

Cutler test site. The Cutler test site was drilled as part of the investigation of factors relating to the cyclic flow. The wells are fully cased and serve as points for collecting water samples and for measuring pressure heads at isolated depths. The termini of the wells are indicated as dots in the Cutler cross sections (Figs. 4, 7 to 10).

Potential throughout the fresh-water salt-water environments. Ground water of uniform density moves in the direction of decreasing fluid potential. Where the density varies, from point to point, measurements of head do not indicate the direction of movement directly. This is illustrated in Figure 7 where the figures at the well-casing termini are daily-average water levels referred to the daily-average bay level (0.0). The concentration of salt water in the casing is the same as that in the aquifer at the casing terminus; the head values, as shown, are the original data.

Obviously, all flow cannot converge upon the sink surrounded by the −0.2-foot contour, and equipotential diagrams constructed from the original basic data are not usable.

In Figures 8 to 10 the equivalent fresh-water heads are shown for high, low, and average tide on September 18, 1958. For the wells that contain salty water, the equivalent head of fresh water has been computed, so that in all cases the heads are the same as if the casing had been filled with fresh water at the time of measurement.

Conversion of environmental salt-water head to fresh-water head in a given well is accomplished by application of the hydrostatic pressure equation:

$$p = \rho g l$$

where p is the pressure at casing terminus, ρ is the density of water in the casing, g is the acceleration due to gravity, l is the measured length of water column above the casing terminus. Equating the right term of the above equation for fresh-water and salt-water columns:

$$\rho_f g l_f = \rho_s g l_s$$

$$l_f = \rho_s l_s / \rho_f$$

where the subscripts f and s refer to fresh water and salt water, respectively. The density of fresh water is assumed to be 1.000.

The following table (using observed daily-average data for the deepest well, 350 feet inland from the shoreline, September 18, 1958, Fig. 10) gives a typical computation:

Salt-water head	+0.60 ft, msl
Depth of casing terminus	−97.9 ft, msl
Length (l_s) of salt-water column	98.50 ft
Density (ρ_s) of casing water (18,000 ppm chloride content)	1.0240
$l_f = \rho_s l_s / \rho_f$	100.86 ft
Fresh-water head (subtract 97.9 ft)	+2.96 ft, msl
Daily-average water level of Biscayne Bay	+0.90 ft, msl
Daily-average fresh-water head referred to Bay level	+2.06 ft

The equipotential lines pass vertically downward from their intersection with the water table and at depth deflect toward the horizontal. In the upper, fresh-water part of the aquifer, the lines indicate the potential of fresh water in a fresh-water environment, and hence indicate comparative potentials. As flow lines must be perpendicular to these equipotential lines, a seaward movement of fresh water is indicated.

In the lower and seaward part of the aquifer the equipotential lines indicate the potential of fresh water in a region occupied by salty water. *Hubbert* [1940, pp. 868–869] has shown that

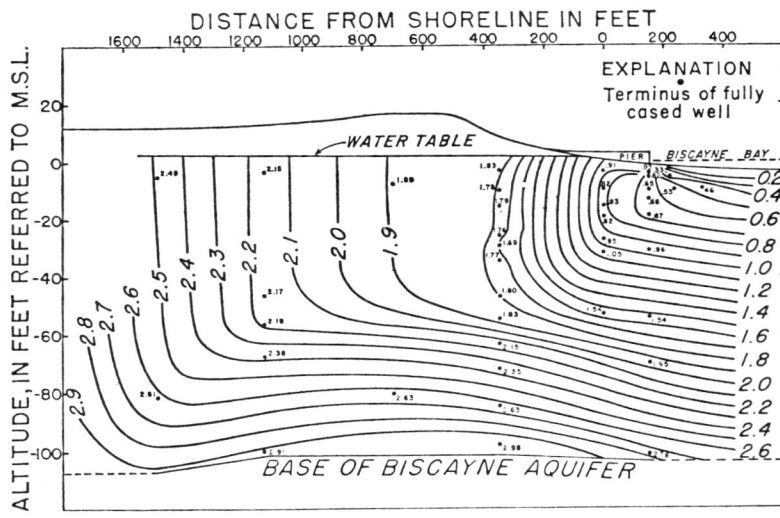

Fig. 9. Cross section through the Cutler area showing equipotential lines in terms of equivalent fresh-water head at 0750 EST (bay low tide), September 18, 1958.

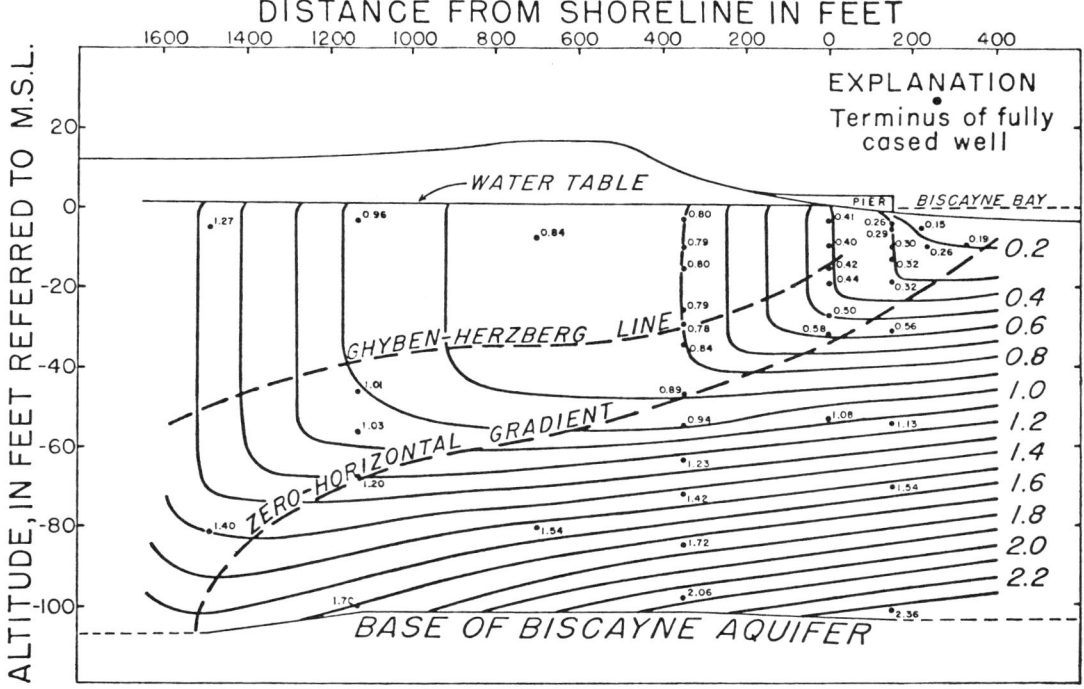

Fig. 10. Cross section through the Cutler area showing equipotential lines in terms of equivalent fresh-water head, average for September 18, 1958.

fresh-water equipotential surfaces in a region occupied by salt water will be horizontal if the salt water is static. The equipotential lines in the salt-water regions of Figures 8 to 10 are not horizontal, but slope inland in Figure 8, seaward in Figure 9, and inland in Figure 10. From this it is concluded that the salt water is not static but must be in motion in the direction of the slope. The slope at high and low tidal stages is indicative of the flow that accommodates the changes of storage taking place in the aquifer at these times. The daily-average equipotential diagram (Fig. 10) indicates that the instantaneous movements occurring throughout the day average out in such a way as to produce a net inland movement of salt water on this date.

The pattern of fresh-water equipotential lines serves as a guide for separating the region of seaward-flowing water from that of the inland-flowing water (Fig. 10). Such a separation is formed by a line passed through the points of horizontality of the individual equipotential lines. Such a line is shown in Figure 10. At all points on this line the water must be flowing vertically upward.

The water above and below the line will have seaward and landward horizontal components of flow, respectively. It is of interest to see where the line is located within the zone of diffusion. Its trace is shown in Figure 4. The location indicates that water containing more than 16,000 ppm of chloride may have a seaward horizontal component of flow.

The salt-water flow cycle. The quantity of inflowing sea water is not continuously balanced by an equivalent seaward discharge of diluted salt water through the upper flow region. For example, rough estimates of the movement of salt under the intrusion conditions shown in Figure 10 indicate that about 20 per cent of the total salt that flows inland discharges seaward through the upper flow region; the remaining 80 per cent stays in the aquifer to increase the volume of salt water in storage and to replace discharged fresh water. Nevertheless, a complete cyclic flow of part of the sea water occurs during the intrusion phase, and this cycle acts as a deterrent to the encroachment of sea water because of return to the sea of part of the inland flow.

Acknowledgments. I am deeply indebted to H. H. Cooper, Jr., M. I. Rorabaugh, and N. D. Hoy, of the U. S. Geological Survey at Tallahassee, Florida, for guidance, discussion of principles, and critical review of the manuscript. My colleagues in the Miami office of the U. S. Geological Survey aided in many ways: Howard Klein reviewed the manuscript critically, C. B. Sherwood and W. F. Lichtler took part in a number of special field studies, and J. E. Hull performed a large part of the field work for the Cutler investigation.

References

Bosworth, R. C. L., The mechanisms of diffusional processes, *Roy. Australian Chem. Inst. J. & Proc.*, 460–482, 1949.

Cooper, H. H., Jr., A hypothesis concerning the dynamic balance of fresh water and salt water in a coastal aquifer, *J. Geophys. Research, 64,* 461–467, 1959.

Day, P. R., Dispersion of a moving salt-water boundary advancing through saturated sand, *Trans. Am. Geophys. Union, 37,* 595–601, 1956.

Eriksson, Erik, A note on the dispersion of a salt-water boundary moving through saturated sand. *Trans. Am. Geophys. Union, 39,* 937–938, 1958.

Hubbert, M. K., The theory of ground-water motion, *J. Geol. 48,* 785–944, 1940.

Kaufman, W. F., and G. T. Orlob, An evaluation of ground-water tracers, *Trans. Am. Geophys. Union, 37,* 297–306, 1956.

Orlob, G. T., and G. N. Radhakrishna, The effect of entrapped gases on the hydraulic characteristics of porous media, *Trans. Am. Geophys. Union, 39,* 648–659, 1958.

Parker, G. G., Geologic and hydrologic factors in the perennial yield of the Biscayne aquifer, *J. Am. Water Works Assoc., 43,* 817–835, 1951.

Rifai, M. N. E., W. J. Kaufman, and D. K. Todd, Dispersion phenomena in laminar flow through porous media, *Progr. Rept. 2,* Canal Seepage Research, Univ. California, Berkeley, 157 pp., 1956.

Wentworth, C. K., Growth of the Ghyben-Herzberg transition zone under a rinsing hypothesis, *Trans. Am. Geophys. Union, 29,* 97–98, 1948.

(Manuscript received February 29, 1960; revised April 29, 1960; presented at the Fortieth Annual Meeting of the American Geophysical Union, Washington, D. C., May 5, 1959).

GEOLOGIC FIELD EVIDENCE SUGGESTING MEMBRANE PROPERTIES OF SHALES

By FREDERICK A. F. BERRY and BRUCE B. HANSHAW
U.S.A.

Some anomalous pressure and salinity data observed within oil- and water-bearing reservoir rocks cannot be explained by prior theories in hydrodynamics and geochemistry. Laboratory evidence has shown that compacted clay minerals act as semipermeable membranes and thereby exhibit osmotic-pressure and saltfiltration effects. Osmotically induced pressure and salt filtering occur in reservoir rocks adjacent to shales presumably serving as semipermeable membranes. Osmotic conditions might result from differences across a shale of salt concentration. Pressure would tend to increase in the reservoir rock on the emergent side of the shale membrane and decrease on the influx side under osmotic conditions. Cross-formational flow through a shale membrane may also cause salt filtering and thereby increase the salinity on the influx side of the membrane.

Three widely separated areas in North America (central Alberta, Canada; San Juan Basin, New Mexico and Colorado; and Wheeler Ridge anticline, San Joaquin Valley, California) have anomalous potentials and salinities that may be explained by the movement of water cross-formationally through shales acting as semipermeable membranes. Pressure and salinity anomalies from other areas possibly may be explained by shale-membrane phenomena.

13

Copyright ©1965 by the American Association of Petroleum Geologists
Reprinted from pages 342-354 and 363-366 of *Am. Assoc. Petroleum Geologists Mem.* **4**:342-366 (1965)

SALINE WATERS OF SEDIMENTARY ROCKS[1]

DONALD E. WHITE[2]
Menlo Park, California

ABSTRACT

Most saline waters of marine sedimentary rocks were probably similar initially to present-day ocean water. Many early diagenetic changes in sediments and waters are related to organic content and bacterial activity; ion exchanges and perhaps some other early changes are inorganic. Diagenetic and later changes in sedimentary rocks cannot be understood without considering the associated fluids, which are mobile and leave little direct and easily interpretable evidence of their changing compositions with time.

Compaction of sediments and escape of interstitial water start at the time of deposition and probably continue for millions of years. The evidence is now convincing that fine-grained sediments behave as semipermeable membranes, permitting selective escape of water and concentrating dissolved components in remaining pore fluids. The initial driving force is lithostatic pressure; after maximum compaction has been attained, salt-filtering may continue under certain circumstances of topography, structure, and lithology, with meteoric water providing the driving energy.

As salinity of the retained pore fluids increases, proportions of some constituents also change, with calcium in particular tending to increase relative to sodium in normal marine sediments. The usual explanation involves liquid-solid cation exchange reactions. This paper suggests that mobilities of individual dissolved constituents differ greatly and that CO_2, B, and perhaps also NH_4^+ and H_2S normally are relatively high in the escaping water. When cation-exchanging clay minerals are tightly compressed, their fixed negative charges evidently repel anions including Cl^-, HCO_3^- and $CO_3^=$. The compositions of thermal and mineral waters of sedimentary rocks, however, strongly indicate that CO_2 does escape through clays, perhaps as uncharged H_2CO_3 molecules. There is a little experimental evidence to indicate that any excess of cations over anions in the fluid passing through the membrane is balanced by a net return flow of H^+ ions. In carbonate-free sediments the system presumably soon comes to a steady state, but if carbonates are in the sediments, they may dissolve in the H^+-enriched environment behind the membrane, thereby maintaining a supply of HCO_3^- and H_2CO_3 for further escape. The chemical evidence suggests that Ca^{++} (mostly from calcite?) is less mobile than Cl^-, probably because of the double charge of Ca^{++}, and is enriched in the retained brine.

Original ocean water in contact with normal sediments evolves to connate (as redefined) or fossil water vastly different from its initial composition, with rates of migration and chemical evolution depending upon its physical and chemical environment. Membrane-filtered and membrane-concentrated types of connate water are recognized, and all gradations can be expected between extremes of these two types.

Chemical and isotopic criteria also clarify the origin of other saline waters in addition to those of normal marine sediments. These other types include connate waters of marine and nonmarine evaporites, several kinds of waters of relatively low salinity that have dissolved evaporites, sulfate and bicarbonate waters of oil fields, and waters driven out of sedimentary rocks during progressive metamorphism. Relatively complete analyses of waters from oil fields and other sources are included to illustrate some principles, but much additional field and experimental study is needed.

Waters escaping from depth into a lower temperature environment will normally exchange dissolved K for Na from the solid phases. In some circumstances the K/Na ratio of a water serves as a crude geothermometer for water temperature.

INTRODUCTION

The origin, hydrocarbon content, lithologic characteristics, and mineral compositions of sedimentary rocks have been studied extensively, but very little attention has been given to the associated waters. Probably every chemical and physical change occurring in a sediment is accompanied

---

in natural waters include Julian Hemley, P.B. Hostetler, Bruce Hanshaw, L.J.P. Muffler, Edgar Bailey, Wayne Hall, A.H. Truesdell, John D. Hem, J.H. Feth, Robert Meade, and D.F. Hewett. Brine samples from a Kings County, Calif. oil-test well and supplementary data were generously provided by Prof. Fred Berry, University of California, Berkeley; four samples from Texas and two from Louisiana were collected by Dr. Marcus Hanna, formerly of Gulf Oil Company; four samples from Alberta, Canada, were collected by Dr. Brian Hitchon, Research Council of Alberta; and nearly all isotope analyses of this report were provided by Professor Harmon Craig, University of California, La Jolla.

[2] Research geologist, U.S. Geological Survey.

---

[1] Read before the 6th Annual Meeting of the Southwestern Federation of Geological Societies and Southwestern Regional Meeting of The American Association of Petroleum Geologists, at Midland, Texas, January 31, 1964. Manuscript received, July 13, 1964. Publication authorized by the Director, U.S. Geological Survey.

The writer is greatly indebted to many individuals with whom the problems and relationships of natural waters have been discussed. Associates of the U.S. Geological Survey who have especially shared interests

by a corresponding change in the associated water. Some of these changes are drastic and others, no doubt, have little significance, but the whole system cannot be understood until we also understand the mobile fluids. This paper outlines some principles that are now reasonably well substantiated and proposes several hypotheses to explain observed compositions and relationships of natural waters.

Subsurface waters have been analyzed extensively for only a few major constituents. Where a dependable water supply is the major objective, analyses are necessary to prove that the water is within desirable limits for drinking or industrial use and free of objectionable contaminants. Many saline waters in sedimentary basins have also been analyzed for major constituents in the course of exploration for oil and gas; during one period in the history of the petroleum industry, compositions of the waters were considered useful for stratigraphic correlation, but in recent years electric logging and micropaleontology have almost entirely replaced correlation by water analyses. Little attention has been given to the minor constituents and origin of saline waters of sedimentary rocks, and to the role of water in the origin, migration, and accumulation of petroleum and natural gas, especially in view of the high value of these associated fluids. The many problems concerned with meaningful collection and analysis of oil-field waters are reviewed by Chave (1960).

In this report the major chemical types of waters in marine sedimentary rocks are first summarized. Essentially all interstitial waters differ from ocean water in chemical composition and in concentration. Changes start immediately at the time of deposition of sediments and continue to occur in response to changing conditions without limit in time. The role of fine-grained sediments is extremely important, not only in providing "salt-sieving" effects but probably also in changing the compositions of escaping and retained fluids. Some dissolved constituents seem to be highly mobile and others relatively immobile.

Summary of major chemical types.—The major chemical types of saline water of sedimentary basins are largely those types that are associated with petroleum and were considered in early years by Rogers (1917), Mills and Wells (1919), Russell (1933), Case and others (1942), and De Sitter (1947). Recent general reviews include those by Chebotarev (1955), Schoeller (1956), Buneev (1956), Zyka (1958), Tageeva (1960), Chave (1960), Engelhardt (1960, 1961), White (1957) and White, Hem, and Waring (1963, p. F9–F10, tables 12–16).

The salinity of water commonly increases with depth below the surface and ranges from much less than that of sea water to about 10 times more. No tendency has been found for salinity to be identical with or even very close to that of sea water. The dominant anion is generally Cl^-, and the proportion of Cl^- to total anions generally increases with salinity. Chebotarev (1955) and others have noted that waters near the surface are commonly dominated by sulfate, with a mean depth of 1,700 ft for Chebotarev's examples. Bicarbonate waters tend to be intermediate both in salinity and depth, with a mean of 2,300 ft for Chebotarev's examples. With rare exceptions, Na^+ is the dominant cation in chloride waters, but Ca^{++} is generally present in larger proportions than in sea water. The relative proportion of Ca^{++} in the cations tends to increase strongly with increasing salinity and commonly increases with depth and age of the rocks (Fig. 1). Many exceptions are known, however, including high-Ca waters in upper Tertiary rocks (this paper) and even waters in Recent sediments of the Black Sea (Shishkina, 1958).

Less well known chemical types of saline water also occur in sedimentary rocks, and some of these are considered later in this report.

Previous explanations for major chemical types.—In an early outstanding paper, Rogers (1917) recognized that oil-field waters are not necessarily high in salinity; sulfate waters are likely to occur near the surface; bicarbonate waters are commonly intermediate in depth; and chloride is generally the dominant anion at depth, in waters most closely associated with petroleum. Our knowledge and understanding of the sulfate and bicarbonate waters of oil fields have increased only slightly since Rogers' study. Sulfate near the surface is derived either from solution of anhydrite and gypsum or from oxidation of pyrite in the rocks; the high bicarbonate content at interme-

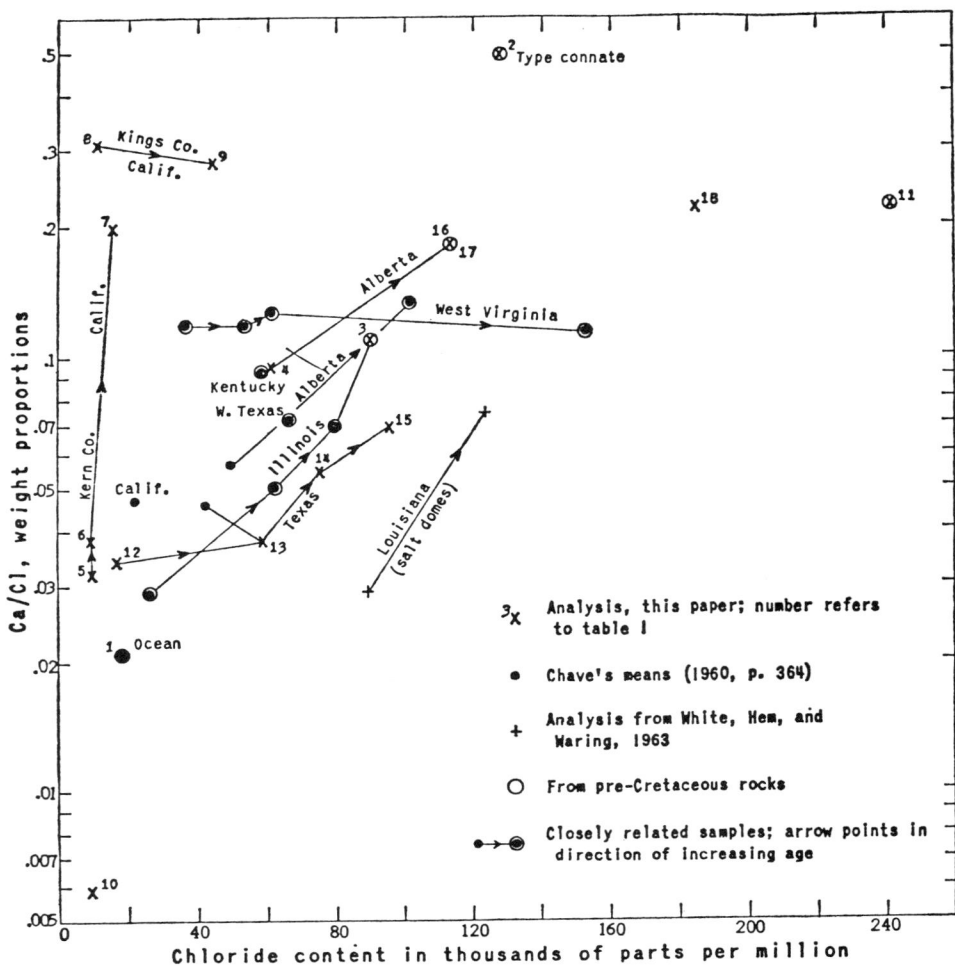

FIG. 1.—Ca/Cl vs. Cl content of some saline waters associated with sedimentary rocks.

diate depths is presumably due to bacterial reduction of sulfate in the organic-rich environment, the oxygen from the sulfate combining with hydrocarbons to form carbonate species and H_2O.

Most attention has been given to the chloride waters, in particular to the great differences in salinity and ion proportions relative to sea water. Several mechanisms to account for these changes have been proposed and these have been reviewed by Russell (1933), De Sitter (1947), Chave (1960), and Engelhardt (1960, 1961). Evaporation and entrainment in expanding subsurface gas (Mills and Wells, 1919) and density stratification due to molecular settling are completely unsatisfactory mechanisms in view of the quantitative requirements (Russell, 1933). A relationship to evaporites is possible for some brines, but high salinities are very common where evidence for evaporites is lacking (Chave, 1960; White, Hem, and Waring, 1963, p. F9).

Major progress in understanding the high salinities started with De Sitter's suggestion (1947) that fine-grained sediments behave as semipermeable membranes or "salt sieves," permitting a selective escape of water and concentrating the dissolved salts in water remaining behind the membrane. Recent studies by Engelhardt (1960), Engelhardt and Gaida (1963), Kemper (1960). McKelvey and Milne (1962), and Bruce Hanshaw (1962) have supplied firm experimental support for the actual effectiveness of the mechanism.

Chebotarev (1955), on the other hand, seems to conclude that the original interstitial water of marine sediments is normally flushed out and is replaced at later times by waters of different compositions, determined largely by rates of flushing. Krejci-Graf and others (1957) conclude that oil-field waters of the Vienna Basin of Austria formed from the original water content of organisms. A mechanism for complete flushing of the original large proportions of interstitial water in the sediments prior to decomposition of the organisms is not evident, and the hypothesis has not been supported by studies of interstitial waters of recent marine sediments.

Bredehoeft and others (1963; see also discussion by Rittenhouse, 1964, and reply by Bredehoeft and others, 1964) accept De Sitter's principle of salt-sieving, and propose that saline waters can also result from concentration of dissolved matter contained in dilute meteoric waters migrating down dip from the outcrops of an aquifer. Near the center of a basin, the water moves upward through confining strata that serve as semipermeable membranes. These authors state (1964) that the saline interstitial water of sedimentary rocks can also be derived from original ocean water, with the energy for continued salt-sieving provided by the hydrostatic pressure of meteoric water in the up-dip parts of an aquifer, as on the flanks of a basin.

In contrast to the Bredehoeft model involving the hydrostatic pressure of meteoric waters, the energy for salt-sieving during initial compaction and metamorphism of sedimentary rocks is provided by gravitational attraction of the solid phases of the overlying rocks. During compaction and metamorphism, fluid pressures on interstitial waters are higher than for simple hydrostatic pressure at a comparable depth.

DEFINITIONS OF SOME TERMS USED IN THIS REPORT

Formation water.—The water present in rocks immediately before drilling (Case, 1955). This term is generally accepted in the petroleum industry as a useful non-genetic term for waters of unknown age and origin.

Meteoric water.—Defined by White (1957, p. 1661) as

> ...water that was recently involved in atmospheric circulation. The age of meteoric ground water is slight when compared with the age of the enclosing rocks, and is not more than a small part of a geologic period.

This definition is very similar to Meinzer's (1923, p. 31) except that age is specifically considered. Meteoric water is normally low in salinity, but a high content of dissolved substances can be related either to extensive evaporation or to solution of salt deposits.

Gorrell (1958) has objected to the term "meteoric" and has proposed "hydrospheric" to take its place. Both terms are inherently ambiguous in regard to ancient or fossil waters that were formerly involved with atmospheric circulation, and the ambiguity of either must be removed by specifically considering the age of a water since its last direct contact with the atmosphere.

Connate or fossil water.—Connate water as defined by White (1957, p. 1661) is

> 'fossil' water that has been out of contact with the atmosphere for at least an appreciable part of a geologic period. It consists of the fossil interstitial water of unmetamorphosed sediments and extrusive volcanic rocks and water that has been driven from the rocks.

The term has been redefined a number of times since it was proposed by Lane (1908), who first implied and then stated (1927; Lane and Alter, 1941) that connate water has remained since burial with the specific rocks in which it occurs, and that its chemical composition has remained unchanged.

Much recent evidence indicates that (1) the original interstitial water of sediments normally undergoes diagenetic changes related to bacterial activity and decomposition of organic matter; (2) inorganic reactions between mineral phases and interstitial water occur during early and late diagenesis and also during metamorphism; (3) compaction always occurs, at least to some extent, in response to sediment load, so water must migrate either locally or extensively in directions of decreasing potential; (4) "salt-sieving" mechanisms, only slightly understood until very recent years, account for changes in salinity of probably all deep waters of sedimentary basins; and (5) other mechanisms, even less well understood, account for changes in chemical composition of saline waters.

Connate water, in Lane's sense, almost certainly does not exist. A brine from Lane's type area for connate water was recently sampled and analyzed in detail (Table I, analysis 2; White, Hem, and Waring, 1963, p. F52-F53). The brine dripped slowly from the back of a new deep entry into the copper-bearing amygdaloidal lava flow of Precambrian age in Houghton County, Michigan. Few, if any, present-day geologists could accept Lane's contention that this brine was "born with" the lava flow at the time of its extrusion, that the brine could have been preserved *in situ* during the copper mineralization, or that it could have failed to react chemically with associated rocks as physical conditions changed during deep burial. Whatever its initial origin may be, this brine clearly must have migrated into the lava flow after eruption, and extensive chemical changes have no doubt occurred.

The term "connate" is commonly used without insistence upon proof of preservation of original composition or proof that the water has not migrated. Some authors, on the other hand, have argued strongly for retention of Lane's original definition, but I am not aware that any alternative genetic terms have been proposed to permit evolution in composition, concentration, and physical association. I prefer to redefine a familiar term as the need becomes apparent, rather than to propose an unfamiliar term.

Connate water, as here defined, generally is similar in age or somewhat younger, since last direct contact with the atmosphere, than the age of its associated rocks. Most connate waters are probably marine in origin and are associated with normal marine sediments, but some are suspected of being connate to marine and nonmarine evaporites. The dilute meteoric water "born with" normal nonmarine sediments is probably never preserved in its original relationships but is normally displaced after burial by other dilute waters and eventually by any saline water (with higher density) that has access to the sediments.

If the concept of "salt-sieving" by semipermeable membranes (discussed in detail in a later section) is accepted, we must also distinguish between connate water on the output side of an effective membrane and the more saline water retained behind the membrane. "Membrane-filtered connate water" is here proposed for the former, and "membrane-concentrated connate water" for the latter. The water of any specific pore space or reservoir rock can be of both types with respect to upstream or downstream membranes, but the water escaping from a complex system constitutes one extreme, and the most concentrated pore solution upstream in the system is an opposite extreme.

In addition, Engelhardt (1961; Engelhardt and Gaida, 1963) has called attention to the layers of adsorbed water immediately adjacent to cation-exchanging clay-mineral grains with fixed negative charges. Some evidence indicates that these layers of water are lower in dissolved salts than the associated water of the larger pore spaces. "Adsorbed connate water" is here suggested as a useful term for this type.

Metamorphic water.—Metamorphic water was defined by White (1957, p. 1662) as

water that is or that has been associated with rocks during their metamorphism.

Metamorphic water may have been in part originally connate, interstitial to or adsorbed on mineral grains, particularly during early stages of metamorphism, but much must be derived from hydrous minerals during metamorphic reconstitution to less hydrous or anhydrous minerals. Connate water is considered to grade into metamorphic water without a sharp or distinct break, just as sedimentary rocks grade through different stages into metamorphic rocks.

EVOLUTION OF INTERSTITIAL WATERS OF NORMAL MARINE SEDIMENTS

The compositions of interstitial waters of recent marine sediments and of deep saline waters of sedimentary basins are reviewed to determine the normal courses of evolution of marine connate waters.

Early diagenetic changes.—Much attention has been given in recent years to the mineral and organic constituents of present-day marine sediments, and a few of these studies are also con-

[*Editors' Note:* Table 1 has been omitted.]

cerned with interstitial waters and gases of these sediments (Emery and Rittenberg, 1952; Emery, Orr, and Rittenberg, 1955; Rittenberg and others, 1955, 1963; Bruevich, 1956a, 1956b; Bruevich and Vinogradova, 1946; Zaitseva, 1956; Lisitsin, 1956; Shishkina, 1955a, 1955b, 1958, 1959a, 1959b, 1960; Shishkina and Bykova, 1962; Chilingar, 1958; Starikova, 1959, 1961; and Siever and others, 1961). These studies indicate that many of the chemical contrasts between oil-field waters and normal ocean water are initiated when the sediments are first deposited.

Fine-grained sediments are very porous and high in water content when first deposited, commonly containing 65 per cent or more of water by volume near the sediment interface. Porosity decreases irregularly downward, depending upon the nature of the sediments and rates of sedimentation. Lisitsin's (1956) median curve from the sediment interface to 100 ft in depth in long cores is in reasonable agreement with Emery and Rittenberg's earlier published curve (1952, p. 753) extending to greater depths in Tertiary oil fields. The compaction processes evidently continue over long periods of time. The moisture content of sediments in the experimental Mohole, for example, indicates porosities of 60 per cent or more to depths of about 140 m below the sediment surface (Rittenberg and others, 1963, Table 2). Calcareous-siliceous ooze of probable middle Miocene age 170 m below the sediment surface still has a porosity of about 50 per cent.

Ocean-bottom sediments that are deposited slowly in direct contact with oxygenated water show little consistent change in composition of associated interstitial water with depth below the sediment surface. Sulfate content of the water of such sediments does not seem to change significantly with depth. Although these sediments commonly contain 0.5 per cent or more of organic carbon, sufficient oxygen evidently diffuses from above the interface to oxidize the most unstable organic components. Sediments from the experimental Mohole (Rittenberg and others, 1963) are evidently of this type, to judge from measured Eh's and calculated sedimentation rates.

Shishkina's studies (1955a, 1955b, 1958, 1959a, 1959b, 1960; Krasintseva and Shishkina, 1959; Shishkina and Bykova, 1962) clarify the behavior of some key elements in the interstitial waters of more rapidly deposited organic-rich sediments, generally with more than 0.5 per cent of organic carbon. Emery and Rittenberg (1952), ZoBell and others (1953), Morita and ZoBell (1955), Bruevich (1956a; 1956b), Zaitseva (1956), and Oppenheimer and Kornicker (1958) have also clarified the relationships in organic-rich sediments.

Without access to sufficient external oxygen, anaerobic bacteria reduce the sulfate of interstitial water as some of the organic matter is oxidized. CO_2 and sulfide, probably with intermediate sulfur species, are products of the reactions. The total dissolved CO_2 species, including HCO_3^- and $CO_3^=$ species, are increased, evidently with little change in pH. The increase in $CO_3^=$ in turn decreases the Ca content of the water (Shishkina, 1955b, 1959a). NH_4^+ from decaying proteins of organisms increases greatly in the interstitial water and also in exchange positions in clay minerals. Sulfide produced by reduction of sulfate is largely fixed by reaction with Fe to form Fe sulfides, but some dissolved H_2S is generally also evident in the sediments (Emery and Rittenberg, 1952; Ostroumov, 1957). Br is commonly enriched slightly in the water, presumably from decomposition of bromine-concentrating organisms. B also increases somewhat in the interstitial water. The B content of fine-grained marine sediments (Harder, 1959, 1960, 1961a, 1961b) is so much greater than average continental abundance that marine shales constitute the major boron reservoir of the crust. How much of the increase in B takes place while the clays are in suspension in sea water and how much is a diagenetic addition by diffusion from sea water above the sediment interface are not yet evident, but both factors are probably important. Na^+ and K^+ show no consistent trends with depth but tend to increase slightly with increasing HCO_3^-, and Mg tends to decrease downward but not as rapidly as Ca. The decrease in Mg is probably related at least in part to slow formation of chlorite rather than dolomite (Chave, 1960, p. 365–367) or to fixation of Mg in the fine-grained mixed-layer micas. An irregular increase in dissolved SiO_2 with depth, attaining concentrations of 10 to 60 ppm in comparison to a normal content of about 7 ppm in deep ocean water (Emery and Rittenberg, 1952, p. 795) is probably related to dissolution of the more soluble parts of diatoms and radiolaria.

Shishkina (1958, 1959b) has also described an apparent diagenetic formation of interstitial sodium-calcium-chloride water at shallow depths within the sediments of parts of the Black Sea. This is a type of water that normally does not form until long after the early diagenetic stages. As in the organic-rich marine environment, $SO_4^=$ decreases, NH_4^+ increases considerably, and the Br/Cl ratio increases slightly with depth. Ca increases notably but HCO_3^- (determined as alkalinity) does not increase; presumably the excess CO_2 resulting from bacterial activity reacts with Ca and is precipitated as carbonate. The source of the excess Ca is not at all clear, but it may be related in some way to the special environment of the Black Sea.

The diagenetic chemical changes now known to occur at shallow depths in recent rapidly deposited, organic-rich marine sediments bridge most of the gap in composition between ocean water and many oil-field waters of comparable or slightly lower salinity. Compare, for example, the analysis of ocean water (Table I, analysis 1) with analyses 5, 6, and 10. For each component that has been studied, the observed diagenetic trend in organic-rich sediments is in the direction to be expected if these oil-field waters are indeed connate ocean water. The more saline oil-field waters have additional complications considered in following sections. I am not aware that diagenetic changes in I, Fe, Ba, Sr, and F contents of the interstitial waters of sediments have been looked for. The expected change in iodine, especially, seems almost certain to be found; other possible changes will depend upon the major controlling reactions.

Differences in salinity probably related to semipermeable membranes.—The possibility that fine-grained rocks behave as semipermeable membranes, permitting the passage of some components but not others, has been recognized for many years (Lindgren, 1933, p. 173; Korzhinskii, 1947). De Sitter (1947) seems to have been the first to suggest that some saline brines have been concentrated by a "salt-sieving" action of fine-grained sediments, permitting water molecules to escape but retaining the ions of dissolved salts. The theoretical basis for the hypothesized behavior was considered by Marshall (1949), Wyllie (1948, 1955), McKelvey and others (1957), Kemper (1960), Bernstein (1960), and Engelhardt (1960, 1961; Engelhardt and Gaida, 1963). Recent experimental evidence confirming the effectiveness of the mechanism has been provided, especially by McKelvey and others (1957, 1962), Engelhardt (1961; Engelhardt and Gaida, 1963), Lomtadze (1954), Kozin and Mzhachikh (1958), and Hanshaw (1962). The phenomenon is analogous to the Donnan membrane equilibria studied in particular in connection with organic protein sols (Hartman, 1947, p. 392–400).

According to one explanation for membrane-filtering involving fine-grained sediments, montmorillonite and illite have fixed negative charges that are balanced by adsorbed exchangeable cations. When porosity of a clay-bearing sediment is high and channels between grains are relatively large, water and its dissolved salts pass through the sediment in proportions approximately equal to their abundance on the "input" side of the clay membrane. As compaction of the sediment proceeds, however, spaces between adjacent clay-mineral grains decrease. The fixed negative charges of adjacent grains become so close together that anions in solution on the input side of the membrane are repelled and cannot escape. The adsorbed cations, however, can still move between adjacent exchange sites and through the membrane. An electrical "streaming" potential is set up between the two sides of the membrane, and cations cannot continue to move through the membrane without increasing the electrical imbalance. Uncharged water molecules, however, can pass through the membrane in the direction of decreasing water pressure, thus increasing the salt content on the input side. McKelvey and Milne (1962) found that compressed bentonite membranes with porosities of 31–41 per cent had upstream to downstream salinity ratios ranging from 8/1 to 1.7/1, the differences depending on concentration and other factors. Membranes made from ground and compressed shale with relatively low contents of cation-exchanging clays were not as effective, but

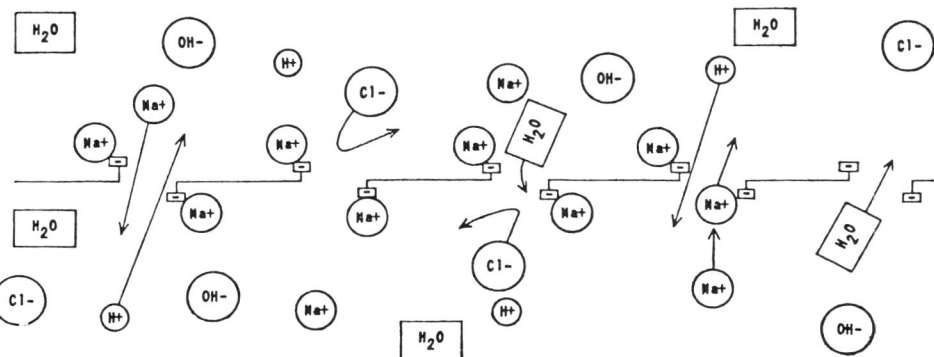

Fig. 2.—Cation-exchanging semipermeable membrane at equilibrium, with Na$^+$, Cl$^-$, and water in initially equal concentrations on both sides. The membrane is assumed to be permeable to cations and uncharged water molecules, but charged anions are repelled by fixed negative charges.

concentration ratios were 1.3 and 1.7, thus confirming the effectiveness of natural shale in inhibiting movement of the dissolved salts.

Differences in proportions of the dissolved constituents.—In understanding changes that take place in proportion of dissolved constituents of connate waters, let us consider other aspects of the Donnan membrane theory. In the classic Donnan experiment (*see* Hartman, 1947, p. 392–400) the membrane has pores large enough to permit passage of all small ions but is impervious to large ionized colloidal particles; an equilibrium is established between the components on the two sides of a membrane. When a colloidal electrolyte with ionizable Na, such as Congo red, is dialyzed against pure water, an equilibrium is established between H$^+$ and Na$^+$. Some Na$^+$ diffuses through the membrane to the colloid-free side; in main-

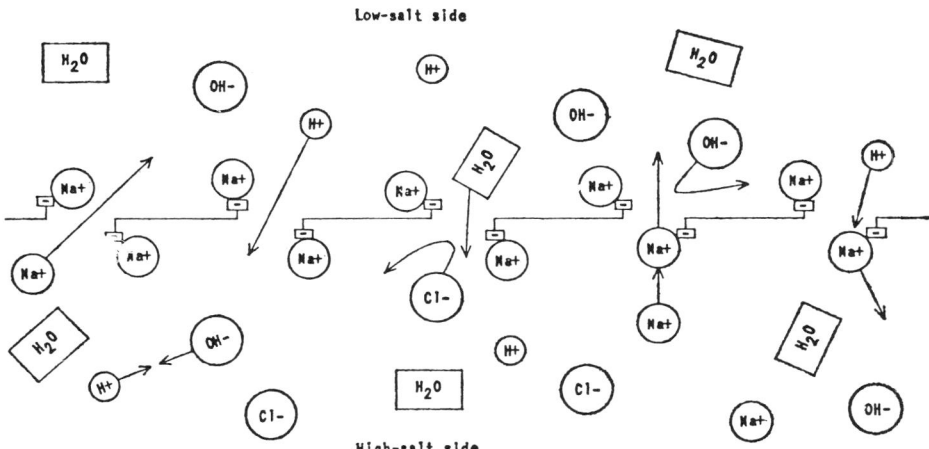

Fig. 3.—Cation-exchanging semipermeable membrane at equilibrium with NaCl initially on only one side of membrane assumed permeable to Na$^+$, H$^+$, and water molecules but impermeable to charged anions. Sodium diffuses to the low-sodium side because of the steep Na$^+$ concentration gradient; membrane is impermeable to Cl$^-$, so an electrical charge imbalance of potential is set up; H$^+$ ions then diffuse to the high-salt side in response to the potential until equilibrium is attained. An excess osmotic pressure develops on the high-salt side due to net addition of H$_2$O; this side has become acid and the low-salt side alkaline.

taining electrical balance, water molecules ionize and a net diffusion of H^+ ions occurs through the membrane to the side of the colloid. The water associated with the Congo red side of the membrane has become slightly acid, and water on the opposite side has become alkaline. This phenomenon is called membrane hydrolysis.

With sediment membranes containing cation-exchanging clays, experimentation in water-NaCl and water-$CaCl_2$ systems has shown that water molecules can pass through the membranes, but movement of the salts is impeded. This has been explained as repulsion of anions by the fixed negative charges of cation-exchanging clays. The effects of clays on inorganic anions may be comparable to those of the Donnan membrane described above, which is impermeable to the large colloidal anions of Congo red. The principles of equilibrium as applied here to sediment membranes are represented diagrammatically in Figures 2 and 3.

The salt solution on the high-Cl side of the membrane of Figure 3 will have an acid pH, and that on the low-Cl side will be alkaline.

Many geologists and geochemists have noted that Cl is greatly dominant over other anions in most brines of high salinity. Some have also noted a tendency for Ca to increase relative to Na as salinity increases (*see*, for example, Engelhardt, 1960, p. 158; 1961, Fig. 4; White, 1957, p. 1673; White, Hem, and Waring, 1963, p. F59). In Figure 1, Chave's (1960, p. 364–365) mean ratios of Ca/Cl from many analyses are plotted against chlorinity. Individual analyses from Table I and a few additional analyses from White, Hem, and Waring (1963) are also plotted in this figure. Two tendencies are evident—in general, in any one province or sedimentary basin, salinity increases with age, and the Ca/Cl ratio, with few exceptions, increases with salinity. Individual basins differ greatly from each other, indicating the existence of local influences that cannot be explained by changes in a common ocean. Kramer (1963), Spiro and Vovk (1961), and some others who have noted high Ca contents of some waters have assumed that Ca was more abundant in the ancient oceans, but the data of Figure 1 prove that such a simple relation to time is untenable. The ratio Ca/Cl rather than Ca was selected as the vertical coordinate because this permits an easy recognition of saline brines that mix near the surface with meteoric water, with little change in proportion of components. The nearly horizontal trends of the West Virginia brines and the least saline of the Texas brines (Fig. 1) may be explained in this way.

Engelhardt (1961) and Chave (1960) have both demonstrated by comparison with ocean water that the high-Ca waters cannot result from simple dolomitization (base exchange) of limestone. Engelhardt has also emphasized that fixation of Mg to form chlorite, with exchange of Na and Ca from the ancestral clay minerals, cannot explain the high-Ca brines. Nor can solution of calcite or gypsum in a closed system account for high Ca in brines in which Cl is the overwhelmingly dominant anion. Engelhardt therefore concludes that, with increasing salinity, Na from the brine must be exchanged for Ca from the solid phases, reversing the usual trend at low salinities. I have found no substantiating experimental evidence that simple base exchange can account for the observed trends of Figure 1; Ca, for example, seems to increase greatly at relatively low salinities in the sampled California waters. Each of the plotted areas or sedimentary basins seems to have a characteristic trend. Although the type or quantity of cation-exchanging clay mineral could be a controlling factor for specific formations, other factors common to a whole basin or large area have considerable appeal.

An alternative explanation based on the compositions and field distribution of natural waters is that fine-grained sediments are not equally permeable to all constituents, but that some have greater mobility than others. The term "mobility" as used here indicates relative ease of escape through fine-grained mineral membranes and does not refer to a strict thermodynamic concept. Intergrain distances may become so narrow at sites of greatest restriction along escape channels that the properties of different constituents become critical. Possible differences that could be important include charge and ionic and molecular radii. Marshall (1949, p. 174–175) and McKelvey and Milne (1962) have suggested that ion-exchange membranes permit different rates of diffusion of cations, and recent developments with natural and synthetic "molecular sieves" and chromatographic methods of analysis suggest that differences among constituents should actually be expected. However, I have found no discussion of specific or relative rates to be expected for individual constituents.

Observed distributions and trends in natural waters become understandable if Na^+ is appreciably more mobile than Ca^{++}. In view of the normally great tendency for Ca^{++} to be adsorbed on cation-exchanging clays, its double charge, and perhaps also the greater radius of its hydrated ion (9.6 A as compared to 5.6 A for Na^+, in data for multihydrated cations, summarized by Grim, 1953, p. 148), it seems reasonable to expect less mobility for Ca^{++} than for Na^+. The cited experimental evidence for salt-sieving and the evidence from natural waters generalized in Figure 1 provide very strong evidence that molecular H_2O is considerably more mobile than Cl ions.

The relative mobilities of some other constituents can be hypothesized if certain oil-field waters with salinities slightly lower than ocean water are assumed to be membrane-filtered connate waters and most brines of high salinity are assumed to be membrane-concentrated connate waters. Waters of Analyses 5, 6, and 10 of Table I are suggested as examples of membrane-filtered waters, and those of Analyses 3 and 4 as membrane-concentrated waters. (Other analyses of Table I not specifically considered here seem to have additional complications, as will be seen.) Tentative conclusions based on this hypothesis, without regard for other possible dominating effects, are that Na^+, HCO_3^-, F^-, I^-, and B (as boric acid?) are considerably more mobile than Cl^-; Mg^{++} is slightly less mobile, and Ca^{++} and SO_4^{--} are much less mobile than Cl^-. No clear trend is evident for K^+, Li^+, NH_4^+, and Br^-. Waters of possible metamorphic origin, discussed in a later section, seem to point to high mobility for Na^+, HCO_3^- and B relative to Cl^-, and also indicate that NH_4^+ and H_2S are relatively mobile. The relative mobilities indicated here are in part consistent with the hypothesis that small single-charged ions are the most mobile, and large double-charged ions the least mobile.

The suggested high mobilities of I^-, HCO_3^-, B, and some form of sulfide, however, demand additional explanation because HCO_3^- and H_3BO_3 are large species with multiple large oxygen atoms, and I^-, S^{--}, and HS^- are also relatively large. A possible factor of major importance suggested by these species is that constituents forming or potentially forming uncharged molecules may be relatively mobile and can pass through or around fixed negative charges on surfaces of closely spaced cation-exchanging clay minerals. A behavior similar to this has already been proved experimentally for H_2O. In near-neutral solutions, uncharged H_3BO_3 is the recognized dominant boron species; H_2CO_3 and H_2S are minor but important species in nearly neutral solutions and perhaps can escape through electrical barriers that repel the more abundant negatively charged species. Undissociated $NaHCO_3$ is a relatively abundant carbonate species in a near-neutral solution high in Na (P. B. Hostetler, oral communication, 1963) and perhaps also escapes through membranes in addition to the uncharged H_2CO_3 molecules assumed in the following discussion. Uncharged iodine (I_2?) is likely to be a minor iodine species in a reducing sedimentary environment, but this element is also the most readily oxidizable of the halogen group. Is it possible that the uncharged iodine molecule, in minor abundance relative to I^- ions, is the mobile species of iodine just as H_2CO_3 and H_2S are the indicated mobile species of CO_2 and sulfide?

An important point to note here is that, as uncharged H_2CO_3 molecules escape from the membrane-concentrated water, more uncharged molecules form by association of HCO_3^- and H^+ in response to the disequilibrium. If iodine molecules are also escaping, new molecules form behind the membrane in equilibrium proportions to the more abundant I^- ions, thus perhaps decreasing slightly the Eh of the environment.

As charged cations and uncharged molecules escape through the membrane, an electrical potential is set up between the two sides of the membrane. Evidence has already been cited from Donnan equilibria experiments (see Figs. 2 and 3 and discussion; Hartmann, 1947, p. 396–398) that H^+ ions from dissociating water flow in the reverse direction in response to the electrical potential, thus balancing the escaping Na^+ ions. This is also illustrated diagrammatically in Figure

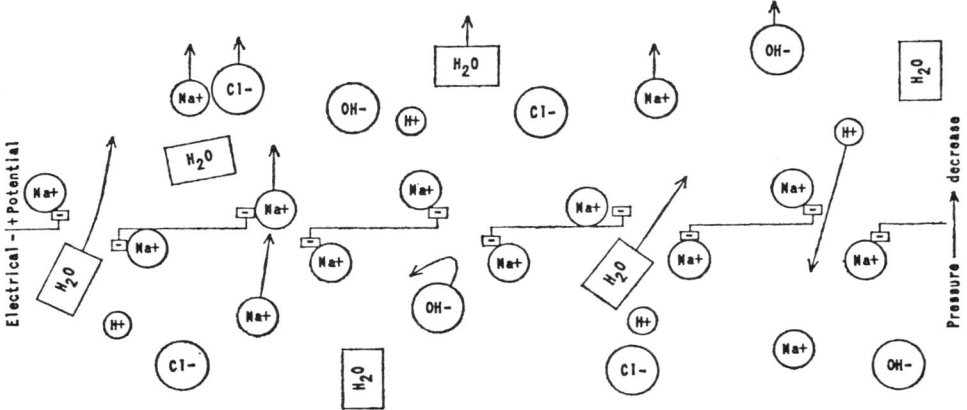

Fig. 4.—Non-equilibrium system with the same initial equality as in Fig. 2, but with a pressure differential continuing to force water through the cation-exchanging membrane impermeable to anions. With a pressure differential, H_2O flows through the membrane, diluting ions on the low-pressure side; with this concentration gradient, Na^+ diffuses through the membrane, setting up an electrical potential. In response to this potential, some H^+ ions flow in a reverse direction similar to that of Fig. 3 until a steady state is attained.

4 for the non-equilibrium situation where a pressure difference and continuing but very slow flow of water are maintained. If any carbonate species (not shown in Fig. 4) were originally in the water, the return flow of H^+ ions would permit H_2CO_3 molecules to form from any HCO_3^- on the input side of the membrane, and these uncharged molecules, by the present hypothesis, can escape through the membrane. New HCO_3^- ions and water molecules form on the output side by reaction with OH^-. This mechanism continues to operate only so long as CO_2 species are available on the input side of the membrane.

If carbonate minerals are present on the input side of the membrane, more H_2CO_3 forms to become available for escape. Ca^{++} ions of low mobility are thus supplied to offset the relatively mobile Na^+ ions escaping through the membrane. This model may account very nicely for the observed increase in Ca/Cl and Ca/Na ratios of many natural brines, here suggested as examples of membrane-concentrated connate waters. The mechanism is illustrated diagrammatically in Figure 5.

In the preceding discussion, all Ca^{++} is assumed to come from carbonates, but another important source during diagenesis and early stages of metamorphism is plagioclase, which can react to form more stable minerals, such as albite, mica, or montmorillonite.

The three waters selected on other chemical evidence as possible examples of membrane-filtered connate waters (Table I, analyses 5, 6, and 10) range in pH from 7.1 to 7.6, in contrast to the two waters suggested as typical membrane-concentrated connate waters (Table I, analyses 3 and 4), with pH's of 6.2 and 6.8. These differences are consistent with predictions based on semipermeable membrane theory. Very few examples of reliable pH of oil-field waters are to be found in the literature, as pointed out by Chave (1960). Present reasoning suggests that Chave was not warranted in arbitrarily discarding *all* analyses with pH's above 7.5 and below 6.5 and many within his accepted range may be equally poor, but his general skepticism is very well founded.

Special field measurements of pH's of saline brines are greatly needed to determine whether the commonly reported range of about 4.5 to 6.5 is real in freshly collected samples, particularly for the more acid part of this range. An important factor not considered by Chave that probably accounts for many laboratory-determined acid pH's involves access of atmospheric oxygen during and after collection of high-Ca brines. Such waters always have very low concentrations of

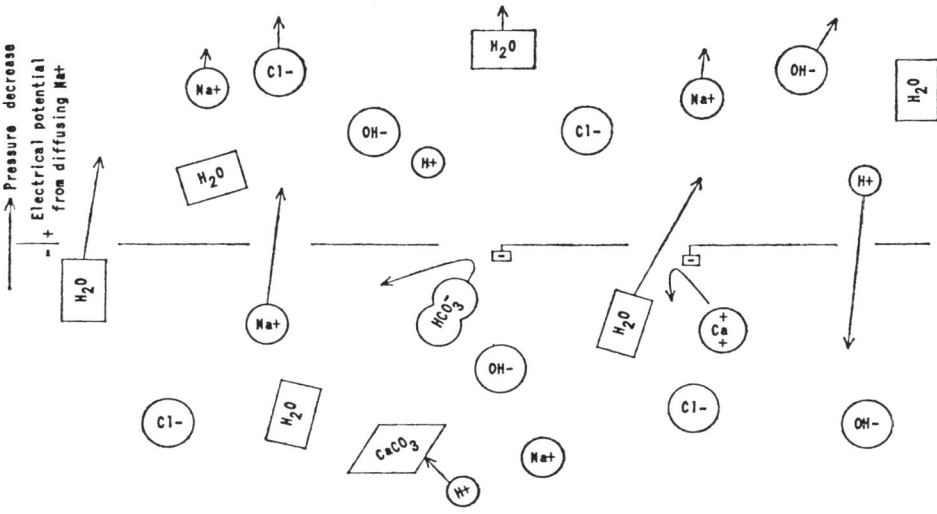

FIG. 5.—Non-equilibrium system with same components as in Fig. 4 but with calcite as a solid phase; the membrane also is assumed less permeable to Ca^{++} than to Na^+, H^+, and all uncharged molecules. With a pressure differential, the low-pressure side is diluted by increase in H_2O, gains Na^+ because of concentration gradient, and loses H^+ in response to electrical potential, thus becoming alkaline. The rare uncharged H_2CO_3 (not shown, to avoid unreasonable proportions) that diffuses through the membrane reacts with part of the excess OH^-. The high-pressure side has lost H_2O and Na^+ but has gained H^+, which is then largely neutralized by reacting with $CaCO_3$ to form Ca^{++} and HCO_3^- ions; salinity and the Ca/Na ratio of the brine thus increase. New H_2CO_3 molecules are formed to offset those that are lost, and are then available in turn to diffuse through the membrane.

buffering CO_2 species and generally also contain significant concentrations of Fe^{++} ions. Oxidation of the ferrous iron produces $Fe(OH)_3$ and H^+ ion according to the reaction

$$Fe^{++} + 2H_2O + \tfrac{1}{2}O_2 + e \rightarrow Fe(OH)_3 + H^+$$

The relative abundance of carbonates and the abundance, composition, and relative instability of clastic plagioclase in a sedimentary pile may determine the rate of increase of Ca in connate waters with time and chlorinity (Fig. 1). Additional factors of possible importance are rate of deposition, depth of burial, porosity and permeability relationships of the immediate environment and of the whole sedimentary pile, and temperature relationships.

The highest Ca/Cl ratio of Figure 1 is the "type connate" water (Table I, analysis 2) from the copper-bearing amygdaloid lava flow of Michigan. Although the detailed origin of this water is not clear, its very high Ca content is likely to be related to abundant unstable calcic plagioclase of the original basaltic lava flow.

The next highest Ca/Cl ratios of Figure 1 are from different depths in a single oil-test well in Kings County, California (Table I, analyses 8, 9), and are from a relatively thick and deeply buried sequence of lower Miocene rocks. The depths of these rocks are three to four times greater than the lower Miocene reservoir rock of the Cymric oil field (Table I, analysis 5; Bailey and others, 1961) from the same major sedimentary basin. The differences in Ca content of these lower Miocene rocks may be related to differences in depth and rates of deposition. I have no data on the mineralogy of the brine reservoir rocks of the two areas, and this should be investigated.

The Leda sand (of local usage), from which sample 9 of Table I was obtained, is especially notable for its tremendously high fluid pressure, which is about 1.7 times the normal hydrostatic pressure for its depth; this approximates the probable lithostatic (geostatic) pressure of the overlying sedimentary rocks and their contained water. The high fluid pressure in turn indicates that the Leda sand and adjacent shales are not yet thoroughly compacted, and that the fluid is sustaining essentially the full weight of the overlying rocks (Hubbert and Rubey, 1959, p. 150-157). The Leda sand is apparently a shoestring sand completely enclosed in all directions by great thicknesses of shale. This is apparently an extreme example of a permeable reservoir completely enclosed in thick, rapidly deposited shale and was originally, and probably still is, relatively high in porosity (and water content) but low in mass permeability. The fluid pressure and average rate of deposition of the sediments are near the upper limits of examples cited by Rubey and Hubbert (1959, p. 169-173). Details of escape routes of the interstitial water of the shale are not known; the Leda sand is very probably serving as a major escape route at depth, but its up-dip termination is evidently surrounded by shale, which is the membrane or "salt sieve" that permits selective escape of water, concentrating the dissolved salts in the sand.

To this point, attention has been focused largely on individual idealized semipermeable membranes assumed to be permeable or impermeable to different constituents. Actual field relationships are far more complicated than this. A normal sedimentary pile contains innumerable lithologic units and subdivisions of units, each with its own gross and detailed membrane characteristics. Probably no shale is a perfect membrane; some anions no doubt continue to escape through the largest of interconnected openings. As compaction continues and the large interconnections decrease in size and new restrictions are formed, each membrane presumably becomes more nearly perfect. In the meantime, however, the salinity on the input side of each membrane has been increasing. The net effect of an innumerable series of imperfect membranes in a sedimentary pile is to decrease the salinity of water in the larger pore spaces of comparable size along the escape routes. These routes will in general trend upward but can locally be downward. The average water in all pore spaces of an individual shale membrane may differ greatly in salinity, depending on the proportion of adsorbed connate water of low salinity (*see* Engelhardt, 1961, and definitions in this report).

When deeper burial ceases and particularly after a sedimentary pile is uplifted and partially eroded, the trend toward decreasing porosity comes to an end and, exceptionally, may even be reversed. The earlier regime in which water is driven out by lithostatic pressure of the overlying rocks is followed by a regime that is dominated by hydrostatic pressure. Meteoric water circulating near the surface will then flush out formerly existing saline water. Bredehoeft and others (1963, 1964) have proposed that the pressure of meteoric water in structurally and topographically high parts of a hydrologic system can provide the energy for continued salt-sieving; salinity in some parts of the system may continue to increase.

Rittenhouse (1964) has questioned whether high Cl brines form by membrane filtration from dilute meteoric waters greatly dominated by components other than Cl, and I share this skepticism. Do interstitial waters of marine sediments continue to increase in salinity after compaction ceases, according to the Bredehoeft model, with salts retained but with meteoric water filtering through to displace the original ocean water? This possibility must be clarified by further experimental, chemical, and isotope studies.

[*Editors' Note:* Material has been omitted at this point.]

CONCLUSIONS

This brief survey concludes that, of the different kinds of saline waters likely to be found in sedimentary rocks, the most abundant are probably connate waters of marine origin. These waters normally migrate during association with sediments, and the waters have changed chemically in response to changes in physical and chemical environments.

Many detailed characteristics of saline waters of sedimentary rocks are best explained by the theory that fine-grained sediments are semipermeable membranes, permitting a selective escape of H_2O, CO_2, Na, B, S, and to lesser extents certain other components relative to Cl and Ca. A somewhat complicated but attractive mechanism is proposed to explain details of behavior.

Experimentation is needed greatly to fortify or to revise the tentative conclusions favored here. If these conclusions are supported by further work, major progress is then to be expected in understanding many mineralogical and chemical changes that occur during diagenesis and metamorphism.

REFERENCES

Bailey, E.H., Snavely, P.D., Jr., and White, D.E., 1961, Chemical analyses of brines and crude oil, Cymric field, Kern County, California: Art. 398 *in* U.S. Geol. Survey Prof. Paper 424-D, p. D306-309.

Banwell, C.J., 1963, Thermal energy from the Earth's crust—Introduction and Pt. 1: New Zealand Jour. Geol. and Geophysics, v. 6, p. 52-69.

Bernstein, Fabian, 1960, Distribution of water and electrolyte between homoionic clays and saturating NaCl solutions, *in* Swineford, Ada, ed., Clays and clay minerals: Natl. Conf. on Clays and Clay Minerals, 8th, Norman, Okla., Proc., p. 122-149.

Bredehoeft, J.D., Blyth, C.R., White, W.A., and Maxey, G.B., 1963, Possible mechanism for concentration of brines in subsurface formation: Am. Assoc. Petroleum Geologists Bull., v. 47, no. 2, p. 257-269.

Bredehoeft, J.D., White, W.A., and Maxey, G.B., 1964, Reply (to Rittenhouse discussion): Am. Assoc. Petroleum Geologists Bull., v. 48, no. 2, p. 236-238.

Bruevich, S.V., 1956a, Chemistry of sediments of Okhotsk Sea: Akad. Nauk SSSR, Inst. Okeanol. Trudy, v. 17, p. 41-132.

—— 1956b, The vertical distribution of the biogenous elements in the bottom solutions of the Okhotskoe More (Sea of Okhotsk): Acad. Sci.U.S.S.R. Proc., Geochem. Sec. v. 3, no. 1-6, p. 121-124 (English ed.).

—— and Vinogradova, E.G., 1946, Biogenic elements in sediment solutions of the northern, middle, and southern parts of the Caspian Sea: Akad. Nauk SSSR, Doklady, v. 54, no. 5, p. 419-422.

Buneev Aleksandr Nikolaevich, 1956, Osnovy gidrogeokhimii mineralnykh vod osadochnykh otlozhenii: Tsentral. Inst. Kurort, Moscow, 227 p.

Case, L.C., 1955, Origin and current usage of the term, "connate water:" Am. Assoc. Petroleum Geologists Bull., v. 39, no. 9, p. 1879-1882.

—— Heck, E.T., Fash, R.H., Minor, H.E., Crawford, J.G., Jensen, F.W., Foster, M.D., and Ginter, R.L., 1942, Selected annotated bibliography on oil-field waters: Am. Assoc. Petroleum Geologists Bull., v. 26, no. 5, p. 865-881.

Chave, K.E., 1960, Evidence on history of sea water from chemistry of deeper subsurface waters of ancient basins: Am. Assoc. Petroleum Geologists Bull., v. 44, no. 3, p. 357-370.

Chebotarev, I.I., 1955, Metamorphism of natural waters in the crust of weathering: Geochim. et Cosmochim. Acta, v. 8, nos. 1-2, p. 22-48; no. 3, p. 137-170; no. 4, p. 198-212.

Chilingar, G.V., 1958, Some data on diagenesis obtained from Soviet literature—a summary: Geochim. et Cosmochim. Acta, v. 13, nos. 2-3, p. 213-217.

Craig, Harmon, 1961a, Isotopic variations in meteoric waters: Science, v. 133, no. 3465, p. 1702-1703.

—— 1961b, Standard for reporting concentrations of deuterium and oxygen-18 in natural waters: Science, v. 133, no. 3467, p. 1833-1834.

—— Boato, Giovanni, and White, D.E., 1956, Isotopic geochemistry of thermal waters, chap. 5 *of* Nuclear processes in geologic settings: Natl. Research Council, Comm. Nuclear Sci. Ser. Rept., no. 19, p. 29-38.

Craig, H., Gordon, L.I., and Horibe, Y., 1963, Isotopic exchange effects in the evaporation of water, I. Low-temperature experimental results: Jour. Geophys. Research, v. 68, no. 17, p. 5079-5087.

Crawford, J.G., 1949, Water analyses (characteristics of oil-field waters of Rocky Mountain region), *in* Subsurface geologic methods: Colorado School Mines Quart., v. 44, no. 3, p. 188-210.

De Sitter, L.U., 1947, Diagenesis of oil-field brines: Am. Assoc. Petroleum Geologists Bull., v. 31, no. 11, p. 2030-2040.

Ellis, A.J., and Wilson, S.H., 1960, The geochemistry of alkali metal ions in the Wairakei hydrothermal system: New Zealand Jour. Geology and Geophysics, v. 3, no. 4, p. 593-617.

Emery, K.O., Orr, W.L., and Rittenberg, S.C., 1955, Nutrient budgets in the ocean, *in* Essays in the natural sciences in honor of Captain Allan Hancock: Los Angeles, Univ. Southern Calif. Press, p. 299-309.

Emery, K.O., and Rittenberg, S.C., 1952, Early diagenesis of California basin sediments in relation to origin of oil: Am. Assoc. Petroleum Geologists Bull., v. 36, no. 5, p. 735-806.

Engel, A.E.J., and Engel, C.G., 1960, Migration of elements during metamorphism in the northwest Adirondack Mountains: Art. 212 in U.S. Geol. Survey Prof. Paper 400-B, p. B465-B470.

Engelhardt, Wolf von, 1960, Der Porenraum der Sedimente: Berlin, Springer, 207 p.

―――― 1961, Zum Chemismus der Porenlösung der Sedimente (On the Chemistry of the Pore Solution of Sediments): Upsala Univ. Geol. Inst. Bull., v. 40, p. 189-204.

―――― and Gaida, K.H., 1963, Concentration changes of pore solutions during the compaction of clay sediments: Jour. Sed. Petrology, v. 33, p. 919-930.

Fenner, C.N., 1936, Bore-hole investigations in Yellowstone Park: Jour. Geology, v. 44, no. 2, pt. 2, p. 225-315.

Gorrell H.A., 1958, The importance of subsurface water data in petroleum geology: Canadian Mining and Metall. Bull., v. 51, no. 560, p. 754-758.

Grim, Ralph E., 1953, Clay mineralogy: New York, McGraw-Hill Book Co., Inc., 394 p.

Hall, W.E., and Friedman, Irving, 1963, Composition of fluid inclusions, Cave-in-Rock fluorite district, Illinois, and Upper Mississippi Valley zinc-lead district: Econ. Geology, v. 58, no. 6, p. 886-911.

Hanshaw, Bruce B., 1962, Membrane properties of compacted clays: Harvard University, Cambridge, Mass., Unpub. Ph.D. thesis, 113 p.

Harder, Hermann, 1959, Geochemistry of boron—I, Boron in minerals and igneous rocks; II, Boron in sediments: Akad. Wiss. Göttingen Nachr., Math.-phys. Kl., no. 5, p. 67-122; no. 6, p. 123-183.

―――― 1960, The boron cycle: Internat. Geol. Cong., 21st, Copenhagen 1960, Rept., pt. I, p. 10-13 (in German).

―――― 1961a, Geochemistry of boron—III, Boron in metamorphic rocks and the geochemical cycle: Akad. Wiss. Göttingen Nachr., II. Math.-phys. no. 1, p. 1-26.

―――― 1961b, Incorporation of boron in detrital clay minerals—Experiments explaining the boron content of clay sediments: Geochim. et Cosmochim. Acta, v. 21, nos. 3-4, p. 284-294.

Hartman, R.J., 1947, Colloid chemistry, 2d ed.: Boston, Mass., Houghton Mifflin Co., 572 p.

Hemley, J.J., and Jones, W.R., 1964, Aspects of the chemistry of hydrothermal alteration with emphasis on hydrogen metasomatism: Econ. Geology, v. 59, no. 4, p. 538-569.

Holser, W.T., 1963, Chemistry of brine inclusions in Permian salt from Hutchinson, Kansas: Symposium on Salt, Northern Ohio Geol. Soc., Cleveland, Ohio, Mar. 3, 1962, p. 86-103.

Hubbert, M.K., and Rubey, W.W., 1959, Mechanics of fluid-filled porous solids and its application to overthrust faulting, pt. 1 of Role of fluid pressure in mechanics of overthrust faulting: Geol. Soc. America Bull., v. 70, no. 2, p. 115-166.

Kawaguchi, Hiroshi, 1959, Geochemical investigation of boron in Japanese rocks—II, Boron in Japanese metamorphic rocks: Nippon Kagaku Zasshi, v. 80, p. 151-155.

Kemper, W.D., 1960, Water and ion movement in thin films as influenced by the electrostatic charge and diffuse layers of cations associated with clay minerals surfaces: Soil Sci. Soc. Am. Proc., v. 24, p. 10-16.

Korzhinskii, D., 1947, The filtration effect in solutions and its role in geology: Akad. Nauk SSSR Izv. ser. geol. [1947], no. 2, p. 35-48.

Kozin, A.N., and Mzhachikh, K.I., 1958, Study of the aqueous fluids from formation rocks obtained by squeezing at high pressures: Gosudarst. Vsesoyuz. Issledovatel. i Proekt. Inst. "Giprovostokneft" Trudy, no. 1, p. 110-117.

Kramer, J.R., 1963, History of the composition of sea water—liquid inclusions compared with a chemical equilibrium model [abs.]: Geol. Soc. America Spec. Paper 73, p. 190.

Krasintseva, V.V., and Shishkina, O.V., 1959, problems of the distribution of boron in marine sediments: Akad. Nauk SSSR Doklady, v. 128, p. 815-817.

Krejci-Graf, Karl, Hecht, F., and Pasler, W., 1957, Über Ölfeldwässer des Wiener Beckens: Germany, Geol. Landesanst., Geol. Jb. bd. 74, p. 161-209.

Landergren, S., 1958, On the distribution of boron on different size classes in marine clay sediments: Geol. Fören. Stockholm, Förh. bd. 80, h. 1, no. 492, p. 104-107.

Lane, A.C., 1908, Mine waters and their field assay: Geol. Soc. America Bull., v. 19, p. 501-512.

―――― 1927, Calcium chloride waters, connate and diagenetic: Am. Assoc. Petroleum Geologists Bull., v. 11, no. 12, p. 1283-1305.

―――― and Alter, C.M., 1941, Connate waters recognized in underground circulation [Abs.]: Geol. Soc. America Bull., v. 52, no. 12, pt. 2, p. 1919-1920.

Lindgren, Waldemar, 1933, Mineral deposits, 4th ed: New York, McGraw-Hill Book Co., Inc., 930 p.

Lisitsin, A.P., 1956, Change of moisture content in long cores from the Bering Sea: Akad. Nauk SSSR Doklady, v. 108, no. 2, p. 313-316.

Lomtadze, V.D., 1954, The role of compression of clay deposits in the formation of ground water: Akad. Nauk SSSR Doklady, v. 98, p. 451-454.

Marshall, C.E., 1949, The colloid chemistry of the silicate minerals: New York, Academic Press, Inc., 195 p.

McKelvey, J.G., and Milne, I.H., 1962, The flow of salt solutions through compacted clay, in Swineford, Ada, ed., Clays and clay minerals: Natl. Conf. on clays and clay minerals, 9th, Lafayette, Ind., 1960, Proc., p. 248-259.

McKelvey, J.G., Jr., Spiegler, K.S., and Wyllie, M.R.J., 1957, Salt filtering by ion-exchange grains and membranes: Jour. Phys. Chem., v. 61, p. 174-178.

Meinzer, O.E., 1923, Outline of ground water hydrology, with definitions: U.S. Geol. Survey Water Supply Paper 494, 71 p.

Mills, R.V.A., and Wells, R.C., 1919, The evaporation and concentration of water associated with petroleum and natural gas: U.S. Geol. Survey Bull. 693, 104 p.

Morita, R.Y., and ZoBell, C.E., 1955, Occurrences of bacteria in pelagic sediments collected during the Mid-Pacific expedition: Deep-Sea Research, v. 3, no. 1, p. 66-73.

Ogienko, V.S., 1959, Distribution of bromine in the salt

of the Angara-Lena salt basin and the possibility of finding potassium salts: Geochemistry (Geokhimiya), no. 8, p. 893–900.
Oppenheimer, C.H., and Kornicker, L.S., 1958, Effect of the microbial production of hydrogen sulfide and carbon dioxide on the pH of recent sediments: Inst. Marine Sci. Pub., v. 5, p. 5–15.
Orville, P.M., 1962, Alkali metasomatism and feldspars: Norsk. Geologisk Tidsskrift, v. 42, no. 2 (Feldspar volume), p. 283–316.
——— 1963, Alkali ion exchange between vapor and feldspar phases: Am. Jour. Sci., v. 261, no. 3, p. 201–237.
Ostrouvmov, E.A., 1957, Sulfur compounds in the bottom deposits of the Okhotsk Sea: Akad. Nauk SSSR, Trudy Inst. Okeanol., v. 22, p. 139–157.
Pavlyuchenko, M.M., Medvedeva, A.P., and Mazel, M.I., 1961, The bromine content of the Starobino potassium salts: Trudy Soveshchaniya po Ispol'zovan. i Obogashchen. Kaliinykh Solei Belorussi (Inst. Gen. and Inorg. Chem., Minsk) U.S.S.R., p. 41–47.
Rankama, Kalervo, and Sahama, Th.G., 1950, Geochemistry: Chicago, Ill. Chicago Univ. Press, 912 p.
Rittenberg, S.C., Emery, K.O., and Orr, W.L., 1955, Regeneration of nutrients in sediments of marine basins [Calif.]: Deep-sea research, v. 3, no. 1, p. 23–45.
Rittenberg, S.C., Emery, K.O., Hulsemann, J., Degens, E.T., Fay, R.C., Reuter, J.H., Grady, J.R., Richardson, S.H., and Bray, E.E., 1963, Biogeochemistry of sediments in experimental Mohole: Jour. Sed. Petrology, v. 33, p. 140–172.
Rittenhouse, Gordon, 1964, Discussion of Bredehoeft and others, 1963, Possible mechanism for concentration of brines in subsurface formations: Am. Assoc. Petroleum Geologists Bull., v. 48, no. 2, p. 234–236.
Rogers, G.S., 1917, Chemical relations of the oil-field waters in San Joaquin Valley, California: U.S. Geol. Survey Bull. 653, 119 p.
Rubey, W.W., and Hubbert, M.K., 1959, Overthrust belt in geosynclinal area of western Wyoming in light of fluid-pressure hypothesis, pt. 2 of Role of fluid pressure in mechanics of overthrust faulting: Geol. Soc. America Bull., v. 70, no. 2, p. 167–205.
Russell, W.L., 1933, Subsurface concentration of chloride brines: Am. Assoc. Petroleum Geologists Bull., v. 17, no. 10, p. 1213–1228.
Schoeller, H., 1956, Geochimie des eaux souterraines; application aux eaux des gisements de pétrole: Paris, Société des Editions, p. 1–213.
Schulze, Günter, 1960, Stratigraphic and genetic interpretation of the bromine distribution in the rock-salt deposits of the Zechstein formation of central Germany: Freiberger Forschungsh. C83, 114 p.
Shaw, D.M., 1956, Major elements and general geochemistry, pt. 3 of Geochemistry of pelitic rocks: Geol. Soc. America Bull., v. 67, no. 7, p. 919–934.
Shishkina, O.V., 1955a, The question of silty sea waters: Akad. Nauk SSSR Inst. Okeanol. Trudy, v. 13, p. 94–99.
——— 1955b, Salt content of the waters formed in marine precipitates: Akad. Nauk SSSR Doklady, v. 105, p. 1289–1292.
——— 1958, Chloride-sodium-calcium waters in the Quaternary deposits of the Black Sea: Acad. Sci. U.S.S.R., Geochem. Sec. 115, 116, 117, Proc., p. 65–68.

——— 1959a, The chemical composition of the mud waters of the Pacific Ocean: Akad. Nauk SSSR Inst. Okeanol. Trudy, v. 33, p. 146–164.
——— 1959b, Metamorphism of the chemical composition of the silt waters of the Black Sea, in Perception of the diagenesis of sediments [K Poznaniyu Diageneza Osadkov]: Akad. Nauk SSSR, Sbornik Statei, p. 29–50.
——— 1960, Changes of the salt content of mud water during diagenesis: Akad. Nauk SSSR, Okeanograf. Komissii Trudy, v. 10, no. 2, p. 13–20.
——— and Bykova, V.S., 1962, Chemical composition of mud waters of the Atlantic Ocean: Akad. Nauk SSSR, Morsk. Gidrofiz. Inst. Trudy, v. 25, p. 187–194.
Siever, Raymond, Garrels, R.M., Kanwisher, John, and Berner, R.A., 1961, Interstitial waters of recent marine muds off Cape Cod: Science, v. 134, no. 3485, p. 1071–1072.
Sigvaldason, G.E., 1963, Epidote and related minerals in two deep geothermal drill holes, Reykjavik and Hveragerdi, Iceland: Article 200 in U.S. Geol. Survey Prof. Paper 450-E, p. E77–E79.
Spiro, N.S., and Vovk, Ts.L., 1961, Changes of the salt composition in the world ocean: Trudy Nauch.-Issledovatel. Inst. Geol. Arktiki Ministerstva Geol. i Okhrany Nedr. SSSR, v. 119, no. 2, p. 23–37.
Starikova, N.D., 1959, Organic substance in the liquid phase of the deposits of the Black Sea, in Perception of the diagenesis of sediments [K Poznaniyu Diageneza Osadkov]: Akad. Nauk SSSR., Sbornik Statei, p. 72–91.
——— 1961, Organic matter in the ground waters and its distribution in deep-sea and ocean deposits: Akad. Nauk SSSR Doklady, v. 140, p. 1423–1426.
Tageeva, N.V., 1960, Geochemistry of the formation waters in oil reservoirs, in Problems of hydrology [Problemy Gidrogeologic]: Akad. Nauk SSSR, Mezhdunarod. Assotsiatsii Gidrogeol., Copenhagen, Doklady k Sobraniyu, p. 265–268.
Valyashko, M.G., 1956, Geochemistry of bromine in the processes of salt deposition and the use of the bromine content as a genetic and prospecting criterion: Geochemistry (Geokhimiya), no. 6, p. 570–589.
——— 1957, Physico-chemical conditions of the formation of potassium salt deposits of the past: Geochemistry (Geokhimiya), no. 6, p. 553–565.
——— 1961, Geochemistry of halogenesis: Mezhdunar. Geol. Congr., 21st Sessii, Sb. Tr. Geol. Fak. Mosk. Univ. [Collection of papers of the Geology Dept., Moscow Univ., submitted to the Internatl. Geol. Cong., 21st, Copenhagen, 1960], p. 211–218.
——— 1962, The influence of the chemical composition of sea water in the composition of ground water in sedimentary strata: Geochemistry (Geokhimiya), no. 2, p. 109–115.
White, D.E., 1955, Thermal springs and epithermal ore deposits: Econ. Geology, 50th Ann. Volume, p. 99–154.
——— 1957, Magmatic, connate, and metamorphic waters: Geol. Soc. America Bull., v. 68, no. 12, pt. 1, p. 1659–1682.
——— 1964, Preliminary evaluation of geothermal areas by geochemistry, geology, and shallow drilling; in Geothermal energy, I: United Nations Conf. New Sources Energy, Rome, 1961, Proc., v. 2, p. 402–408.

——— Anderson, E.T., and Grubbs, D.K., 1963, Geothermal brine well—mile-deep drill hole may tap ore-bearing magmatic water and rocks undergoing metamorphism: Science, v. 139, no. 3558, p. 919–922.

——— Hem, J.D., and Waring, G.A., 1963, Chemical composition of subsurface waters, *in* Data of Geochemistry: U.S. Geol. Survey Prof. Paper 440-F, 67 p.

——— and Roberson, C.E., 1962, Sulphur Bank, California, a major hot-spring quicksilver deposit, *in* Petrologic studies: Geol. Soc. America, Buddington Volume, p. 397–428.

——— and Sigvaldason, G.E., 1963, Epidote in hot-spring systems, and depth of formation of propylitic epidote in epithermal ore deposits: Art. 201 *in* U.S. Geol. Survey Prof. Paper 450-E, p. E80–E84.

Wyllie, M.R.J., 1948, Some electrochemical properties of shales: Science, v. 108, no. 2816, p. 684–685.

——— 1955, Role of clay in well-log interpretation, *in* Pask, J.A., and Turner, M.O., eds., Clays and clay technology—Proceeding of first national conference on clays and clay technology: California Div. Mines Bull. 169, p. 282–305.

Zaitseva, E.D., 1956, Ammonium exchange in bottom sediments of the Pacific Ocean: Proc. Acad. Sci. USSR, Geochem. Sec., v. 111, p. 117–119.

ZoBell, C.E., Sisler, F.D., and Oppenheimer, C.H., 1953, Evidence of biochemical heating in Lake Mead mud [Ariz.-Nev.]: Jour. Sed. Petrology, v. 23, no. 1, p. 13–17.

Zyka, V., 1958, Role of oilfield waters in the accumulation and distribution of chemical elements: Acta Geol. Acad. Sci. Hungaricae, v. 5, no. 3–4, p. 435–478.

14

Copyright ©1969 by the Society of Economic Paleontologists and Mineralogists
Reprinted from *Jour. Sed. Petrology* 39:1188-1201 (1969)

DIAGENESIS, CHEMICAL SEDIMENTS, AND THE MIXING OF NATURAL WATERS[1]

DONALD D. RUNNELLS
Department of Geological Sciences, University of Colorado, Boulder, Colorado 80302

ABSTRACT

Mixing of aqueous solutions is an important procedure in chemical laboratories. The chemical effects of mixing in the laboratory include changes in the concentration and electrical properties of a solution, shifts in homogeneous equilibria, and precipitation or dissolution of a solid phase. Similar chemical effects can be expected to occur in nature as a result of the mixing of natural waters, and in some instances these may result in diagenesis. A simple classification yields 14 distinct categories of natural waters with 91 possible combinations of mixing of pairs of such waters.

Examples are known of the precipitation of hydroxides and carbonates as a result of the mixing of chemically dissimilar surface waters. Documented examples of subsurface mixing and diagenesis are few. However, because of the great variety of chemically dissimilar subsurface waters, plus the existence of several types of potential fields which cause the movement of subsurface waters, mixing and diagenesis in the subsurface may be a common and significant geologic process.

The solubility of rock forming minerals is a non-linear function of such independent variables as salinity, partial pressure of gases, temperature, and so on. The dissolution of calcium carbonate as a result of "Mischungskorrosion" can be viewed in the light of the non-linearity of the solubility curve. Experimental data on the solubility of calcium carbonate and calcium sulfate as a function of added salts indicate that mixing of solutions which differ only in their content of dissolved electrolytes may cause either the precipitation or dissolution of the rock-forming minerals. Other common rock-forming minerals, with the exception of quartz at low temperatures, should exhibit similar behavior.

INTRODUCTION

This paper offers the hypothesis that the mixing of natural waters is an important process of diagenesis and chemical sedimentation. The hypothesis is presented in general terms, in the belief that the necessary testing of the concept will be done in the field by geologists who are not necessarily specialists in geochemistry. Experimental data from the chemical literature are used to construct the solubility curves that comprise the basis of the hypothesis. Examples of natural mixing and associated diagenesis and chemical sedimentation are offered in support of the concept.

The term "diagenesis" is used throughout this paper in accord with Pray and Murray (1965, p. 1): "In its broadest sense, diagenesis encompasses those natural changes which occur in sediments and sedimentary rocks between the time of initial deposition and the time—if ever—when the changes created by elevated temperature, or pressure, or by other conditions can be considered to have crossed the threshold into the realm of metamorphism." For purposes of this discussion it is adequate to arbitrarily define the upper limits of diagenesis as roughly equivalent to the conditions of temperature and lithostatic pressure encountered in the deepest modern oil wells, on the order of about 200°C at 1500 atmospheres, corresponding to depths of about 20,000 to 25,000 ft (gradient of 1°C/100 ft and 100 psi/100 ft for water-saturated sedimentary rocks; Levorsen, 1967, p. 415, 394). A more rigorous definition of diagenesis involves the concept of the zeolitic facies of regional metamorphism, named by Fyfe, Turner, and Verhoogen (1958). The zeolitic facies bridges the transition from diagenesis to the greenschist facies of metamorphism.

LABORATORY MODELS OF MIXING

When a chemist mixes aqueous reagents in the laboratory, he has one of three purposes in mind. First, he may wish to change the concentration and electrical properties of a solution by adding distilled water or an inert electrolyte. Second, he may intend to initiate a homogeneous reaction in solution, such as the formation of a colored complex or the oxidation and reduction of dissolved species. Third, he may wish to precipitate or dissolve a solid phase. Other objectives, such as the determination of electrochemical or thermochemical parameters, depend on one or more of the preceding processes. These three laboratory procedures can serve as models for the chemical effects of mixing of waters in nature. In particular, the third model of mixing, involving a solid phase, is analogous

[1] Manuscript received November 5, 1968; revised March 3, 1969.

to diagenesis and chemical sedimentation in nature.

NATURAL WATERS AVAILABLE FOR MIXING

Just as the chemist has many aqueous reagents available for his experiments, so are there many natural waters with different chemical and physical properties. It would be irrelevant to attempt a detailed classification here, but in order to demonstrate the many possibilities for natural mixing of dissimilar waters it is useful to introduce a few simple subdivisions. It is implicit that most of the waters will differ in chemical and (or) physical properties.

One of many possible frameworks of classification which can be used is the location of the waters with respect to the surface of the earth, summarized in table 1 as folows: meteoric (above the surface), surface, and subsurface. Further breakdown of the surface waters is possible into at least the four major categories of lakes, streams, swamps, and oceans. The subsurface waters are logically subdivided at least into vadose (above the water table Meinzer, 1923) and phreatic (below the water table). Although meteoric waters may differ greatly in chemical composition (Carroll, 1962), no subdivisions of this category are proposed.

Because of the fundamental importance of the chemistry of the waters involved in diagenesis, it is useful to characterize surface and subsurface waters at least as fresh or saline. A distinction is therefore made between fresh lake water and saline lake water, fresh phreatic water and saline phreatic water, and so on. The exact point of separation of fresh from saline is not relevant.

Additional subdivisions of phreatic waters should be made in recognition of the distinct differences in chemical composition that may obtain as a function of the genesis of subsurface waters. The two broad categories of normal formation water and ascending thermal water are therefore suggested. Papers by White (1957a, b) and White, Hem, and Waring (1963) illustrate the differences in chemical and physical properties that may obtain between phreatic waters of meteoric or connate origin and those which may be of volcanic, magmatic, or metamorphic origin. Most phreatic waters encountered in wells are of connate or recent meteoric origin. Although this distinction is certainly not clearcut, it is adequate for present purposes.

This simple classification results in 14 separate categories of water, summarized in table 1. Although specific samples of water from several of the categories may be virtually identical, many will differ significantly. There are 91 com-

TABLE 1.—*Classification of natural waters*

I. Meteoric waters
II. Surface waters
A. Lakes
1. Fresh
2. Saline
B. Streams
1. Fresh
2. Saline
C. Swamps
1. Fresh
2. Saline
D. Oceans
III. Subsurface waters
A. Vadose
1. Fresh
2. Saline
B. Phreatic
1. Normal formation water
a. Fresh
b. Saline
2. Ascending thermal water
a. Fresh
b. Saline

binations of mixing of any pair of waters in table 1. Obvious examples include the mixing of fresh stream water with ocean water, saline swamp water with fresh stream water, meteoric water (fresh) with saline lake water, and ascending saline thermal water with fresh phreatic formation water. Three or four waters may also mix.

Additional subdivisions, with a resulting increase in the possible combinations of mixing could be justified on the basis of specific chemical composition, such as saline sodium chloride water or fresh calcium bicarbonate water. Quantitative considerations of temperature and pressure might also be introduced. However, the only point to be made here is that there are a great many different aqueous solutions available for mixing in nature, with manifold possibilities for chemical reaction.

EXAMPLES AND EFFECTS OF MIXING OF SURFACE WATERS

In order to cause diagenesis, the mixing of natural waters must result in a solution which is either undersaturated or supersaturated with respect to one or more mineral phases. This is the natural analog of the third model of mixing in the laboratory, mentioned above. If minerals are precipitated, they may either be the same as those already present in the adjacent framework of rock and sediment or they may be new. Depending on the textures involved and the background and bias of the investigator, such processes may be termed replacement, alteration, cementation, leaching, recrystallization, mineralization, and so on. Several textural changes may take place simultaneously, with one or more minerals being dissolved while others are being precipitated.

There are a few documented examples of the precipitation of minerals as the result of mixing of natural surface waters. An interesting case was described by Theobald, Lakin, and Hawkins (1962) for the mixing of two small streams in Summit County, Colorado. One of the streams is strongly acidic, with a pH of about 4, and is rich in dissolved sulfate, silica, aluminum, and heavy metals. The unusual chemical character of this water results from the oxidation of disseminated pyrite, with attendant acid leaching of the igneous and metamorphic rocks in its drainage basin. The other stream has a pH of about 8 and contains smaller quantities of dissolved components. Mixing of the two waters results in the precipitation of flocculent white aluminum hydroxide and basic aluminum sulfate downstream from their junction. This is easily explained on the basis of the change in solubility of aluminum hydroxide as a function of pH, as discussed in a following section. Raymahashay (1968) has recently described a similar situation from Yellowstone Park, in which a precipitate of poorly-crystallized silica plus aluminum-iron hydroxides or hydrated aluminum silicates is formed at the junction of streams of acid and alkaline waters derived from thermal springs. Zelenov and others (1965) mention that the mixing of acidic waters of volcanic origin with sea water, with the production of precipitates of aluminum and iron hydroxides, is a common occurrence in volcanic areas of the world. It was suggested long ago (von Gumbel, 1878) that manganese in deep sea sediments is derived from submarine hydrothermal springs, and it is reasonable to expect that mixing of sea water with hydrothermal waters from submarine springs does take place.

Baranov (1956) concludes that concretions of calcite and magnesite and other carbonate sediments in lakes of the Kulundin Steppe probably result from the mixing of waters of different composition and temperature. Sedel'nikov (1959) found that the mixing of calcium bicarbonate waters from the Caspian Sea with magnesium sulfate waters in the Gulf of Kara-Bogaz results in the precipitation of a sequence of different salts along the path of mixing. Tsurikova and Tsurikov (1966) conclude from field geochemical studies that the mixing of river and sea water may cause either the precipitation or the dissolution of calcium carbonate, depending on the chemical composition and concentration of the river water. They state that in seven out of nine instances of influx of river water at the head of the Taganrog Gulf in the Sea of Azov, the chemistry of the water indicates that precipitation of calcium carbonate is taking place. In contrast, according to Tsurikova and Tsurikov (1966), the mixing of sea water and very fresh river water increases the solubility of calcium carbonate.

Other examples of mixing of surface waters are easily imagined. One would be the dilution of surface waters by meteoric water, corresponding to the first laboratory model of mixing described above. A trivial consequence of this type of mixing is that it may permit the dissolution of additional minerals, such as salts from the floor of a playa lake. A more interesting consequence is that simple dilution may cause significant shifts in homogeneous equilibria, such as the formation or breaking of dissolved complexes or the oxidation and reduction of dissolved species. For example, Eardley (1938) found that the pH of water from Great Salt Lake rose from 7.4 to 8.5 upon dilution with six or seven volumes of distilled water. Such a change in pH may result from a redistribution in solution of one or more hydrogen-bearing and hydroxyl-bearing species, such as HCO_3^-, HS^-, $Mg(OH)^+$, $MgHCO_3^+$, and many others. Silicates, carbonates, and most other rock-forming minerals are sensitive to changes in pH, so that simple dilution of a surface water by rain water could theoretically cause acid-base alteration of associated sediments.

Similarly, introduction of dissolved oxygen into an oxygen-poor environment will tend to shift the prevailing equilibrium between such species as SO_4^{-2}–HS^-, NO_3^-–NO_2^-, and Fe^{+2}–Fe^{+3}, although kinetic factors may prevent attainment of equilibrium (Morris and Stumm, 1967). Minerals which incorporate these species in their structure may undergo diagenesis as a result of changes in the relative proportions of the ions in solution. Obvious examples of mixing with probable changes in the content of dissolved oxygen include the mixing of fresh stream water with stagnant lake water and of meteoric water with organic-rich swamp water.

EXAMPLES AND EFFECTS OF MIXING OF SUBSURFACE WATERS

It is more difficult to find examples in the literature of mixing and associated diagenesis in the subsurface, probably because workers in the field have not generally been aware of the possible importance of this process. Back and Hanshaw (1965, p. 81) do suggest that precipitation of calcite and other minerals may result from the mixing of two waters, neither of which was capable of forming a precipitate before mixing, such as encroaching sea water and fresh phreatic water near the coast. They also mention (p. 81) that mixing of waters may be important in the development of caliche and beach rock. Unfortunately Back and Hanshaw (1965) give neither a reason nor supporting evidence for their suggestions. However, from recent work in the Soviet Union (Voroshilov, Dzhalilov, and Samedov, 1968) it does in fact appear that the subsurface mixing of sea water with formation waters in oil fields may cause cementation by calcium and magnesium carbonates. The Russian authors (1968) found that chemical analyses of waters from the formations in the Surakhan and Karachakhura oil fields indicate an influx of sea water with possible precipitation of salts, chiefly magnesium and calcium carbonates.

An artificial but instructive example of the type of cementation that may take place as a result of mixing of waters in nature, and of which we may at present be totally ignorant, is offered by the results of water-flooding operations in the Wilmington oil field near Long Beach, California (Gates and Garaway, 1965). Large scale injection of sea water into the producing horizons in the Wilmington field caused the precipitation of troublesome quantities of barium sulfate scale in the production wells. Gates and Garaway (1965) found that the scaling resulted from the mixing of sulfate-rich sea water with barium-rich formation water. As an interesting sidelight, Sawkins (1966) suggests that the precipitation of barite in the ore deposits of the Northern Pennines in England may have resulted from the mixing of hydrothermal waters with barium-rich formation waters.

In view of the fact that encroachment by sea water into coastal aquifers is common, it would be surprising not to find that diagenesis occurs at various localities near the coast as a result of mixing in the subsurface. This would be unquestionable if the formation water involved could react with sea water to form insoluble precipitates, such as calcium carbonate or barium sulfate. However, as will be shown in following sections of this paper, simple changes in salinity due to mixing may also cause the diagenesis of adjacent rock and sediment.

Although published descriptions of subsurface mixing and associated diagenesis are sparse, indirect evidence virtually dictates that the process must operate in nature. In fact, it must be a relatively common process. Consider first the wide variety of chemically dissimilar waters which exist in the subsurface, as documented in the compilation of analyses by White, Hem, and Waring (1963). Subsurface waters exist in which the predominant dissolved component is sodium chloride, calcium chloride, sodium sulfate, calcium sulfate, sodium bicarbonate, magnesium sulfate, calcium bicarbonate, and so on, with mixtures in between and an enormous range of total salinities. An interesting, but not unusual example of chemical variation with depth was determined by Kravtzov (in Kamensky, 1958) in the Donetz Basin, as follows: calcium-bicarbonate water near the surface, superseded successively at depth by sodium-bicarbonate water, sodium-sulfate water, sodium-bicarbonate water, sodium-bicarbonate-chloride water, and possibly saline sodium-chloride water. Unless the intervening rocks act as impermeable barriers, mixing must take place at each interface between zones of chemically dissimilar water, with attendant opportunity for diagenesis.

Bulk movement of water in the subsurface occurs in response to potential fields (Hubbert, 1953). The potential fields may be created by differences in elevation, pressures due to the weight of overlying fluids and rocks, secondary cementation of the pores in the rock (Levorsen, 1967), differences in temperature, osmotic pressures across semipermeable membranes (DeSitter, 1947; Hanshaw, 1962; White, 1965), and other miscellaneous effects such as earthquakes and chemical and biochemical reactions (Levorsen, 1967). Even in stable basins, some mixing of subsurface waters will take place. Roberts (1962) (quoted in Baker, 1967, p. 322–323) states: "In the history of a basin fluid system, the influence of compaction pressures must gradually give way to the influences of artesian pressures, although some influence of both may always exist. The net result is a more or less blended pressure system. The system is never dead, even in the most quiescent basin. There is always fluid exchange occurring between the subsurface and the atmosphere and (or) hydrosphere. Any equilibrium attained is always dynamic and vulnerable to the strengthening or weakening of any part of the system. Significant lateral and vertical pressure gradients have been observed even in the most stable basins where data are available." Baker (1967, p. 323), presents a diagram (from Roberts, 1962) showing the hypothetical mixing of waters entering a basin with waters being driven from the basin as a result of compaction.

Documented examples of subsurface mixing, without investigation of possible associated diagenesis, have been published by a few workers. Henningsen (1962) collected and analyzed water from wells in the Cretaceous Trinity aquifers of central Texas. He showed that recharge into the Trinity aquifers consists chiefly of two chemically distinct masses of water draining areas of different lithology. As the waters move basinward in the subsurface, they mix to form a third distinct mass which changes gradually with continuing migration. Further mixing occurs where the Trinity aquifers cross the Balcones fault system, with water being introduced along the fault system from other formations. Another example is that published by Popov and Goldshteyn (1957) in which they describe large hydrodynamic systems in central Asia with a central zone of ascending saline water, surrounded by peripheral zones of descending fresh' water. A hydrochemical facies (Back, 1960, 1966) of carbonate-sulfide water is developed in the region of mixing between the ascending and descending masses. Popov and Goldshteyn (1957) suggest that the mixed zone is a favorable locus for the deposition of certain metallic ores. From the description, one might suspect that bacterial action is important in the zone of mixing, with anaerobic bacteria utilizing dissolved organic matter and sulfate in their metabolism, releasing dissolved sulfide and bicarbonate. This mechanism has been suggested by Feely and Kulp (1957) for the deposits of native sulfur and limestone caprock associated with salt domes.

The mixing of fresh phreatic water with encroaching sea water near the coast is common and well known (Turner and Foster, 1934; Foster, 1942; Kohout, 1960; Columbus, 1965; Upson, 1966). Recent drilling in the ocean floor on the Blake Plateau has revealed the presence of artesian fresh water and a zone of mixing with sea water about 40 to 50 miles east of the coast of Florida (Corwin and Bradley, 1966).

An interesting example of mixing of several different waters has been discussed by Bjorklund (1958) and Motts (1968) for the area near Carlsbad, New Mexico. The groundwater pumped from wells in the area results from the mixing of the following: potable water from the Limestone aquifer of the Guadalupe Mountains, saline water seeping downward from Lake Avalon, seepage water from the Pecos River, and return irrigation water.

Other lines of evidence also show that there is significant movement of subsurface brines, with attendant opportunities for mixing and reaction. For example, saline springs and seeps are common in many areas of the world. Irelan and Mendieta (1962) report that natural springs and seeps of brine with salinities up to 300,000 parts per million account for as much as 50 percent of the dissolved load of the Brazos River in Texas, and natural springs of sodium chloride brine with concentrations up to 190,000 parts per million are described from western Oklahoma by Leonard and Ward (1962).

An interesting example of the probable mixing of brines in the deep subsurface comes from the work of McNeal (1965) in the Permian Basin of western Texas. McNeal points out that his maps of the potentiometric surface and salinity of water in the Permian Delaware sand suggest vertical movement through fault zones.

To summarize this section: in view of the great variety of chemically different subsurface waters and the many potential fields that cause the movement of such waters, it is likely that mixing is a relatively common phenomenon in the subsurface. Further, in every case in which dissimilar waters mix, some sort of chemical change must take place. In many cases, the chemical changes may involve only homogeneous equilibria in solution, perhaps reflected by a change in pH, the oxidation of dissolved iron, or the formation of dissolved complexes. In other cases, as shown by the examples cited previously, mixing may cause the dissolution or precipitation of minerals.

EXAMPLES OF MIXING IN THE FORMATION OF ORE DEPOSITS

Many published reports cite the mixing of waters as a probable or possible factor in the formation of hydrothermal ore deposits. From theoretical considerations, Garrels (1941) and Helgeson (1964) point out that the dilution of a concentrated hydrothermal solution of alkali halides should cause the precipitation of complexed metals. Similarly, Dickson and Tunell (1958) present experimental results showing that dilution of an aqueous solution of Na_2S saturated with HgS will cause the precipitation of HgS. Dickson and Tunell (1958) also state that mixing of an acid solution with the solution of Na_2S will break the aqueous sulfide complexes and cause the precipitation of HgS. Shawe (1966) concludes that the distribution of uranium, vanadium, and selenium in the "roll" ore bodies of the Colorado Plateau is compatible with earlier hypotheses that the precipitation of the ore took place at the interface between two dissimilar subsurface waters.

With regard to the mixing of waters and the precipitation of gangue minerals in hydrothermal deposits, Segnit, Holland, and Biscardi (1962) conclude (p. 1330) that it is difficult to account for the widespread deposition of calcite from hydrothermal solutions by simple cooling and loss of pressure and vapor along a normal geothermobarometric gradient. As one of several alternative suggestions for the cause of the extensive precipitation of calcite from hydrothermal solutions, the writers state (p. 1330): "Mixing of hydrothermal fluids with ground waters may result in large-scale precipitation of calcite. The importance of such a mechanism seems doubtful, however, except at small depth."

Hall and Friedman (1963) investigated the chemical and isotopic composition of fluid inclusions in the fluorspar deposits of the Kentucky-Illinois district, and their data were best explained as the result of mixing of hydrothermal waters with connate brines similar in composition to those of the Illinois Basin. In a recent discussion of textures and fluid inclusions in sphalerite from Pine Point, Northwest Territories, Roedder (1968) concludes (p. 447–448): "In particular, it is suggested that an upward flow of hot, metal-bearing brines mixed with a small amount of relatively fresh, cold surface waters in the vicinity of the ore deposit. This mixing may explain the precipitation of the ore, the range in salinity and homogenization temperatures, and the minute, regular color banding seen in some of the sphalerite." Also using Pine Point as an example, Jackson and Beales (1967) conclude that many ore deposits of the Mississippi Valley type are best explained as the result of mixing of saline connate waters expulsed during compaction with sulfide-rich waters indigenous to carbonate reservoir rocks. In his discussion of the chemistry of fluid inclusions and its bearing on the origin of Mississippi Valley type ore deposits, Sawkins (1968, p. 935) states: "The data are consistent however with a genetic model involving the mixing of small amounts of relatively high potassium saline solutions containing sulfate with average connate brines."

In summary, with regard to ore deposits, Holland (1967, p. 431) says: "Mixing offers many delightful opportunities for chemical reactions and is a particularly tantalizing mechanism, because its operation is so difficult to prove or disprove in the evolution of any particular hydrothermal system." Holland's remarks regarding the role of mixing in the deposition of hydrothermal ore deposits are certainly appropriate, because in the study of ore deposits we are hard-pressed to define unequivocally the composition of the hydrothermal solution, the temperature and pressure of mineralization, or, especially, the hydrodynamics of the system. In most cases, the hydrothermal fluid is available only as micro-samples from fluid inclusions, and the temperature, pressure, and hydrodynamics have long since changed or ceased to function. In sharp contrast, in a study of the role of mixing in diagenesis and chemical sedimentation, we often have the opportunity to collect large samples of both the liquid and the rock or sediment involved, and to investigate the hydrodynamics of the environment.

SOLUBILITY CURVES AND DIAGENESIS

Diagenesis or chemical sedimentation can occur as the result of mixing only if the resulting mixed water is undersaturated or supersaturated with respect to one or more minerals. In order to be able to predict whether or not this will happen we must know how the solubility of the minerals of interest varies as a function of the chemical and physical parameters which distinguish the waters from each other. For example, the solubility of calcite is a non-linear function of the partial pressure of carbon dioxide gas in the co-existing vapor phase, as shown in figure 1. We can therefore predict (Bögli, 1964) that the mixing of two waters, such as A and B in figure 1, both at equilibrium with calcite but in contact with different partial pressures of carbon dioxide, will result in a locus of mixed waters, undersaturated with respect to calcite, between points A and B. In order to again reach saturation, the mixed waters along the locus A-B in figure 1 must dissolve additional calcite and move up to the equilibrium solubility curve. This is the basis of the "Mischungskorrosion" effect of Bögli (1964). Mischungskorrosion has been applied to the excavation of caves in limestone, but Thrailkill (1968) lists several reasons why it may actually be relatively ineffective in nature. Whether or not field observations eventually prove Mischungskorrosion to be an effective agent in the excavation of caves, the simple fact that the solubility curve is non-linear suggests that we should seek other systems with similar non-linear solubility curves as possible models of the diagenetic effects of mixing in nature. This is the basis of the hypothesis offered here.

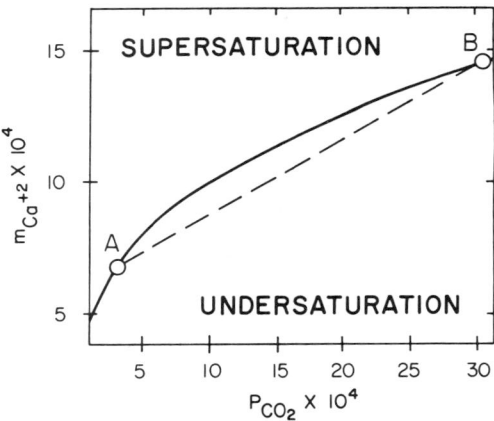

FIG. 1.—Solubility of calcite as a function of an independent variable, which is the partial pressure of carbon dioxide gas in this case. Modified slightly from Thrailkill (1968).

PRODUCTION OF POROSITY THROUGH MIXING OF SODIUM CHLORIDE WATERS

Figure 2 shows the equilibrium solubility of calcite at various temperatures as a function of the content of dissolved sodium chloride in the water. The content of sodium chloride in sea water is indicated, as is the concentration of sodium chloride which would be equivalent to the total ionic strength (0.7) of sea water. From the fact that the curves in figure 2 are concave downward, it can be predicted that mixing of two waters, both saturated with calcite at the same temperature but containing different amounts of dissolved sodium chloride, should cause undersaturation and leaching by the mixed water. Therefore, if kinetics allow, porosity will be created in the adjacent rock or sediment if calcite is available.

The general shape of the curves in figure 2 is to be expected whenever a neutral, non-reactive salt is introduced into a saturated solution of another salt. "Salting-in" takes place with the initial additions of the neutral electrolyte, and "salting-out" occurs at high concentrations. The exact nature of the interaction among ions and solvents, which governs the shape of solubility curves, continues to be the subject of intensive research in physical chemistry. Briefly, the initial increase in solubility with the addition of a neutral electrolyte reflects a decrease in the thermodynamic activity of the dissolved ions as a result of increased mutual electrostatic interactions, and more of the solid must dissolve to maintain a constant ion activity product in solution. At higher concentrations of dissolved salts, the major factors are a decrease in the dielectric constant of water and the removal of free water due to solvation of the ions (Lee, 1959). At higher concentrations of added electrolyte, the relative activity of the dissolved species is enhanced, and the solid is precipitated in order to maintain a constant ion activity product in solution. Geochemists will recognize that this behavior is described by the variation in the values of the ion activity coefficients, as illustrated in figures 2.14 and 2.15 of Garrels and Christ (1965).

Note that the solubility curves in figure 2 become more nearly linear with increasing temperature, as is characteristic of most similar systems. We therefore expect that the diagenetic effects of mixing will decrease with rising temperature.

The extent of diagenesis possible through mixing is not trivial in geologic terms. Consider waters A and B in figure 2, arbitrarily mixed in equal volumes to yield water C. In order to reach the equilibrium curve, the mixed water must dissolve calcite and move from C to C^1. In

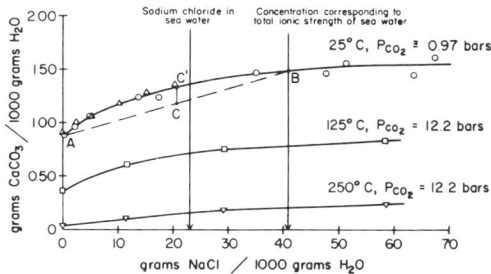

Fig. 2.—Solubility of calcite as a function of the content of a neutral electrolyte (sodium chloride) in solution.

this particular example, the quantity of calcite which will be dissolved at equilibrium is about 0.17 grams for each 1000 grams of mixed water. Assuming a specific gravity of 2.5 for a limestone with a porosity of about 10 percent, each 1000 grams of mixed water would therefore be able to leach about 0.06 to 0.07 cc of limestone wallrock. If such a process operates at maximum efficiency, the passage of about 100 liters of mixed water through a cube of limestone 10 cm on a side would increase the pore space from 100 cc to about 107 cc. It is probably not unreasonable to assume a high efficiency for the leaching of calcite by very slowly moving subsurface waters. Harris and Matthews (1968) have shown that the dissolution of aragonite and the reprecipitation of calcite by percolating vadose waters in the carbonate sediments of Barbados, West Indies, proceeds with an efficiency of more than 90 percent. Although the processes involved are not entirely equivalent, the work by Harris and Matthews (1968) indicates that the rate of response of the carbonates to changes in the state of saturation of the waters is rapid relative to the rate of movement of the water through the rock, and it is to be expected that waters in the deep subsurface will move much more slowly than the percolating vadose waters in Barbados. Neither is the assumption of the passage of 100 liters of water through a 10 cm cube of limestone unreasonable. If the porosity of the limestone is 10 percent, there will be 100 cc of water present in the 10 cm cube at saturation. The example of 100 liters of mixed water amounts to about 100,000 cc, so that a column of water (1000 × 10 cm) 10,000 cm long would have to pass through the 10 cm cube of limestone; this is a length of only 100 m. Hanshaw, Back, and Rubin (1965) found an average rate of movement of 7 m per year for water in a carbonate aquifer in central Florida. This would result in a period of time of only about 14 years for the 100,000 cc in our example. If we increase the volume of mixed water in the example by an order of magnitude, to 1000 liters, and reduce the rate of flow found by Hansaw, Back, and Rubin (1965) by an order of magnitude, to 0.7 m per year, the time required for passage of the mixed water is still only 1400 years. With 100 percent efficiency of leaching the passage of 1000 liters of the mixed water in figure 2 would increase the porosity of the 10 cm cube of limestone from 10 percent to 16 percent; this is geologically significant.

Figure 3 shows the solubility of calcium sulfate in water with added sodium chloride. Some of the experimental data have been recalculated for plotting in figure 3. As expected for an added neutral electrolyte, the solubility curves for both gypsum and anhydrite are again concave toward the axis of the independent variable, with a pronounced maximum at low temperatures. Mixing of such waters, as represented by points A and B in figure 3, will cause undersaturation and should lead (kinetics allowing) to leaching of calcium sulfate from the adjacent rock or sediment. Mixing of waters A and B in figure 3 in the proper proportions to produce water C will result in undersaturation of about 1.9 grams of calcium sulfate per 1000 grams of mixed water or about 0.66 cc of anhydrite (sp gr about 2.9). Mixing of waters D and E in figure 3 would result in a much larger degree of undersaturation (as at point F). Experimental results by Marshall and Slusher (1968) show that the shape of the solubility curves for calcium sulfate at various temperatures in dilutions and concentrates of sea water is essentially identical to that shown in figure 3 for pure sodium chloride, but the absolute solubility is significantly higher in sea water at elevated temperatures due to the formation of complex ions.

The increase in linearity of the solubility curves with increasing temperature is more marked for anhydrite than for calcite. It is clear from figure 3 that mixing of sodium chloride solutions saturated with anhydrite will produce little diagenesis as the temperature approaches 200°C. Although 325°C is beyond the temperatures of concern here, it is interesting to note from figure 3 that at 325°C the solubility curve for anhydrite as a function of dissolved sodium chloride is slightly concave upward. Mixing of waters which lie on such a curve will result in supersaturation. It is also interesting to observe in figure 3 that the solubility curves at various temperatures cross at high salinities; this indi-

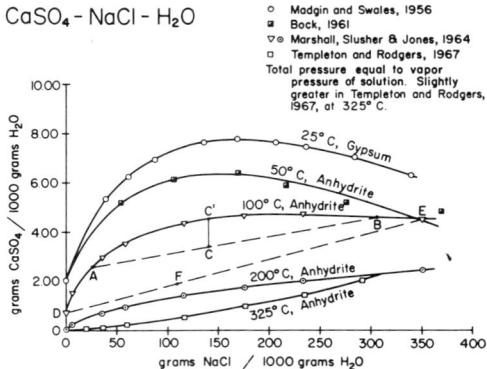

FIG. 3.—Solubility of calcium sulfate minerals as a function of the content of a neutral electrolyte (sodium chloride) in solution.

cates that the solubility of anhydrite, which increases with falling temperature at low salinities, will decrease with falling temperature at salinities above the points of intersection.

Therefore, as a rule of thumb, we can predict from figures 2 and 3 that the mixing of sodium chloride waters in the presence of limestone, gypsum, anhydrite, and probably dolomite, will result in undersaturation and the production of porosity. The greater the difference in content of dissolved neutral salts in the two waters, the greater will be the extent of leaching of the adjacent rock or sediment as a result of the mixing.

It is important to point out that the sodium chloride waters involved in figures 2 and 3 are geologically realistic. As shown by the lines in figure 2, mixing of waters with a content of sodium chloride equal to or less than that of sea water will cause a significant degree of undersaturation. With reference to the higher salinities in figure 3, White, Hem, and Waring (1963) list several analyses of oil field brines in which the principal dissolved salt is sodium chloride, with total dissolved solids up to 323 grams per 1000 grams of water. Many saline lakes are also rich in sodium chloride. Great Salt Lake is chiefly a sodium chloride water, with a concentration about eight times that of sea water (Whitehead and Feth, 1961). Figures 2 and 3 are therefore directly applicable to many natural waters.

Because subsurface mixing has not previously been recognized as a potentially important diagenetic process, specific examples of such diagenesis are difficult to document. However, in a recent publication, Vernon (1969, p. 8) states: "The information developed from numerous wells drilled in the search for oil and gas indicates that the artesian system of Florida consists of fresh and brackish waters that rest upon and have depressed a dynamic, pulsating body of heavily mineralized salty water. An enormously cavernous area with broadly developed transmissivities has been formed approximatey along the contact of the two bodies of water. Dense dolomite rock forms the walls of these caverns...." This is precisely the natural situation which would be predicted from a consideration of the mixing of sodium chloride waters in the presence of limestone and probably dolomite, as shown in figure 2 and discussed above. Although he does not so state specifically, Kohout (1965) implies that these deep Florida waters are similar in composition to sea water. Vernon (1969, p. 13) says that the salinity of the water ranges up to ten times that of sea water.

PRECIPATION DUE TO MIXING

If the solubility curve of a mineral is concave upward as a function of added salts, we can predict that mixing of saturated waters will cause supersaturation, with the potential for precipitation and cementation. This is illustrated in figure 4, which shows the solubility of gypsum as a function of added calcium chloride. The presence of a common ion, in this case calcium, depresses the solubility of gypsum. Mixing along line A-B in figure 4 will result in supersaturation. Gypsum must be precipitated if the mixed water is to again achieve equilibrium saturation.

Subsurface brines rich in calcium chloride are relatively common (White, Hem, and Waring, 1963, table 13). In the Michigan Basin, for example, calcium chloride is the predominant dissolved component in many of the subsurface brines, with total dissolved solids up to 399 grams per liter (Graf and others, 1966, table 2). Bedded gypsum and anhydrite are also common in the Michigan Basin. If such waters mix, calcium sulfate may be precipitated.

Unfortunately, experimental data are apparently not available for the solubility of anhydrite as a function of the concentration of calcium chloride, with the exception of the determinations by Templeton and Rodgers (1967) at 250° and 300°C. Figure 4 shows that at 250°C the solubility of anhydrite increases with increasing dissolved calcium chloride; at this temperature the increase in solubility due to the formation of aqueous complexes (Templeton and Rodgers, 1967) overwhelms the decrease due to the common ion effect.

The examples given previously of the precipitation of aluminum and iron hydroxides as a result of the mixing of surface waters are now easily understood in terms of the shape of the solubility curves. Substances like $Al(OH)_3$ and $Fe(OH)_3$ are amphoteric; that is, they exhibit both acidic and basic properties, passing through a minimum in solubility as a function of pH. (See for example the graphs in Black, 1967, p. 278–279.) Mixing of two waters, both saturated with respect to the hydroxides but with different values of pH, will produce a mixed water which falls above the equilibrium solubility curve and which must therefore precipitate the solid phase to again achieve equilibrium. In each of the examples of the precipitation of hydroxides cited earlier, the precipitation does in fact result from the mixing of waters with significant differences in initial pH. The theoretical phase diagram on p. 356 (fig. 10.2) of Garrels and Christ (1965) suggests that kaolinite may also be precipitated as the result of mixing of waters of different initial pH.

In this case, as a rule of thumb, we can predict that the mixing of calcium-chloride waters in the presence of gypsum or anhydrite will cause precipitation and cementation. As shown in the next section, the same is true for calcite (and probably dolomite) at low concentrations of calcium chloride. Amphoterix substances will tend to be precipitated if saturated waters of differing pH are mixed. And in the most general terms we can also say that diagenesis will occur in other types of rocks which are particularly pH-sensitive, such as carbonates, if the mixing of waters results in significant changes in pH.

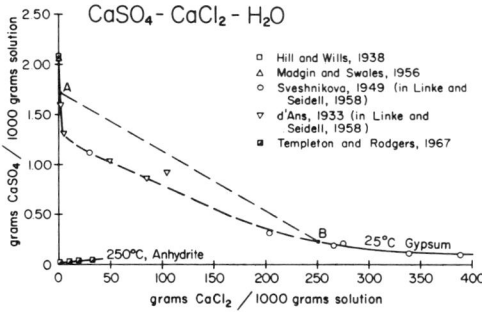

FIG. 4.—Solubility of calcium sulfate minerals as a function of an added salt with the common ion calcium. Note that the units of concentration are slightly different than in the preceeding figures.

DISSOLUTION AND PRECIPATATION DUE TO MIXING

If the solubility curve of a rock-forming mineral exhibits portions which are both concave upward and concave downward, then mixing can lead to either undersaturation or supersaturation, depending on the portion of the curve (salinity of the waters) involved. This is illustrated in figure 5, in which the solubility of gypsum and anhydrite is plotted as a function of added sodium sulfate. With increasing concentrations of added sodium sulfate, the solubility of the minerals first decreases, as expected due to the common ion effect, and then increases. The increase in solubility at higher concentrations is in part due to electrostatic effects in the solution, but in this system the formation of such dissolved complex ions as $CaSO_4^\circ$ and $NaSO_4^-$ is probably also important. At 250°C (fig. 5), the solubility of anhydrite as a function of added sodium sulfate and sodium chloride is consistent with the formation of the complexes $CaSO_4^\circ$, $Ca(SO_4)_2^{-2}$, and $Ca(SO_4)_3^{-4}$ (Templeton and Rodgers, 1967, fig. 8), and the common ion effect is unimportant.

In figure 5 mixing of waters A and B or A and E will result in supersaturation and possible precipitation of anhydrite cement. Mixing of similar waters at 25°C will cause supersaturation with respect to gypsum. In contrast, mixing of waters B and D or E and D in figure 5 will result in undersaturation and possible leaching of anhydrite. Continuous mixing in the direction of the arrows along line D-A would first cause undersaturation, followed by equilibrium at point E, with supersaturation from E to A. One must wonder if some of the diagenetic textures observed and interpreted by sedimentary petrographers as multiple and separate stages of precipitation and leaching may not represent continuous mixing of waters of differing salinity, as along the path D-A in figure 5.

Waters which are predominantly sodium sulfate in composition are not common in nature, but they do occur (White, Hem, and Waring, 1963, table 14; Swarzenski, 1968).

The behavior of calcite in the presence of calcium chloride is exactly analogous, at least for the one set of experimental conditions for which

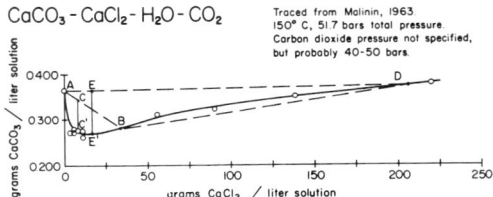

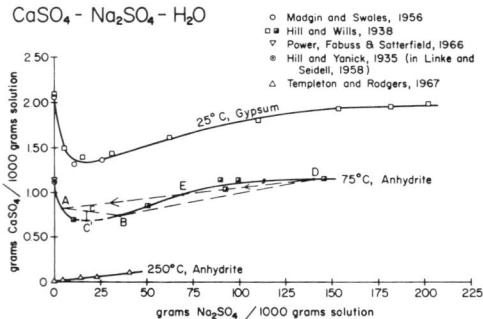

Fig. 5.—Solubility of calcium sulfate minerals as a function of added sodium sulfate.

Fig. 6.—Solubility of calcite as a function of added calcium chloride. Units of concentration are slightly different than in preceeding figures.

data are available, as shown in figure 6. The initial decrease is due to the common ion effect, and the increase at higher concentrations of calcium chloride is attributed by Malinin (1963) to a decrease in the mean activity coefficient of calcium bicarbonate in solution. The minimum in the curve represents the point of balance between the opposing effects (Malinin, 1963). The possibilities for diagenesis due to mixing along the curve in figure 6 are similar to those described above for calcium sulfate in the presence of dissolved sodium sulfate. Either leaching or cementation could occur, depending on the relative proportions of the waters involved. As noted earlier, calcium chloride brines are relatively common in the subsurface, and it is interesting to speculate on the patterns of diagenetic alteration which might be produced in carbonate rocks by the mixing of such waters. The important and perplexing problem of the origin of patchy calcite cement in quartz sandstone might also be fruitfully investigated from the point of view of mixing of calcium chloride or similar waters.

Summarizing, we can predict that the mixing of waters which have an ion in common with the rock-forming minerals will probably cause supersaturation and precipitation at low concentrations, with undersaturation and leaching being the result of mixing at higher concentrations of the added salt. Continuous mixing of varying proportions of the waters may result in sequential stages of leaching and precipitation.

INITIAL DEPARTURES FROM EQUILIBRIUM

It should be emphasized that while the preceding discussion implies that the waters involved in the mixing are initially at equilibrium with the minerals involved, this need not be the case. In figure 7, for example, waters which fall below the hypothetical curve and are therefore undersaturated with respect to the mineral phase, such as A and B, may become supersaturated through mixing, as at point C. Similarly, waters D and E, both supersaturated with re-

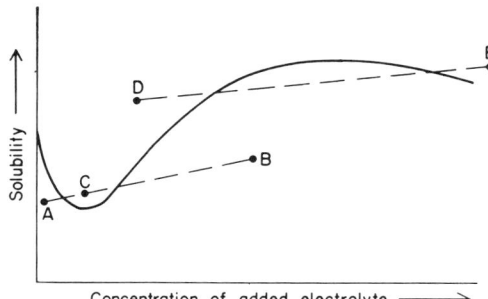

Fig. 7.—Hypothetical solubility curve showing how mixing of undersaturated waters (points A and B) may result in a supersaturated water (point C), and how mixing of supersaturated waters (points D and E) may result in undersaturation.

spect to the hypothetical mineral involved, may mix to form a series of undersaturated waters along the line D-E (below the saturation curve).

Again, the development of patchy calcite cement in quartz sandstone might be investigated in light of this phenomenon, in which waters initially undersaturated with respect to calcite can become supersaturated through mixing, as at point C in figure 7. (See also fig. 6.)

OTHER MINERALS

Leaching or precipitation of a mineral will tend to take place as a result of mixing whenever the solubility of the mineral is a non-linear function of added salts. Therefore, the processes discussed above in terms of calcium sulfate and calcium carbonate will probably hold true for any mineral which dissolves to form free ions in solution, although the probable effects for less soluble minerals will be correspondingly smaller. This includes the feldspars, pyroxenes, amphiboles, and others, but it does not include quartz. At low temperatures quartz dissolves to form the uncharged aqueous species H_4SiO_4 (Alexander, Heston, and Iler, 1954), and the solubility of quartz at low temperatures is little affected by added electrolytes (Holland, 1967). However, at 700°C and 4 kilobars pressure the solubility curve of quartz as a function of added potassium chloride is strongly concave downward (Anderson and Burnham, 1967), so that mixing will result in undersaturation. Similar experimental data are lacking for most other silicates.

OTHER INDEPENDENT VARIABLES

Although the content of added salts has been emphasized throughout this discussion as the independent variable of greatest interest, mixing of waters in which other parameters differ may also be important. For example, mixing of waters of different temperature, pressure, pH, content of dissolved organics, partial pressure of gasses (Mischungskorrosion), and so on, may also cause changes in the state of saturation. Only if the solubility of a mineral is a straight line function of a given independent parameter will mixing result in a water which remains at equilibrium with the adjacent solids. The principle is of broad application.

CONCLUSIONS AND SUMMARY

Little study has heretofore been given to the possible role of mixing of natural waters in diagenesis and chemical sedimentation, but indirect evidence strongly suggests that it should be an important process. The variety of chemically dissimilar waters in nature is large, and the opportunities and possible combinations for mixing are numerous.

One obvious result of mixing is the precipitation of the components of relatively insoluble minerals, such as barite and the carbonates. Amphoteric substances, such as aluminum and iron hydroxides, will also be precipitated as a result of mixing. Less obvious are the possible diagenetic effects of the mixing of waters which differ only in salinity. Depending on the shape of the solubility curves of a mineral as a function of added salts, mixing can result in either supersaturation or undersaturation, with attendant possibilities for diagenesis. It can be predicted that mixing of waters in which the predominant dissolved salt is neutral and nonreactive will result in undersaturation with respect to the rock forming minerals. In contrast, if the dissolved salts have an ion in common with the rock-forming minerals, mixing will cause supersaturation at low salinities and, in some instances, undersaturation at higher salinities.

The hypothesis offered here must be tested by geologic observations. In addition, solubility curves are needed for the common rock-forming minerals in waters with added mixed salts, more nearly approximating natural surface and subsurface waters. Experiments should also be conducted to determine if kinetic factors allow the predicted changes to take place within geologically reasonable periods of time. It has been assumed in this discussion that after mixing the waters will approach solution equilibrium with the solid at a geologically meaningful rate.

ACKNOWLEDGEMENTS

My thanks to an unnamed reviewer for his constructive criticism. Gratitude is also expressed to F. J. Lucia, R. C. Murray, R. M. Lloyd, and R. J. Dunham for introducing me to the study of diagenesis when I was with Shell

Development Company. The first draft of this paper was completed during the tenure of a Faculty Summer Fellowship at the University of California at Santa Barbara. Miss Carol Boyce drafted the figures.

REFERENCES

ALEXANDER, G. B., HESTON, W. M., AND ILER, H. K., 1954, The solubility of amorphous silica in water: Jour. Phys. Chem., v. 58, p. 453–455.
ANDERSON, G. M., AND BURNHAM, C. W., 1967, Reactions of quartz and corundum with aqueous chloride and hydroxide solutions at high temperatures and pressures: Amer. Jour. Sci., v. 265, p. 12–27.
BACK, WILLIAM, 1960, Origin of hydrochemical facies of ground-water in the Atlantic Coastal Plain, in Internat. Geol. Congr., Geochemical Cycles: Intern. Geol. Congr., 21st, Copenhagen, Proc., pt. 1, p. 87–95.
———, 1966, Hydrochemical facies and groundwater flow patterns in northern part of Atlantic Coastal Plain: U.S. Geol. Survey, Prof. Paper 498-A, 42 p.
———, AND HANSHAW, B. B., 1965, Chemical geohydrology, p. 49–109 in Chow, V. T., ed., Advances in hydroscience. v. 2, Academic Press, New York, 288 p.
BAKER, E. G., 1967, A geochemical evaluation of petroleum migration and accumulation, p. 299–329 in Nagy, B., and Colombo, U., eds., Fundamental aspects of petroleum geochemistry. Elsevier Publ. Co., Amsterdam, 388 p.
BARANOV, K. A., 1956, The origination of carbonates and the chemical composition of natural brines in Kulundin Steppe: Izvest. Akad. Nauk S.S.S.R., Ser. Geol., no. 5, p. 92–98.
BJORKLUND, L. J., 1958, Flow analysis of Carlsbad Springs, Eddy County, New Mexico: U.S. Geol. Survey, Open File Report, 25 p.
BLACK, A. P., 1967, Electrokinetic characteristics of hydrous oxides of aluminum and iron, p. 274–300 in Faust, S. D. and Hunter, J. V., eds., Principles and applications of water chemistry. John Wiley and Sons, Inc., New York, 643 p.
BOCK, E., 1961, On the solubility of anhydrous calcium sulphate and of gypsum in concentrated solutions of sodium chloride at 25°C, 30°C, 40°C, and 50°C: Canad. Jour. Chem., v. 39, p. 1746–1751
BÖGLI, ALFRED, 1964, Mischungskorrosion—ein Beitrag zum Verkarstungsproblem: Erdkunde, Bd. 18, p. 83–92.
CARROLL, DOROTHY, 1962, Rainwater as a chemical agent of geologic processes—a review: U.S. Geol. Surv., Water Supply Paper 1535-G, 18 p.
COLUMBUS, NATHAN, 1965, Viscous model study of sea water intrusion in water table aquifers: Water Resources Res., v. 1, p. 318–323.
CORWIN, GILBERT, AND BRADLEY, EDWARD, Sea bottom tapped for fresh water: Undersea Technology, January, p. 59–60.
D'ANS, J., 1933, Die Lösungsgleichwicht der Systeme der Salze ozeanischer Salzlagerungen, p. 669 in Linke, W. F., and Seidell, Atherton, eds., 1958. Solubilities or inorganic and metal-organic compounds. D. van Nostrand, Inc., Princeton.
DESITTER, L. V., 1947, Diagenesis of oil-field brines: Amer. Assoc. Petrol. Geol. Bull., v. 31, p. 2030–2040.
DICKSON, F W., AND TUNELL, GEORGE, 1958, Equilibria of red HgS (cinnabar) and black HgS (metacinnabar) and their saturated solutions in the systems HgS-Na$_2$S-H$_2$O and HgS-Na$_2$S-Na$_2$O-H$_2$O from 25°C. to 75°C. at 1 atmosphere pressure: Amer. Jour. Sci., v. 256, p. 654–679.
EARDLEY, A. J., 1938, Sediments of Great Salt Lake: Amer. Assoc. Petrol. Geol. Bull., v. 22, p. 1305–1411.
ELLIS, A. J., 1963, The solubility of calcite in sodium chloride solutions at high temperatures: Amer. Jour. Sci., v. 261, p. 259–267.
FEELY, H. W., AND KULP, J. L., 1957, The origin of Gulf Coast salt dome sulfur deposits: Amer. Assoc. Petrol. Geol. Bull., v. 41, p. 1802–1853.
FOSTER, M. D., 1942, Base-exchange and sulphate reduction in salty ground waters along Atlantic and Gulf Coasts: Amer. Assoc. Petrol. Geol. Bull., v. 26, p. 838–851.
FREAR, G. L., AND JOHNSTON, JOHN, 1929, The solubility of calcium carbonate (calcite) in certain aqueous solutions at 25°: Amer. Chem. Soc. Jour., v. 51, p. 2082–2093.
FYFE, W. S., TURNER, F. J., AND VERHOOGEN, JEAN, 1958, Metamorphic reactions and metamorphic facies. Geol. Soc. Amer. Mem. 73, Waverly Press, Baltimore, 259 p.
GARRELS, R. M., 1941, The Mississippi Valley type lead zinc deposits and the problem of mineral zoning: Econ. Geol., v. 36, p. 729–744.
———, AND CHRIST, C. L., 1965, Solutions, minerals, an dequilibria. Harper and Row, New York, 450 p.
GATES, G. L., AND GARAWAY, W. H., 1965, Oil well scale formation in waterflood operations using ocean brines, Wilmington, California: U.S. Bur. Mines Rept. Invest. 6658, 28 p.
GRAF, D. L., MEENTS, W. F., FRIEDMAN, IRVING, AND SHIMP, N. F., 1966, The origin of saline formation waters, III: calcium chloride waters: Illinois State Geol. Survey, Circ. 397, 60 p.
HALL, W. E., AND FRIEDMAN, IRVING, 1963, Composition of fluid inclusions, Cave-in-Rock fluorite district, Illinois and upper Mississippi Valley zinc-lead district: Econ. Geol., v. 58, p. 886–911.
HANSHAW, B. B., 1962, Membrane properties of compacted clays: Ph.D. Thesis, Harvard University, 113 p.
———, BACK, WILLIAM, AND RUBIN, MEYER, 1965, Radiocarbon determinations for estimating groundwater flow velocities in Central Florida: Science, v. 148, p. 494–495.
HARRIS, W. H., AND MATTHEWS, R. K., 1968, Subaerial diagenesis of carbonate sediments: efficiency of the solution-reprecipitation process: Science, v. 160, p. 77–79.
HELGESON, H. C., 1964, Complexing and hydrothermal ore deposition. Pergamon Press, the Macmillan Co., N.Y., 128 p.
HENNINGSEN, E. R., 1962, Water diagenesis in Lower Cretaceous Trinity aquifers of Central Texas: Baylor Geol. Studies, Bull. 3, 37 p.

HILL, A. E., AND YANICK, N. S., 1935, Ternary systems: XX. Calcium sulfate, ammonium sulfate, and water: Amer. Chem. Soc. Jour., v. 57, p 645–651, in Linke and Seidell, eds., 1958; Solubilities of uninorganic and metal-organic compounds. D van Nostrand Inc., Princeton.

———, AND WILLS, J. H., 1938, Ternary systems: XXIV, Calcium sulfate, sodium sulfate and water: Amer. Chem. Soc. Jour., v. 60, p. 1647–1655.

HOLLAND, H. D., 1967, Gangue minerals in hydrothermal deposits, p. 382–436 in Barnes, H. L., ed., Geochemistry of hydrothermal ore deposits. Holt, Rinehart, and Winston, Inc., New York, 670 p.

HUBBERT, M. K., 1953, Entrapment of petroleum under hydrodynamic conditions: Amer. Assoc. Petrol. Geol. Bull., v. 37, p. 1954–2026.

IRELAN, BURDGE, AND MENDIETA, H. B., 1962, Chemical quality of surface waters in the Brazos River Basin, Texas: U.S. Geol. Survey, Prof. Paper 450-B, p. 129–130.

JACKSON, S. A., AND BEALES, F. W., 1967, An aspect of sedimentary basin evolution: the concentration of Mississippi Valley-type ores during late stages of diagenesis: Canad. Petrol. Geol. Bull., v. 15, p. 383–433.

KAMENSKY, G. N., 1958, Hydrochemical zoning in the distribution of underground water: Symposium of Ground Water Proc., Calcutta, 1955, Publ. 4, p. 281–292.

KOHOUT, F. A., 1960, Cyclic flow of salt water in the Biscayne aquifer of southeastern Florida: Jour. Geophys. Res., v. 65, p. 2133–2141.

———, 1965, A hypothesis concerning cyclic flow of salt water related to geothermal heating in the Floridan aquifer: N. Y. Acad. Sci. Trans., Sec. 2, v. 28, p. 249–271.

LEE, T. S., 1959, Chemical equilibrium and the thermodynamics of reactions, p. 185–275 in Kolthoff, I. M., Elving, P. J., and Sandell, E. B., eds., Treatise on analytical chemistry. The Interscience Encyclopedia, Inc., New York, 809 p.

LEONARD, A. R., AND WARD, P. E., 1962, Use of Na/Cl ratios to distinguish oil-field from salt-spring brines in western Oklahoma: U.S. Geol. Survey, Prof. Paper 450-B, p. 126.

LEVORSEN, A. I., 1967, Geology of petroleum. 2nd ed., revised by F. A. F. Berry. W, H. Freeman and Co., San Francisco, 724 p.

LINKE, W. F., AND SEIDELL, ATHERTON, 1958, Solubilities of inorganic and metal-organic compounds. v. 1, 5th ed., D. van Nostrand, Inc., Princeton, N.J., 1487 p.

MADGIN, W. M., AND SWALES, D. A., 1956, Solubilities in the system $CaSO_4$-$NaCl$-H_2O at 25°C and 35°: Jour. Appl. Chem., v. 6, p. 482–487.

MALININ, S. D., 1963, An experimental investigation of the solubility of calcite and witherite under hydrothermal conditions: Geochemistry (Transl.), no. 7, p. 650–667.

MARSHALL, W. L., AND SLUSHER, RUTH, 1968, Aqueous systems at high temperature. Solubility to 200°C of calcium sulfate and its hydrates in sea water and saline water concentrates, and temperature-concentration limits: Jour. Chem. Engin. Data, v. 13, p. 83–93.

———, AND JONES, E. V., 1964, Solubility and thermodynamic relationships for $CaSO_4$ in $NaCl$-H_2O solutions from 40° to 200°C 0 to 4 molal NaCl: Jour. Chem. Engin. Data, v. 9, p. 187–191.

MCNEAL, R. P., 1965, Hydrodynamics of the Permian Basin, p. 308–326 in Young, Addison, and Galley, J. E., eds., Fluids in subsurface environments. Amer. Assoc. Petrol. Geol., Mem. 4, Tulsa, 414 p.

MEINZER, O. E., 1923, Outline of ground-water hydrology, with definitions: U.S. Geol. Survey, Water Supply Paper 494, 71 p.

MORRIS, J. C., AND STUMM, WERNER, 1967, Redox equilibria and measurements of potentials in aquatic environments, in Equilibrium concepts in natural water systems, p. 270–285 in Gould, R. F., ed., Adv. in Chem. Ser. 67. Amer. Chem. Soc., 344 p.

MOTTS, W. S., 1968, The control of groundwater occurrence by lithofacies in the Guadalupian Reef Complex: Geol. Soc. America Bull., v. 79, p. 283–298.

POPOV, A. I., AND GOLDSHTEYN, R. I., 1967, Hydrologic zoning of hydrostatic systems as a mineralizing factor in the stratal cover of central Asia: Dokl. Akad. Nauk S.S.S.R., v. 176, Earth Science Section (Transl.), p. 118–120.

POWER, W. H., FABUSS, B. M., AND SATTERFIELD, C. N., 1966, Transient solute concentrations and phase changes of calcium sulfate in aqueous sodium chloride: Jour. Chem. Engin. Data, v. 11, p. 149–154.

PRAY, L. C., AND MURRAY, R. C., eds., 1965, Dolomitization and limestone diagenesis, a symposium. Soc. Econ. Paleontologists and Mineralogists, Spec. Publ. 13, Tulsa, 180 p.

RAYMAHASHAY, B. C., 1968, A geochemical study of rock alteration by hot springs in the Paint Pot Hill area, Yellowstone Park: Geochim. et Cosmochim. Acta, v. 32, p. 499–522.

ROBERTS, W. H. III, 1962, Unpublished data (quoted in Baker, 1967).

ROEDDER, EDWIN, 1968, Temperature, salinity, and origin of the ore-forming fluids at Pine Point, Northwest Territories, Canada, from fluid inclusion studies: Econ. Geol., v. 63, p. 439–450.

SAWKINS, F. J., 1966, Ore genesis in the North Pennine ore-field in light of fluid inclusion studies: Econ. Geol., v. 61, p. 385–401.

———, 1968, The significance of Na/K and Cl/SO_4 ratios in fluid inclusions and subsurface waters with respect to the genesis of Mississippi Valley-type ore deposits: Econ. Geol. v. 63, p. 935–942.

SEDEL'NIKOV, G. S., 1959, Hydrochemical relations of salt deposition in the Gulf of Kara-Bogaz: Freiberger Forschung., A-123, p. 166–174.

SEGNIT, E. R., HOLLAND,, H. D., AND BISCARDI, C. J., 1962, The solubility of calcite in aqueous solutions-I. The solubility of calcite in water between 75° and 200° at CO_2 pressure up to 60 atm: Geochim. et Cosmochim. Acta, v. 26, p. 1301–1331.

SHAWE, D. R., 1966, Zonal distribution of elements in some uranium-vanadium roll and tabular ore bodies on the Colorado Plateau: U.S. Geol. Survey, Prof. Paper 550-B p. 169–175.

SHTERNINI, E. B., AND FROLOVA, E. V., 1945, Compt. Rend. Sci. U.S.S.R., v. 47, p. 544 in Linke, W. F., and Seidell, Atherton, eds., 1958, Solubilities of inorganic and metal-organic compounds. D. van Nostrand, Inc., Princeton.

SVESHNIKOVA, V. N., 1949, Izvest. Sektova Fiz-Khim. Anal., Inst. Obshchei i Neorg. Khim., Akad. Nauk S.S.S.R., v. 17, p. 345–350; p. 699 *in* Linke, W. F., and Seidel, Atherton, eds., 1958, Solution of inorganic and metal-organic compounds. D. van Nostrand, Inc., Princeton.

SWARZENSKI, W. V., 1968, Fresh and saline ground-water zones in the Punjab region, West Pakistan: U.S. Geol. Survey Water Supply Paper 1608–I, 24 p.

TEMPLETON, C. C., AND RODGERS, J. C., 1967, Solubility of anhydrite in several aqueous salt solutions between 250°C and 325°C: Jour. Chem. Engin. Data, v. 12, p. 536–547.

THEOBALD, P. K., LAKIN, H. W., AND HAWKINS, D. B., 1963, The precipitation of aluminum, iron and manganese at the junction of Deer Creek with Snake Creek in Summit County, Colorado: Geochim. et Cosmochim. Acta, v. 27, p. 121–132.

THRAILKILL, JOHN, 1968, Chemical and hydrologic factors in the excavation of limestone caves: Geol. Soc. America Bull., v. 79, p. 19–46.

TSURIKOVA, A. P., AND TSURIKOV, V. L., 1966, Precipitation of calcium carbonate and changes in salinity during mixing of waters: Khim. Protessy Moryakh Okeanakh, Akad. Nauk S.S.S.R., p. 12–18; Chem. Abstrs., v. 67, 57187d, p. 5393.

TURNER, S. F., AND FOSTER, M. D., 1934, A study of salt-water encroachment in the Galveston area, Texas: Amer. Geophys. Union. Trans., v. 15, p. 432–435.

UPSON, J. E., 1966, Relationships of fresh and salty ground water in the northern Atlantic Coastal Plain of the United States: U.S. Geol. Survey, Prof. Paper 550–C, p. 235–243.

VERNON, R. O., 1969, The geology and hydrology associated with a zone of high permeability (Boulder Zone) in Florida: Soc. Mining Engineers, A.I.M.E., preprint 69–AG–12.

VON GUMBEL, G., 1878, Ueber die im Stillen Ozean auf dem Meersgrunde Vorkommenden Manganknollen: Sitz. Berichte d.k. Bayerischen Akad. d. Wissenschaften München, Matem.-Physik. Klasse, p. 189–209.

VOROSHILOV, E. A., DZHALILOV, T. I., AND SAMEDOV, N. K., 1968, Precipitation of salts during a change in the chemical composition of stratal waters during a reaction: Azerb. Neft. Khoz, v. 47, p. 28–30; Chem. Abstrs., v. 69, 30004 W, p. 2814.

WHITE, D. E., 1957a, Thermal waters of volcanic origin: Geol. Soc. America Bull., v. 68, p. 1637–1658.

———, 1957b, Magmatic, connate, and meetamorphic waters: Geol. Soc. America Bull., v. 68, p. 1659–1682.

———, 1965, Saline waters of sedimentary rocks, p. 342–366 in Young, Addison, and Galley, J. E., eds., Fluids in subsurface environments. Amer. Assoc. Petrol. Geol. Mem. 4, Tulsa, 414 p.

———, HEM, J. D., AND WARING, G. A., 1963, Chemical composition of subsurface waters, *in* Fleischer, Michael, ed., Data of geochemistry, Chapt. F, 6th ed. U.S. Geol. Survey, Prof. Paper 440, 67 p.

WHITEHEAD, H. C., AND FETH, J. H., 1961, Recent chemical analyses of waters from several closed-lake basins and their tributaries in the western United States: Geol. Soc. America Bull., v. 72, p. 1421–1426.

ZELENOV, K. K., ZOTOV, A. V., MAKSAREVA, T. S., AND FOKEEV, V. M., 1965, Characteristics of the neutralization of acid solutions by sea water: Izvest. Vysshikh Uchebenykh Favedenii, Geologia i. Razvedke, v. 8, p. 120–123. Transl. by Assoc. Tech. Services. Glen Ridge, N.J.

15

Copyright ©1976 by The Geological Society of America
Reprinted from *Geol. Soc. America Bull.* **87**:1301–1316 (1976)

Hydrogeochemistry of Bermuda: A case history of ground-water diagenesis of biocalcarenites

L. N. PLUMMER   *U.S. Geological Survey, National Center, Mail Stop 432, Reston, Virginia 22092*
H. L. VACHER   *Department of Geology, Washington State University, Pullman, Washington 99163*
F. T. MACKENZIE   *Department of Geological Sciences, Northwestern University, Evanston, Illinois 60201*
O. P. BRICKER   *Maryland Geological Survey, 33rd and Charles St., Baltimore, Maryland 21218*
L. S. LAND   *Department of Geological Sciences, University of Texas, Austin, Texas 78712*

ABSTRACT

Bermuda is composed of relatively young skeletal limestones currently undergoing diagenesis by the ground water passing through them. The saturated zone consists of separate fresh-water bodies laterally surrounded and underlain by extensive brackish aureoles, in which the meteoric water is mixed with sea water. The meteoric water enters the aquifer after passing through the soil or through marshes (outcrops of the ground-water bodies), in each case causing an influx of CO_2 to the saturated zone.

Examination of the ground-water chemistry enables mapping of (a) the extent of mixing of meteoric ground water and sea water; (b) P_{CO_2}; (c) the extent of saturation with calcite and aragonite; (d) concentration of Sr; and (e) the amount of calcium and magnesium derived from the limestones.

It is concluded that three processes control the chemistry of Bermudian ground water: (a) generation of elevated CO_2 partial pressures in soils and marshes; (b) dissolution of metastable carbonate minerals (principally aragonite); and (c) mixing with sea water. Bermuda ground water apparently approaches a steady state of aragonite dissolution (at slight subsaturation) and concurrent precipitation of calcite cement. Large Sr/Ca ratios in the ground water indicate that the dissolution of aragonite is incongruent. Dissolution is most pronounced near the marshes where CO_2 content is highest. Mixing with sea water is not significant in controlling calcite saturation.

Only small amounts of magnesium enter the ground water by incongruent dissolution of magnesium calcite, an apparently slow process on the time scale of passage of the ground water through the saturated zone. All of the waters are well undersaturated with respect to dolomite. It is estimated that the present rate of recrystallization of aragonite to calcite is about 0.32 cm^3 of aragonite to calcite per m^3 of the saturated zone per year. At the present rate of chemical weathering, 360 m^3 of the saturated zone is lost each year through solution and transported to the sea by ground water.

INTRODUCTION

Bermuda is a well-known group of limestone islands located 1,000 km east of Cape Hatteras. The limestones are skeletal calcarenites, principally eolian, and are of Pleistocene age. They are currently undergoing diagenesis by percolation of meteoric water through them.

Traditionally, the approach to understanding the diagenetic processes active in exposed limestones is essentially petrographic — that is, study of the mineralogy and cement fabric of variously altered Pleistocene limestones with due consideration of *their* chemical and isotopic information. Largely on account of accessibility, the petrographic effort has been concentrated on the vadose zone. Work of this nature has had a long history in Bermuda (Friedman, 1964; Gross, 1964; Land and others, 1967; Ristvet, 1971; Lafon and Vacher, 1975) and has been reviewed by Bathurst (1971, Chap. 8).

From petrographic analysis of middle Pleistocene biosparites in Bermuda, Land (1970) concluded that phreatic meteoric diagenesis is a more rapid process than vadose-meteoric diagenesis. This conclusion has been upheld by study of Pleistocene limestones of other islands (Jamaica by Land; Barbados by Matthews and co-workers), so that now the view is that, insofar as meteoric diagenesis is concerned, "the phreatic zone, not the vadose zone, is of vast importance in the diagenesis of carbonate rocks" (Land, 1973, p. 412).

In this paper, we examine in some detail the hydrochemistry of the Bermudian phreatic zone in order to clarify the chemical processes that produce the observed and documented phreatic diagenesis. That is, in studying the Bermudian ground water — its partial pressure of CO_2, the concentration of Sr, the amount of Ca and Mg over that due to mixing with sea water, and the extent of saturation with calcite and aragonite — we ask the question: "What is this water doing, now, to the rocks through which it is passing?"

HYDROGEOLOGIC FRAMEWORK

Geology

Limestones. Bermuda is a limestone-capped volcanic pinnacle. The exposed portion of the cap is a complex mosaic of Pleistocene strandline- and shoreline-dune calcarenites and interlacing paleosols. The limestones, deposited during interglaciations, are composed of cemented, sand-sized fragments of organisms that lived offshore at the time the exposed beach and dune limestones were deposited. Details of the mosaic and interpretation of the depositional history and oscillating sea levels that produced the mosaic are given by Bretz (1960), Land and others (1967), and Vacher (1971, 1973).

Three rock-stratigraphic units make up the exposed limestones (Fig. 1). The three formations are disjunct time-stratigraphically. Each formation is bracketed by paleosols which also occur within the formations. Each formation, by volume, is predominantly of eolian facies, but each also contains one or more marine tongues. In order of increasing age, the formations are Paget Formation (Vacher, 1973), Belmont Formation (Land and others, 1967), and Walsingham Formation (Land and others, 1967).

The three limestones differ lithologically, largely on account of the varying extent of their diagenetic alteration. Land and others (1967) illustrate the progressive change (with time) of a number of lithologic features, including mineralogy, cementation, and porosity. Ristvet (1971) gives a detailed account of the stratigraphic variation of mineralogy (compare with Vacher, 1973, Fig. 5). Figure 2, compiled from the data of

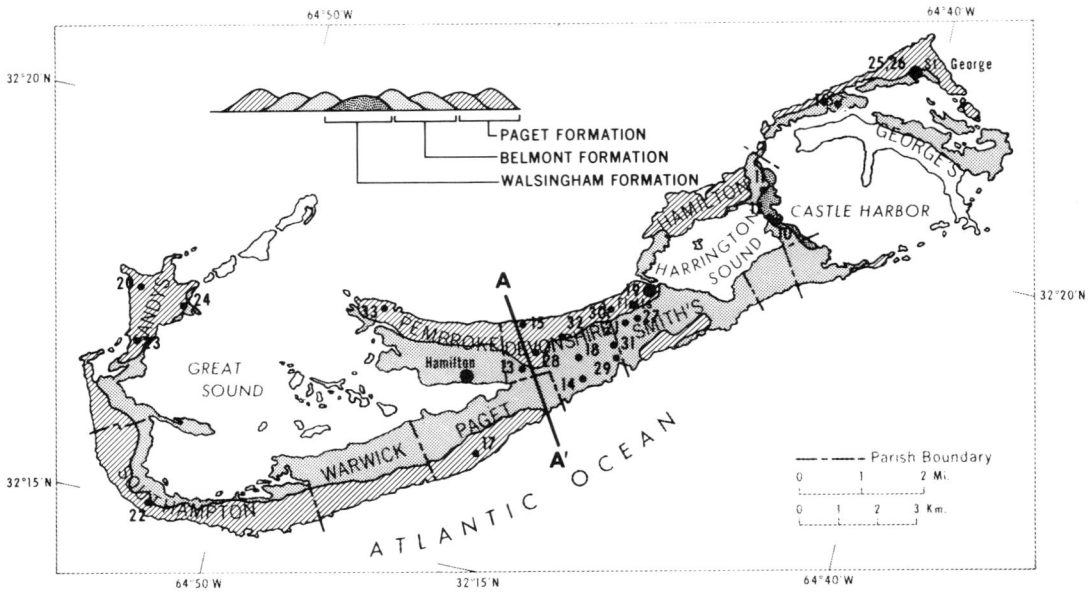

Figure 1. Geologic map of Bermuda, showing the distribution of the three major formations (from Vacher, 1974) and the location of water samples keyed to Tables 2 and 3.

Ristvet (1971), compares the mineralogy of the lithologic units as a function of geologic age. Recent sediments from beaches contain about 35 percent aragonite by weight and about 50 percent Mg calcite by weight. The mode mole percent $MgCO_3$ in the magnesian calcites of recent sediments is about 14 (Ristvet, 1971). With increasing geologic age, the aragonite content of the Paget Formation remains similar to that of recent sediments and the amount of magnesian calcite decreases, forming low-magnesian calcites. Although the total amount of magnesian calcite in the Paget Formation decreases with age, the mode mole percent $MgCO_3$ in the magnesian calcites remains relatively constant: 14.6 mole percent $MgCO_3$ for the upper Paget member (Southhampton Formation of Land and others, 1967) and 12.7 mole percent $MgCO_3$ in the lower Paget member. The Belmont Formation contains little or no Mg calcite and approximately 12 percent aragonite by weight. The Walsingham Formation is composed essentially of calcite. All the samples collected by Ristvet are from the present vadose zone; the samples from the Paget Formation have probably had little contact with the phreatic zone, but many of the samples from the Belmont and Walsingham Formations have had past contact with the saturated zone. Lafon and Vacher (1975) discuss the implications of Ristvet's stratigraphically controlled data to the rate of vadose alteration in the mineralogic composition of a pile of sand consisting originally of Mg calcite and aragonite.

On account of the varying extent of diagenesis of the three formations, the pore-space characteristics also differ from formation to formation (Vacher, 1974). The Paget is essentially a pile of well-sorted calcarenitic sand; the permeability is related principally to intergranular porosity. The Belmont, in which nearly all of the original Mg calcite has been converted to calcite (Ristvet, 1971), has considerable moldic porosity (Land and others, 1967), and, more importantly, it contains many pencil-sized solution channels in the present-day saturated zone (observable in excavations below water table). The Walsingham is cavernous, and hence ground-water flow is essentially unrestricted. As shown below, this variation in permeability exerts a strong influence on the distribution of fresh and brackish ground water that now occupies the limestone cap — and hence, the present-day diagenetic environment within the saturated zone.

Geomorphology. The topography of Bermuda reflects the origin and subsequent diagenetic history of the limestones. They formed as shoreline dunes and were later modified by chemical erosion (Bretz, 1960; Mackenzie, 1964a, 1964b; Land and others, 1967; Vacher, 1973). Of fundamental importance is the parallelism of rock and morphostratigraphic units (Vacher, 1973). The implication is that the geochem-

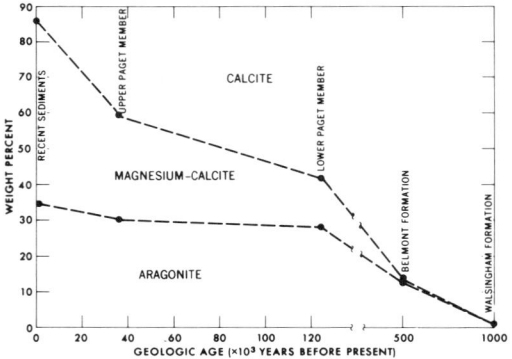

Figure 2. Relative weight percent of calcite, Mg calcite, and aragonite in Bermuda limestones as a function of geologic age (compiled from the data of Ristvet, 1971).

ical processes attending the passage of water through the Pleistocene calcarenites affects the evolution of both the limestone lithology and the surface landscape simultaneously.

In Younger Bermuda (Sayles, 1931), the morphostratigraphic equivalent to the Paget Formation (Vacher, 1973), the topography is largely depositional. The eolianite ridges of the Paget consist of individual dune-shaped hills representing large accretionary mounds that merged laterally as they grew landward from the source beaches.

In Older Bermuda, which includes the Belmont and Walsingham Formations, "the subdued, rolling topography is in striking contrast to the dune topography of the coasts" (Sayles, 1931, p. 445; see also Vacher, 1973, Fig. 3). Although sedimentary-structural patterns in the Belmont are like those in the Paget, the patterns are less concordant with topography than in the Paget. A striking departure of the present topography from that of the initial, depositional topography occurs in the topographic lows formed by onlap of Paget dunes onto the windward slopes of the Belmont eolianites. In these areas, the initial lows have been deepened by chemical erosion. The windward-topset cross-beds of the Belmont eolianites are breached, erosion having penetrated into the underlying foresets; as a result, the Paget-Belmont soil now occurs some distance up the slope toward the crest of the Paget eolianite. In some areas, the implied deepening of the initial topographic lows amounts to as much as 15 m (Vacher, unpub. data).

The Walsingham is characterized by cave-related karst described by Bretz (1960). Apparently, much of the karst formed on the Walsingham prior to deposition of the Belmont. In some areas, the Belmont eolianites onlap a karst topography, and locally Belmont deposits breach caves in the Walsingham.

In summary, the Bermudian landscape is a solution-modified dune topography; cave-related karst seems principally inherited from early, pre-Belmont events.

Soils. The importance of Bermuda soils in the diagenetic alteration of the eolianites is discussed by Plummer (1970). His calculations show that meteoric water passing through sandy soils may dissolve considerable amounts of calcium carbonate before entering the vadose zone. Depending upon rates of CO_2 production, calcite dissolution, and various hydrologic factors, meteoric water passing through sandy soils may be either slightly subsaturated or supersaturated with calcite as it enters the underlying eolianites (Plummer, 1970). Because the metastability of magnesian calcites increases with increasing mole percent $MgCO_3$ in solid solution (Plummer and Mackenzie, 1974), soil waters close to saturation with calcite alter only the most metastable phases in the mineral assemblaged first and have lesser corrosive effects on the relatively more stable aragonite and low-magnesian calcites as they percolate through the vadose zone. Meteoric water passing through nonsandy red soils and peat marshes are subsaturated with calcite and have considerable capacity to dissolve calcium carbonate. Thus the composition of meteoric water reaching the phreatic zone depends, in part, on the type of soil through which it passed and on the mineralogy of the eolianites along the vadose zone pathways to the saturated zone.

Many diagenetic features are observed at the bases of modern and Pleistocene soils. These include (a) soil base, a 1- to 3-cm-thick, well-cemented layer of fine-grained limestone at the soil-rock contact; (b) stringers of well-cemented calcarenite extending down foreset bedding planes from the lower contact of the soil; (c) breccia consisting of 2- to 18-cm-long pieces of eolianite beds; (d) scattered black phosphate nodules; (e) chalky limestone characterized by moldic porosity and a nodular weathering surface occurring as lenses at the soil base or as wedges penetrating cracks in the underlying eolianite; (f) soil pipes (in the paleosols), carrot-like structures, up to 1.5 m in diameter, which extend a maximum of 3 m into the underlying rocks; and (g) shells of land snails, commonly whole in the modern and accretionary soils, but highly comminuted in terra-rossa paleosols. In addition, soils commonly contain rhizoconcretions. In the lower part of some outcrops of the sub-Belmont soil, stringers of euhedral calcite crystals occur. A 1-m-thick zone beneath the soil is characteristically different from the underlying rocks; in some areas, this zone is tightly cemented and contains abundant calcite-crystal veins; in other areas, the zone is almost friable owing to leaching by percolating meteoric waters. Thus, the base of the soil (that is, the lower portion of the soil itself, the soil-rock contact, and a 1-m zone below the soil) shows the effects of both dissolution and precipitation of calcium carbonate.

TABLE 1. BERMUDIAN CLIMATIC DATA

Annual	J	F	M	A	M	J	J	A	S	O	N	D	
1,463	113.3	115.3	117.3	100.6	116.8	106.4	111.8	137.2	133.9	161.6	130.0	118.6	Rainfall (mm)
360.7	39.7	37.8	36.2	27.6	29.0	22.3	18.8	24.0	24.8	36.9	30.7	32.9	Hours of rain
46	54	55	48	37	35	37	42	48	45	49	50	51	No. of rain days (>0.01 in.) (%)
2.1	2.3	2.4	2.6	2.5	2.8	2.0	1.4	1.6	1.7	2.2	2.2	2.0	Duration of rain on rain days (hr)
4.06	2.85	3.05	3.24	3.64	4.03	4.77	5.94	5.72	5.41	4.38	4.24	3.61	Intensity* (mm/hr)
	4.4	4.3	5.7	7.0	8.1	7.6	7.6	6.5	6.6	6.0	4.9	4.6	Maximum dry periods (<0.01) (days)
	17.5	19.9	18.3	15.8	13.4	11.9	10.8	10.3	12.1	14.4	15.9	17.1	Average wind speed (mph)
43	61	75	66	50	31	22	15	18	31	39	53	61	Days with mean 10-min. wind >25 mph (%)
10	21	27	19	9	2	1	0	2	5	5	11	11	Days with gale (%)
21.1	17.2	16.7	17.2	18.3	21.1	24.9	26.1	26.7	25.6	23.3	20.6	18.4	Temperature (°C)
58	49	49	53	58	59	60	70	67	64	55	55	49	Hours of sunshine (% of possible)
14.4	11.3	10.2	10.8	12.3	15.0	17.9	20.0	20.0	18.8	16.2	13.4	11.6	Moisture content of air (g/m³)
	68.6	82.6	108.0	124.5	137.2	143.5	163.8	157.5	130.8	99.2	80.0	69.9	Potential evapotranspiration† (mm)

Note: Data in monthly averages.
Sources: Macky (1957).
* Calculated by monthly rain/hours of rain.
† Calculated by Penman from Macky's data.

General Hydrology

Climate. About 100 yr of meteorologic data have been compiled in the Bermuda area (Macky, 1957). These data are summarized in Table 1.

Long-term averages of monthly precipitation and temperature show that the climate is rather constant through the year. On average, the rainfall is 12 cm/month; the range of monthly averages is 6 cm. The yearly average of temperature is 21°C and the range of monthly averages, 10°C.

There is marked seasonality, however, in duration and frequency of rainfall, percentage of possible hours of sunlight, wind speed, and number of days with high or gale-force winds. Winter months (October through March) have relatively frequent rains of low intensity and long duration, with correspondingly short dry periods between rainfalls; in winter, wind speed and number of gale days are high, and the number of sunshine hours and humidity are low. Summer months, on the other hand, have frequent rains of greater intensity and shorter duration, wind speed and number of gale days are low, and humidity and percentage of sunshine are high.

The balance between rainfall and potential evapotranspiration also varies seasonally. Potential evapotranspiration (PE, the amount of water which would be used by plants if sufficient water were continually available) has been calculated from Macky's meteorological compilations by Penman (for the Bermuda Dept. Agriculture and Fisheries; I. W. Hughes, 1975, personal commun.) using the technique of Penman (1963). Penman's monthly PE figures are included in Table 1 and shown in comparison with rainfall (R) in Figure 3. The seasonality of the R-PE excess is apparent in the figure.

Water Budget. The annual excess of rainfall over potential evapotranspiration is on the order of 10 cm (Table 1). Thus, in areas where water is continually available for plant consumption (that is, the marshes), there is a net water surplus of about 10 cm per yr.

In nonmarshy areas, the annual actual evapotranspiration (AE) is less than the annual PE, because PE exceeds R during the summer months. That is, the plants at times do not get the water they need, and, therefore, changes in soil moisture must be taken into account in order to estimate the amount of water passing downward from the soil (Penman, 1963). A first-order estimate of the annual soil-water excess has been made (Vacher, 1974), using a Thornthwaite-type water-balance inventory (Thornthwaite and Mather, 1955; see also Strahler and Strahler, 1973). The calculation is an accounting technique for the amount of water in the soil: the accounting takes rainfall as a monthly income, PE as a monthly expenditure, and makes the assumption that soil-moisture content can neither exceed field capacity (at which point the excess over field capacity drains downward under gravity) nor be less than wilting point (at which point the plants cannot withdraw water from the soil to satisfy their consumptive needs). The difference between field capacity and wilting point — the "plant available water" — is taken as 7.5 cm.

The resulting budget is summarized in Figure 4. The results suggest that recharge through soils (that is, the soil-water excess) is about 18 cm/yr and occurs seasonally, from early November through March. From the end of the recharge season to late June, the soil moisture decreases from field capacity to wilting point, and from September to the beginning of the recharge season in November, the soil moisture builds up again to field capacity.

The results of Figure 4 can be considered only rough approximations because of the assumptions used. However, as discussed by Vacher (1974), the estimated annual recharge through soils is consistent with other hydrologic evidence such as size and geometry of the fresh ground-water lenses, permeability of the Paget and Belmont limestones as estimated from dampening of tidal oscillations, and regional steady-state drawdown by large-scale extraction.

Surface Water. There is no surface runoff in Bermuda — no streams or sheet wash. Water does not puddle on natural surfaces, even after intense summer or prolonged winter rainfalls. The infiltration capacity of the soil and underlying permeable calcarenites seems, always, to exceed rainfall rates.

Inland surface waters are limited to lakes, ponds, and marshes. These bodies are essentially outcrops of the ground-water lenses, and they occur where inter-eolianite topographic lows extend below regional ground-water levels. Chemistry and tidal oscillations in the surface water bodies are similar to the surrounding ground water.

Recharge and Geochemical Interpretation. Water arrives at the saturated zone of the limestones by two main routes, each with its own geochemical heritage:

(a) That which enters through the marshy "outcroppings" of ground water essentially bypasses the vadose zone of the limestones. The net annual contribution from this source is given above as about 10 cm/yr. It is likely that the quantity of water that enters the saturated zones of the limestones from the peat-filled bedrock depression below the marshes is more than the (net) annual sum of monthly R-PE and is closer to the gross sum of 25 cm, over the seven months that R exceeds PE. During the other five months, the marsh acts as an evapotranspiration-driven pump that pulls the ground water out of the limestones to make up the R-PE deficit. The fate of rainwater falling on the marsh is even more complicated than that surmised from the monthly averages, as even in the summer months, heavy rainfalls cause a mounding of the water table that dissipates, in part due to spreading into the adjoining limestones.

(b) That which percolates downward from the soil zone during periods of soil-water excess (Fig. 4) passes through the aerated zone of the limestones. Some percolates as vadose seepage (Thrailkill, 1968), along wetting fronts as illustrated in laboratory experiments by Smith (1967). The rest passes through the aerated calcarenites, more rapidly as vadose flow (Thrailkill, 1968), along cracks which are most common in Belmont and Walsingham Forma-

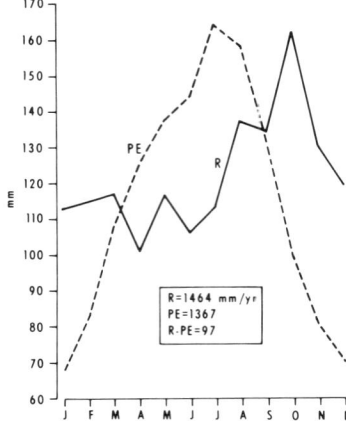

Figure 3. Graphs showing the run of rainfall (R) and potential evapotranspiration (PE), as calculated by Penman. Monthly averages are plotted. Yearly totals are in the box.

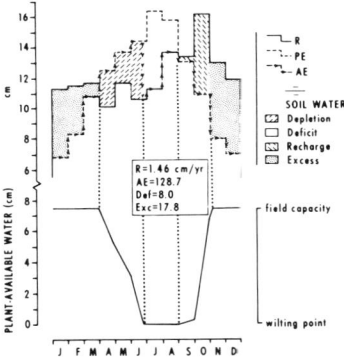

Figure 4. Thornthwaite-type soil-water budget, showing soil-moisture conditions as they change through the year. The budget is calculated from R and PE values of Figure 3 and an assumed 7.5-cm excess of field capacity over wilting point. The results of the model suggest that the yearly soil-water excess is about 18 cm; this amount percolates downward through the aerated zone to recharge the lens.

tions. Infiltration via the soil zone is about 20 times that occurring through the marshes (Vacher, 1974).

To a first approximation, the recharge is uniform geographically. This complicates the concept of residence time. It is impossible to say, for example, that ground water sampled at a particular locality has been in the saturated zone for an explicitly stated period of time, as the sample represents a mixture of the waters that have arrived between the point of collection and the water-table divide. This problem is inherent in geochemical interpretation of nearly any water-table aquifer and is similar, in principle, to the interpretation of residence time of water in a water catchment drained by streams. A parameter that can be calculated and does have physical meaning pertains to the time required to flow from one particular locality to another, along a flow line. Of particular interest is the time for flow between the area of the water-table divide to the point of discharge at the shoreline, as this time represents the maximum time that any water is in the Bermudian saturated zone. The longest residence times are for water that flows southward from near the water-table divide of the eastern portion of Devonshire lens (described below). Preliminary calculations (Vacher, unpub. data) indicate that this time is about 300 yr.

GROUND WATER

Distribution of Fresh and Brackish Ground Water

The distribution of fresh and brackish ground water in Bermuda has been mapped by Vacher (1974). This mapping has documented the island-wide, geographic variation of salinity and, in the area of the Devonshire lens (the most extensive body of fresh ground water), the spatial variation of salinity.

The mapping is based on determination of conductivity (proportional to total dissolved solids) on two kinds of samples: (1) Water from pumped household wells distributed throughout the island. The household wells are normally drilled 10 ft, or less, into the saturated zone and are pumped, probably considerably less than 400 gpd, for flushing water, gardening use, and filling swimming pools. (2) Water from government observation wells drilled through the fresh-water column and into the underlying brackish zone. The samples were obtained by use of a thief sampler. More than 300 household wells were sampled in order to determine the geographic distribution of fresh and brackish ground water. Seventeen deep wells comprise the Devonshire lens observation-well network, from which the three-dimensional variation is determined.

The results are presented here in terms of percentage sea water, or relative salinity, defined by Bear and Todd (1960) to be

$$\epsilon = \frac{C - C_f}{C_s - C_f}$$

where C is the measured conductivity (of a mixture of fresh and salty water), C_f is the conductivity of the unmixed fresh water, and C_s is the conductivity of undiluted sea water. Use of this conversion rests on the premise that the TDS of the brackish water is determined primarily by the process of mixing "fresh" water and sea water, a conclusion that is demonstrated by the chemical data of this paper and supported by the overall structure of the transition zone that underlies the Devonshire lens (Fig. 5). C_s was determined from sampled sea water. Selection of a value for C_f followed interpretation of conductivity profiles measured in the deep observation wells. The profiles consistently showed an upper layer of constant-conductivity water and then at greater depths a rapid rise with increasing depth. C_f was selected as the conductivity of this upper layer, which was approximately 1 percent that of undiluted sea water. The same value of C_f was used for all conversions, the implicit assumption being that variations in the TDS of the fresh water are overwhelmed by the addition of the very much saltier sea water, even in relatively small amounts.

Use of relative salinity greatly facilitates treatment of the transition zone. Carrier (1959) and Bear and Todd (1960) have shown from dispersion models of the transition zone between the fresh-water lens of an island and underlying sea water that the increase in relative salinity with depth can be described by an error function; that is, if Z is depth, then

$$\epsilon\,(Z) = 1/2\left[1 - erf\frac{Z - \bar{Z}}{2\,\sigma}\right]$$

where

$$\bar{Z} = (Z)_{\epsilon\,=\,50\%}$$

and

$$\sigma = 1/2\left[(Z)_{\epsilon\,=\,84.1\%} - (Z)_{\epsilon\,=\,15.9\%}\right].$$

Thus the set of values of relative salinity determined at various depths in a borehole can be plotted on probability paper, the regression line drawn, and the value of Z corresponding to any relative salinity of interest interpolated easily. Of special interest is the depth corresponding to a relative salinity of 50 percent, as this datum corresponds to the theoretical interface (Todd and Meyer, 1971) treated by the Ghyben-Herzberg relation (Todd, 1959).

Figure 6 shows two relative-salinity contours (1 and 25 percent) determined from the island-wide sampling. Five fresh-water bodies are defined, the largest of which is the Devonshire lens, which covers an area about twice the combined total of the others. The fresh-water bodies can be visualized as nuclei of the unmixed fresh water within a highly mixed, brackish-water lens. The fresh-water areas represent only about 20 percent of Bermuda.

Figure 7 shows the three-dimensional geometry of the island lens in the central parishes. Figure 7A shows the depth below water table of the iso-surface of 50 percent relative salinity; this map indicates the thickness of the Ghyben-Herzberg lens, bounded by the water table and the fresh-water–sea-water interface, that would occur in Bermuda, if there were no mixing about the interface. Figure 7B shows the depth below water table of the iso-surface of 1 percent relative salinity; this figure indicates the thickness of the fresh-water nu-

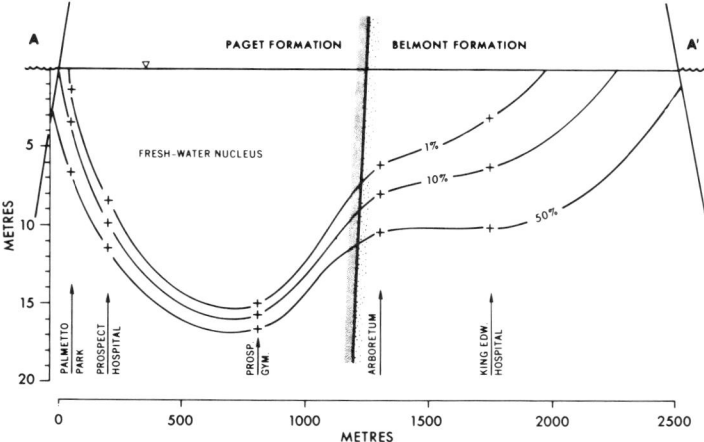

Figure 5. Hydrogeologic cross section through Devonshire lens (located on Figs. 1 and 6) showing iso-lines of relative salinity, ϵ (from Vacher, 1974). The section shows the effect of the distribution of the two rock units (Paget and Belmont) on the thickness of the fresh-water nucleus ($\epsilon<1$ percent) and the transition zone (1 percent $<\epsilon<50$ percent). The Paget-Belmont contact dips northward at about 15°; vertical exaggeration is 64.

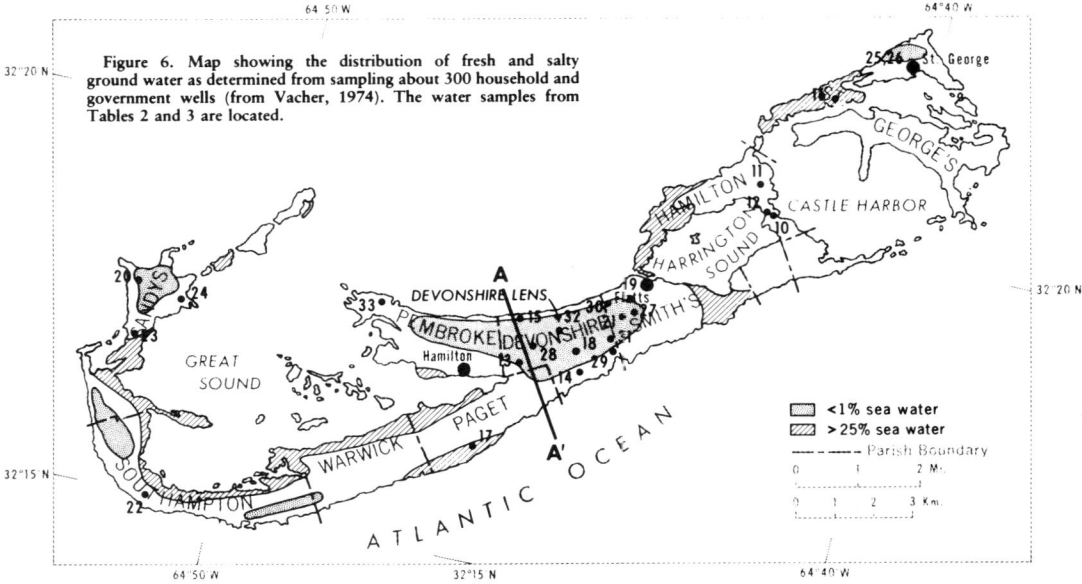

Figure 6. Map showing the distribution of fresh and salty ground water as determined from sampling about 300 household and government wells (from Vacher, 1974). The water samples from Tables 2 and 3 are located.

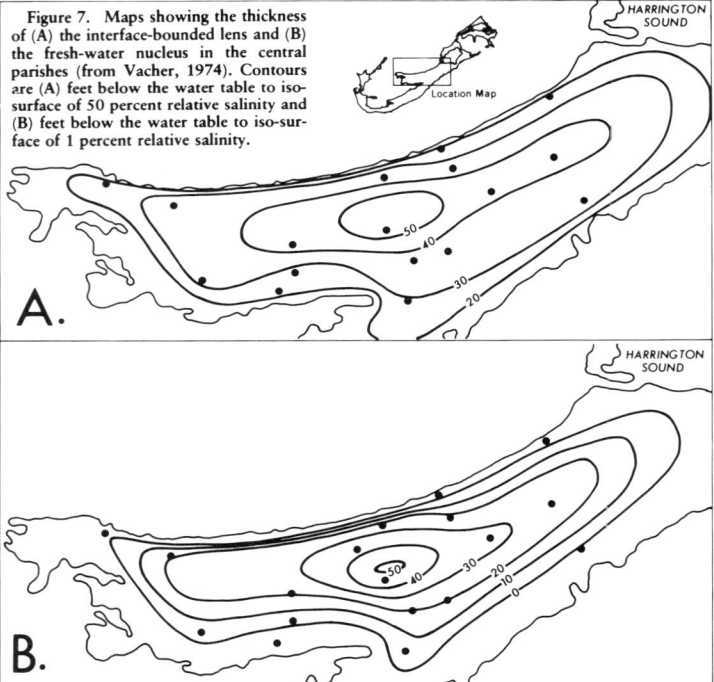

Figure 7. Maps showing the thickness of (A) the interface-bounded lens and (B) the fresh-water nucleus in the central parishes (from Vacher, 1974). Contours are (A) feet below the water table to iso-surface of 50 percent relative salinity and (B) feet below the water table to iso-surface of 1 percent relative salinity.

cleus contained within the interface-bounded lens. Figures 7A and 7B provide complete three-dimensional determination of the salinity variation in the ground water of the central parishes. For at any locality, the Z (ϵ = 1 percent) and Z (ϵ = 50 percent) can be used to calculate σ of the error function and hence the depth below water table of any other relative salinity of interest. In this way, the maps of Figure 7A and 7B, which show depth below water table of two selected iso-surfaces of relative salinity, can be converted to maps showing the position of any other iso-surface of relative salinity.

Geologic Control of Fresh-Water Distribution

As shown in Figure 6, the geographic variation of relative salinity is not axisymmetric as would be the case if the island were homogeneous, that is, composed of a single rock unit. Comparison of Figures 1 and 6 shows that the location of the fresh-water bodies is influenced by the occurrence of a broad band of Paget Formation.

The nature of the geologic control is displayed visually by hydrogeologic cross sections of the Devonshire lens. The cross section shown in Figure 5 slices through the Devonshire lens where the thickness of fresh water is the greatest on the island (Fort Prospect). The three iso-lines of relative salinity (1, 10, and 50 percent) are plotted directly from transition-zone profiles determined at the five observation wells indicated on the figure. The Paget-Belmont contact is shown; its location is known from surface mapping, and it was encountered (sub-Paget soil) in drilling the Arboretum well.

Figure 5 shows that the distribution of geologic units affects both the interface-bounded lens and the transition zone.

The interface-bounded lens attains its maximum thickness in Paget rocks. Southward, as it crosses the Paget-Belmont contact, the midline of the transition zone rises abruptly and levels off to a lesser depth in Belmont rocks.

The transition zone thins inland from each shoreline. However, for a given distance from the shoreline, the transition zone is thinner in Paget rocks than in Belmont rocks such that, as the contact is crossed, the iso-lines of relative salinity markedly diverge southward.

Thus the location of the bulk of the fresh ground water is related to the occurrence of the Paget Formation, because the interface-bounded lens tends to swell in that unit, and the thickness of the transition zone, in general, tends to be less there. This thickening and thinning of the interface-bounded lens and the transition zone are the hydraulic result of relatively low permeability Paget rocks along the north shore and more highly permeable Belmont rocks along the south shore (Vacher, 1974).

WATER CHEMISTRY

Methods

Water from 20 wells throughout Bermuda was sampled between fall 1969 and late summer 1970. Some additional samples were collected in the summer of 1971. The wells sampled are located on Figures 1 and 6 where the index numbers correspond to those of Tables 2 and 3. In addition to 54 samples of ground water, the data (Table 2) include analyses of rainwater, soil pore water, and vadose drip water from caves.

Various techniques were used (ultracentrifuge, squeezing) to obtain soil pore waters. Most attempts yielded about 3-ml samples on which analyses for calcium, magnesium, and alkalinity were made. The pH was taken *in situ*, except for sample number 5 (Tables 2, 3), which was measured on the squeezed water. More complete analyses of soil pore waters (2 and 3) were made from 15- to 20-ml samples squeezed from red soils collected during a heavy rain. All pore waters are from red soils containing varying amounts of organic matter and calcite. Most of these soils contained about 25 percent pore water by weight when collected. Equilibria calculations for the more complete soil pore-water analyses (2 and 3) were made using the program WATEQ (see below), and for the remaining partial soil pore-water analyses, a constant ionic strength, 0.01, similar to analysis 2, was assumed. The equilibria for these waters were then computed using constant activity coefficients of 0.81 for Ca^{++} and Mg^{++}, 0.90 for HCO_3^-, and 0.68 for CO_3^{--}, and ignoring ion pair formation. Most of the ground water was sampled from domestic wells that penetrate 10 ft, or less, into the saturated zone. Ground-water samples were taken only after continuous pumping resulted in a steady (lowest) temperature.

Calcium and magnesium were determined by EDTA titration; alkalinity by potentiometric titration with .01N HCl; chloride by titration with 0.01N $AgNO_3$, using an Orion Chloride electrode as the end point indicator; and potassium and strontium by flame photometry. Temperature and pH were recorded in the field, the latter using a Photovolt meter on expanded scale and combination electrode.

Sulfate was estimated from the chloride content using the average molar ratio of SO_4^{--}/Cl^- in surface sea water of 0.052 (Goldberg, 1957), and sodium was estimated by charge balance. The computed charge balance sodium to analytical chloride ratio in the average Bermuda ground water is 0.84, not unlike the Na^+/Cl^- ratio in sea water (.85). Sodium determined in the Watlington water (28) by flame photometry differs from the charge balance sodium by only 2.5 percent. The estimated sulfate and sodium concentrations were used only for the purpose of estimating ionic strength and are not reported in Table 2.

Calculation of ion activities in solution was made on an IBM 360/65 using a FORTRAN IV version of the aqueous model, WATEQ, of Truesdell and Jones (1973, 1974). Reference thermodynamic data at 25°C were computed to field temperatures by means of the van't Hoff equation, and in some cases, from analytical expressions derived by Truesdell and Jones (1973, 1974) from experimental data in the literature (Table 4). The distribution of aqueous species was solved by a successive approximation routine involving the simultaneous solution of a set of mass action and mass balance equations (Garrels and Thompson, 1962; Garrels and Christ, 1965) that represent the aqueous model. Individual ion activity coefficients of charged species were computed from the extended Debye-Hückel equations of Truesdell and Jones (1973, 1974), and activity coefficients of neutral species were calculated from the relation $\gamma_i^0 = 10^{0.11}$ where I is the ionic strength of the solution. Twenty-seven aqueous species of calcium, magnesium, sodium, potassium, chlorine, sulfur, and carbon were considered in the calculations (Table 4).

The saturation indices for aragonite, calcite, and dolomite (Table 3) were computed from the calculated activities of Ca^{++}, Mg^{++}, and CO_3^{--} in the waters. Saturation indices (Langmuir, 1971) are defined as the log of the ratio of the computed ion activity product of the mineral to the appropriate thermodynamic equilibrium constant, SI = log IAP/K. Saturation indices less than zero indicate subsaturation, and values greater than zero correspond to supersaturation. The partial pressure of CO_2 (Table 3) in the waters was computed from the carbonate equilibria. Considering the uncertainties in the analyses and equilibrium calculations, the maximum uncertainty in the saturation indices is about ±0.1 SI units.

Results

Rainwater. Bermuda rainwater was sampled in order to account for extra-island contributions to the water chemistry. Our data indicate that Bermuda rainwater is near equilibrium with atmospheric CO_2 pressures, and it contains approximately 0.07 percent sea water. If the sea-water-derived salts in fresh ground water (~1.0 percent) were derived from rainwater, an evapotranspiration rate of 93 percent is indicated. This result is only slightly higher than that suggested by Figure 4 (88 percent).

Soil Pore Water. The soil pore water sampled is characterized, in general, by high CO_2 pressures, nearly two orders of magnitude greater than atmospheric, and apparent supersaturation with respect to calcite and dolomite. The two samples collected during a heavy rain (samples 2 and 3) have lower CO_2 pressures and are subsaturated with respect to calcite and dolomite. Sample 3 was taken from a low depression near sea level on the grounds of the Biological Station where no fresh-water lens is known. The pore water in this soil has a high sea-water content.

The pH of the pore water is largely a function of the calcite content of the soils. Sample 8 has an unusually low pH, 5.26, and contains no detectable calcite. The presence of bicarbonate and the very low content of calcium and magnesium in sample 8 indicate that silicate reactions may also be important in the soils. Unfortunately, no analyses for silica have been made.

Cave Water. The cave water, collected from the roofs of caves in the Walsingham and Belmont Formations (Frantz, 1971), are examples of vadose water percolating through open fractures. In general, the CO_2 pressures are an order of magnitude lower than both the soil pore water and the ground water, and the dissolved calcium carbonate is about 50 percent of that found in soil pore water and ground water. The decreased CO_2 and calcium carbonate content of the cave water indicates that along the vadose pathways sampled, secondary calcite is cementing open fractures as the CO_2 escapes.

TABLE 2. ANALYTICAL DATA

Sample no.	Well name	Date	Temperature (°C)*	pH	Calcium (Ca)	Magnesium (Mg)	Bicarbonate (HCO$_3$)	Chloride (Cl)	Potassium (K)	Strontium (Sr)	Type of water†
1		8- -71	..	6.22	0.0	0.0	4.9	14.2	0.0	0	1
2		7-15-71	..	7.54	62.5	25.3	148.9	265.9	..	0	2
3		7-15-71	..	7.35	90.2	212.8	341.7	5779	..	1.4	2
4		7- -68	..	7.24	60.9	51.1	431.3	..	..	..	2
5		7- -68	..	7.22	107.0	0.0	213.5	..	..	..	2
6		7- -68	..	7.17	140.3	31.2	225.7	..	..	..	2
7		7- -68	..	6.88	344.7	131.3	183.0	..	..	..	2
8		7- -68	..	5.26	0.0	0.0	122.0	..	..	..	2
9		7- -68	..	6.61	52.1	48.6	237.9	..	..	..	2
10	Gov't. Quarry	8- 4-70	..	7.86	50.1	8.8	87.6	111.0	..	..	3
11	Crystal Cave	8- 4-70	..	7.80	50.9	10.7	155.0	98.7	..	..	3
12	Lemmington Cave	8- 4-70	..	7.77	60.5	8.3	147.5	77.1	3.0	..	3
13a	Arboretum	9- 5-69	24.2	7.26	96.7	8.5	235.7	220.0	6.6	6.0	4
13b	do.	10-17-69	24.5	7.24	88.7	11.1	235.7	223.5	6.8	6.2	4
13c	do.	11-26-69	22.5	7.44	83.5	13.0	232.6	220.0	6.3	5.2	4
13d	do.	5-12-70	21.0	7.16	82.1	16.4	232.0	202.2	5.9	6.6	4
13e	do.	8- 4-70	27.0	7.35	79.8	10.2	239.3	156.1	5.0	..	4
14a	Ashley Hall	9- 5-69	24.7	7.39	107.0	35.9	264.5	425.9	11.8	2.35	4
14b	do.	10-17-69	23.5	7.29	117.8	27.8	252.3	472.1	12.3	2.47	4
14c	do.	11-26-69	19.5	7.21	105.4	35.9	250.5	504.1	13.1	1.12	4
14d	do.	5-12-70	22.0	7.16	114.5	47.9	253.5	504.1	13.6	2.10	4
15a	Breakers	9- 5-69	27.0	7.21	83.5	15.3	239.3	159.6	5.9	5.8	4
15b	do.	5-12-70	22.0	7.25	81.0	14.8	222.2	163.2	3.7	4.3	4
15c	do.	8- 4-70	26.0	7.07	79.0	13.4	233.2	113.5	3.5	..	4
16a	Cepheid	10-28-69	22.0	7.51	377.7	1182	211.4	17132	342.3	..	4
16b	do.	11-25-69	22.0	7.32	381.1	1177	202.6	17054	341.0	..	4
16c	do.	1-20-70	21.0	7.55	383.6	1154	200.0	16979	339.4	..	4
17a	Coral Beach	9- 5-69	25.3	7.26	205.3	310.5	281.3	4649	105.9	5.7	4
17b	do.	11-26-69	18.5	7.35	219.1	326.0	289.1	4677	118.7	4.8	4
17c	do.	5-12-70	21.5	7.16	225.3	325.1	279.5	4696	116.0	5.8	4
17d	do.	8- 4-70	27.0	7.21	221.6	341.5	289.4	4758	121.1	..	4
18	Correias	5-12-70	22.5	6.94	95.9	29.3	279.8	437.8	10.6	1.5	4
19a	De Couto	9- 5-69	22.8	7.06	86.6	8.9	303.4	127.7	6.6	3.7	4
19b	do.	10-17-69	22.5	7.10	78.9	5.9	305.3	156.1	6.9	2.9	4
19c	do.	5-12-70	22.5	7.03	103.3	25.4	324.3	255.5	13.6	3.6	4
19d	do.	8- 4-70	22.5	7.08	107.5	17.8	323.6	149.0	13.8	..	4
20a	Faries	9- 5-69	27.8	7.28	116.3	26.0	264.5	330.0	22.8	6.6	4
20b	do.	10-17-69	24.5	7.33	115.1	24.7	262.6	333.6	23.5	5.1	4
20c	do.	11-26-69	18.5	7.33	115.4	22.7	262.0	319.4	23.3	5.6	4
20d	do.	8- 4-70	25.5	7.30	107.5	32.8	265.0	280.3	25.0	..	4
21a	Hinson's Hall	9- 5-69	23.8	7.04	107.1	14.7	260.7	156.1	7.2	..	4
21b	do.	10-17-69	23.5	7.10	109.1	4.3	260.1	220.0	7.3	2.7	4
21c	do.	11-26-69	23.0	7.04	76.6	9.2	251.6	173.8	5.3	2.5	4
21d	do.	5-12-70	23.0	7.12	93.9	16.0	265.0	202.2	7.2	3.1	4
21e	do.	8- 4-70	23.5	7.14	96.3	12.9	265.6	156.1	8.2	..	4
22a	Olivera's	9- 5-69	33.0	7.04	184.7	263.5	591.6	3895	123.4	5.2	4
22b	do.	10-17-69	25.5	7.07	186.1	227.6	553.5	3966	126.0	4.2	4
22c	do.	11-26-69	22.5	6.89	186.7	280.2	536.8	3945	131.0	4.3	4
23a	Packwood	9- 5-69	28.5	7.50	17.0	3.0	47.6	78.0	.6	.0	4
23b	do.	10-17-69	25.0	7.70	19.1	1.9	54.3	99.3	1.2	1.9	4
23c	do.	11-26-69	20.5	7.74	21.5	1.8	58.6	78.0	.8	1.8	4
23d	do.	8- 4-70	25.2	7.26	16.0	1.1	50.0	9.6	.8	..	4
24	Simmons	8- 4-70	26.0	7.05	146.4	12.9	417.7	113.5	3.5	..	4
25	St. Georges A	10-24-69	23.0	7.17	84.3	28.9	323.2	468.6	11.6	6.7	4
26	St. Georges B	10-24-69	22.0	7.28	112.2	55.3	278.0	869.9	23.3	5.2	4
27a	Terceira's	9- 5-69	23.0	7.06	138.0	18.5	329.8	287.4	16.8	..	4
27b	do.	10-17-69	24.0	7.45	97.1	14.8	238.2	397.5	15.0	2.3	4
27c	do.	11-26-69	17.0	7.89	71.1	36.6	151.5	390.3	15.3	2.7	4
27d	do.	8- 4-70	22.5	6.92	128.0	17.8	339.0	280.4	17.5	2.5	4
28	Watlington	7-31-71	..	7.19	129.7	48.7	294.6	752.9	15.0	5.3	4§
29	Devonshire well	1- -71	..	7.42	104.5	51.7	201.6	790.9	15.8	6.3	4
30	Penhurst	1- -71	..	7.32	64.9	13.3	316.9	131.6	2.6	3.7	4
31a	St. Brendan's	12- -70	..	7.32	54.9	19.6	243.0	182.0	3.6	3.4	4
31b	do.	12- -70	..	7.52	65.7	18.5	226.5	177.7	3.6	3.5	4
31c	do.	12- -70	..	7.58	79.8	19.0	162.4	222.8	4.5	3.9	4
32	Devonshire Marsh	7-13-71	..	6.90	55.1	8.3	114.8	150.7	3.0	1.1	4
33	Admiralty House	2- -71	..	7.42	38.1	8.2	204.5	130.2	2.6	3.5	4

Note: Analytical data in ppm.
* If no value, 25°C has been assumed in equilibria calculations.
† Type of water: 1. rain water, 2. soil pore water, 3. cave water, 4. ground water.
§ Sea-water contamination; sampled during prolonged pumping.

TABLE 3. PERCENT SEA WATER, IONIC STRENGTH, LOG P_{CO_2}, SATURATION, AND SOURCE-ROCK CALCULATIONS

Sample no.	Well name	Percent sea water*	Ionic strength†	Log§ P_{CO_2}	Saturation indices #			Source rock** (mmoles/litre)	
					Aragonite	Calcite	Dolomite	Calcium	Magnesium
1		0.07	..	−3.3	..	..	..	..	..
2		1.3	0.013	−2.36	−0.17	−0.06	−0.12	1.43	0.31
3		27.4	.200	−1.89	−.27	−.16	.48	..	..
4		..	.01	−1.63	−.04	.07	.55	..	..
5		..	.01	−1.84	.13	.24	..	..	..
6		..	.01	−1.75	.15	.26	.29	..	..
7		..	.01	−1.58	.13	.24	.48	..	..
8		..	.01	−.14	..	..	..	..	..
9		..	.01	−1.20	−.85	−.74	−1.09	..	..
10	Gov't. Quarry	.53	.007	−2.90	−.10	.00	−.36	1.20	.06
11	Crystal Cave	.47	.007	−2.60	.08	.19	.09	1.22	.17
12	Lemmington Cave	.37	.007	−2.58	.11	.22	−.03	1.47	.13
13a	Arboretum	1.05	.013	−1.89	−.07	.04	−.59	2.30	−.26
13b	do.	1.06	.013	−1.86	−.12	−.01	−.54	2.10	−.16
13c	do.	1.05	.013	−2.08	.02	.12	−.18	1.97	−.08
13d	do.	.96	.013	−1.81	−.29	−.18	−.70	1.95	.12
13e	do.	.74	.011	−1.95	.01	.12	−.26	1.91	−.01
14a	Ashley Hall	2.02	.022	−1.98	.10	.21	.34	2.45	.30
14b	do.	2.24	.023	−1.90	.00	.11	−.03	2.70	−.16
14c	do.	2.39	.024	−1.85	−.20	−.10	−.29	2.37	.09
14d	do.	2.39	.024	−1.78	−.17	−.07	−.13	2.60	.58
15a	Breakers	.76	.011	−1.81	−.11	.00	−.35	2.00	.19
15b	do.	.77	.011	−1.91	−.19	−.09	−.54	1.94	.16
15c	do.	.54	.010	−1.68	−.29	−.18	−.74	1.91	.39
16a	Cepheid	78.8	.583	−2.34	.06	.16	1.26	1.41	2.44
16b	do.	78.4	.579	−2.17	−.14	−.03	.86	1.65	2.50
16c	do.	78.4	.576	−2.41	.07	.17	1.26	1.84	1.01
17a	Coral Beach	21.85	.166	−1.89	−.05	.05	.71	2.74	−.08
17b	do.	21.85	.256	−2.01	−.03	.07	.71	3.05	.48
17c	do.	22.07	.169	−1.81	−.17	−.07	.43	3.21	.42
17d	do.	22.35	.172	−1.82	−.04	.07	.75	3.09	.89
18	Correias	2.07	.021	−1.51	−.40	−.30	−.72	2.17	.00
19a	De Couto	.61	.011	−1.58	−.21	−.10	−.81	2.10	.01
19b	do.	.74	.012	−1.62	−.21	−.11	−.97	1.89	−.19
19c	do.	1.21	.017	−1.53	−.18	−.07	−.38	2.45	.34
19d	do.	.71	.013	−1.58	−.09	.02	−.37	2.61	.32
20a	Faries	1.56	.018	−1.84	.10	.21	.17	2.73	.16
20b	do.	1.58	.018	−1.91	.09	.20	.12	2.70	.10
20c	do.	1.51	.018	−1.95	.00	.10	−.14	2.72	.06
20d	do.	1.33	.017	−1.87	.06	.17	.21	2.54	.58
21a	Hinson's Hall	.74	.012	−1.62	−.19	−.08	−.65	2.59	.18
21b	do.	1.04	.014	−1.69	−.14	−.04	−1.10	2.61	−.44
21c	do.	.82	.012	−1.64	−.36	−.27	−1.05	1.82	−.10
21d	do.	.96	.013	−1.70	−.19	−.08	−.55	2.24	.10
21e	do.	.74	.012	−1.72	−.13	−.03	−.55	2.32	.01
22a	Olivera's	18.32	.144	−1.29	.13	.25	1.08	2.61	.08
22b	do.	18.66	.147	−1.40	.04	.14	.87	2.61	.45
22c	do.	18.55	.146	−1.25	−.20	−.09	.40	2.64	.62
23a	Packwood	.37	.004	−2.79	−1.14	−1.04	−2.43	.39	−.09
23b	do.	.47	.005	−2.94	−.86	−.75	−2.11	.43	−.20
23c	do.	.37	.004	−2.98	−.80	−.70	−2.11	.50	−.15
23d	do.	.05	.002	−2.52	−1.34	−1.23	−3.22	.40	.02
24	Simmons	.54	.014	−1.42	.18	.29	−.10	3.59	.22
25	St. Georges A	2.22	.023	−1.68	−.17	−.06	−.20	1.86	−.11
26	St. Georges B	4.12	.027	−1.84	−.12	−.02	.05	2.36	−.13
27a	Terceira's	1.36	.018	−1.55	−.02	.09	−.32	3.30	−.04
27b	do.	1.88	.019	−2.08	.08	.19	−.06	2.22	−.46
27c	do.	1.85	.018	−2.77	.07	.18	.42	1.58	.43
27d	do.	1.33	.018	−1.40	−.19	−.08	−.64	3.05	−.05
28	Watlington	3.56	.033	−1.74	−.02	.09	.14	2.85	−.06
29	Dovenshire Well	3.74	.024	−2.13	−.05	.06	.21	2.21	−.06
30	Penhurst	.62	.011	−1.81	−.02	.09	−.13	1.55	.19
31a	St. Brendan's	.86	.012	−1.93	−.21	−.11	−.27	1.28	.30
31b	do.	.84	.012	−2.16	.03	.14	.12	1.55	.27
31c	do.	1.06	.012	−2.36	.03	.13	.03	1.88	.17
32	Dovenshire Marsh	.71	.008	−1.76	−.78	−.67	−1.72	1.30	−.08
33	Admiralty House	.62	.007	−2.09	−.30	−.19	−.67	.89	−.03

* Computed from the chloride content relative to the chloride content of Ferry Reach (Bermuda) sea water (Berner, 1966).
† Computed in program WATEQ using estimated sulfate and sodium values.
§ Computed from carbonate equilibria.
# Saturation index (SI) = log IAP/K. For reference, values of log K used (at 25°C) are −8.305, −8.412, and −17.00, respectively, for aragonite, calcite, and dolomite.
** Computed by subtraction of expected calcium and magnesium contents by dilutions of Ferry Reach (Bermuda) sea water (Berner, 1966) from analyzed values. Expected dilution values were computed from analyzed chloride using the molar ratios from Ferry Reach: Ca^{++}/Cl^- = .018, Mg^{++}/Cl^- = .098.

TABLE 4. THERMOCHEMICAL DATA

Reaction	$\Delta H_r^{\circ *}$	Log $K_{(25)}$†
$CaOH^+ = Ca^{++} + OH^-$	−1.19	−1.40
$CaSO_4^\circ = Ca^{++} + SO_4^{--}$	−1.65	−2.309
$CaCO_3^\circ = Ca^{++} + CO_3^{--}$	−3.13	−3.20
$MgOH^+ = Mg^{++} + OH^-$	−2.14	−2.60
$MgSO_4^\circ = Mg^{++} + SO_4^{--}$	−4.92	−2.238
$MgHCO_3^+ = Mg^{++} + HCO_3^-$	−10.37	−0.928
$MgCO_3^\circ = Mg^{++} + CO_3^{--}$	−0.058	−3.398
$NaSO_4^- = Na^+ + SO_4^{--}$	−2.229	−0.226
$Na_2SO_4^\circ = 2Na^+ + SO_4^{--}$	2.642	−1.512
$NaHCO_3^\circ = Na^+ + HCO_3^-$	..§	0.250
$NaCO_3^- = Na^+ + CO_3^{--}$	−8.911	−1.268
$Na_2CO_3^\circ = 2Na^+ + CO_3^{--}$	..§	−0.672
$NaCl^\circ = Na^+ + Cl^-$	..§	1.602
$KCl^\circ = K^+ + Cl^-$	..§	1.585
$H_2O = H^+ + OH^-$	13.345	−13.998
$CaMg(CO_3)_2$ (dolomite) $= Ca^{++} + Mg^{++} + CO_3^{--}$	−8.29	−17.00
$CaCO_3$ (calcite) $= Ca^{++} + CO_3^{--}$	Log $K(T)$# = 13.870 − 0.04035T − 3059./T	
$H_2CO_3^\circ = H^+ + HCO_3^-$	Log $K(T)$** = 14.8435 − 0.032786T − 3404.71/T	
$HCO_3^- = H^+ + CO_3^{--}$	Log $K(T)$†† = 6.498 − 0.02379T − 2902.39/T	
$KSO_4^- = K^+ + SO_4^{--}$	Log $K(T)$ = −3.106 + 673.6/T	
$HSO_4^- = H^+ + SO_4^{--}$	Log $K(T)$ = 5.3505 − 0.0183412T − 557.2461/T	
$CO_2 + H_2O = H_2CO_3^\circ$	Log $m_{H_2CO_3^\circ}$ = Log P_{CO_2} − 14.0184 + 0.015264T + 2385.73/T − I[0.84344 − 0.004471T + 0.00000666T²]§§	

Note: The thermodynamic data, except where indicated, have been adopted from the recent compilation of Truesdell and Jones (1973) where reference to the original data sources is given. The $CaHCO_3^+$ complex has been ignored for reasons discussed by Plummer and Mackenzie (1974).
* Standard enthalpy of reaction (Kcal/mole).
† Log K for the reaction at 25°C.
§ No value of ΔH_r° is known. Log $K_{(25)}$ has been used at all temperatures considered.
# Jacobson and Langmuir (1974).
** Harned and Davis (1943).
†† Harned and Scholes (1941).
§§ Analytical expression for log $m_{H_2CO_3^\circ}$ as a function of P_{CO_2}, temperature, T°K, and ionic strength of the solution, I (Truesdell and Jones, 1973). This relation has been used in computing calcite solubility as a function of temperature and CO_2 partial pressure.

Ground Water. Bermuda ground water is a mixture of calcium bicarbonate water and a sea-water component. The "sea water" content in the ground water sampled (Table 3) ranges from 0.61 percent (19a) to 78.8 percent (16a), estimated from the chloride content relative to the chloride content of Ferry Reach (Bermuda) sea water (Berner, 1966). While "sea-water" contents of approximately 1 percent can be accounted for by rainfall-evaportranspiration processes, "sea-water" contents greater than 1 percent are attributed to subsurface mixing processes. The molar Mg^{++}/Cl^- ratio in the average ground water, 0.105, is slightly greater than that ratio in the average Ferry Reach sea water of Berner (1966), 0.098.

In order to characterize the hydrochemical processes that determine the composition of Bermuda ground water, we now examine in some detail the dissolved rock contribution of calcium, magnesium, and strontium to the ground water, the spatial extent of ground-water composition within the Devonshire lens, and the factors that control the saturation state of calcite in the ground water.

Source Rock Calculation

Sea Water Subtraction. The rock contribution of calcium and magnesium to the ground-water chemistry has been estimated by subtraction of sea water from the ground water. Using the chloride content of the ground water and the molar Ca^{++}/Cl^- and Mg^{++}/Cl^- ratios (0.018 and 0.098) of Ferry Reach sea water (Berner, 1966), the calcium and magnesium content derived from sea water was estimated and subtracted from the analysis. This calculation assumes that all sources of calcium and magnesium, other than rock sources (for example, rainfall, sea spray, sea-water mixing), have molar Ca^{++}/Cl^- and Mg^{++}/Cl^- ratios similar to that of Ferry Reach sea water. The residual is an estimate of the contribution of calcium and magnesium to the ground water from the rock. The results indicate that the average rock contribution to the ground water (excluding 16, 17, and 22) is dominated by calcium (and bicarbonate); 2.06 mmoles/l Ca^{++} and 0.05 mmoles/l magnesium. The charge balances reasonably well against the average bicarbonate content of 3.95 mmoles/l, indicating that most of the water chemistry can be accounted for by calcium-carbonate dissolution and addition of "sea water."

Diagenetic Magnesium. A plot of magnesium versus chloride (Fig. 8) for each low-salinity ground water shows a trend parallel and near values expected by dilution of Ferry Reach sea water, which again demonstrates that most of the magnesium in the ground water is derived from sea water.

However, many of the waters do contain significant rock contributions of magnesium. Waters from wells that pass through and (or) into the Paget Formation are distinguished from wells in the Belmont Formation (Fig. 8). With few exceptions, the magnesium content of waters from Paget wells is greater than that of sea-water dilution. Waters collected from the roofs of caves in the Walsingham Formation are also enriched in magnesium, because the vadose zone overlying the caves is in part Paget Formation. This "rock-derived" magnesium in the water can be attributed largely to dissolution of Mg-calcite skeletal fragments in the vadose and phreatic zone of the Paget Formation. Waters from wells in the Belmont Formation fall on both sides of the sea-water dilution curve (Fig. 8); the scatter reflects, in part, the analytical error for magnesium. Although we cannot rule out the possibility of local enrichments of Mg calcites in the Belmont Formation, it is suggested that waters from the Belmont Formation containing "rock-derived" magnesium may have pathways, at least in part, through the Paget Formation. The relatively low content of "rock-derived" magnesium in Bermuda ground water reflects the amount of dissolution of Mg-calcite skeletal grains along vadose-zone pathways and in the phreatic zone, the kinetics of dissolution of magnesian calcite, and the rate of flushing of the aquifer.

Without detailed sampling of waters percolating through the unsaturated zone, it is not possible to determine the relative amounts of vadose-zone versus phreatic-zone diagenesis; only the net result is reflected in the ground-water chemistry. We can, however, use kinetic arguments to examine the limiting case of alteration of magnesian calcites in the phreatic zone alone, a case that may well apply to reactions occurring in parts of the Devonshire lens along the north shore where the ground water is in the Paget Formation (Fig. 5).

We consider the dissolution of 14 mole percent Mg calcite in the saturated zone as a function of time. Since the ground water is near equilibrium with calcite, the Mg-calcite grains presumably dissolve incongruently to calcite. The rate of release of magnesium during incongruent dissolution of Mg-calcite grains is parabolic (Plummer and Mackenzie, 1974). The destruction of Mg calcite can, therefore, be described by an integrated form of the parabolic rate law

$$C = kt^{\frac{1}{2}}$$

where C is moles of $Ca_x Mg_{(1-x)} CO_3$ destroyed per cm^2 of surface exposed, k is a constant expressed in moles of $Ca_x Mg_{(1-x)} CO_3$ destroyed per cm^2 per $sec^{\frac{1}{2}}$, and t is time in seconds. Preliminary work on the rates of incongruent dissolution of Mg calcites as a function of P_{CO_2} (Plummer, 1974) indicates that for the incongruent dissolution of 14 mole percent Mg calcite at $10^{-1.9}$ atm CO_2, k is near 10^{-9} moles destroyed/cm^2 $sec^{\frac{1}{2}}$. Estimates of the age of the phreatic water in the Devonshire lens from hydrologic consideration (Vacher, unpub. data) range up to 300 yr, depending upon lateral position within the aquifer. Taking 100 yr as the average age of ground water in the Devonshire lens, and 0.6 mmoles/litre (Fig. 8) as the maximum rock contribution of magnesium over that period of time, we can estimate average surface areas and amounts of Mg calcite in the saturated zone required to account for the observed magnesium content of the water. After 100 yr, the rock contribution of magnesium to the saturated zone would be 7.9×10^{-3} mmoles/litre of ground water per cm^2 of 14 mole percent Mg calcite in the saturated zone. In other words, an average surface area of 76 cm^2 of 14 mole percent Mg calcite in contact with one litre of ground water is required to account for rock contribution of 0.6 mmoles/litre of magnesium in 100 yr.

This estimate is not unreasonable. Taking account of grain size, porosity, and the extent of cementation of the saturated-zone calcarenites, 76 cm^2 surface area is consistent with bulk rock analyses that indicate several percent magnesium calcite by weight. Thus, in general, the results suggest that there is little rock-derived magnesium in the Bermudian ground waters, because residence time of the ground water is relatively short, the rate of release of magnesium by incongruent dissolution of Mg-calcite grains is slow, and the abundance of magnesian calcites in the saturated zone is low.

Diagenetic Strontium. All of the waters analyzed contain large amounts of residual strontium (Table 2). The average Bermuda ground water contains 5.4 ppm strontium, which is 40 times greater than the strontium in a 1.3 percent sea-water dilution. The significantly large amounts of strontium indicate that incongruent dissolution of aragonite is a major diagenetic process currently affecting the Bermudian saturated zone. Similar results have been documented

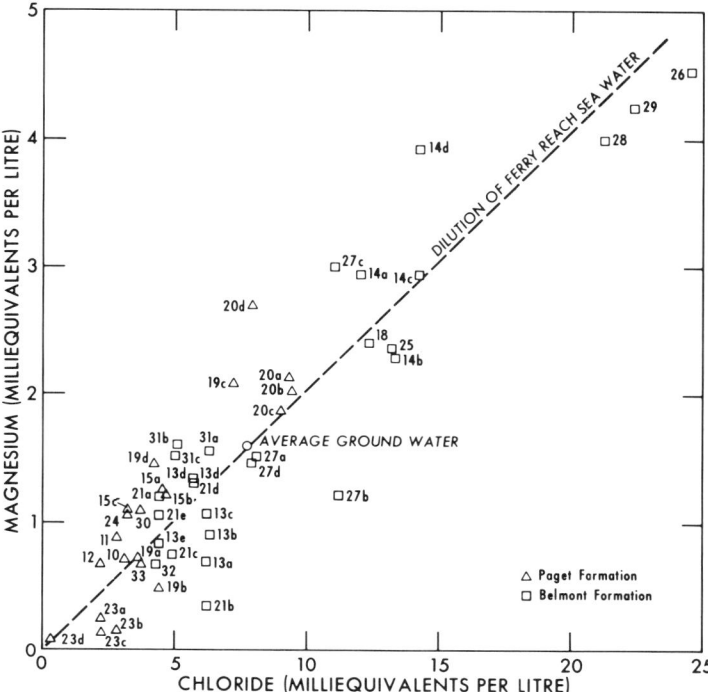

Figure 8. Plot of magnesium versus chloride content for each low-salinity well compared with the dilution curve for Ferry Reach (Bermuda) sea water. Most phreatic water from the younger Paget Formation contains small amounts of "rock-derived" magnesium.

by Harris and Matthews (1968) for the solution-precipitation process on Barbados.

The ratio $m_{Sr^{++}}/m_{Ca^{++}}$, 0.03, in the average Bermuda ground water is approximately three times that common in skeletal aragonite grains (Kinsman, 1969). This result suggests that about three times the amount of dissolved $CaCO_3$ in the ground water is dissolved as aragonite skeletal grains and precipitated as calcite cement during an average residence time interval in the saturated zone.

It is difficult to assign a rate for the recrystallization of aragonite to calcite without knowing how much of the strontium in the ground water is derived from alteration of aragonite in the unsaturated zone. We can, however, obtain a maximum estimate of the present rate of recrystallization in the saturated zone by assuming that all of the strontium in the ground water has been derived from diagenesis of the saturated zone. Taking 20 percent as the average porosity of the saturated zone, 100 yr as an estimate of the average residence time of fresh water in the saturated zone, and 2.1 mmoles/litre as the average amount of rock-derived calcium in the ground water, the Sr/Ca ratio in the ground water indicates that the present rate of recrystallization of aragonite to calcite is about 0.32 cm^3 of aragonite to calcite per m^3 of the saturated zone per year.

In order to examine this rate for the recrystallization of aragonite to calcite further, we have assumed that the recrystallization process involves two steps: aragonite dissolution and calcite precipitation. Depending on which step is rate limiting, two models are possible.

If aragonite dissolution is rate limiting, the ground water should be subsaturated with aragonite. To a first approximation, the rate of dissolution of aragonite (and subsequent precipitation of calcite) is proportional to the amount of aragonite (A) present at time t, that is,

$$\frac{dA}{dt} = kA$$

where k is the first-order rate constant for the process. Taking the average amount of aragonite in the saturated zone today to be about 10 percent, each m^3 of the saturated zone (20 percent porosity) contains about 80,000 cm^3 of aragonite, which indicates that k is about 4×10^{-6} yr^{-1}. The amount

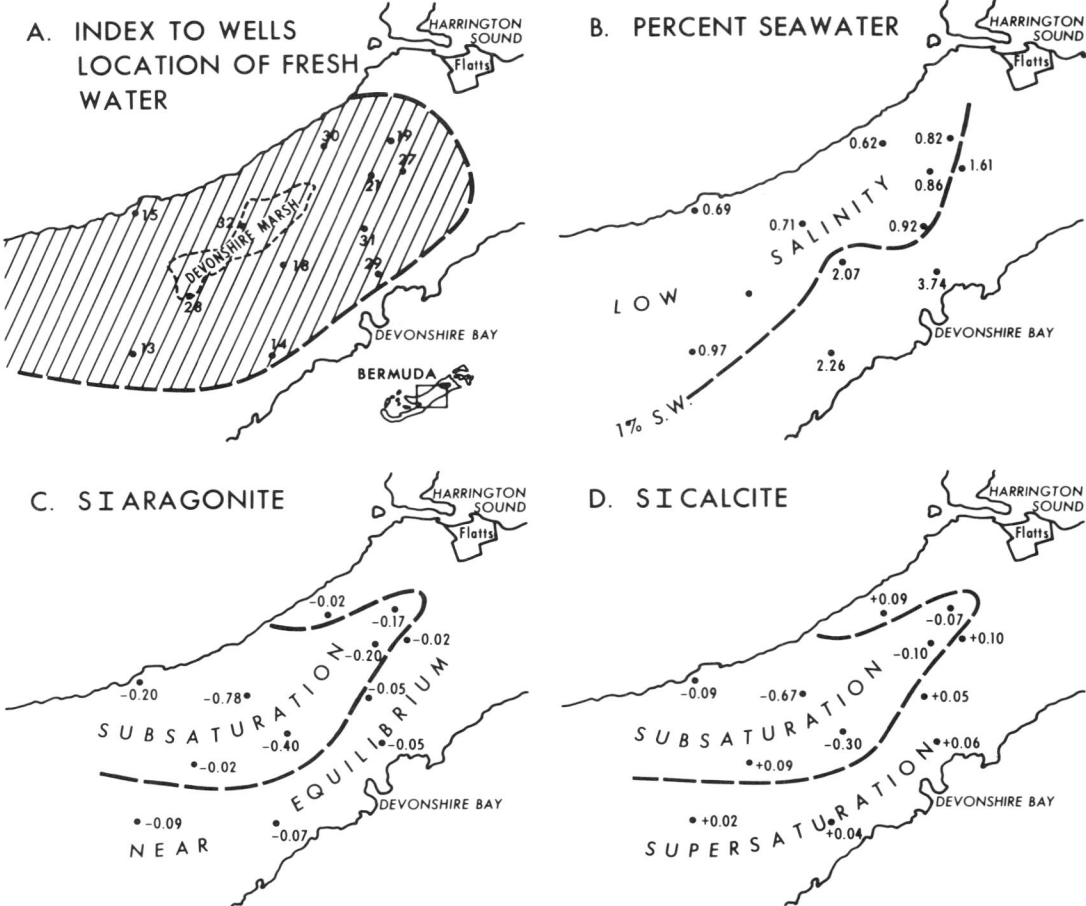

Figure 9. Geochemistry of the Devonshire lens. A. Index to wells and location of fresh water. B. Percentage of sea water. C. Saturation index of aragonite (SI). D. Saturation index (SI) of calcite.

of aragonite remaining (in percent of the mineral assemblage) after t years of steady-state diagenesis in the saturated zone is

$$A = A_0 e^{-kt}$$

where A_0 is the percent of aragonite in unaltered sediment, taken to be 35 percent (Fig. 2). This expression indicates that about 310,000 yr of continuous fresh-water diagenesis could produce the present mineral assemblage containing about 10 percent aragonite. Furthermore, approximately 900,000 yr of steady-state diagenesis are required to produce the assemblage of aragonite in the present Walsingham Formation (1 percent).

If, for the second model, calcite precipitation is the slow step, the ground water should be saturated with aragonite, and the rate of recrystallization should be independent of the amount of aragonite present, provided that amount is in sufficient excess that the first model does not become rate limiting. Because the ground water maintains an approximately constant supersaturation with calcite (aragonite saturation), to a first approximation the rate of recrystallization of aragonite to calcite is constant. Our estimate of 0.32 $cm^3 m^{-3} yr^{-1}$ then requires 625,000 yr of steady-state diagenesis to recrystallize an original sediment of 35 percent aragonite to a limestone containing 10 percent aragonite, and about 825,000 yr are required to produce a limestone equivalent to the Walsingham Formation (1 percent aragonite).

Both models considered for the recrystallization of aragonite to calcite predict results that are reasonable on the time scale of Bermuda diagenesis. Further conclusions concerning the validity of either model must allow consideration for the actual amount of time that Bermuda limestones have been in contact with a saturated zone and for the fact that many of the samples used to construct Figure 2 are from the unsaturated zone (where the rate of recrystallization must be slower because of intermittent recharge). We conclude that within the uncertainties of the calculations, the present rate of recrystallization of aragonite to calcite in the saturated zone is approximately 0.32 cm^3 of aragonite to calcite per m^3 of the saturated zone per year.

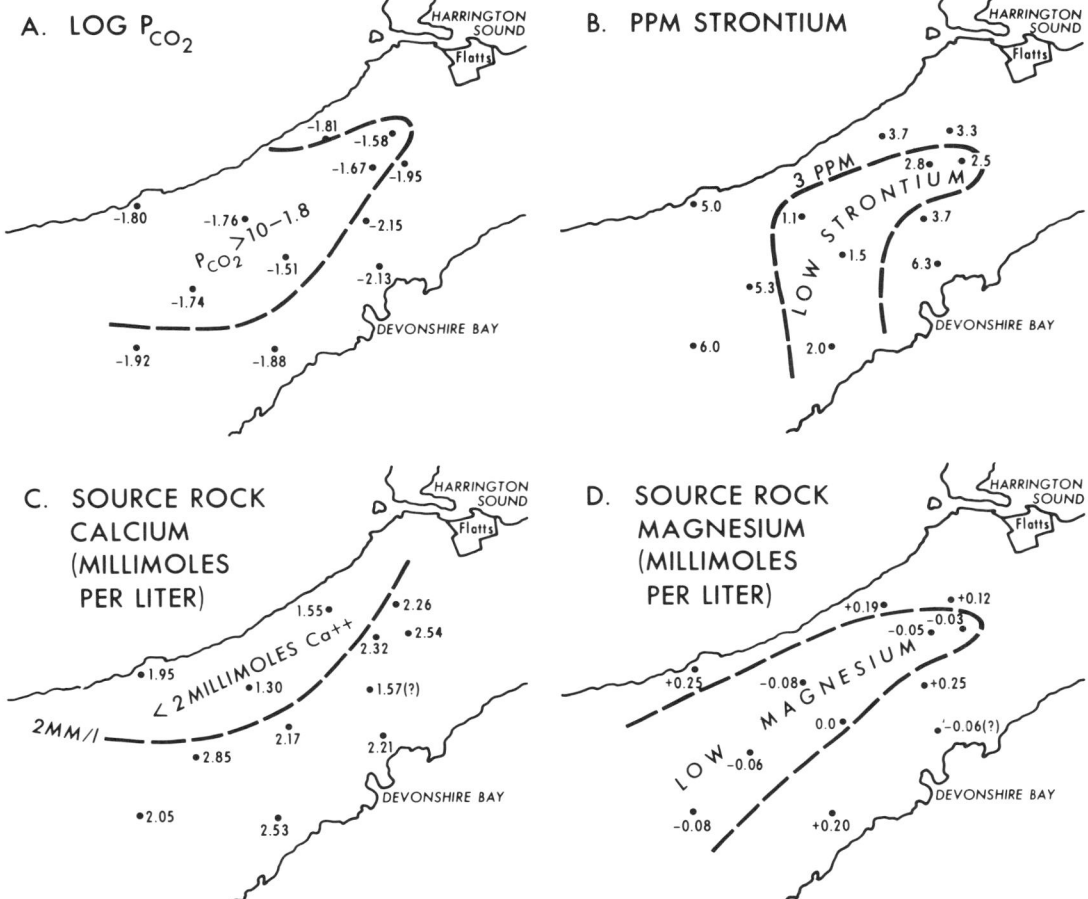

Figure 10. Geochemistry of the Devonshire lens. A. Log P_{CO_2}. B. Strontium ppm. C. "Rock-derived" calcium. D. "Rock-derived" magnesium.

Geochemistry of the Devonshire Lens

Our data are complete enough to allow us to map the water chemistry near the top of the saturated zone within the Devonshire lens in Devonshire Parish. An index to the wells used to map the water chemistry is given in Figure 9A.

The analytical data mapped in Figures 9 and 10 support the conclusions that ground-water flow radiates outward in the saturated zone from the vicinity of Devonshire marsh (which is located near the thickest part of the fresh-water lens), and that the marsh-CO_2 system is an important source of CO_2 for reactions taking place in the vicinity of Devonshire marsh and throughout the Devonshire lens.

In the vicinity of Devonshire marsh, there are low-salinity waters coincident with subsaturation with respect to calcite and aragonite (Figs. 9B through 9D), high CO_2 partial pressures, relatively low strontium concentrations, and low values of source-rock calcium and magnesium (Figs. 10A through 10D). Toward the south shore and farther to the east, the salinity increases, and the waters are near equilibrium with respect to calcite. There is a tendency toward lower CO_2 partial pressures along the south shore, presumably due to evasion. These lower CO_2 partial pressures may coincide with phreatic cementation in that part of the Devonshire lens. The strontium content increases in all directions from Devonshire marsh in the phreatic zone, presumably indicating the dissolution of aragonite. A low strontium well near the south shore (Fig. 10B) may indicate an area where previous diagenetic alteration of aragonite has been more intense.

The fact that the higher salinity waters are more saturated with respect to calcite than are the lower salinity waters suggests that sea-water mixing may not be a significant factor in controlling calcite saturation. (The control of calcite saturation is discussed in detail in the next section of this paper). The higher saturation along the south shore may not result from increased concentration of calcium (or alkalinity), as shown by a plot of the source rock contribution of calcium to the phreatic water (Fig. 10C), but rather by a small loss of

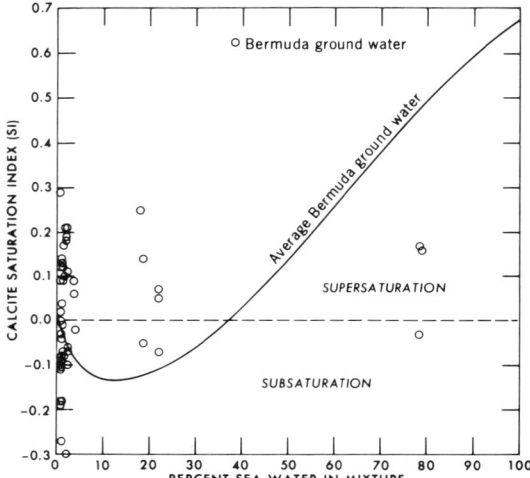

Figure 11. Theoretical calculation of the effect on calcite saturation of mixing sea water with the average Bermuda ground water. Comparison of the theoretical mixing curve with the computed saturation state in Bermuda ground water indicates that mixing with sea water is not a dominant process in determining the saturation state of Bermuda ground water.

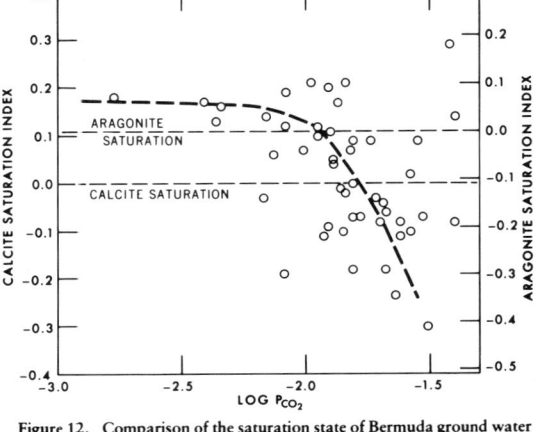

Figure 12. Comparison of the saturation state of Bermuda ground water with the computed log P_{CO_2}. The dashed line suggests a likely trend to subsaturation with higher P_{CO_2}.

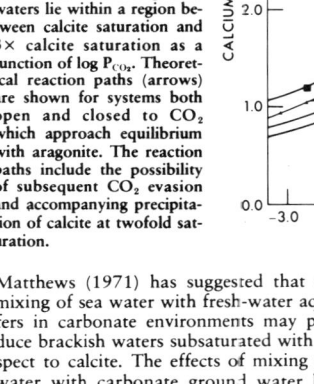

Figure 13. Total calcium in solution as a function of log P_{CO_2} (at 25°C) for waters in equilibrium with calcite and aragonite, and waters twofold and threefold supersaturated with respect to calcite. Most Bermuda ground waters lie within a region between calcite saturation and 3× calcite saturation as a function of log P_{CO_2}. Theoretical reaction paths (arrows) are shown for systems both open and closed to CO_2 which approach equilibrium with aragonite. The reaction paths include the possibility of subsequent CO_2 evasion and accompanying precipitation of calcite at twofold saturation.

dissolved CO_2 (Fig. 10A). The loss of CO_2 might be explained by an expected increase in residence time for waters along the south shore fringe of the Devonshire lens, which would allow more time for CO_2 to evade. In addition, being farther from the marsh, the supply of CO_2 to the ground water is probably less along the south shore.

Control of Calcite Saturation

Many factors may affect the saturation state of calcite in Bermuda ground water, including variability in the soil P_{CO_2}, the flux of CO_2 in the phreatic zone from the soils and marshes, evasion of CO_2 from the ground water along fractures and through areas of minimal soil cover, the kinetics of dissolution and precipitation of carbonate minerals, and the effects of mixing with sea water. We now attempt to determine if any one factor or process contributes significantly to the saturation state of the ground water with respect to calcite.

Sea-Water Mixing. The mixing of two waters saturated and (or) supersaturated with respect to calcite at differing CO_2 pressures can, in many cases, produce waters subsaturated with respect to calcite (Mischungskorrosion) (Bogli, 1964; Howard, 1966; Thrailkill, 1968; Badiozamani, 1973; Plummer, 1975; Wigley and Plummer, 1976). Runnells (1969) has shown that, depending upon the rate of change of mineral solubility as a function of change in ionic strength, it is possible to mix subsaturated waters to produce supersaturated waters and vice versa. The effect can be seen largely in the changes in the ion activity coefficients as a function of mixing.

Matthews (1971) has suggested that the mixing of sea water with fresh-water aquifers in carbonate environments may produce brackish waters subsaturated with respect to calcite. The effects of mixing sea water with carbonate ground water has been quantified by Plummer (1975). These calculations show that as sea water mixes with carbonate ground water, the mixtures become increasingly subsaturated with respect to calcite and that with continued mixing, the subsaturation decreases, so that mixtures containing large amounts of sea water are supersaturated.

To test whether the extensive mixing of Bermuda ground water with sea water (documented on Figs. 5 through 7) is a controlling influence on the state of calcite saturation, we have calculated the saturation index of calcite (at 25°C) as a function of sea water mixing with the average Bermuda ground water (Fig. 11), using the charge-balance scheme of Plummer (1975). The results show that mixtures containing from 0.5 to 37.5 percent sea water by volume are subsaturated with respect to calcite; the greatest subsaturation (−.13 SI) occurs in mixtures containing 13 percent sea water (Fig. 11) (assuming that no mineral dissolution or precipitation follows mixing). Mixing beyond 37.5 percent sea water results in supersaturation.

The observed saturation index of calcite in Bermuda ground water is compared with the theoretical saturation curve in Figure 11. There is considerable scatter in the points, and if any relationship exists between sea-water content and calcite saturation index, it is opposite to that predicted by the mixing calculations. Therefore, processes other than mixing with sea water must be more significant in determining the degree of saturation with respect to calcite in Bermuda ground water.

Nonequilibrium Effects. A plot of calcite and aragonite saturation indices against log P_{CO_2} in the ground water shows a trend to subsaturation with increasing P_{CO_2} (Fig. 12). As P_{CO_2} decreases, the slope of the trend decreases, approaching approximately a constant degree of supersaturation with respect to calcite and aragonite.

If the fresh-water diagenetic alteration of metastable aragonite to stable low-magnesium calcite on Bermuda were governed solely by equilibrium processes, no point would lie above the line for calcite saturation (Fig. 12). Moreover, if the kinetics of the diagenetic process were rapid relative to the age of the ground water, all water should be in equilibrium with calcite. The observed relationship between saturation index and log P_{CO_2} (Fig. 12) indicates that the chemical evolution of Bermuda ground water is controlled not by equilibrium processes, but more likely by nonequilibrium dissolution-precipitation reactions coupled with processes involving fluxes of CO_2 to and from the ground water.

The presence of subsaturated water at high P_{CO_2} indicates that the rate of aragonite (and calcite) dissolution, as equilibrium is approached, is relatively slow on the residence-time scale of Bermuda ground water. We can draw an analogy only to work on the dissolution of calcite (Terjesen and others, 1961; Berner and Morse, 1974; Plummer and Wigley, 1976), indicating that foreign ions (for example, PO_4^{3-}) adsorb onto calcite surfaces from the surrounding fluid, significantly inhibiting the rate of dissolution near equilibrium. Phosphate levels in 15 analyses of Bermuda ground water range from 0.005 to 0.094 ppm. Similar inhibitions may affect the rate of aragonite dissolution near equilibrium, accounting for the observed subsaturation; however, there is no simple systematic relationship between phosphate content and aragonite (or calcite) saturation index.

From a thermodynamic standpoint, and accepting the fact that calcite cement grows from supersaturation (Wollast, 1971), we expect the water composition to approach saturation with the most soluble phase present, which for Bermuda is aragonite, and possibly, high magnesium calcite. From the Mg-calcite stability data of Plummer and Mackenzie (1974), we calculate that the average Bermuda ground water can be saturated with $Ca_{0.95}Mg_{0.05}CO_3$. Magnesium calcites containing more than 5 mole percent $MgCO_3$ dissolve irreversibly in most of the Bermuda ground water. Bermuda ground water does not come to saturation with higher magnesium calcites either because they are essentially absent in the saturated zone or, if present, because of their slow rate of incongruent dissolution relative to the rapid flushing of the aquifer. From solubility considerations, the relatively slow rate of incongruent dissolution of Mg calcites, and the source-rock calculations for magnesium, it is concluded that magnesium-calcite dissolution contributes little to determine the saturation state of Bermuda ground water.

Thus, it appears that, in general, Bermuda ground water approaches a steady-state condition of aragonite dissolution (at slight subsaturation) and precipitation of calcite cement from the accompanying supersaturation with calcite.

P_{CO_2} Effects. Figure 12 shows that waters at lower P_{CO_2} are supersaturated with aragonite, as well as calcite. The solubility of $Ca_{0.95}Mg_{0.05}CO_3$ is not sufficient to account for the oversaturation with respect to aragonite. This supersaturation probably results from loss of CO_2 from the ground water to the atmosphere through the permeable vadose zone in areas of incomplete soil cover, such as hilltops and rocky areas near the shoreline. As P_{CO_2} decreases in the ground water by evasion of CO_2, the saturation index of calcite approaches a constant value (SI ≅ 0.2; 1.6 times saturation) which may coincide with the level of supersaturation required for the crystal growth of calcite. As CO_2 evades from the ground water, calcite precipitation may buffer the activity product of $CaCO_3$.

Theoretical reaction paths for the evolution of carbonate ground water are shown in Figure 13 which gives total calcium in solution as a function of P_{CO_2}. Figure 13 shows that most Bermuda ground water lies within a region defined by calcite saturation and three times calcite saturation as a function of P_{CO_2}. Theoretical reaction paths for aragonite (and calcite) dissolution in systems both closed and open to CO_2 at initial P_{CO_2} values of $10^{-1.5}$ and $10^{-1.0}$ are shown. In a closed system, dissolution of $CaCO_3$ is not accompanied by additional input (or loss) of CO_2 from the soil zone. Along an "open" reaction path, CO_2 from the soil zone is continually added to the ground water as $CaCO_3$ dissolves, buffering the ground-water P_{CO_2} at a value equal to that for the soil zone.

Because the vadose zone is permeable to CO_2 and the marshes occur within the aquifer, there is an apparent circulation of CO_2 within the ground-water system. Near marshes, which extend deep into the aquifers, there is an inflow of CO_2 to the ground water. Farther from the marshes, some CO_2 is probably lost to the atmosphere through the porous vadose zone, as is shown above for the Devonshire lens. As the computed reaction paths show (Fig. 13), most Bermuda ground water could be produced by $CaCO_3$ dissolution approaching aragonite saturation with subsequent CO_2 evasion and calcite precipitation at two times calcite saturation in either a soil–ground-water system open to CO_2 at $10^{-1.5}$ atm or closed at $10^{-1.0}$ atm CO_2.

CONCLUDING REMARKS

Bermuda ground waters are essentially mixtures of sea water and calcium bicarbonate waters resulting from the dissolution of aragonite at greater than atmospheric P_{CO_2}. Strontium to calcium ratios in the ground water, three times that in skeletal aragonite, point to the significance of the recrystallization of aragonite to calcite. It is estimated that approximately 0.32 cm³ of aragonite is recrystallized to calcite in each m³ of the saturated zone each year, a rate that appears reasonable on the time scale of Bermuda diagenesis. Most of the magnesium in the ground water is derived from sea water, the remaining magnesium being accounted for by the low abundance of magnesium calcites in the saturated zone, the slow rate of incongruent dissolution of magnesium calcite, and relatively short residence times for phreatic waters.

Important differences in the water chemistry and mineral saturation are observed spatially within the Devonshire lens. Sea water mixing is probably not significant in contributing to the observed saturation state of the ground water. It is concluded that nonequilibrium dissolution and precipitation reactions coupled with variable fluxes of CO_2 to and from the ground water are important in accounting for the observed saturation. The persistence of subsaturation at high P_{CO_2} indicates that the dissolution of aragonite (and calcite) at small subsaturation is relatively slow on the time scale of Bermuda ground water. Waters supersaturated with calcite and aragonite probably result from CO_2 evasion. The upper limit to supersaturation with calcite may be near twofold saturation and buffered by growth of calcite cement.

Bermuda's aquifers are partially open to CO_2. Peat marshes penetrating a fresh-water lens put in a significant flux of CO_2 to the ground water; farther from the marshes, there may be a net loss of CO_2 to the atmosphere.

In a steady-state lens, essentially all of the recharge and dissolved $CaCO_3$ is lost to the ocean annually. Approximately 2.5×10^9 litres of water pass through the fresh-water lenses of Bermuda and flow into the sea each year (Vacher, 1974). This volume of water carries with it 5.2×10^5 kg of calcium derived from dissolution of the island's limestone bedrock. In terms of volume, 360 m³ of the island (assuming 20 percent porosity) is lost each year through solution and transported to the sea by ground water.

ACKNOWLEDGMENTS

We thank Mrs. F. Bickley for assistance in collection and analysis of ground-water samples. The Public Works Department of Bermuda kindly granted permission to sample government wells. We also wish to thank the many individuals in Bermuda who allowed us to sample their private water supplies. The manuscript was improved by the constructive criticism of William Back, Isaac Winograd, D. C. Thorstenson, and L. A. Heindl. This work was supported by funds from Northwestern University, the Bermuda Government, and National Science Foundation Grant GA-32120.

REFERENCES CITED

Badiozamani, K., 1973, The Dorag dolomitizations model: Application to the Middle Ordovician of Wisconsin: Jour. Sed. Petrology, v. 43, p. 965–984.

Bathurst, R.G.C., 1971, Carbonate sediments and their diagenesis: Amsterdam, Elsevier, 620 p.

Bear, J., and Todd, D. K., 1960, The transition zone between fresh and salt waters in coastal aquifers: California Univ. Water Resources Center Contr. no. 29, 156 p.

Berner, R. A., 1966, Chemical diagenesis of some modern carbonate sediments: Am. Jour. Sci., v. 264, p. 1–36.

Berner, R. A., and Morse, J. W., 1974, Dissolution kinetics of calcium carbonate in sea water: IV. Theory of calcite dissolution: Am. Jour. Sci., v. 274, p. 108–134.

Bogli, A., 1964, Mischungskorrosion: Ein Beitrag zum Verkarstungsproblem: Erdkunde, v. 18, p. 83–92.

Bretz, J. H., 1960, Bermuda: A partially drowned late mature Pleistocene karst: Geol. Soc. America Bull., v. 71, p. 1729–1754.

Carrier, G. F., 1959, The mixing of groundwater and sea water in permeable subsoils: Jour. Fluid Mechanics, v. 4, p. 479–488.

Frantz, J. D., 1971, Bermudian ground and cave waters: Bermuda Biol. Station for Research, Spec. Pub. 7, p. 47–55.

Friedman, G. M., 1964, Early diagenesis and lithification of carbonate sediments: Jour. Sed. Petrology, v. 34, p. 777–813.

Garrels, R. M., and Christ, C. L., 1965, Solutions, minerals and equilibrium: New York, Harper & Row, 450 p.

Garrels, R. M., and Thompson, M. E., 1962, A chemical model for seawater at 25°C and one atmosphere total pressure: Am. Jour. Sci., v. 260, p. 57–66.

Goldberg, E. D., 1957, Biogeochemistry of trace elements, in Hedgpeth, J. W., ed., Treatise on marine ecology and paleoecology, Vol. 1: Geol. Soc. America Mem. 67, p. 345–358.

Gross, M. G., 1964, Variations in the $^{18}O/^{16}O$ and $^{13}C/^{12}C$ ratios of diagenetically altered limestone in the Bermuda Islands: Jour. Geology, v. 72, p. 170–194.

Harned, H. S., and Davis, R., Jr., 1943, The ionization constant of carbonic acid in water and the solubility of carbon dioxide in water and aqueous salt solutions from 0 to 50°: Am. Chem. Soc. Jour., v. 65, p. 2030–2037.

Harned, H. S., and Scholes, S. R., Jr., 1941, The ionization constant of HCO_3^- from 0 to 50°: Am. Chem. Soc. Jour., v. 63, p. 1706–1709.

Harris, W. H., and Matthews, R. K., 1968, Subaerial diagenesis of carbonate sediments: Efficiency of the solution-reprecipitation process: Science, v. 160, p. 77–79.

Howard, A. D., 1966, Verification of the mischungskorrosion effect: Cave Notes, v. 8, p. 9–12.

Jacobson, R. L., and Langmuir, D., 1974, Dissociation constants of calcite and $CaHCO_3^+$ from 0 to 50°C: Geochim. et Cosmochim. Acta, v. 38, p. 301–318.

Kinsman, D.J.J., 1969, Interpretation of Sr^{+2} concentrations in carbonate minerals and rocks: Jour. Sed. Petrology, v. 39, p. 486–508.

Lafon, G. M., and Vacher, L., 1975, Diagenetic reactions as stochastic processes: Application to the Bermudian eolianites, in Whitten, E.H.T., ed., Quantitative studies in the geological sciences: Geol. Soc. America Mem. 142, p. 187–204.

Land, L. S., 1970, Phreatic versus vadose meteoric diagenesis of limestones: Evidence from a fossil water table: Sedimentology, v. 14, p. 175–185.

———1973, Holocene meteoric dolomitization of Pleistocene limestones, North Jamaica: Sedimentology, v. 20, p. 411–424.

Land, L. S., Mackenzie, F. T., and Gould, S. J., 1967, The Pleistocene history of Bermuda: Geol. Soc. America Bull., v. 78, p. 993–1006.

Langmuir, D., 1971, The geochemistry of carbonate ground waters in central Pennsylvania: Geochim. et Cosmochim. Acta, v. 35, p. 1023–1046.

Mackenzie, F. T., 1964a, Bermuda Pleistocene eolianites and paleowinds: Sedimentology, v. 3, p. 51–64.

———1964b, Geometry of Bermuda calcareous cross-bedding: Science, v. 144, p. 1449–1450.

Macky, W. A., 1957, The rainfall of Bermuda: Bermuda Meteorol. Office Tech. Note 8, 58 p.

Matthews, R. K., 1971, Diagenetic environments of possible importance to the explanation of cementation fabric in subaerially exposed carbonate sediments. in Bricker, O. P., ed., Carbonate cements: Baltimore, Johns Hopkins Press, p. 127–132.

Penman, H. L., 1963, Vegetation and hydrology: Farnham Royal, England, Commonwealth Agr. Bur., Commonwealth Bur. Soils, Tech. Comm. no. 53, 124 p.

Plummer, L. N., 1970, Soils and the diagenesis of Bermuda carbonate sands: Bermuda Biol. Station for Research Spec. Pub. 7, p. 56–98.

———1974, Rates of dissolution of Ca-Mg carbonate minerals [abs.]: EOS (Am. Geophys. Union Trans.), v. 55, p. 699.

———1975, Mixing of sea water with calcium carbonate ground water, in Whitten, E.H.T., ed., Quantitative studies in the geological sciences: Geol. Soc. America Mem. 142, p. 219–238.

Plummer, L. N., and Mackenzie, F. T., 1974, Predicting mineral solubility from rate data: Application to the dissolution of magnesian calcites: Am. Jour. Sci., v. 274, p. 61–83.

Plummer, L. N., and Wigley, T.M.L., 1976, The dissolution of calcite in CO_2-saturated solutions at 25°C and 1 atmosphere total pressure: Geochim. et Cosmochim. Acta, v. 40, p. 191–202.

Ristvet, B. L., 1971, The progressive diagenetic history of Bermuda: Bermuda Biol. Station for Research Spec. Pub. 9, p. 118–157.

Runnells, D. D., 1969, Diagenesis, chemical sediments, and the mixing of natural waters: Jour. Sed. Petrology, v. 39, p. 1188–1201.

Sayles, R. W., 1931, Bermuda during the ice age: Am. Acad. Arts and Sci. Proc., v. 66, p. 381–467.

Smith, W. O., 1967, Infiltration in sands and its relation to groundwater recharge: Water Resources Research, v. 3, no. 2, p. 539–555.

Strahler, A. N., and Strahler, A. H., 1973, Environmental geoscience: Interaction between natural systems and man: Santa Barbara, Calif., Hamilton Publishing Co., 511 p.

Terjesen, S. G., Erga, O., Thorsen, G., and Ve, A., 1961, Phase boundary processes as rate determining steps in reactions between solids and liquids: The inhibitory action of metal ions on the formation of calcium bicarbonate by the reaction of calcite with aqueous carbon dioxide: Chem. Eng. Sci., v. 14, p. 277–289.

Thornthwaite, C. W., and Mather, J. R., 1955, The water balance: Publications in Climatology, v. 8, p. 1–86.

Thrailkill, J., 1968, Chemical and hydrologic factors in the excavation of limestone caves: Geol. Soc. America Bull., v. 79, p. 19–46.

Todd, D. K., 1959, Groundwater hydrology: New York, John Wiley & Sons, Inc., 336 p.

Todd, D. K., and Meyer, C. F., 1971, Hydrology and geology of the Honolulu aquifer: Am. Soc. Civil Engineers Proc., Jour. Hydraulics Div., v. 97, no. HY2, p. 233–255.

Truesdell, A. H., and Jones, B. F., 1973, WATEQ, a computer program for calculating chemical equilibria in natural waters: Natl. Tech. Info. Service P.B. 220464, 73 p.

———1974, WATEQ, a computer program for calculating chemical equilibria in natural waters: U.S. Geol. Survey Jour. Research, v. 2, p. 233–248.

Vacher, H. L., 1971, Late Pleistocene sea level history: Bermuda evidence [Ph.D. thesis]: Evanston, Ill., Northwestern Univ., 153 p.

———1973, Coastal dunes of younger Bermuda, in Coates, D. R., ed., Coastal geomorphology: Binghamton, N.Y., State Univ. New York, Binghamton, p. 355–391.

———1974, Groundwater hydrology of Bermuda: Hamilton, Government of Bermuda, Public Works Dept., 87 p.

Wigley, T.M.L., and Plummer, L. N., 1976, Mixing of carbonate waters: Geochim. et Cosmochim. Acta, v. 40 (in press).

Wollast, R., 1971, Kinetic aspects of nucleation and growth of calcite from aqueous solutions, in Bricker, O. P., ed., Carbonate cements: Baltimore, Johns Hopkins Press, p. 264–273.

Manuscript Received by the Society May 27, 1975
Revised Manuscript Received February 23, 1976
Manuscript Accepted April 7, 1976
Contribution No. 669, Bermuda Biological Station for Research, St. George's, Bermuda

16

Copyright ©1978 by the Society of Economic Paleontologists and Mineralogists
Reprinted from Jour. Sed. Petrology **48**:489–501 (1978)

PRECIPITATION OF BEACHROCK CEMENTS: MIXING OF MARINE AND METEORIC WATERS VS. CO_2-DEGASSING[1]

JEFFREY S. HANOR
Department of Geology,
Louisiana State University,
Baton Rouge, Louisiana 70803

ABSTRACT: The origin of calcite, Mg-calcite, and aragonite cements formed near the water table and in the intertidal zone of tropical and subtropical beach sediments has been the subject of extended debate. Following Field (1919), it is proposed here that much of this cement is precipitated as a consequence of loss of CO_2 from carbonate-saturated beach groundwaters. Mass-transport calculations support the further proposal that vertical fluid dispersion in the phreatic zone resulting from tidal oscillation of the water table is sufficient to induce degassing of CO_2 from a seaward-flowing groundwater. Loss of CO_2 is further enhanced by tidal pumping of the gas phase in the vadose zone across the sediment-atmosphere interface. As sediment porosity is lowered by precipitation of cement, the ability of the groundwater system to degas and form new cements is reduced. As long as the system remains thermodynamically open with respect to CO_2 and close to saturation with respect to calcite or aragonite, however, it remains an unlikely site for the precipitation of dolomite.

The hypothesis that degassing alone is sufficient to cause supersaturation and cementation is supported by an experimental study of degassing of mixed beach and marine waters from St. Croix, U.S. Virgin Islands, in which 30 μm-thick, low-Mg calcite crusts were formed which closely resemble natural water-table cements. The waters were spiked with Hg_2Cl_2 to retard biological involvement. The maximum observed rate of precipitation of calcite in solutions in direct contact with the atmosphere was 10^{-9} moles cm^{-2} sec^{-1}, which is a rate sufficient to indurate a beach sediment within a period of 12 hours. Thermodynamic calculations indicate that precipitation could not have been induced by mixing of marine and meteoric waters, as has been proposed by others, but that loss of CO_2, a process independent of mixing, was necessary to cause supersaturation.

INTRODUCTION

The precipitation of calcium carbonate cements near the water table and in the intertidal zone of Recent tropical and subtropical beach sediments is a wide-spread and rapid process. Study of the factors which control the distribution and rate of formation of these cemented zones is of importance to general problems of early carbonate diagenesis.

While a succession of independent studies has been made by various workers concerning the occurrence of water-table and beach-rock cements, there has been a remarkable divergence of view regarding the factors which cause these cements to precipitate. Commonly invoked processes include evaporation of seawater, precipitation induced by mixing of waters of diverse composition, and vague types of biological involvement. Current summaries of past studies are provided by Bathurst (1975) and Milliman (1974).

The purpose of this paper is to show that degassing of groundwater, a simple mechanism which has received less attention in the study of the diagenesis of beach sediments than the processes listed above, may be responsible for producing many of the water-table and beach-rock cements which have been described in the literature. In 1919, Field hypothesized that the escape of CO_2 from carbonate-saturated groundwater could be responsible for the formation of beach rock at Loggerhead Key, Dry Tortugas, Florida. While this explanation was severely criticized by Daly (1924), it was briefly re-invoked by Russell (1970) in his last published work on beach rock. Thorstenson, *et al.* (1972) have subsequently demonstrated by experiment the feasibility of inducing cementation above and below the water table by degassing of carbonate-saturated waters.

As part of a study of the hydrogeochem-

[1] Manuscript received 3/31/77; revised 8/5/77.

istry and diagenesis of carbonate beaches on St. Croix, an experimental study and theoretical calculations have been employed to evaluate the relative roles of mixing and degassing of beach waters in the formation of beach cements. During experimentally controlled loss of CO_2 from beach groundwaters, calcite crusts were formed which resemble some natural cements. The conditions under which these crusts formed provide experimental support for the hypothesis that degassing is a sufficient mechanism to induce carbonate precipitation in many beach environments.

This paper will discuss the properties of the experimentally precipitated crusts and evaluate on a quantitative basis the relative importance of water mixing versus degassing in production of beach cements on St. Croix. Finally, I will discuss the geological and geochemical factors which could lead to degassing in a generalized beach setting.

EXPERIMENTAL STUDY

Purpose

The island of St. Croix, U.S. Virgin Islands (Fig. 1), has been the site of on-going studies of diagenesis of Recent carbonate beach sediments (Boleneus, 1972; Hanor and Moore, 1974; Moore and Hanor, 1974; Moore, 1977; Hanor, this paper). It has been observed that during dry periods on St. Croix, brackish groundwaters high in carbonate alkalinity, dissolved calcium, and dissolved CO_2 develop in beach sediments. We have postulated (Hanor and Moore, 1974; Moore and Hanor, 1974) that during periods of high rainfall, these masses of brackish water are displaced seaward, where they mix with normal marine water. Because the probable zone of water mixing is also the site of the occurrence of beach rock cements (Fig. 2), it is of interest to know if the mixing of two dissimilar water types or other processes associated with the physical transport or chemical modification of groundwater could induce the precipitation of carbonate cement. With this in mind an experiment was run on freshly collected natural waters from St. Croix to evaluate possible controls on carbonate precipitation. Factors to be evaluated included: mixing, loss of CO_2 by degassing, and biological effects.

Study Area

The beach selected for intensive study is in Boiler Bay on the northeast coast of St. Croix (Fig. 1). The beach sediments here are composed primarily of carbonate clastics derived from a seaward reef tract and shallow bay, with minor amounts of terrigenous clastics derived from a basement composed of Cretaceous turbidites. A soil zone has developed landward on these sediments, and the back beach area supports a variety of salt-tolerant grasses, shrubs, and trees (Forman, 1974). Figure (2) shows a generalized cross-section through the beach. Water wells have been emplaced and are sampled at intervals to monitor beach water chemistry. Rainfall

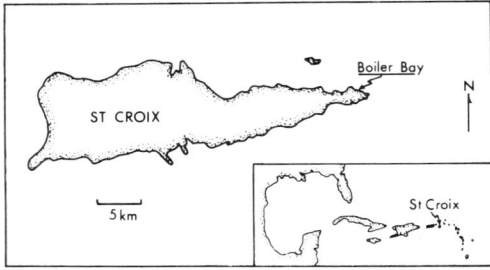

FIG. 1.—Sketch map of the island of St. Croix showing the location of the study area, Boiler Bay. Inset shows the location of St. Croix.

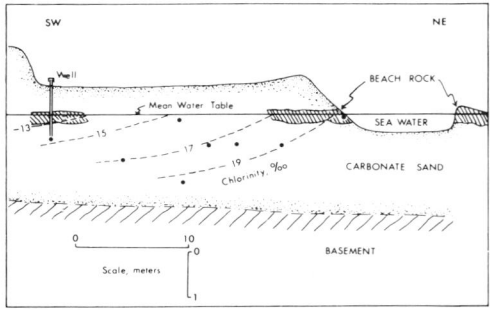

FIG. 2.—Generalized cross-section of beach in western Boiler Bay, St. Croix, showing salinity regime in August 1976 (solid circles are control points) and distribution of sediment types. Unconsolidated carbonate sands overlie a Cretaceous basement. Intertidal and water-table cemented zones are indicated by ruled areas. Groundwater used in experiment (Fig. 5) was pumped from the well shown in the back beach.

averages 91 cm/year. During the rainy season, generally August through January, freshwater infiltrates into the beach sediments. During dry months, beach ground waters become progressively more saline and approach seawater salinities. There is, thus, no permanent, steady-state freshwater lens in the Ghyben-Herzberg sense, but, rather, a continuously-evolving salinity regime. The tidal pattern is markedly diurnal and usually has a range of 30–35 cm. (Adey, 1975). Tidal fluctuations of the ground water table are attenuated landward.

Carbonate-cemented zones occur at the water table both in the back beach area and present intertidal zone, as shown in Figure 2. A presumably earlier generation of beach rock occurs several meters seaward of the present shoreline. The water-table cements in the back beach are dominantly low-Mg calcite, while the beach rock cements to the seaward are composed of magnesian calcite and aragonite (Boleneus, 1972; Moore, 1977).

Water Sampling and Analysis

The wells in Boiler Bay were sampled in early August 1976. Field procedure included pumping the well to remove stagnant water above the well screen. pH measurements were made *in-situ* or on samples immediately drawn up from depth. These measurements were conducted in the early morning when air and groundwater temperatures were equivalent, thus minimizing difficulties in calibrating the pH meter. Alkalinity was determined by Gran titration. An aliquot of water sample was filtered and then pickled with HNO_3 for later major element analysis. A second aliquot, intended for nutrient analysis, was immediately frozen for transport back to LSU. The composition of back beach water is given in Table 1. In the month preceeding the sampling, rainfall on the eastern end of St. Croix was only 0.4 cm (R. Holt, personal communication, 1976), and chlorinities in the back beach had risen to 14°/oo. The calculated P_{CO_2} of the water is $10^{-1.3}$, compared with atmospheric values of approximately $10^{-3.5}$. The high P_{CO_2} value for the water is thought to be the result of high rates of *in-situ* biological respiration of organic material washed down from the soil layer above.

TABLE 1.—*Composition of beach groundwater at Boiler Bay, August 1976, and seawater of 19°/oo chlorinity*

Species	Ground Water ppm	Sea Water ppm
Na	7250	10500
K	140	380
Mg	1075	1350
Ca	710	400
Sr	11.8	8.0
Cl	13556	19000
SO_4	1650	2710
alk	435	135
PO_4	0.1	—
pH	(6.83)	(8.20)

Mixing Experiment

Within one hour of collection, aliquots of back beach groundwater sampled from the back beach were mixed with varying quantities of surface seawater to simulate mixing of groundwater and normal marine water. Another set of aliquots was mixed with distilled water to simulate mixing of brackish beach water with fresh rain waters. A duplicate series of water mixtures was prepared and spiked with Hg_2Cl_2 to retard biological activity, particularly the *in-situ* production of CO_2. The mixtures were agitated vigorously and stored in narrow-necked 1000-ml glass bottles open to the atmosphere. The compositions of water mixtures prepared are listed in Table 2. Alkalinity and pH measurements were made of samples taken periodically from the upper 1 cm of solution.

Formation of Precipitates

It was observed that calcite precipitates had formed within a period of one day at

TABLE 2.—*Summary of water mixing experiments*

(Volume %)				Calcite Precipitate	
Ground Water	Sea Water	Distilled Water	Hg_2Cl_2 Treatment	Time of Appearance (hours)	Amount
100	0	0	yes	12*	+++
			no	12	+++
75	25	0	yes	24*	++
			no	24	++
50	50	0	yes	36	+
			no	48	+
25	75	0	yes	—	0
			no	—	0
0	100	0	yes	—	0
			no	—	0
91	0	9	no	24*	++
50	0	50	no	36*	++
9	0	91	no	—	0

*SEM photo of precipitate shown in Figure 4.

the surface of the solutions of some of the bottles. Precipitates subsequently developed in other bottles at the solution-air interface. The precipitates were allowed to develop until the end of the field session a few days later, at which time they were filtered and returned to LSU for detailed study. Evaporative loss of water during the duration of the experiments is estimated to be a factor of 0.01 or less, as determined by observation of water levels within the bottles. Table 2 lists approximate time of first appearance and quantity of precipitate formed. Precipitates formed both in Hg_2Cl_2-free solutions and in treated solutions.

The precipitates consist of flat, crust-like aggregates of crystals (Fig. 3). Individual crystals range from 15 to 30 μm in width and occur in aggregates 0.5 to 1.5 mm in width and 30 to 50 μm in thickness. As will be shown, the crusts formed in response to loss of CO_2 from the water mixtures and a consequent increase in the state of supersaturation with respect to calcite (Fig. 5). Supersaturation was greatest subjacent to the air-water interface and this is where initial precipitation of calcite took place. The crust-like aggregates were held at the interface by surface tension throughout the course of the experiment. The crusts are flat on their upper side while their lower sides have rhombic terminations which projected downward into the solution from which they were precipitated.

FIG. 3.—SEM photograph of top and bottom surfaces of low-Mg calcite crusts formed by degassing a 3:1 mix of beach groundwater and seawater. Flat surfaces were subjacent to solution-atmosphere interface, rhombic terminations projected downward into solution

Composition and Morphology of Precipitates

The mineralogy of the crusts was determined to be low-Mg calcite on the basis of crystal morphology, chemical composition, and infra-red spectrometry. The magnesium content of the crusts was ascertained by SEM non-dispersive spectrometry, using dolomite and high-Mg calcite as standards, to be below 5 mole % $MgCO_3$. Strontium was below the detection limit of the system, which is estimated to be approximately 1000 ppm at the conditions of operation.

The size, morphology, and crystallographic orientation of individual calcite crystals within the aggregates vary with differences in solution composition. Because of the interest in the relation between morphology of calcium carbonate crystals and environment of precipitation (Folk, 1974), a brief discussion of the morphology of the precipitates will be given here.

Calcites precipitated from unmixed beach groundwater are 30 μm across and are bounded by the steep rhombohedral form $\{40\bar{4}1\}$ (Fig. 4). Single crystals are bounded by numerous offset $40\bar{4}1$ faces and display scores of parallel terminations at the c-axes. The net effect is to give individual crystals a blocky appearance and a height-to-width ratio of 1:1, much lower than the 3.6:1 height-to-width ratio that would be displayed by a simple, six-sided rhombohedron bounded by $\{40\bar{4}1\}$. The crystallographic orientation of individual crystals appears to be random.

Crystals precipitated from beach water-distilled water mixtures average 30 μm in width and are again dominated by $\{40\bar{4}1\}$ (Fig. 4). Individual crystals show multiple c-axis terminations, and there is a pronounced preferred orientation of the c-axes perpendicular to the air-water interface. Some crystals along the margins of the aggregates are oriented parallel to the interface and have a height-to-width ratio of 2:1.

In contrast to the crystals described above, calcite crystals precipitated from beach water-seawater mixtures are smaller, 15 μm, and show a different morphology. They are dominated by the $\{10\bar{1}1\}$ common rhombohedral form and are approximately equant (1:1). A few crystals display $40\bar{4}1$ termina-

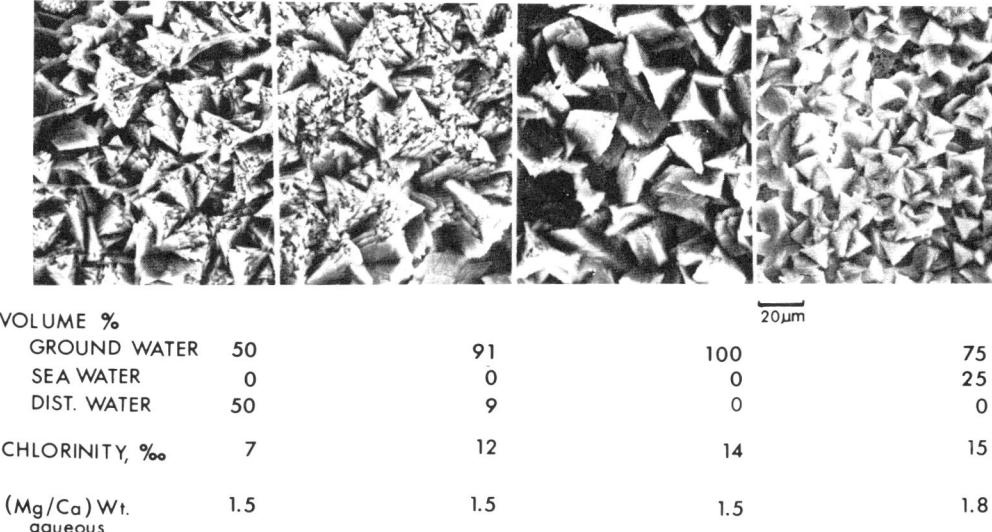

VOLUME %				
GROUND WATER	50	91	100	75
SEA WATER	0	0	0	25
DIST. WATER	50	9	0	0
CHLORINITY, ‰	7	12	14	15
(Mg/Ca) Wt. aqueous	1.5	1.5	1.5	1.8

FIG. 4.—SEM photographs showing variation in crystal morphology as a function of solution composition for cacite crusts produced by degassing mixtures of beach groundwaters with distilled water and seawater. All crusts are apparently low-Mg calcite. Photos are taken perpendicular to original solution-atmosphere interface. All photos at same scale.

tions at one end and $10\bar{1}1$ at the other. There is some preferential orientation of c-axes perpendicular to the interface.

It is possible to estimate rates of precipitation of the calcite. A calcite crust 30 μm thick with a porosity of 0.5 would contain 4×10^{-5} moles of $CaCO_3$ per cm^2 area of crust. The most rapid precipitation of calcite in the experiments occurred for beach groundwater in which crusts of the properties described above formed within a period of 12 hours. This yields a probable maximum rate of local precipitation of approximately 9×10^{-10} moles $CaCO_3$ cm^{-2} s^{-1}. The amount of precipitate formed in each of the bottles did not appear to increase significantly after a day or so of first appearance. The above rate of precipitation would be sufficient to indurate a sediment within a 12-hour period. Actual rates of precipitation in natural beach settings are probably less because of the necessity for diffusive transport of CO_2 through overlying sediments.

The carbonate aggregates formed during degassing and mixing closely resemble in crystal size, morphology, and composition natural calcite crusts formed as water-table cements in Boiler Bay (See figures in Boleneus, 1972; Moore, 1977) and in other beach areas. Beach rock cements, on the other hand, are usually high-Mg calcite and/or aragonite and only rarely low-Mg calcite (Ward, 1975). It is beyond the scope of this paper to speculate on the controls on cement mineralogy and composition. Bulk solution composition, rate of precipitation, and epitaxial growth controls all could conceivably play a role. This aspect of beach rock formation, however, is of secondary importance to the main question at hand here: what is the ultimate cause of the precipitation of cement? The present series of experiments demonstrate the following:

1. Crust-like precipitates formed from pure, unmixed beach groundwater at the air-water interface, indicating that neither mixing of different waters nor unusual substrates were necessary requirements for their formation.

2. The time of first appearance of the precipitate and quantity of precipitate formed decreased with increasing proportion of seawater or distilled water, suggesting that mixing inhibited precipitation.

3. Precipitates formed both in waters

treated with Hg_2Cl_2 and untreated waters, suggesting biologic enhancement or retardation of precipitation in this experiment was not a significant factor.

Controls on Saturation State

The mixing of carbonate waters is geochemically complex and is characterized by non-linear variations in saturation state. In theory, mixtures of solutions saturated with respect to calcium carbonate may be either undersaturated or supersaturated depending on external physical conditions and the compositions of the end-member waters (O'Neil, 1974; Wigley and Plummer, 1976).

Based on the known composition of the waters used in the mixing experiment, it is possible to estimate by calculation the saturation state of the solutions as a function of mixing and external P_{CO_2}. Calculations made in this study employ an ion-pairing model (Hanor, 1969) similar to that employed in other saturation programs (Truesdell and Jones, 1974). It is recognized here that there may be fundamental difficulties concerning the theoretical validity of the ion-pairing model for aqueous solutions (Whitfield, 1975), but the calculations which result from the assumption of inorganic complexing appear to be sufficiently accurate for the purposes of this paper.

In calculating saturation state in mixed waters, the total concentrations of all major dissolved components, including alkalinity, were assumed to be conservative. Compositions of the end-member beach water and seawater are given in Table 1. Using the ion-pairing model, values for P_{CO_2}, total dissolved carbonate, and free-ion activity products were calculated for selected mixtures of these waters for assumed values of pH. From this information it is possible to determine saturation behavior during mixing both in 1) *closed systems,* where there is no exchange of CO_2 and total dissolved carbonate is conserved, and in 2) *open systems,* where P_{CO_2} is externally controlled.

Figure 5 shows calculated values of the ion activity product, $IAP = (a_{Ca}2+) \times (a_{CO_3}2-)$, as a function of P_{CO_2} and degree of mixing between beach ground water and normal seawater. Two additional contours are shown which represent compositions of waters in thermodynamic equilibrium with

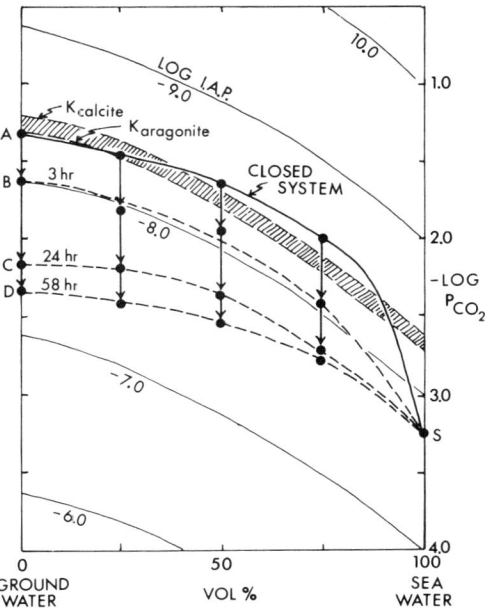

FIG. 5.—Plot of log IAP, $(a_{Ca^{2+}} \times a_{CO_3^{2-}})$, versus $-\log P_{CO_2}$ and percent mixing of beach groundwater and seawater. Line A-S shows the composition of mixed waters initially closed with respect to CO_2. Intermediate compositions are undersaturated with respect to calcite and aragonite. Lines B-S, C-S, and D-S, show measured composition and progressive supersaturation of mixed waters (solid circles) open to the atmosphere after 3 hours, 24 hours, and 58 hours, respectively. All waters, A-S through D-S, are supersaturated with respect to dolomite.

calcite and aragonite. Line A-S, represents water compositions produced by mixing end-member waters in systems closed to CO_2. This line also represents the initial composition of waters produced in the experiment prior to their degassing.

As can be seen from Figure 5 groundwater from the back beach is approximately saturated with respect to aragonite. Seawater is several times supersaturated with respect to both calcite and aragonite. Waters produced by progressively mixing seawater with back beach water in a closed system show a minimum in IAP at approximately 30% seawater, and a range of intermediate waters is undersaturated with respect to both calcite and aragonite. A similar relation, not shown, occurs for waters produced during closed-system mixing of groundwater with distilled water initially in equilibrium with the atmosphere. These calculations are consistent with several previous studies (*e.g.*, Bögli, 1964; O'Neil, 1974; Wigley and Plummer, 1976) which have shown that mixing of calcite-saturated waters of different P_{CO_2} can produce undersaturated waters of intermediate composition. This has been termed the "ΔP_{CO_2} effect" by Wigley and Plummer (1976). Although mixing of waters of diverse composition has been invoked as a mechanism for precipitating beach rock cement (Schmaltz, 1971), it is more likely that this mechanism alone would produce waters of *lower* saturation state, less capable of precipitation and possibly capable of dissolution.

Both the mixed and the end-member waters used in the experiment were open to the atmosphere and those with a high initial P_{CO_2} lost dissolved CO_2 by transport of the gas across the solution-air interface. This CO_2 loss was reflected by a progressive increase with time in pH of the surface water in each bottle and eventually by the formation of a calcite precipitate at the interface in some of the bottles. We can see from Figure 5 that if back beach water were brought into equilibrium with a normal atmospheric P_{CO_2} of approximately $10^{-3.5}$, it would be over a hundred times supersaturated with respect to calcite and aragonite. The degree of supersaturation at which calcite actually began to precipitate in the experiments is probably less than this figure, but its exact value for each water cannot be determined from the present experiment. Alkalinity and pH measurements were made of waters collected from the upper 1 cm of solution in each bottle and this data has been used to calculate the change in P_{CO_2} and saturation state of surface solutions with time shown by curves A-S through D-S in Figure 5. Actual P_{CO_2} values in the uppermost few tens of microns of solution were probably significantly less and the state of supersaturation prior to the onset of precipitation possibly greater at any given time than the depth-integrated values shown in Figure 5. Although not shown in Figure 5, all mixtures, A-S through D-S, are supersaturated with respect to dolomite.

Mixing Versus Degassing

Recent work on the thermodynamic saturation state of mixed carbonate waters has shown that differences in P_{CO_2}, ionic strength, and temperature of calcite-saturated end-members lead to waters of intermediate composition which are undersaturated with respect to calcite (O'Neil, 1974; Plummer, *et al.*, 1976; Wigley and Plummer, 1976). Of all these factors, P_{CO_2} and salinity are particularly variable in a beach groundwater system, and it is improbable that supersaturated waters can be produced simply by mixing of sea water with most beach groundwaters. Degassing of CO_2 or some other mechanism would be necessary to cause precipitation of carbonate.

I have shown that loss of CO_2 from St. Croix waters is sufficient to produce calcite crusts similar to some water-table cements observed in beach sediments. In the next section I will discuss factors which could result in degassing in a general beach setting.

CONTROLS ON CO_2 CONTENT OF BEACH GROUNDWATERS

Introduction

Figure 6 is a generalized representation of the hydrologic regime of an unconfined coastal aquifer. Groundwater flows seaward in response to a potential gradient and discharges into the ocean in a zone at and just below sea level. Because of tidal fluctuations in sea level, the top of the groundwater table, the phreatic surface, oscillates vertically. The amplitude of oscillation decreases landward. Mixing of groundwater and seawater occurs as a result of fluid dispersion around fixed sediment particles, and a zone of water of

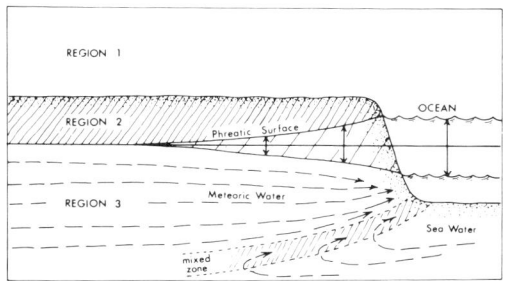

FIG. 6.—Idealized cross-section of an unconfined coastal aquifer dominated by a seaward flux of meteoric water. Region 1 is the atmosphere. Region 2, the vadose zone, is sediment dominated by gas-filled pore space. Region 3, the phreatic zone, is sediment dominated by water-filled pore space. The interface between regions 2 and 3 oscillates vertically in response to tidal fluctuations.

intermediate composition is formed. At high seaward groundwater fluxes and low tidal amplitudes, however, the interface between seawater and groundwater may be sharply defined (Bear, 1972).

An idealized vertical profile through the beach is characterized by three regions: *Region 1,* the atmosphere; *Region 2,* the vadose zone, sediment with air-filled pore space; *Region 3,* the phreatic zone, sediment with waterfilled pore space. On St. Croix and in many other tropical coastal regions, a soil zone is developed on beach sediments landward from the immediate shore area. Carbon dioxide is produced in the sediment column by plants living in the soil zone and by bacterial decay of organic material washed down into the beach groundwater. CO_2 concentrations in the system are lowest in Region 1, the atmosphere, and CO_2 thus continuously diffuses upward in response to this concentration gradient (Fig. 8).

The vertical variation in CO_2 with time within each of the two regions of the sediment column is due primarily to three variables. The first is vertical dispersive transport, which is the result both of molecular diffusion caused by concentration gradients and of fluid dispersion or mixing indused by vertical advective flow of the sediment atmosphere and groundwater column up and down through a fixed framework of sedimentary particles.

The second is bulk advective transport, caused primarily by tidal oscillation. In Region 3, the phreatic zone, the time-integrated transport by this process will be zero. In Region 2, the vadose zone, however, tidal pumping will push CO_2-rich air up across the sediment-air interface and pull CO_2-poor air back in its place, resulting in the net upward transport of CO_2. Advection of the gas phase in the vadose zone could further be caused by stochastic variations in barometric pressure and by diurnal changes in temperature.

Finally, variations in CO_2 can be caused by *in-situ* production or consumption of CO_2. This process at St. Croix is envisioned as being primarily a function of biological production of CO_2 by respiration. Rate of respiration in turn, however, may be a function of the rate of oxygen or nutrient influx into the system. Secondary *in-situ* controls of CO_2 concentration in the vadose zone may include absorption or desorption of CO_2 gas from thin films of water on mineral grains.

Each of the three factors above is influenced by variations in porosity. In a carbonate beach, porosity can vary as a result of solution and precipitation of solids and because of the downward eluviation of fine-grained particles generated in the soil zone. The most critical variation in porosity, however, may occur in the vadose zone, where transient, but significant variations in porosity can occur as a result of partial saturation of air-filled pores by rainwater and by groundwater left behind during tidal oscillation.

Fluid Dispersion as a Transport Mechanism

Mixing occurs when parts of a fluid are physically divided and recombined during flow around fixed particles in a porous sediment. This mixing, or fluid dispersion, tends to even out differences in fluid composition. In a beach, for example, a highly stratified water column could eventually be homogenized simply by continually moving it up and down over short distances through a fixed framework of solid grains. As we will see, the vertical velocity, V (cm sec^{-1}), of the groundwater table relative to the sediment-air interface is an essential factor in estimating the relative importance of fluid dispersion in the vertical transport of CO_2. It can be readily shown that for simple diurnal tidal oscillations, such as those that characterize the shores of St. Croix, the average absolute velocity $\bar{V} = 2.3 \times 10^{-5} h$, where h (cm) is the total tidal range (*i.e.*, twice the amplitude of the oscillation).

Dominick, *et al.,* (1973) have measured the variation in groundwater level in a series of wells in a tropical carbonate beach on Grand Cayman Island. They found that tidally-induced oscillations are attenuated in amplitude landward. I have taken their field data and have calculated the ratio of maximum observed range of groundwater level to maximum tidal range for two different tidal oscillations. This ratio is plotted as a function of distance back from the shore line (Figure 7). At approximately 20 meters landward from the shore line at low tide, groundwater oscillations have half the amplitude of open

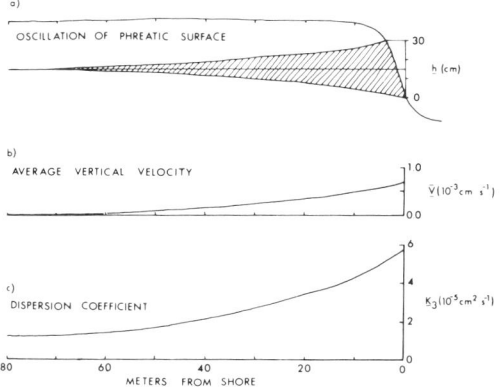

FIG. 7.—Vertical transport parameters in a carbonate beach of the type studied on Grand Cayman by Dominick, et al. (1973), subjected to a 30-cm tidal range. a) shows magnitude of oscillation of phreatic surface as a function of distance from the shore. b) is the integrated average absolute vertical velocity of beach groundwater. c) shows the significant increase in the dispersion coefficient, K_3, for dissolved CO_2 in the phreatic zone from the back beach to near shore area. It is proposed that the increase in K_3 can result in degassing of CO_2 from the phreatic zone and the precipitation of carbonate cements.

tidal oscillations. At 70 meters, variations in water level are within the error of field measurement. Although these data are specific to the beach studied by Dominick, et al., they provide some idea of the order of magnitude of water table oscillation that could be expected in similar carbonate beaches.

Figure 7 shows the landward variation in average vertical velocity of the groundwater table for a hypothetical carbonate beach which has the physical properties of the Grand Cayman beach and which is subjected to a simple diurnal tide with a total tidal range of 30 cm. This type of tide is typical of many coastal areas of the Caribbean. Maximum vertical velocities of the groundwater table are 1.1×10^{-3} cm sec^{-1} at the shore line and decrease landward to an estimated 1.1×10^{-5} cm sec^{-1} or less at 70 meters.

The dispersion coefficient, K, is a measure of the ease with which compositional differences within a fluid phase in a porous medium can be equalized by a combination of molecular diffusion and fluid dispersion. Experimental studies of fluids flowing through unconsolidated porous media have shown that the combined effects of diffusion and dispersion can be expressed by the relation (Fried and Canbarnous, 1971):

$$K/D_o = (D_s/D_o) + \alpha (Vd/D_o)^m \quad (1)$$

where

D_o = Molecular diffusion coefficient in free medium (cm^2 sec^{-1})
D_s = Apparent or sediment diffusion coefficient corrected for tortuosity, L, where $D_s = D_o/L^2$ (Stoessell and Hanor, 1975)
V = Fluid velocity (cm sec^{-1})
d = Dispersivity or spherical grain diameter (cm)
α, m = Dimensionless constants

Beach sands are texturally complex, and few data on their hydrologic properties are presently available. I will consider an idealized case for homogeneous medium composed of packed spheres of uniform size. Let $d = 0.1$ cm and $D_s/D_o = 0.7$ (Perkins and Johnston, 1963). D_o for CO_2 in air is 1×10^{-2} and in pure water 1.9×10^{-5} cm^2 sec^{-1} (Horne, 1969). Let $V = \bar{V}$, the average absolute vertical velocity of the groundwater table at various distances for the beach shown in Figure 7. For values of (Vd/D) in the range of interest, I have assumed $\alpha = 0.5$ and $m = 1.2$ (see Fried and Canbarnous 1971). For this idealized case then, values of K_2 and K_3 for Regions 2 and 3 respectively, can be calculated as a function of distance back from the shoreline.

It can be readily shown that fluid dispersion of the gas phase in Region 2, the vadose zone, is negligible compared with molecular diffusion and that K_2 is nearly constant at 7×10^{-3}. Dispersion in the aqueous phase in Region 3, the phreatic zone, however, though negligible landward, is highly significant in areas of the beach near the shoreline (Fig. 6). At seventy meters inland, where fluctuations of the groundwater table are slight, dispersion is negligible, and K_3 approaches the values of the apparent molecular diffusion coefficient, $D_s = 1.3 \times 10^{-5}$. At twenty meters inland, K_3 has increased to 3.4×10^{-5}. At the shoreline, $K_3 = 5.8 \times 10^{-5}$. There is thus over a four-fold lateral increase in K_3 from back beach to shore.

The important conclusions that can be derived from the above calculations are that 1) dispersion as a mechanism of net mass transport will be more significant in the aqueous phase of the phreatic zone than in the gas phase of the vadose zone, and 2) the potential exists even with a modest tidal range for a significant increase in the dispersion coefficient in the phreatic zone as one moves from the back beach to the shoreline. It is likely that the above calculation represents a minimum value for the increase in K_3. In a real beach sediment, tortuosities (L) and dispersivities (d) are probably greater than the values assumed for the homogeneous packed spheres. From equation 1, this would result in a much larger increase in K with V.

Factors Which Could Cause Degassing

A formal mathematical description of the vertical mass transport and loss of CO_2 from a beach system has been developed (Hanor, in prep.). A qualitative treatment will be used for the conceptual model developed in this paper.

In August 1976, a series of soil atmosphere samples were taken in Boiler Bay (Hanor, in preparation). The concentration profile in Region 2 of the back beach is approximately linear, and CO_2 values range from 1.2×10^{-5} moles liter^{-1} of gas phase at the sediment-air interface to 2.0×10^{-4} moles liter^{-1} near the top of the groundwater table (Figure 8). The measured value in Region 3 that is shown in Figure 8 comes from the screen level of a beach well and is 3.6×10^{-2} moles liter^{-1} of aqueous phase. It can be shown from basic laws of mass transport that most of the *in-situ* production of CO_2 in this profile is occurring in Region 3, the phreatic zone, and that the CO_2 profile should be convex downward as shown in curve t_1 in Figure 8. An estimate of the ratio of the dispersion coefficients, K_2/K_3 (Hanor, in prep.) yields the limiting slope for the CO_2 profile in the uppermost part of Region 3.

If the beach waters were subjected to significant tidal oscillation, fluid dispersion would become an important transport mechanism in the phreatic zone and K_3 would increase (Eqn. 1). As shown in the previous section, however, fluid dispersion would not

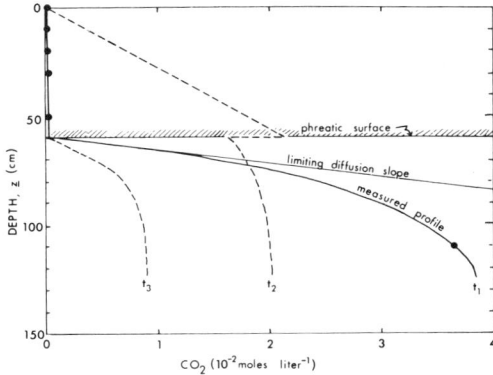

FIG. 8.—Probable distribution of CO_2 with depth for the back beach area in Boiler Bay in August, 1976 (solid circles, line t_1). Line t_3 is a hypothetical profile showing effects of degassing induced by tidal oscillation. Line t'_2 is a possible profile at an intermediate time (see text). The assumed offset in CO_2-concentrations at the phreatic surface reflects the preferential partitioning of the CO_2 molecule into the gas phase at equilibrium.

be as significant in the vadose zone, and, as a result, K_2/K_3 would decrease. At an extended period of time later, the CO_2 profile in Region 3 would appear as shown in curve t_3 in Figure 8, with a lowered limiting slope at the interface and lowered CO_2 concentration at any given depth below the interface. Even with rates of *in-situ* production of CO_2 constant with time, there would result a degassing of the phreatic zone with the decrease in CO_2 concentration at depth being approximately inversely related to the increase in K_3. Final CO_2 concentrations in the immediate vicinity of the interface, however, would decrease much less.

As vertical fluid dispersion becomes important in the phreatic zone, tidal pumping should become increasingly significant in the overlying vadose zone. The time-integrated effect of tidal pumping should be to shift the intercept of the CO_2-profile and all points at depth within the phreatic zone toward lower values.

If all of these waters in the phreatic zone were initially close to saturation with respect to calcium carbonate, the degassing caused by an increase in fluid dispersion and tidal pumping would be highly effective in causing supersaturation and possible precipitation.

If the increase in fluid dispersion with time is rapid, then there will be a series of transient, intermediate CO_2-profiles of the type indicated diagrammatically by profile t_2 in Figure 8. The CO_2 concentrations in the waters just below the interface at the time of increased tidal fluctuation could actually increase, giving these waters the capacity to dissolve carbonate. As degassing progresses, the CO_2 concentration of these waters will begin to decrease and possible supersaturation and precipitation at the interface will ensue.

There are other factors which could influence the CO_2-content of waters in the phreatic zone. These include mixing with underlying marine waters, changes in the rate of *in-situ* production of CO_2, changes in the height of the overlying sediment column, and changes in sediment porosity. By simply moving a parcel of groundwater to an area where depth to the water table is less, for example, the path length for diffusion in the vadose zone would be shortened and some degassing could take place. If the pore space in the vadose zone were partially saturated by water, as could occur during rain, the decrease in air-filled porosity should retard upward diffusion and cause an increase in the CO_2-content of the phreatic zone. A decrease in porosity, however, in addition to inhibiting the upward diffusion of CO_2 will also inhibit the downward diffusion of O_2, necessary for the aerobic production of CO_2 in Region 3. The longterm effect, then, could be to consume CO_2 by reaction with sulfide and ammonia during anaerobic respiration (Berner, 1971). This alone could cause the precipitation of carbonate.

In summary, as a parcel of beach groundwater flows seaward in an unconfined aquifer, several factors combine to increase the efficiency of vertical mass transport and potential degassing. These include, in probable order of importance for many beaches:

1. A significant increase in the dispersion coefficient for dissolved CO_2 in groundwater as induced by even modest tidal oscillation.
2. Tidal pumping across the sediment-atmosphere interface.
3. A decrease in the average depth to the water table.
4. An increase in porosity from back beach sediments containing eluviated soil particles to better-sorted forebeach sands.

It is likely that the specific contribution of each of the above will vary significantly from one beach to another.

CONCLUSIONS

It is highly probable that carbonate cements in Recent beach sediments can be produced simply by the seaward migration of CO_2-rich, carbonate-saturated groundwaters into shore areas where degassing can take place. Following Matthews (1971), all that ultimately is required is that the waters have sufficient residence time in the landward areas to acquire a high P_{CO_2} and dissolved calcium and carbonate and sufficient residence time in the forebeach areas that degassing can take place before the waters are discharged into the oceans. Such a sequence would be logical in tropical and subtropical coastal areas, where seasonal fluctuations in rainfall could cause episodic variations in groundwater discharge (Hanor and Moore, 1974). In the previous section, it has been shown that even coastal areas subjected to a modest tidal range of 30 cm, oscillation of the watertable in forebeach areas could greatly increase dispersive transport of CO_2 and result in degassing and supersaturation of groundwaters. Mixing of water types appears to be incidental to precipitation, but may influence cement chemistry and morphology (Figure 4).

The precipitation of beach cements is less likely in areas where high groundwater discharge results in low residence time and less likelihood of degassing in forebeach sediments. In discussing the occurrence of beachrock on Samoa, for example, Daly (1924, p. 138) describes a beach where beachrock is absent and where, ". . . the natives are able to get potable water at low tide by scooping out holes in the sand just in front of the lapping waves. . . ." Rather than agreeing with Daly's contention that this is a refutation of Field's (1919) degassing hypothesis for the origin of beachrock, I would accept this as a possible example of an area where groundwater fluxes are too high to permit *in-situ* build-up of CO_2 and/or degassing.

It is not necessary to have a permanent fresh-water lens for the formation of beach cements. In fact, the groundwater that may be involved in degassing and precipitation could be brackish or saline (Fig. 2). What is necessary, however, is that the overall annual seaward flux of groundwater be positive, so that continuous or episodic degassing according to the mechanism proposed above could take place. In highly arid regions, where coastal sediments are characterized by the continuous landward flow and deep refluxing of seawater, the degassing hypothesis proposed here would not be applicable.

In the degassing hypothesis, it is possible to make a case for the formation of beachrock being a self-limiting process both in terms of the relative volume of pore space which can be occluded by cement and of the maximum thickness of the cemented zone which can develop. When cement is precipitated, sediment porosity is decreased, and this decreases the ability of the system to degas. New beachrock may form in an area where older beachrock has been removed, but once porosity has again been reduced by precipitation, cementation may cease. Watertable cements, formed landward from the shore in the vicinity of the water table may also be precipitated by degassing of groundwaters derived from further inland. This would seem most likely in coastal areas subjected to large tidal ranges, where significant oscillations of the water table are propagated a considerable distance inland.

The precipitation of dolomite appears to be characteristic of some geologic environments where there is mixing of marine and meteoric waters. As proposed by Badiozamani (1973), this would most likely occur where mixing in systems closed with respect to CO_2 produces waters undersaturated with respect to calcite and aragonite, but supersaturated with respect to dolomite (for example, line A-S in Fig. 5). However, many beach systems, such as that on St. Croix, are thermodynamically open with respect to CO_2 and tend to maintain equilibrium with respect to calcite or aragonite (Hanor, in prep.). It is less likely that dolomite will precipitate here than in the closed systems postulated by Badiozamani.

The stable isotope geochemistry of beachrock and water-table cements should provide additional insight into sources of carbonate and mechanisms of precipitation. Available δC^{13} values for cements of this type show a wide range, $-40°/oo$ to $+4°/oo$ (Moore, 1973; Roberts and Whelan, 1975). It should be recognized that although many beach groundwaters are probably depleted in C^{13} relative to seawater because of the input of carbon derived from the oxidation of organic matter or of methane, preferential fractionation of C^{12} into the overlying vadose zone atmosphere and its loss by diffusion could cause progressive enrichment of C^{13} in a static groundwater mass.

The formation of beach rock is a complex geological process. I have considered in this paper only one aspect of the problem, the cause of precipitation of beach rock cements. A wide variety of precipitation mechanisms continue to be invoked in the literature. Of these, precipitation due to mixing of different water types, as we have seen, is probably not significant. Recent studies (Davies and Kinsey, 1973) have cast doubt on the direct biological precipitation of beach rock cements. Indirect biological precipitation, however, may be possible in anerobic environments where bacterial processes can consume CO_2 (Berner, 1971). Calcium carbonate can be precipitated out of sea water during evaporation. Continued evaporation of beach groundwaters, however, will produce a denser fluid which should reflux seaward at depth, rather than precipitate carbonate at the ground-water table or in the intertidal zone. Degassing of beach groundwaters has not received much detailed attention as a precipitation mechanism, but, as this paper hopes to demonstrate, would appear to be a phenomenon that could account for many of the known occurrences of beach rock.

ACKNOWLEDGMENTS

Acknowledgment is made to the Donors of the Petroleum Research Fund, administered by the American Chemical Society for the support of this research through grants 6031-AC2 and 7791-AC2-C. I would like to thank the staff of the West Indies Laboratory, Fairleigh Dickinson University, for their generous help during the field portion of this study. Virginia Colten, William DeLoach, Arthur Johnson, and Margie Thomas did most of the laboratory analyses. I am grateful o Lynn M. Walter for many useful discussions and for reviewing this manuscript.

REFERENCES

ADEY, W., 1975, The algal ridges and coral reefs of St. Croix, their structure and Holocene development: Atoll Res. Bull. 187, 67 p.

BATHURST, R. G. C., 1975, Carbonate sediments and their diagenesis (2d ed.): New York, Elsevier, 658 p.

BADIOZAMANI, K., 1973, The dorag dolomitization model—application to the Middle Ordovician of Wisconsin: Jour. Sed. Petrol., v. 43, p. 965–984.

BEAR, J., 1970, Two-liquid flows in porous media: Advances in Hydro-science, v. 6, New York, Academic Press, p. 142–252.

BERNER, R. A., 1971, Bacterial processes effecting the precipitation of calcium carbonate in sediments: In Bricker, O. P. (ed.) Carbonate cements, Johns Hopkins Univ. Press, p. 247–251.

BÖGLI, A., 1964, Mischungskorrosion: ein Beitrag zum Verkastungsproblem: Erkunde, v. 18, p. 83–92.

BOLENEUS, D., 1972, Diagenesis of some Holocene, intertidal carbonate sands, St. Croix, U.S. Virgin Islands: Louisiana State Univ., M.S. Thesis, 115 p.

DALY, R. A., 1924, The geology of American Samoa: Carnegie Inst. of Wash. Pub. 340, p. 93–143.

DAVIES, P. J. AND D. W. KINSEY, 1973, Organic and inorganic factors in Recent beach rock formation, Heron Island, Great Barrier Reef: Jour. Sed. Petrology, v. 46, p. 952–966.

DOMINICK, T. F., B. WILKINS, AND H. H. ROBERTS, 1973, A one-dimensional mathematical model for beach ground-water fluctuations: LSU Coastal Studies Inst. Tech. Report 152, p. 1–78.

EMERSON, J., 1975, Chemically enhanced CO_2 gas exchange in a eutrophic lake: A general model: Limnol. and Oceanog., v. 20, p. 743–753.

FIELD, R. M., 1919, Investigations regarding calcium carbonate oozes at Tortugas and the beach rock at Loggerhead Key: Carnegie Inst. Wash., Yearbook, v. 18, p. 197–198.

FOLK. R. L., 1974, The natural history of crystalline calcium carbonate: effect of magnesium content and salinity: Jour. Sed. Petrol., v. 44, p. 40–53.

FRIED, J. J., AND M. A. CANBARNOUS, 1971, Dispersion in porous media: In Advances in Hydro-science, v. 7, New York, Academic Press, p. 170–282.

FORMAN, R. T. T., 1974, An introduction to the ecosystems and plants on St. Croix: In Multer, H. G., and Gerhard, L. C., (eds.) Guidebook to the geology and ecology of some marine and terrestrial environments, St. Croix, U.S. Virgin Islands, West Indies Lab. Special Pub. 5, p. 201–238.

HANOR, J. S., 1969, Barite saturation in seawater: Geochim. et Cosmochim. Acta, v. 33, p. 894–898.

———, AND C. H. MOORE, 1974, Hydrogeochemistry of a carbonate beach area, St. Croix: West Indies Lab. Special Pub. 6, p. 10.

HORNE, R. A., 1969, Marine chemistry: New York, Wiley-Interscience, 568 p.

MATTHEWS, R. K., 1971, Diagenetic environments of possible importance to the exploration of cementation fabric in subaerially exposed carbonate sediments: In Bricker, O. P. (ed.) Carbonate cements, Johns Hopkins Univ. Press., p. 127–132.

MILLIMAN, J. D., 1974, Marine carbonates: New York, Springer-Verlag, 375 p.

MOORE, C. H., 1973, Intertidal carbonate cementation, Grand Cayman, West Indies: Jour. Sed. Petrol., v. 43, p. 591–602.

———, 1977, Beach rock origin: Some geochemical, mineralogical, and petrographic considerations: In Walker, H. J. (ed., R. J. Russell Symposium volume, Louisiana State Univ. (in press).

———, AND J. S. HANOR, 1974, Boiler Bay beach rock St. Croix: West Indies Lab. Special Pub. 5, 71–75.

O'NEIL, T. J., 1974, Chemical interactions due to mixing of meteoric and marine waters in a Pleistocene reef complex, Reo Bueno, Jamaica: Lousiana State Univ., M.S. Thesis, 186 p.

PERKINS, T. K., AND O. C. JOHNSTON, 1963, A review of diffusion and dispersion in porous media: Soc. Petrol. Engineers Jour., v. 1, p. 70–84.

PLUMMER, L. N., H. L. VACHER, F. T. MACKENZIE, O. P. BRICKER, AND L. S. LAND, 1976, Hydrogeochemistry of Bermuda: A case history of groundwater diagenesis of biocalcarenites: Geol. Soc. Amer. Bull., v. 87, p. 1301–1316.

ROBERTS, H. H., AND T. WHELAN, 1975, Methane-derived carbonate cements in barrier and beach sands of a subtropical delta complex: Geochim. et Cosmochim. Acta, va. 39, p. 1085–1089.

RUSSELL, R. J., 1970, Florida beaches and cemented water-table rocks: LSU Coastal Studies Inst. Tech. Rept. 88, 53 p.

SCHMALZ, R. G., 1971, Formation of beach rock at Eniwetok Atoll: In Bricker, O. P. (ed.) Carbonate cements, Johns Hopkins Univ. Press, Md p. 17–24.

STOESSELL, R. K., AND J. S. HANOR, 1975, A nonsteady state method for determining diffusion coefficients in porous media: Jour. Geophys. Res., v. 80, p. 4979–4982.

THORSTENSON, D. C., F. T. MACKENZIE, AND B. L. RISTVET, 1972, Experimental vadose and phreatic cementation of skeletal carbonate sand: Jour. Sed. Petrology v. 42, p. 162–167.

TRUESDELL, A. H., AND B. F. JONES, 1974, WATEQ, a computer program for calculating chemical equilibria of natural waters: U.S. Geol. Survey Jour. Res. v. 2, p. 233–248.

WARD, W. C., 1974, Collector's guide to carbonate cement types, northeastern Yucatan Peninsula: In Weidie, A. E. (ed.) Field Seminar on water and carbonate rocks of the Yucatan Peninsula, Mexico, New Orleans Geol. Soc., p. 176–198.

WHITFIELD, M., 1975, An improved specific interaction model for seawater at 25° C and 1 atmosphere total pressure: Marine Chem., v. 3, p. 197–213.

WIGLEY, T. M. L., AND L. N. PLUMMER, 1976, Mixing of carbonate waters: Geochim. et Cosmochim. Acta, v. 40, p. 989–998.

Part III

THE EQUILIBRIUM APPROACH

Editors' Comments
on Papers 17 Through 22

17 HEM
Excerpt from *Equilibrium Chemistry of Iron in Ground Water*

18 HANSHAW, BACK, and RUBIN
Carbonate Equilibria and Radiocarbon Distribution Related to Groundwater Flow in the Floridan Limestone Aquifer, U.S.A.

19 THRAILKILL
Chemical and Hydrologic Factors in the Excavation of Limestone Caves

20 LANGMUIR
The Geochemistry of Some Carbonate Ground Waters in Central Pennsylvania

21 BRICKER and GARRELS
Mineralogic Factors in Natural Water Equilibria

22 TRUESDELL and JONES
WATEQ, A Computer Program for Calculating Chemical Equilibria of Natural Waters

Development of the equilibrium approach had a revolutionary impact on chemical hydrogeology. This approach had its beginnings in the late 1940s and early 1950s in the exploration of the Colorado plateau for uranium ores that brought to the attention of the geologic profession many of the concepts of chemical thermodynamics. The first of these concepts was the Nernst equation relating electromotive force, pH, temperature, and dissolved species (Krumbein and Garrels, 1952; McKelvey et al., 1955; Evans and Garrels, 1958).

By 1960, Garrels (1960) and Hem (1959) had completed laboratory investigations studying the speciation of dissolved iron and iron phases in relation to the Eh-pH concept. Paper 17 is a synthesis of these studies and later work. Based largely on this work, the first application of redox to groundwater was concerned with high iron concentration in the Maryland coastal plain aquifers (Barnes and Back, 1964a; Back and Barnes, 1965).

Two review articles appeared in 1960 that summarized the Eh-pH relationships in natural environments: the first by Baas Becking et al. (1960), and the second by Sato (1960) on application of the redox concept to oxidation of sulfide ores. The construction, application, and interpretation of Eh-pH diagrams with respect to geologic environments are discussed by Garrels and Christ (1964), Cloke (1966), and Stumm and Morgan (1970).

It is ironic that the entire field of aqueous geochemistry had its beginnings in a concept as controversial as redox. This controversy is concerned with (1) whether the conceptual potential should be referred to as "Eh" symbolizing the electromotive force of the half-cell reaction or should be "pe" symbolizing the negative logarithm of the concentration of an assumed aqueous electron which perhaps does not exist; (2) the validity of field measurements; and (3) whether the redox concept is applicable to nonequilibrium systems. Nevertheless, general agreement exists that the concept of oxidation potential can be used effectively to establish fields of stability for aqueous and solid phases as demonstrated by Hem (Paper 17). The redox concept has been corroborative in studies of mine drainage problems and of well-casing corrosion and encrustation (Barnes and Clarke, 1969; Clarke and Barnes, 1969). Edmunds (1973) measured oxidation potentials in the field that were correlative with the distribution and behavior of trace and minor constituents in groundwater in the Lincolnshire limestone in eastern England. Because of the difficulty of measuring oxidation potentials in the field and their controversial significance, much thought has been given to the possibility of identifying and using species couples of carbon, nitrogen, or sulfur to determine redox conditions as discussed by Thorstenson (1970).

Chemical thermodynamics was first applied in 1957 to carbonate equilibrium in groundwater in a term paper for a seminar of which R. M. Garrels was the professor (Back, 1961). This paper was based partly on Paper 5 by Langelier; a series of papers giving the dissociation constants of carbonic acid and the solubility product of calcite, which were well known in sanitary engineering but were not part of the geologic literature; and the framework provided by Garrels and Dreyer (1952). Field application of this approach resulted in several publications of which Paper 18 is an example.

During the early 1960s, the validity of the equilibrium model in natural systems was examined by Hall (1963) and Hsu (1963) who compared chemical character of groundwater with calculated values based on mineral solubilities; by Hostetler (1964) who studied the degree of saturation of water with reference to carbonate minerals that had various amounts of magnesium and calcium; by Barnes and Back (1964b) who determined the solubility product for dolomite from chemical analyses of groundwater; and by Barnes (1965) who studied the water from a small stream in eastern California to determine the causes of the control on travertine deposition from streams. Later work included that by Cherry (1968) who studied the mineral equilibria of gypsum in brackish waters and by Back and Hanshaw (1970) who compared the results of their investigation in Florida to a similar geologic terrane of the Yucatan Peninsula. Examples of the application of carbonate equilibria to regional aquifer systems are given by Cherry (1972), Paper 19, and Paper 20.

Research on silicate geochemistry has proceeded along lines somewhat different from that of carbonate geochemistry because of the difficulty in obtaining free energy values and, therefore, mineral solubilities. In studies of silicates the emphasis has been on mass-balance concepts and activity diagrams as in Bricker and Garrels (Paper 21); Mackenzie and Garrels (1966); Mackenzie et al. (1967); Garrels (1967); Garrels and Mackenzie (1967); Bricker et al. (1968); Cleaves et al. (1970, 1974); and Harriss and Adams (1966). Excellent background for these papers can be obtained by studying chapter four in Krauskopf (1967) that reviews the basic chemical reactions of silicate weathering.

Because of the quantity and complexity of the calculations required in these regional geochemical studies, they could not have been undertaken without the development of computer programs. The first such model and program, called IONIC ACTIVITY, was prepared by Fred Sower and Ivan Barnes in 1962; it calculated the departure from equilibrium for only three minerals—calcite, dolomite, and gypsum. Current programs have the data for more than one hundred minerals. A great number of programs are now available, each of which has a slightly different chemical model. Two commonly used are SOLMNEQ (Kharaka and Barnes, 1973), and WATEQ by Truesdell and Jones (Paper 22). Thermodynamic properties of minerals for many of the models were obtained from Robie and Waldbaum (1968). Recently, a group of scientists active in developing aqueous models and programs prepared an evaluation of the approaches, assumptions, and limitations inherent in them (Nordstrom et al., 1979).

REFERENCES

Baas Becking, L. G. N., I. R. Kaplan, and D. Moore, 1960, Limits of the Natural Environment in Terms of pH and Oxidation-Reduction Potentials, *Jour. Geology* **68:**243-284.

Back, W., 1961, Calcium Carbonate Saturation in Ground Water from Routine Analysis, *U.S. Geol. Survey Water-Supply Paper 1535-D,* pp. 1-14.

Back, W., and I. Barnes, 1965, Electrochemical Potential and Iron-Bearing Waters Related to Ground-Water Flow Patterns, *U.S. Geol. Survey Prof. Paper 498-C,* pp. 1–16.

Back, W., and B. B. Hanshaw, 1970, Comparison of Chemical Hydrogeology of the Carbonate Peninsulas of Florida and Yucatan, *Jour. Hydrology* **10:**330–368.

Barnes, I., 1965, Geochemistry of Birch Creek, Inyo County, California, A Travertine Depositing Creek in an Arid Climate, *Geochim. et Cosmochim. Acta* **29:**85–112.

Barnes, I., and W. Back, 1964a, Geochemistry of Iron-Rich Ground Water of Southern Maryland, *Jour. Geology* **72:**435–447.

Barnes, I., and W. Back, 1964b, Dolomite Solubility in Ground Water, *U.S. Geol. Survey Prof. Paper 475-D,* pp. 179–180.

Barnes, I., and F. E. Clarke, 1969, Chemical Properties of Ground Water and Their Corrosion and Encrustation Effects on Wells, *U.S. Geol. Survey Prof. Paper 498-D,* 58p.

Bricker, O. P., A. Godfrey, and E. T. Cleaves, 1968, Mineral-Water Interaction during the Chemical Weathering of Silicates, *Advances in Chemistry* **73:**128–142.

Cherry, J. A., 1968, Chemical Equilibrium between Gypsum and Brackish and Slightly Saline Waters at Low Temperatures and Pressures, *Chem. Geology* **3:**239–247.

Cherry, J. A., 1972, Geochemical Processes in Shallow Groundwater Flow Systems in Five Areas in Southern Manitoba, Canada, *Internat. Geologic Congress, 24th, Montreal, Canada,* Section 11, pp. 208–221.

Clarke, F. E., and I. Barnes, 1969, Evaluation and Control of Corrosion and Encrustation in Tube Wells of the Indus Plains, West Pakistan, *U.S. Geol. Survey Water-Supply Paper 1608-L,* 59p.

Cleaves, E. T., A. Godfrey, and O. P. Bricker, 1970, Geochemical Balance of a Small Watershed and its Geomorphic Implications, *Geol. Soc. America Bull.* **81:**3015–3032.

Cleaves, E. T., D. Fisher, and O. P. Bricker, 1974, Chemical Weathering of Serpentine in the Eastern Piedmont of Maryland, *Geol. Soc. America Bull.* **85:**437–444.

Cloke, P. L., 1966, The Geochemical Application of Eh-pH Diagrams, *Jour. Geol. Education* **14:**140–148.

Edmunds, W. M., 1973, Trace Element Variations across an Oxidation-Reduction Barrier in a Limestone Aquifer, *Proceedings of the Symposium on Hydrogeochem. and Biogeochemistry, vol. 1 Tokyo 1970,* The Clarke Company, Washington, D.C., pp. 500–526.

Evans, H. T., Jr., and R. M. Garrels, 1958, Thermodynamic Equilibria of Vanadium in Aqueous Systems as Applied to the Interpretation of The Colorado Plateau Ore Deposits, *Geochim. et Cosmochim. Acta* **15:**131–149.

Garrels, R. M., 1960, *Mineral equilibria at low temperature and pressure,* Harper & Bros., New York, 254p.

Garrels, R. M., 1967, Genesis of some Ground Waters from Igneous Rocks, in *Researches in Geochemistry,* vol. 2, P. H. Abelson, ed., Wiley New York, pp. 405–420.

Garrels, R. M., and C. L. Christ, 1964, Solutions, minerals, and equilibria, Harper and Row, New York, 450p.

Garrels, R. M., and R. M. Dreyer, 1952, Mechanism of Limestone Replacement at Low Temperatures and Pressures, *Geol. Soc. America Bull.* **63:**325-379.

Garrels, R. M., and F. T. Mackenzie, 1967, Origin of the Chemical Compositions of some Springs and in Lakes, *Advances in Chemistry* **67:**222-242.

Hall, F. R., 1963, Calculated Chemical Composition of Some Sulfate-Bearing Waters, *Internat. Assoc. Sci. Hydrology, Pub. No. 64*, pp. 7-15.

Harriss, R. C., and J. A. Adams, 1966, Geochemical and Mineralogical Studies on the Weathering of Granitic Rocks, *Am. Jour. Sci.* **264:**146-173.

Hem, J. D., 1959, Study and Interpretation of the Chemical Characteristics of Natural Water, *U.S. Geol. Survey Water-Supply Paper 1473*, 363p.

Hostetler, P. B., 1964, The Degree of Saturation of Magnesium and Calcium Carbonate Sediments in Natural Waters, *Internat. Assoc. Sci. Hydrology, Commission of Subterranean Waters Pub. 64*, pp. 34-49.

Hsu, K. J., 1963, Solubility of Dolomite and Composition of Florida Ground Waters, *Jour. Hydrology* **1:**288-310.

Kharaka, Y., and I. Barnes, 1973, SOLMNEQ: Solution-Mineral Equilibrium Computations, *National Technical Information Service (NTIS)*, PB 215/899, 82p.

Krauskopf, K. B., 1967, *Introduction to Geochemistry*, McGraw-Hill, New York, 721p.

Krumbein, W. C., and R. M. Garrels, 1952, Origin and Classification of Chemical Sediments in Terms of pH and Oxidation-Reduction Potentials, *Jour. Geology* **60:**1-32.

McKelvey, V. E., D. L. Everhart, and R. M. Garrels, 1955, Origin of Uranium, *Econ. Geology* **50:**464-533.

Mackenzie, F. T., and R. M. Garrels, 1966, Chemical Mass Balance between Rivers and Oceans, *Am. Jour. Sci.* **264:**507-525. (Reprinted as Paper 6 in *Geochemistry of Water*, Y. Kitano, ed., Benchmark Papers in Geology, vol. 16, Dowden, Hutchinson & Ross, Stroudsburg, Pa., pp. 120-138 and as Paper 5 in *Sea Water*, J. Drever, ed., Benchmark Papers in Geology, vol. 45, Dowden, Hutchinson & Ross, Stroudsburg, Pa., pp. 97-115.)

Mackenzie, F. T., R. M. Garrels, O. P. Bricker, and F. Bickley, 1967, Silica in Sea Water: Control by Silica Minerals, *Science* **155:**1404-1405. (Reprinted as Paper 6 in *Sea Water*, J. Drever, ed., Benchmark Papers in Geology, vol. 45, Dowden, Hutchinson & Ross, Stroudsburg, Pa., pp. 116-117.)

Nordstrom, D. K. et al., 1979, A Comparison of Computerized Models for Equilibrium Calculations in Aqueous Systems, in *Chemical Modeling in Aqueous Systems*, E. A. Jenne, ed., American Chemical Society, Washington, D.C., pp. 857-892.

Robie, R. A., and D. R. Waldbaum, 1968, Thermodynamic Properties of Minerals and Related Substances at 299.15°K (25.0°C) and One Atmosphere (1.013 Bars) Pressure and at Higher Temperatures, *U.S. Geol. Survey Bull. 1259*, 256p.

Sato, M., 1960, Oxidation of Sulfide Ore Bodies, Geochemical Environments in Terms of Eh and pH, *Econ. Geology* **55:**928-961.

Stumm, W., and J. J. Morgan, 1970, *Aquatic Chemistry*, John Wiley & Sons, New York, 583p.

Thorstenson, D. C., 1970, Equilibrium Distribution of Small Organic Molecules in Natural Waters, *Geochim. et Cosmochim. Acta* **34:**745-770.

17

Copyright ©1967 by John Wiley & Sons, Inc. Publishers
Reprinted by permission from pages 625-637 and 642-643 of Principles and Applications of Water Chemistry, S. D. Faust and J. V. Hunter, eds., Wiley, New York, 1967, 643p.

EQUILIBRIUM CHEMISTRY OF IRON IN GROUND WATER

John D. Hem
U. S. Geological Survey
Menlo Park, California

Differences in solid mineral assemblages and impaired circulation of water often bring about chemical variation in ground water. The solution of iron may be favored at some places whereas the precipitation of iron may be favored at other places. Mixing of water from all the water saturated material to which a well is open may yield a water that is objectionably high in dissolved iron. At the same time, encrustation in the well may interfere with water movement. Exposure of reduced iron minerals, such as pyrite, to an oxidizing condition or of oxidized iron minerals, such as limonite, to reducing conditions, brings large amounts of iron into solution. If bicarbonate species are present, the solubility of iron may be controlled by the solubility of siderite under conditions where both pyrite and limonite are unstable. Large amounts of iron can be present in solution at equilibrium with siderite.

Equilibrium relations have explained the occurrence of iron in ground waters in portions of the Atlantic Coastal Plain and Mississippi Embayment geologic regions of the United States. These relations also aid in explaining encrustation of wells with siderite in certain irrigated areas of West Pakistan. Proper well construction and operation techniques may aid in avoiding problems resulting from iron-bearing ground water.

APPLICABILITY OF EQUILIBRIUM CONCEPTS

The chemical behavior of iron in underground waters often seems complex and puzzling. For example, two adjacent wells may yield water of similar composition except that one is high in iron content whereas the other is low or the iron content may fluctuate with time in an apparently irrational way. Besides the problem of treatment of the pumped water to decrease the iron content, some wells present additional problems of declining yields caused by encrustation of iron compounds on well screens or at other openings where water enters the well.

A sufficient understanding to explain the observed behavior of iron in ground water and to predict its occurrence can be obtained from suitable applications of principles of chemical equilibrium Difficulties in sampling and in extrapolating laboratory results to field conditions prevent completely rigorous application of the equilibrium model but it has been generally useful in explaining field observations.

In this discussion the physical and chemical features of ground-water systems will be outlined in a general way and illustrated with specific examples. Wider application of chemical principles in planning well construction and operation offers promise of coping more effectively with problems related to the high iron content of some ground water.

GROUND-WATER CIRCULATION

Water at and near the land surface flows in response to imposed energy potentials. That portion of the circulating water which is present in the zone where interconnected pores and other openings of rock strata are saturated completely, is commonly termed ground water. It moves from areas of recharge or replenishment toward areas of discharge in response to hydraulic head or potential. The bodies of rock capable of transmitting and yielding water are termed aquifiers. That portion of the ground-water body which extends to a depth of a few thousand feet is of greatest interest to man as a source of readily recoverable water although locally it contains undesirable quantities of dissolved ions.

In general, recharge to the ground-water body represents water from precipitation moving downward through the soil and subsoil or rock to the saturated zone. Ground water moves at a relatively slow rate through the saturated section following a pattern that often is represented by a simple flow net. When this general picture is examined in more detail, however, it becomes evident that water-bearing rocks transmit water in a non-uniform way. Flow is more rapid through the coarse-grained, well-sorted detrital sediments and slower where the grains are fine or where interstices are partly closed as by cementation. Movement of water generally occurs more freely along the bedding surfaces of rocks than across them. Crystalline rocks may transmit water only along fractures whereas carbonate rocks may transmit water principally through channels enlarged by solution. Vertical circulation of ground water often is considerably impaired. In an artesian system, the drastic impairment of vertical movement confines ground water and causes development of the pressure that is the distinguishing feature of the artesian aquifer.

Ground water may discharge naturally where the zone of saturation intersects the land surface. It may be withdrawn at a well where the hydraulic potential is artificially lowered and water movement toward the well thereby is induced.

A well generally penetrates a considerable distance into the aquifer wherefrom the water represents a mixture of that moving in various parts of the adjacent saturated zone. The mixture will contain a larger portion of water from the more permeable zones than from the less permeable ones. As the head around the well declines with continued pumping, movements of water considerably different from the initial natural flow patterns will take place.

Operation of the pump may change conditions in the water in the well bore by introducing air or by causing dissolved gases to be released. The latter effect could cause an increase in pH of the solution when CO_2 is released. Even when it is not being pumped, the well bore provides a short circuit through which waters from different zones that do not naturally contact each other, may have opportunity for intimate mixing.

GEOCHEMICAL SYSTEMS IN GROUND-WATER AQUIFERS

In the usual ground-water system a relatively large surface area of solid minerals is exposed to slowly circulating water. A tendency toward establishment of an equilibrium between dissolved ions and solid minerals is promoted by these conditions. Non-homogeneous composition of the aquifer, however, can be expected to influence the chemical composition of the circulating water as well as the flow patterns that control

with aquifer materials. The nature of solid mineral alteration products and the composition of the solution in an aquifer are controlled by the total environment. For example, near sources of recharge the ground water contains dissolved oxygen and CO_2 replenished by contact with the atmosphere and soil air. Under these conditions carbonates tend to dissolve and oxidized iron minerals are generally stable. In sections of the saturated zone far separated from the air, especially where organic matter may be present to consume any dissolved oxygen, reducing conditions are to be expected wherein reduced iron minerals will be stable. Transitional zones and the seasonal rise and fall of the water table impose oxidizing conditions on reduced minerals and vice versa. Thus, iron-bearing minerals are attacked and parts of the saturated zone may at times contain water high in dissolved iron.

Wells obviously may penetrate strata containing water of differing composition. Solutions entering the well bore, however, are mixed together in the water discharged through the pump. Hence, composition of water samples from a well that withdraws water from two or more zones of differing chemical composition may not help much in understanding the geochemistry of the ground water. Usually, in studying an area, wells can be found which are constructed so that they intercept only shallow or only deep ground water. In such places, some insight into the changes in composition with depth can be obtained.

THE Eh- pH DIAGRAM

The actual solubility of iron under different conditions at equilibrium can be calculated conveniently with the results shown graphically by means of the Eh-pH or stability-field diagram. Such diagrams are prepared using the Nernst equation and other chemical thermodynamic relationships and data. The procedures are described by Garrels (1) and have been used by the writer in other reports. The two variables, Eh and pH, can be used to define areas of "stability" for compounds whose free energies are known if other pertinent conditions are specified. For iron oxide or hydroxide species, the only other variable is the dissolved iron activity. It is assumed the conditions of significance are those in which water itself is stable at $25°C$ and 1 atm pressure.

A diagram of iron-species stability fields for a total dissolved iron activity of 2×10^{-7} molal was given by Hem and Cropper (2). If additional concentration lines are shown, the diagram can be used as an indication of the solubility of iron in the presence of the assumed solids. The indicated solids are those stable under the specified conditions of pH and Eh.

Natural waters contain other solutes, notably dissolved CO_2 and S species, which are involved in precipitation or solution of other solid forms of iron. To represent clearly the effects of these solutes in a two-dimensional diagram, however, requires that a constant single activity for each be used. The results of different amounts of dissolved CO_2 and S species can be represented by using different diagrams each with a single activity of these constituents.

To demonstrate the principal factors involved in establishing equilibrium iron solubility, three Eh-pH diagrams have been prepared and are shown in Figures 1, 2, and 3. A constant activity of 10^{-4} molal of sulfur was used (equivalent to about 9.6 mg/l SO_4^{-2}). One diagram is for a total dissolved CO_2 species activity of 10^{-4} molal, one for 10^{-3} molal, and one for 10^{-2} molal. These are equivalent to HCO_3^- activities at pH 8.2 of 6.1, 61, and 610 mg/l, respectively. Iron solubility lines were drawn for each at molalities of 10^{-5}, 10^{-4}, 10^{-3}, 10^{-2}, 10^{-1}, 10^0, and 10^1. The areas on the diagram where iron solubility is below 10 molal were identified by using distinctive patterns for the species $Fe(OH)_3$ (representing limonite or other hydrated ferric oxide minerals), $FeCO_3$ (siderite), $Fe(OH)_2$, and FeS_2 (pyrite). Increasing or decreasing the sulfur activity would have relatively little effect on the positions of the pyrite boundaries. (See equations 23-1 to 23-18).

The positions of the boundaries were determined from the Nernst Equation and the mass law. Equilibrium constants and standard potentials were calculated from standard free-energy values for the species considered. The free-energy data were obtained from Latimer (3).

The fundamental equations are:

$$\ln K = \frac{-\Delta F^O}{RT} \qquad (23\text{-}19)$$

$$E^O = \frac{-\Delta F^O}{n \mathcal{F}} \qquad (23\text{-}20)$$

$$Eh = E^O + \frac{RT}{n \mathcal{F}} \ln \frac{\text{oxid}}{\text{red}} \qquad (23\text{-}21)$$

where $\ln K$ = natural log of equilibrium constant
ΔF^O = standard free-energy change in reaction

R = gas constant
T = temperature in degrees Kelvin
E^O = standard potential
n = number of electrons exchanged per ion of reactant reduced
$\mathcal{F}$ = Faraday constant
Eh = redox potential
$\ln \frac{\text{oxid}}{\text{red}}$ = natural log of mass-law statement of the activities of reactants and products

The European sign convention is used in these calculations; that is, increasingly positive potentials are considered to represent increasingly oxidizing conditions.

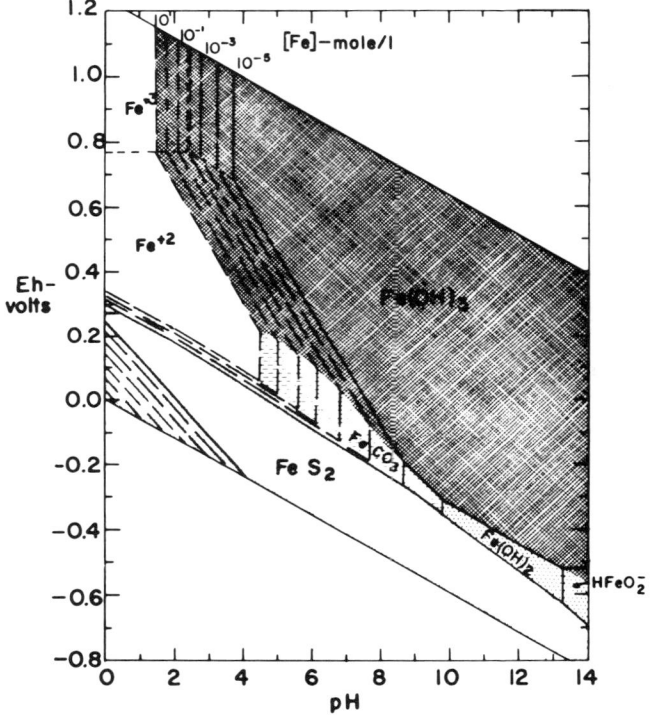

FIGURE 1 SOLUBILITY OF IRON IN RELATION TO pH AND Eh AT 25°C AND 1 ATM. TOTAL DISSOLVED SULFUR 10^{-4}M; BICARBONATE SPECIES 10^{-4}M.

IRON

Chemical equilibria considered in making the diagrams include:

$$Fe^{+3} + e^- = Fe^{+2} \tag{23-1}$$

$$Fe^{+3} + H_2O = FeOH^{+2} + H^+ \tag{23-2}$$

$$Fe(OH)_{3(c)} + 2H^+ = FeOH^{+2} + 2H_2O \tag{23-3}$$

$$Fe(OH)_{3(c)} + 3H^+ = Fe^{+3} + 3H_2O \tag{23-4}$$

$$Fe(OH)_{3(c)} + 3H^+ + e^- = Fe^{+2} + 3H_2O \tag{23-5}$$

$$Fe(OH)_{3(c)} + HCO_3^- + 2H^+ + e^- = FeCO_{3(c)} + 3H_2O \tag{23-6}$$

$$Fe(OH)_{3(c)} + H^+ + e^- = Fe(OH)_{2(c)} + H_2O \tag{23-7}$$

$$Fe(OH)_{3(c)} + e^- = HFeO_2^- + H_2O \tag{23-8}$$

$$FeCO_{3(c)} + H^+ = Fe^{+2} + HCO_3^- \tag{23-9}$$

$$FeCO_{3(c)} + 2H_2O = Fe(OH)_{2(c)} + CO_3^{-2} + 2H^+ \tag{23-10}$$

$$FeCO_{3(c)} + 2SO_4^{-2} + 17H^+ + 14e^- = FeS_{2(c)} + 8H_2O + HCO_3^- \tag{23-11}$$

$$Fe(OH)_{2(c)} = HFeO_2^- + H^+ \tag{23-12}$$

$$Fe(OH)_{2(c)} + 2SO_4^{-2} + 18H^+ + 14e^- = FeS_{2(c)} + 10H_2O \tag{23-13}$$

$$Fe^{+2} + 2SO_4^{-2} + 16H^+ + 14e^- = FeS_{2(c)} + 8H_2O \tag{23-14}$$

$$Fe^{+2} + 2HSO_4^- + 14H^+ + 14e^- = FeS_{2(c)} + 8H_2O \tag{23-15}$$

$$FeS_{2(c)} + 4H^+ + 2e^- = Fe^{+2} + 2H_2S\,(aq) \tag{23-16}$$

$$Fe^{+2} + 2S_{(c)} + 2e^- = FeS_{2(c)} \tag{23-17}$$

$$HFeO_2^- + 2SO_4^{-2} + 19H^+ + 14e^- = FeS_{2(c)} + 10H_2O \tag{23-18}$$

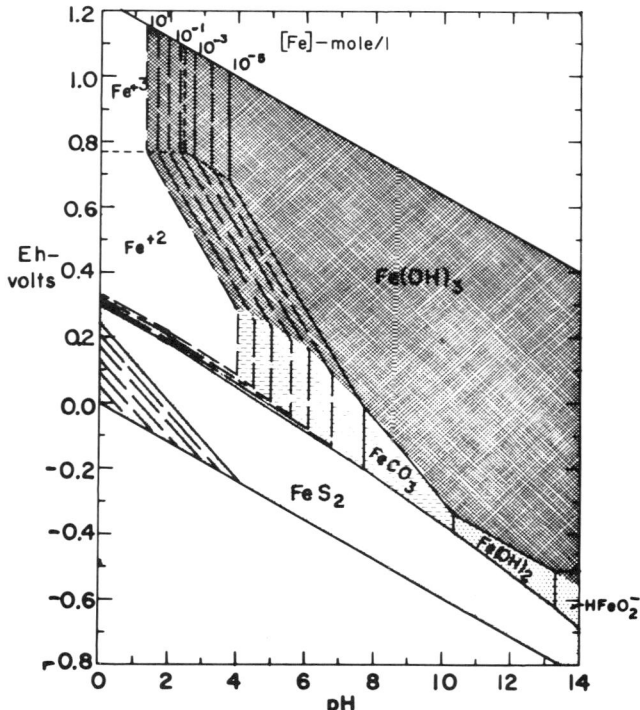

FIGURE 2 SOLUBILITY OF IRON IN RELATION TO pH AND Eh AT 25°C AND 1 ATM. TOTAL DISSOLVED SULFUR 10^{-4}M; BICARBONATE SPECIES 10^{-3}M.

From Figure 1 it can be seen that iron concentrations below 0.5 mg/l are to be expected over a wide area of the $Fe(OH)_3$ and FeS_2 regions. The conditions in an aerated water or a water in a strongly reduced environment, near neutral pH, are favorable for only very small dissolved-iron contents. Within the siderite region of the diagram, however, iron becomes relatively soluble at pH values between 5.0 and 9.0.

Figure 1 indicates that the solubility of iron at pH 7.0 is a little less than 10^{-3}M (around 50 mg/l) between Eh values of -0.14 and +0.03. The HCO_3^- activity assumed for this diagram is 10^{-4}M or about 6 mg/l.

IRON 633

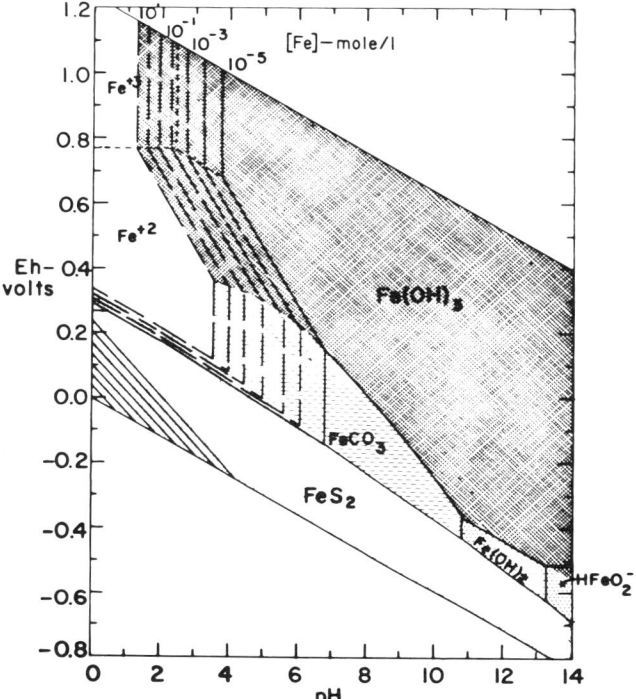

FIGURE 3   SOLUBILITY OF IRON IN RELATION TO pH AND Eh AT 25°C AND 1 ATM. TOTAL DISSOLVED SULFUR 10^{-4}M; BICARBONATE SPECIES 10^{-2}M.

Figure 2 shows what happens when all other parameters remain constant but HCO_3^- activity becomes 10^{-3}M (61 mg/l). Here the solubility of iron at pH 7.0 is decreased 10-fold to near 5 mg/l, but the Eh range over which this value would occur is widened from -0.15 to +0.08 v. In Figure 3 where the HCO_3^- activity is 610 mg/l, a little less than 0.5 mg/l of iron could remain in solution at pH 7.0 whereas Eh values range still more widely from -0.16 to -0.12 v. Between the stability regions for $Fe(OH)_3$ and pyrite there lies an area of relatively high iron solubility. This area may be reached either by reduction of ferric species or oxidation of pyrite. Within this region iron solubility is controlled only by pH and HCO_3^- activity and is independent of Eh because the iron is in the same oxidation state in solution, Fe^{+2}, as in the solid. If CO_2 is absent or present in very small amounts, the solubility of iron in the region between the pryite and $Fe(OH)_3$ fields is even higher than in Figure 1.

APPLICATIONS TO NATURAL CONDITIONS

The specific capacity of a well, that is, the rate at which water can be pumped per foot of lowering of the head often tends to decrease with increasing length of service. In many areas the relative shortness of the useful life of wells is a significant economic problem. The decline in yield commonly results from deposits of $CaCO_3$, iron oxide, and sometimes other materials around screen openings, in the gravel pack, or on the surface of water-yielding rock if the well is uncased. Specific instances of encrustation have been described for wells in New Jersey by Linn (4) and at Lansing, Michigan, by Erickson and Wright (5). A study of well stimulation by Koenig gives some idea as to the widespread nature of the problem and methods used for coping with it (6).

Precipitates containing iron could form in wells through mixing of waters having incompatible chemical characteristics or as a result of changes in pH or Eh resulting from the act of pumping. These precipitates may be associated also with growth of microbiota occurring in the well.

The mass-law statement of the solubility equilibrium for siderite is:

$$\frac{[HCO_3^-][Fe^{+2}]}{[H^+]} = K = 0.46 \qquad (23-22)$$

In the solubility equilibrium for calcite, the equilibrium constant is considerably greater:

$$\frac{[HCO_3^-][Ca^{+2}]}{[H^+]} = K = 97 \qquad (23-23)$$

A solution at equilibrium with both solids simultaneously would have an activity of Ca^{+2} about 200 times that of Fe^{+2}. Two solutions which started with an equal supply of dissolved CO_2 and reached equilibrium in separate environments, one where siderite was present and one where calcite was present, would have very different pH values. When the two waters were mixed, a solution strongly supersaturated with respect to siderite would generally result.

To demonstrate, assume solutions I and II both of which are continually exposed to a gas phase with a partial pressure of CO_2 of 10^{-2} atm. Solution I enters a stratum containing siderite and no calcite. By means of simultaneous equations for siderite solubility, CO_2 dissociation, and ionic balance it can be calculated that, at equilibrium, the pH value would be 6.57, activity of HCO_3^- $10^{-3.28}$ molal or 32 mg/l, and activity of Fe^{+2}, $10^{-3.64}$ molal or 13 mg/l. Figure 4 is a graphical representation of conditions which will occur at equilibrium in the presence of siderite.

Solution II, exposed to the same partial pressure of CO_2, enters a stratum in which calcite is plentiful. Using similar calculations it can be shown that at equilibrium the pH value of this solution would be 7.37, activity of HCO_3^-, $10^{-2.48}$ molal or 202 mg/l, and activity of Ca^{+2}, $10^{-2.90}$ molal or 50 mg/l.

In calculations of this type, the mass-law equations are based on thermodynamic concentrations and the ionic balance equations are based on stoichiometric concentrations. The activity coefficients for the ions were assumed as a first approximation to be unity from which the calculated results were used to compute an ionic strength of the solution. This value in turn was used with the Debye-Hückel equation to calculate approximate activity coefficients for the ions. The equilibrium calculation was repeated with these approximate coefficients to obtain a more exact set of values. The process was repeated until final values of satisfactory accuracy were obtained. This recycling calculation procedure is described by Garrels (1).

If now solutions I and II are intercepted by a well and thereby are mixed in equal proportions, the stoichiometric concentrations (calculated from the activity values) are:

Solution	HCO_3^-	Fe^{+2}	Ca^{+2}	I
I	33	15	-	0.00097
II	215	-	68	0.0051
Mixture	124	7.5	34	0.003

The final activities of the mixture, before any chemical reaction occurs, are $[HCO_3^-]$ = 118 mg/l or $10^{-2.71}$ molal, and $[Fe]$ = 6.0 mg/l or $10^{-3.97}$ molal. In the partial pressure of CO_2 remains constant, the mixture will still be $10^{-3.50}$ molal with respect to H_2CO_3. The pH values of the mixture can be

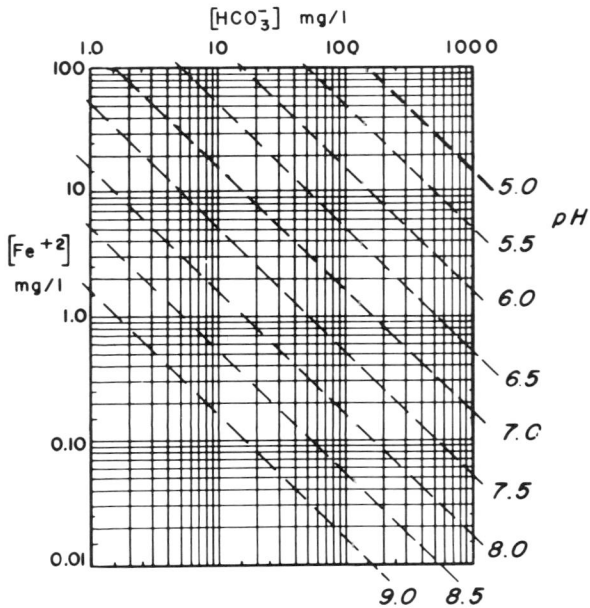

FIGURE 4   EQUILIBRIUM SOLUBILITY OF FERROUS IRON IN THE PRESENCE OF SIDERITE; 25°C, 1 ATM.

calculated from the dissociation constant for H_2CO_3 and the activities which equals 7.14.

From Figure 4 it is evident that this mixture will be unstable with respect to siderite at pH 7.14. In the presence of 118 mg/l HCO_3^- activity, only about 1.0 mg/l of iron can be retained in solution at equilibrium rather than the 6.0 mg/l present.

The conditions assumed for this calculation were simplified to include an assumption of constant partial pressure for CO_2. In ground-water systems such a condition is probably not usually attained. If no gas phase is present, however, the components of the mixture would probably differ more widely from each other and produce a still more unstable combination.

Of course, if the redox potential rises owing to aeration, the stable solid will be the even less soluble $Fe(OH)_3$. Both the carbonate precipitation and oxidation reactions are slow compared with reactions that can be conveniently studied in the laboratory unless the solutions are grossly out of equilibrium. It should be borne in mind, however, that the encrustation builds up over months and years and in the course of time may effectively seal the well.

Besides effects of mixing, the act of pumping itself may tend to change the composition of water in and around a well. As a result of turbulence and head losses, some CO_2 may be released from solution, which tends to raise the pH of the effluent above the level it would otherwise have had. This release of CO_2 implies the generation of bubbles of gas at precipitation sites. Positive evidence that this is occurring is difficult to obtain. Probably the release of CO_2 during pumping is seldom an important effect insofar as the behavior of iron is concerned. An effect of greater potential importance is the tendency for O_2 to be introduced from the air thus bringing about a higher Eh in the upper part of the saturated zone near the well, and especially in the dewatered zone, than would normally occur.

[Editors' Note: Material has been omitted at this point.]

ACKNOWLEDGEMENT

Publication was authorized by the Director, U.S. Geological Survey, Washington, D.C.

LITERATURE CITED

1. Garrels, R.M., Mineral Equilibria, Harper and Bros, New York, N.Y., (1960).

2. Hem, J.D. and W.H. Cropper, U S Geol Survey Water-Supply Paper 1459-A, (1959).

3. Latimer, W.M., Oxidation Potentials, 2nd Ed, Prentice-Hall, New York, N.Y., (1952).

4. Linn, G.L.E., J Am Water Works Assoc, 46, 534 (1954).

5. Erickson, C.R. and R.C. Wright, J Am Water Works Assoc, 49, 817 (1957).

6. Koenig, L., J Am Water Works Assoc, 52, 333 (1960).

7. Turcan, A.N., Jr. and S.W. Fader, Louisiana Dept of Pub Works, Water Resources Pamphlet 6, (1959).

8. Jones, P.H., et al, U S Geol Survey Water-Supply Paper 1364, (1956).

9. Barnes, I. and W. Back, J Geol, 72, 435 (1964).

10. Lawrence, R.E. and R.H. Hess, J Am Water Works Assoc, 55, 1081 (1963).

11. Clarke, F.E. and I. Barnes, Preliminary Evaluation of Corrosion and Encrustation Mechanisms in Tube Wells of Indus Plains, West Pakistan, U S Geol Survey Open-File Report (1964).

12. Suter, M., J Am Water Works Assoc, 54, 371 (1962).

13. Broom, M.E., Personal Communication.

14. Barnes, I., Personal Communication.

DISCUSSION

DR. RIEMAN: We will discuss this briefly if there are questions or comments.

DR. MORGAN: I wonder if one of the geochemists present would like to comment on the possible accuracy of the free energy value for siderite? Some have been concerned with that.

MR. HEM: This is an old value.

DR. D. LANGMUIR (U. S. Geological Survey): The siderite free energy value quoted by Latimer is based on a 1918 solubility study. A comparison of the activity product of ferrous iron times bicarbonate for a number of New Jersey Coastal Plain ground waters flowing in aquifers which contain recent siderite, suggests at least the approximate validity of Latimer's free energy value. However, another experimental determination of siderite stability in the laboratory would be highly desirable. In any case a slight revision of the free energy value for siderite probably won't affect Mr. Hem's over-all picture significantly.

18

Reprinted from *Hydrology of Fractured Rocks,* vol. 1, Proceedings of the Dubrovnik Symposium, October, 1965, International Association of Scientific Hydrology-UNESCO, 1965, pp. 601-614

CARBONATE EQUILIBRIA AND RADIOCARBON DISTRIBUTION RELATED TO GROUNDWATER FLOW IN THE FLORIDAN LIMESTONE AQUIFER, U.S.A.[1]

Bruce B. HANSHAW, William BACK and Meyer RUBIN
U.S. Geological Survey, Washington D.C., U.S.A.

SUMMARY

　　The principal artesian aquifer of Florida is composed predominantly of Tertiary limestone with lesser amounts of dolomite and gypsum. A north-south geohydrologic section through the piezometric high depicts the spatial and temporal changes of chemical character of water. The major area of recharge is characterized by calcium bicarbonate type water, by low concentrations of total dissolved solids and of sulfate ions, by undersaturation with respect to calcite and dolomite, and by young C^{14} ages. Concentration of dissolved solids and sulfate increases markedly as a function of length of flow path and residence time in the aquifer. Radiocarbon ages indicate that within different parts of the system velocities range from 2 to 12 meters per year.
　　Away from the recharge area the water is supersaturated with respect to calcite and dolomite. Throughout most of the area the water has not reached equilibrium with aragonite and gypsum. Supersaturation of the water with respect to calcite tends to buffer the bicarbonate concentration. This study shows that the geochemistry of the water conforms well with the hydrologic history of the system.

RÉSUMÉ

L'équilibre des carbonates et la répartition du radiocarbone par rapport à l'écoulement de la nappe aquifère calcaire de Floride, aux États-Unis d'Amérique

　　Le principal aquifère artésien de Floride se compose surtout de calcaires tertiaires auxquels se mêlent, en quantités moindres, des dolomies et du gypse. Une coupe géohydrologique nord-sud suivant le maximum piézométrique fait apparaître les changements dans l'espace et dans le temps du caractère chimique de l'eau. Dans l'aire principale d'alimentation de la nappe, l'eau est caractérisée par son contenu en bicarbonate de calcium, par de faibles concentrations en corps solides dissous et en ions sulfates, par une sous-saturation en dolomie et en calcite et par des carbones 14 d'âge récent. La concentration en solides dissous et en sulfate s'accroît notablement avec l'accroissement de la distance parcourue par les eaux et de la durée de séjour dans la nappe. L'âge du radiocarbone indique que, dans différentes parties du système, la vitesse d'écoulement est de l'ordre de 2 à 12 mètres par an.
　　Loin de l'aire d'alimentation, l'eau est sursaturée en dolomie et en calcite. Dans la plus grande partie de l'aire, l'eau n'a pas atteint son équilibre en aragonite et en gypse. La sursaturation de l'eau en calcite tend à jouer le rôle de tampon à l'égard de la concentration en bicarbonate. Cette étude montre que la géochimie de l'eau suit l'histoire hydrologique du système.

INTRODUCTION

　　The purpose of this paper is to describe the application of geochemical principles to a reasonably well known hydrologic environment to determine which solid phases play the most important part in controlling the chemistry of water as it moves through the aquifer system. Use of basic principles of physical chemistry show how chemical equations express solution and precipitation of common carbonate and evaporite minerals. A chemical equilibrium model aids in understanding geologic processes such as formation of secondary permeability by solution, precipitation of minerals in the aquifer, cementation, and recrystallization. Isotopic investigations may be used in the study of a hydrologic system, to determine the various origins of water, residence time

---

[1] Publication authorized by Director, U.S. Geological Survey.

of water, velocity of groundwater movement and rates of chemical reactions. A combined study of general water chemistry, mineral saturation of the water, and distribution of radiocarbon isotopes is helpful to identify the principal areas of recharge and to identify which geochemical processes control the chemical character of water. This approach is also useful in predicting which, if any, chemical changes may occur as a result of imposing such stresses as increased pumping or artificial recharge on the hydrologic system.

The regional flow pattern in the hydrologic system in central Florida is controlled largely by the two piezometric highs shown in figure 1. A north-south line of wells through the major piezometric high was chosen for the study of changes in aqueous

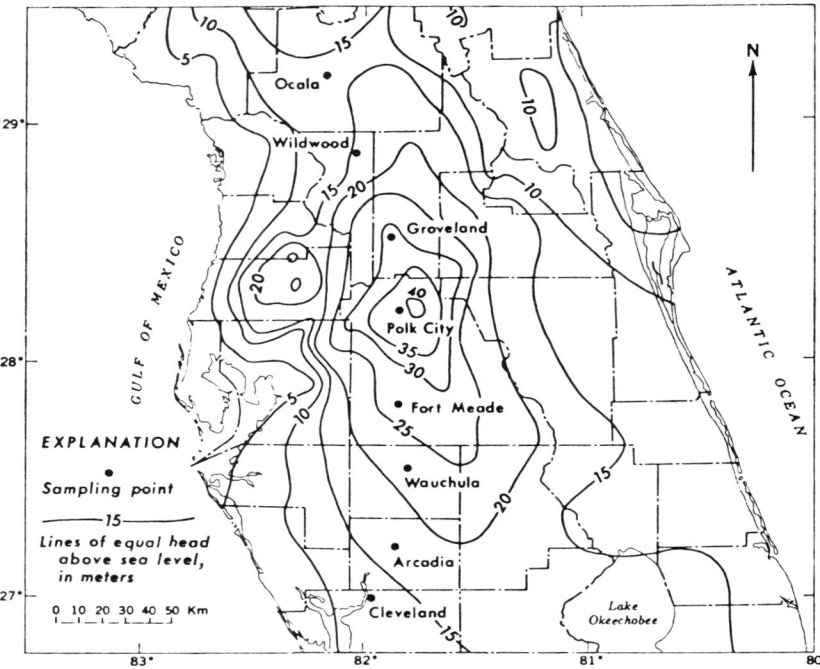

Fig. 1 — Piezometric map of the principal artesian aquifer of central Florida showing location of sample points used on cross-sections (after Stringfield, 1936).

chemistry as the water flows from a recharge area toward deeper parts of the aquifer. The principal artesian aquifer of Florida consists of a series of limestones of Tertiary age which contain minor amounts of dolomite, disseminated quartz sand, gypsum, and anhydrite. The limestones are interconnected and function as a hydrologic unit. The eight wells sampled during this study (fig. 1) range in depth from approximately 80 to 300 metres and are completed in limestone that ranges in age from middle Eocene to Miocene. The principal artesian aquifer is overlain by confining beds of Miocene age, chiefly clay of the Hawthorn Formation (Pride, Meyer, and Cherry, 1961; Le Grand and Stringfield, 1965). The major piezometric high has the shape of a north-south elongated dome; water flows down gradient at right angles to the piezometric contours.

This investigation is part of a continuing study of the chemistry of the principal artesian limestone aquifer of Florida. The complete study includes about 50 wells covering the centre of the State, an area of roughly a half million square kilometres. The final report will include mineral equilibrium studies of solid phases in addition to those presented here and investigations of the isotopes of hydrogen, carbon, oxygen and sulphur.

CHEMICAL CHARACTER OF WATER

The chemical character of the water reflects the combined effect of chemical activity between water and limestone and the flow pattern within the aquifer. Within the study area most of the water (Wildwood, Groveland, Polk City and Fort Meade)

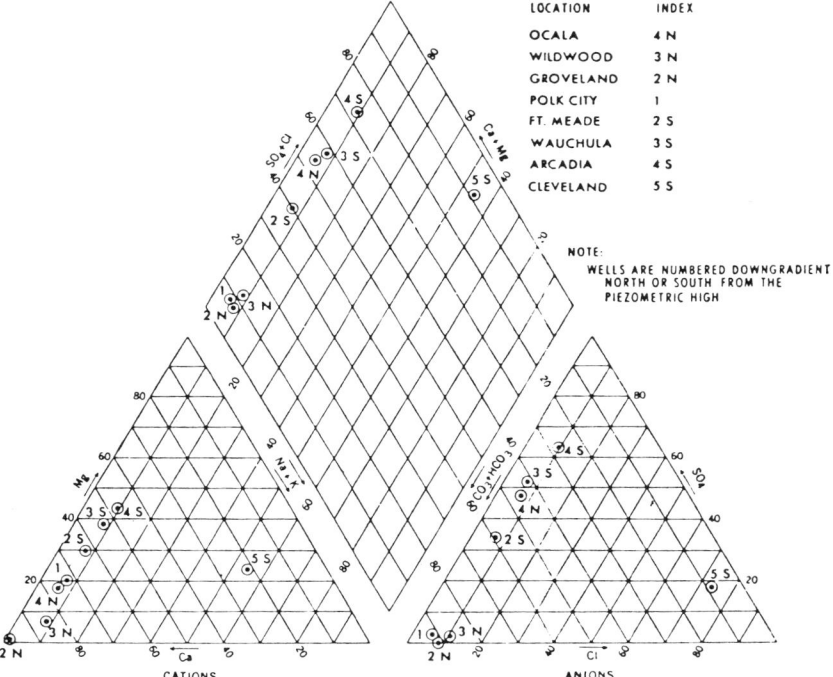

Fig. 2 — Chemical composition of ground water from the principal artesian aquifer of central Florida.

is of the $CaHCO_3$ type; the magnesium and sulphate content increases downgradient so that a mixed type of water is produced by the wells at Ocala and Arcadia, and a $CaSO_4$ type water is produced by the well at Wauchula. The well at Cleveland produces water of NaCl type. The chemical character of water in the area of study is shown in table 1 and is represented in figure 2 as a trilinear plot of the percentages of each major constituent to the total anions or cations in milliequivalents per litre.

The Polk City well is close to the highest part of the piezometric surface (fig. 1) and water in this part of the aquifer flows radially downgradient. The area near Polk

City is a principal area of recharge, and the Groveland and Wildwood areas also contribute substantial quantities of water to that portion of the aquifer system north and westward of the Polk County high (Stringfield, 1936, p. 151). Between the area of recharge and Ocala, the principal change in composition is in the anion facies, which changes from predominantly HCO_3 to a mixture that is predominantly $SO_4 + HCO_3$ (table 1).

A simple progressive change in type of water occurs as water moves from Polk City toward the Cleveland area, indicating only minor recharge south of Polk City. Between Polk City and Arcadia there is a progressive increase in the Mg:Ca ratio in the cations and a similar increase of SO_4:HCO_3 ratio in the anions. A major change in aqueous chemistry occurs between Arcadia and Cleveland. At Arcadia the hydrochemical facies is $Mg+Ca$, SO_4 whereas water produced by the Cleveland well is a NaCl type water.

Similar patterns of chemical change away from the principal recharge area are shown by the constituents which are discussed in the following paragraphs, and by the isotope measurements and mineral saturation studies which are discussed in the two following sections. The TDS (total dissolved solids) content of the water in the study area increases north and south away from the area of recharge, and in particular, from the area of recharge at Polk City (fig. 3a). The Polk City-Wildwood area is characterized by water with less than 200 mg/l TDS. South of the Polk City recharge area, water from the aquifer exhibits a progressive increase in TDS to a maximum of 1600 mg/l at Cleveland. Northward, the TDS increases to 420 mg/l.

A similar pattern is observed for the concentration of sulphate ions in the groundwater (fig. 3b). In the Polk City-Wildwood area the sulphate is less than 10 mg/l and increases to 148 mg/l at Ocala. South of the Polk City area the sulphate content increases progressively as far as Arcadia and decreases between Arcadia and Cleveland.

Similar reversals in other chemical parametres occur between Arcadia and Cleveland. These may be due in part to mixing of waters from different geohydrologic environments, as reflected in the change in hydrochemical facies from $Mg+Ca$, SO_4 to NaCl. South of the Cleveland area, water from the principal artesian aquifer becomes increasingly more saline and approaches the concentration of ocean water. It is possible also that water produced by the well at Cleveland includes a significant amount of water from the Hawthorn Formation because the well is not cased through the Hawthorn Formation. The quality of water in the Hawthorn Formation does not affect the well at Polk City because the well is near the piezometric high and the limestone aquifer is recharged only by percolation downward through the Hawthorn Formation, whose water has essentially the same chemical character as that of the Floridan aquifer. However, at Cleveland, water from the Hawthorn is probably quite different from water in the Floridan aquifer and from water in the Hawthorn at Polk City.

The bicarbonate content of water from the aquifer is relatively uniform throughout the area of study (fig. 4a). In the Polk City-Wildwood area the bicarbonate content is somewhat lower than elsewhere, indicating that this is an area of major recharge. The uniform bicarbonate content reflects a buffering mechanism by the solid carbonate phases of the aquifer rocks.

To summarize the chemical data presented so far, an area of significant recharge between Polk City and Wildwood is identified by downgradient changes in hydrochemical facies, and by changes in concentrations of total dissolved solids and of sulphate ion.

RADIOCARBON CONCENTRATIONS

Results of radiocarbon measurements made on the bicarbonate ion in water from each well are provided in table 2 and figure 4b. The measurements are part of a continu-

TABLE 1

Standard chemical analyses of water from the Floridan aquifer. Wells are listed from north to south

Well location	Open interval, meters	Temperature °C	Milligrams per liter										Dissolved solids, residue at 180°C	Field pH, ±0.02
			SiO_2	Ca^{+2}	Mg^{+2}	Na^+	K^+	Field HCO_3^-	SO_4^{-2}	Cl^-	F^-	NO_3^-		
Ocala 4	30-115	24.5	10	96	15	7.8	1.0	175	148	11	0.3	1.6	420	7.50
Wildwood 2	39-82	23.8	10	51	2.6	4.7	0.2	150	3.2	8.0	0.2	3.8	158	7.59
Groveland	40-180	23.7	11	42	4.1	3.6	0.5	143	1.6	6.5	—	0.1	148	7.80
Polk City	6-172	23.8	0.2	34	5.6	3.2	0.5	124	2.4	4.5	0.1	0.1	138	8.00
Fort Meade	127-291	26.6	16	56	17	6.1	0.7	163	71	9.0	0.4	0.1	272	7.75
Wauchula	114-245	25.4	18	66	29	8.3	2.0	168	155	10	0.7	—	392	7.69
Arcadia	100-151	26.3	31	106	60	21	3.7	206	344	28	2.2	—	762	7.44
Cleveland	39-152	26.7	18	114	82	283	9.6	145	216	655	0.9	0.1	1600	7.51

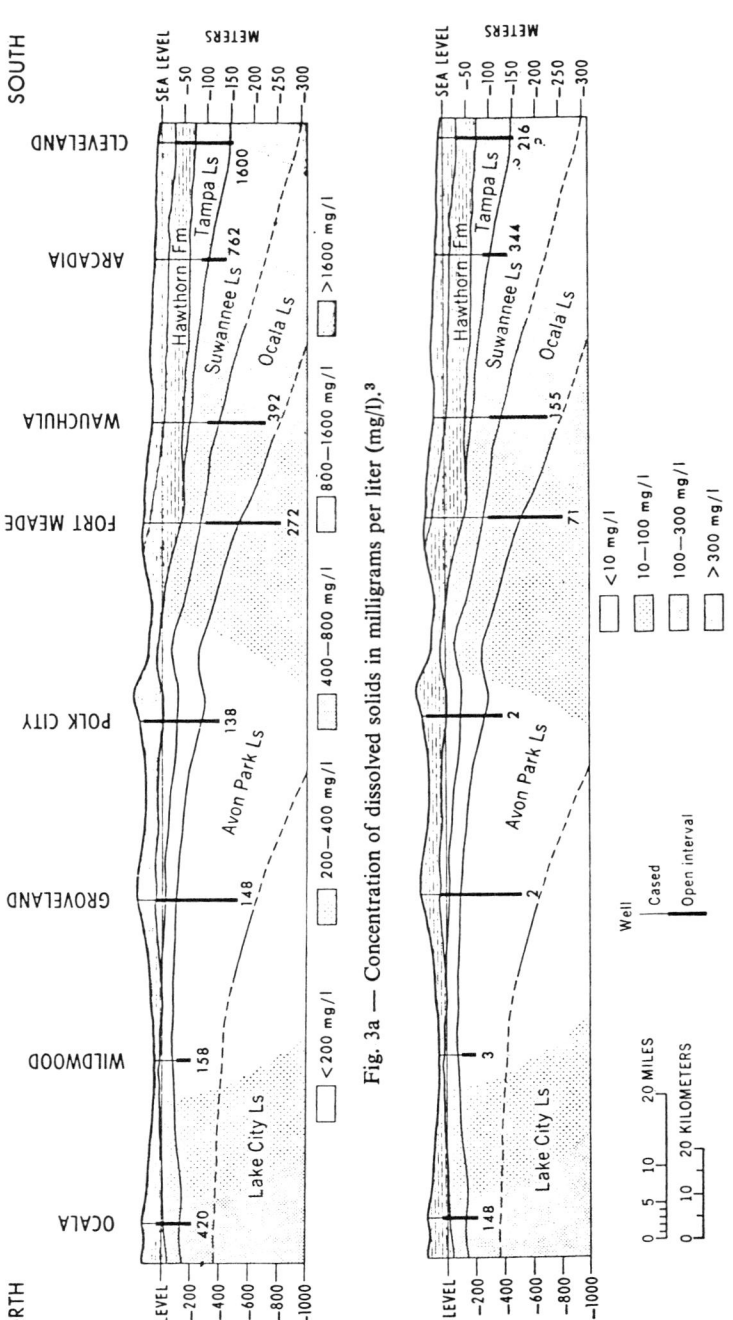

Fig. 3a — Concentration of dissolved solids in milligrams per liter (mg/l).[3]

Fig. 3b — Concentration of sulfate ion in milligrams per liter.

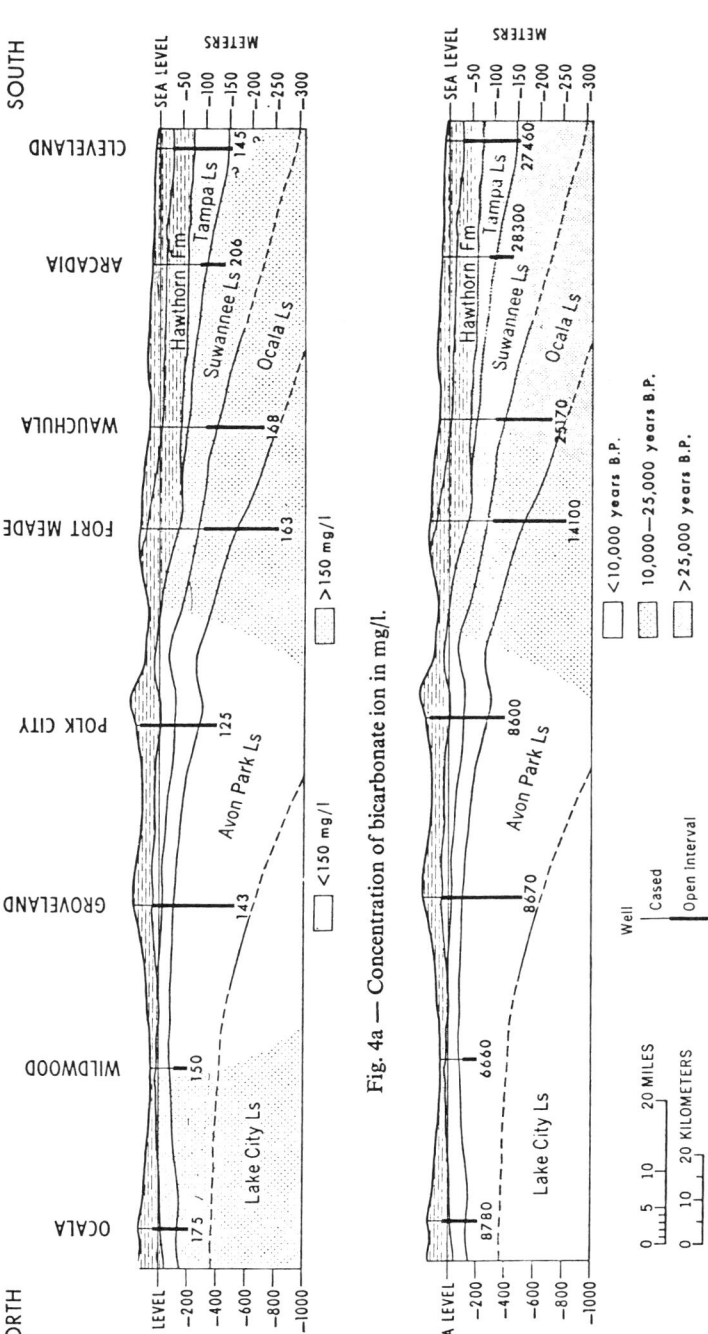

Fig. 4a — Concentration of bicarbonate ion in mg/l.

Fig. 4b — Radiocarbon ages in apparent years before present (B.P.).

ing study of the isotopic composition of water and rock samples representative of the aquifer, that will include an investigation of the isotopes of hydrogen, carbon, oxygen, and sulphur.

The results of C^{14} determinations have been published elsewhere (Hanshaw, Back and Rubin, 1965) for the group of wells from Polk City south to Cleveland. Because there is no evidence of significant recharge south of Polk City, differences in C^{14} ages between wells are assumed to be due to time of travel and are used to calculate apparent velocities of groundwater flow. Velocities range between two and 12 metres per year (table 2). Net velocities between Polk City and wells to the south calculated from hydrologic equations are essentially in agreement with the radiocarbon-determined velocities (table 2). Although apparent radiocarbon ages will not give absolute residence time, age difference between two wells will approximate relative residence time.

MINERAL EQUILIBRIUM

Measurements of departure from chemical equilibrium between groundwater and the minerals which comprise an aquifer aid in delineating principal areas of recharge and in predicting areas subject to solution or deposition of minerals. Equilibrium studies also increase our knowledge of the minerals that control the chemistry of water and the changes that may occur with time or because of the application of a stress upon the hydrologic system.

The minerals considered in this study are those most common in the principal artesian aquifer: calcite, aragonite, dolomite, gypsum, and anhydrite. The thermodynamic model used in this study was developed by Back (1961) and has been applied to a study of calcium carbonate saturation of groundwater in Florida (Back, 1963). Each of the minerals is handled thermodynamically in the same manner; gypsum will be used as an example. The chemical equation describing the solution of gypsum is

$$CaSO_4 \cdot 2H_2O = Ca^{+2} + SO_4^{-2} + 2H_2O \qquad (1)$$

and the equilibrium constant, K_{eq}, for gypsum K_{gyp}, is

$$K_{gyp} = \frac{\alpha_{Ca^{+2}} \alpha_{SO_4^{-2}} \alpha_{H_2O}^2}{\alpha_{CaSO_4 \cdot 2H_2O}} \qquad (2)$$

where α is activity. By definition, the activity of a solid phase is equal to one; for dilute water the activity of water may also be considered equal to one, hence equation (2) becomes

$$K_{gyp} = \alpha_{Ca^{+2}} \alpha_{SO_4^{-2}} \qquad (3)$$

The equilibrium constant was calculated from the Gibbs free energy values by standard thermochemical methods (see Garrels, 1960, pp. 6-18). The K_{eq} given in the heading of table 3 is at 25°C; the K_{eq} of the minerals at the temperature of the groundwater was calculated by use of the van't Hoff equation.

Several steps are required to go from standard chemical analyses such as those in table 1 to activities of single ions. (See Garrels, 1960, pp. 3-42, Back, 1963.) The calculated activities of calcium and sulphate ions are used in equation (3) to obtain the ion activity product, K_{iap}. The K_{iap} is compared with the equilibrium constant to determine the departure from equilibrium of the water with respect to gypsum. If the ratio, $K_{iap}: K_{gyp}$, is equal to one (or 100%) the water is saturated with respect to gypsum; a ratio less than one (less than 100%), the water is undersaturated and if the ratio is greater than one, the water is supersaturated. From the form of equation (3) it is seen that the K_{iap} for gypsum and anhydrite are the same.

The principals for determining carbonate mineral saturation are identical. However,

TABLE 2

Radiocarbon measurements and apparent velocities of groundwater determined from C^{14} or hydrologic data

Well location	USGS Lab No.[1]	δC^{14} ±20‰ [2]	Apparent age yrs	Calculated apparent velocities		
				Between wells indicated	$v_{hydrologic}$[3] m/yr	$v_{C^{14}}$ m/yr [4]
4N. Ocala	W-1441	−665	8780			
3N. Wildwood	W-1445	−564	6660			
2N. Groveland	W-1440	−660	8670			
1. Polk City	W-1442	−657	8600			
				1 and 2S	10	8
2S. Fort Meade	W-1439	−827	14100			
				2S and 3S	7	2
3S. Wauchula	W-1444	−956	25170			
				3S and 4S	5	12
4S. Arcadia	W-1438	−970	28300			
				4S and 5S	5.5	—
5S. Cleveland	W-1443	−967	27460			
				1 and 3S	9	4
				1 and 4S	7	5
				1 and 5S	7	7

[1] U.S. Geological Survey, Radiocarbon Laboratory number.
[2] $C^{14} = [(R_{sample} - R_{standard})/R_{standard}] \, N$.
    Where R is the ratio of C^{14} to C^{12} and N is 1000.
[3] Based on $v = ki/p$ (Hanshaw, Back and Rubin, 1965).
[4] Based on (differences in distance/difference in ages).

determination of saturation of carbonate minerals requires accurate determination of both *pH* and bicarbonate in the field. This is necessary because the equilibrium constant for calcite is

$$K_{calc} = \alpha_{Ca^{+2}} \alpha_{SO_3^{-2}} \qquad (4)$$

whereas the major dissolved carbon species is commonly bicarbonate ion at the pH of most groundwater. Therefore it is necessary to calculate αCO_3^{-2} from the relationship

$$K_{HCO_3} = \frac{\alpha_{CO_3^{-2}} \alpha_{H^+}}{\alpha_{HCO_3^-}} \qquad (5)$$

and $\alpha_{CO_3^{-2}}$ is sensitive to changes in α_{H^+}(pH). As shown above for gypsum and anhydrite, the K_{lap} for calcite and aragonite is the same.

Departure of the groundwater from equilibrium with respect to gypsum and anhydrite is shown in table 3 and figure 5a. The entire study area contains water undersaturated with respect to both gypsum and anhydrite. However, water in the Polk City-Wildwood area is most undersaturated. This is related to the low sulphate ion concentration in water of that area. The paucity of sulphate may be caused either by lack of deposition of gypsum and anhydrite in this area, or—perhaps more likely—by the great amount of recharge and the consequent removal of sulphate minerals by solution during the millennia since the rocks were elevated to where they could be dissolved by percolating waters.

It is suggested that the progressive decrease in undersaturation of water with respect to gypsum and anhydrite south of Polk City is an expression of distance of travel time from the recharge area and residence time (fig. 4b) in the aquifer system.

In table 3 and on all the figures representing the results of the mineral equilibrium study (fig. 4b-6b) an apparent reversal of trend is observed in water from the Cleveland well. This reversal may be the result of mixing as mentioned above.

In the study area the groundwater is undersaturated with respect to aragonite (fig. 5b) but is most undersaturated in the principal area of recharge—the Wildwood-Polk City area.

Results of the calcite saturation investigation (fig. 6a and table 3) are generally similar to, yet significantly different from the picture developed for the aragonite study. (Compare fig. 6a and 5b.) All but one well—that at Wildwood—produce water supersaturated with respect to calcite. However, the Groveland-Polk City area contains water which is less supersaturated than elsewhere; the generally low degree of saturation at Wildwood probably indicates the influence of significant amounts of recharge on the chemistry of groundwater.

An important but unanswered question is, "How does water in a limestone aquifer remain supersaturated with respect to the major solid phase present in the aquifer?" First, perhaps the analytical results may be slightly in error and the supersaturation may be more apparent than real; only a slight excess of calcium or carbonate ion is necessary to achieve the percentages of supersaturation shown in table 3. However, water from wells used in this study has been sampled and analysed twice at different times of the year with similar results. Thus we believe that supersaturation is real and not an implied false value resulting from analytical error.

Secondly, this study was conducted using the values of free energy of pure, ordered phases of $CaCO_3$ to calculate equilibrium constants. However, many limestones are not pure $CaCO_3$ but contain as much as several per cent magnesium in the calcite. This departure from a pure phase increases the free energy of the phase which, in turn, increases the value of the equilibrium constant. Results obtained in this study are within the limits of the possibility that the groundwater is in equilibrium with a magnesian calcite.

A third and intriguing possibility (which is also compatible with the hypothesis that the water is approximately in equilibrium with a magnesian calcite phase) is that

TABLE 3

A comparison of K_{eq} and K_{iap} for gypsum, anhydrite, aragonite, calcite and dolomite

Well location	Gypsum $\log K_{eq} = -4.61$		Anhydrite $\log K_{eq} = -4.43$		Aragonite $\log K_{eq} = -8.14$		Calcite $\log K_{eq} = -8.34$		Dolomite $\log K_{eq} = -17$	
	$\log K_{iap}$	% satn	$\log K_{iap}$	% satn	$\log K_{iap}$	% satn	$\log K_{iap}$	% satn	$\log K_{iap}$	% satn
Ocala 4	−5.80	7	−5.80	4	−8.22	82	−8.22	129	−17.02	96
Wildwood 2	−7.61	0.1	−7.61	0.05	−8.40	52	−8.40	83	−17.88	13
Groveland	−7.99	0.04	−7.99	0.02	−8.29	67	−8.29	106	−17.38	42
Polk City	−7.89	0.05	−7.89	0.03	−8.24	77	−8.24	122	−17.04	92
Fort Meade	−6.29	2	−6.29	1	−8.18	94	−8.18	149	−16.66	220
Wauchula	−5.94	5	−5.94	3	−8.20	87	−8.20	138	−16.53	293
Arcadia	−5.49	13	−5.49	8	−8.21	87	−8.21	138	−16.44	362
Cleveland	−5.77	7	−5.77	4	−8.33	68	−8.32	107	−16.55	279

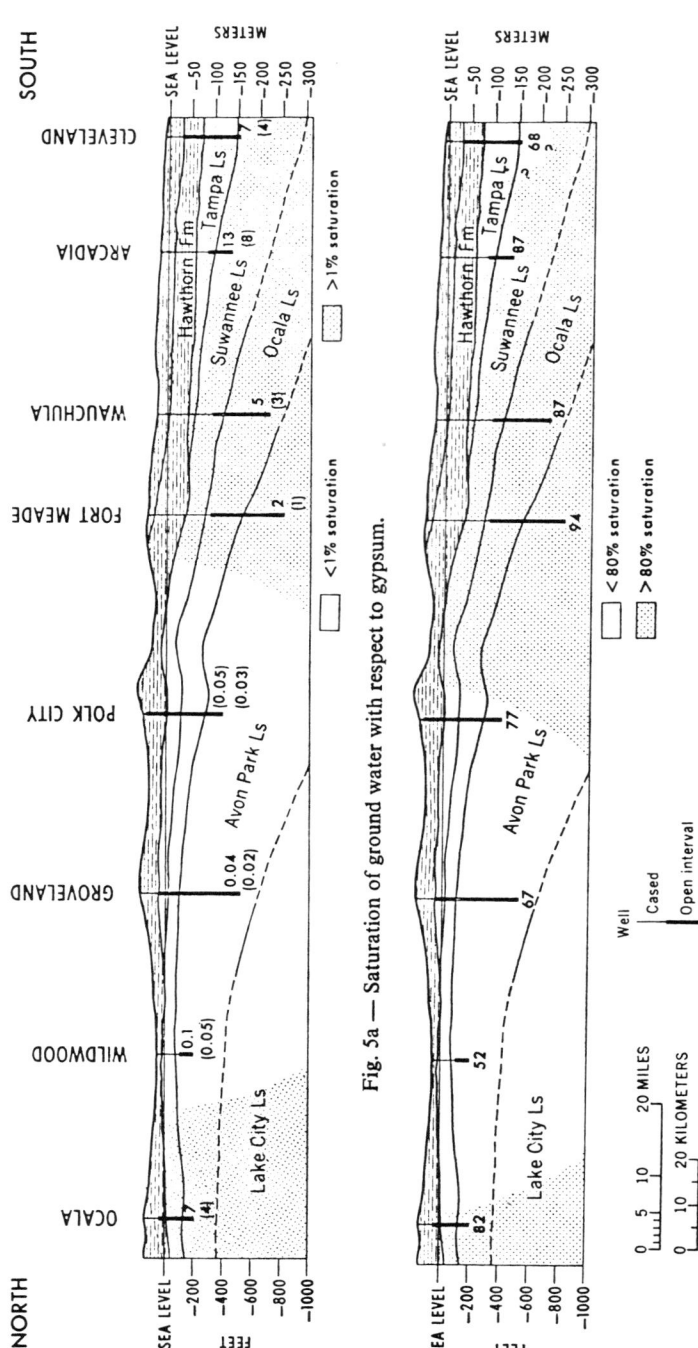

Fig. 5a — Saturation of ground water with respect to gypsum.

Fig. 5b — Saturation of ground water with respect to aragonite.

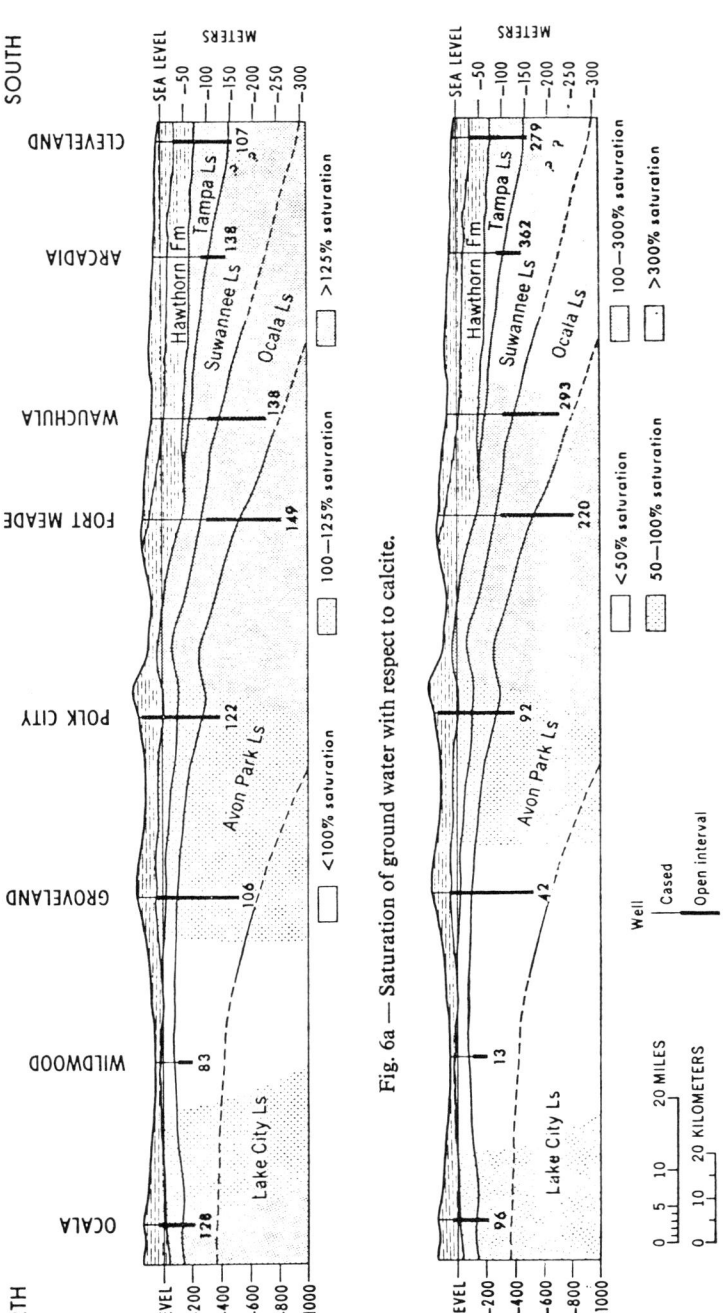

Fig. 6a — Saturation of ground water with respect to calcite.

Fig. 6b — Saturation of ground water wit respect to dolomite.

water will maintain supersaturation with respect to pure calcite, possibly for kinetic reasons, until saturation with respect to aragonite is reached. Once equilibrium is attained, aragonite may precipitate and later invert to calcite. In the ocean and in closed basin environments, aragonite, in preference to calcite, is the usual phase to precipitate out of solution. Perhaps, in groundwater also, aragonite and not calcite is the control on carbonate equilibrium.

Although the water between Ocala and Polk City is undersaturated with respect to dolomite, the water from the wells at Ocala and Polk City is nearly at equilibrium. The departure from equilibrium at Wildwood and Groveland suggests this area is one of major recharge to the aquifer. In this study, a commonly accepted value 1×10^{-17} was used as the K_{eq} for dolomite. The relatively high degree of supersaturation may be the result of choosing a K_{eq} which is too high; perhaps a value of 2 or 3×10^{-17}, as suggested by Hsu (1963) and by Barnes and Back (1964), is more nearly correct.

Lack of equilibrium with respect to dolomite in the Ocala-Polk City area is controlled by the low magnesium content of groundwater in this region. Away from the area of major recharge the calcium content increases by a factor of two or, at most, three. The magnesium content, however, increases by a factor of from three to 30 (table 1 and fig. 2). The increase in magnesium content probably is caused by the greater solubility of a magnesium calcite compared to pure calcite. The concentration of calcium ion is maintained at a rather constant value because the groundwater in nearly all the study area is supersaturated with respect to calcite.

CONCLUSIONS

This investigation confirms previous studies (Stringfield, 1936, p. 151) that major recharge to the Floridan aquifer may occur not only in high but also in some lower areas of the piezometric surface. Principal areas of recharge in the Floridan aquifer are indicated by low total dissolved solids, low sulphate and magnesium ion content, high C^{14} concentrations (low apparent ages), and by degree of undersaturation with respect to various solid phases. As water moves away from the recharge area it increases in total dissolved solids, sulphate and Mg:Ca ratio. In addition, C^{14} decreases in a systematic manner allowing calculation of apparent groundwater velocities. A study of mineral equilibrium indicates that groundwater increases in saturation away from recharge areas and may become supersaturated with respect to some solid phases.

REFERENCES

BACK, William, 1961, Calcium carbonate saturation in ground water, from routine analyses: *U.S. Geol. Survey Water-Supply Paper* 1535-D, 14 p.
—, 1963, Preliminary results of a study of calcium carbonate saturation of ground water in central Florida: *Internat. Assoc. Sic. Hydrol.*: 8, 43-51.
BARNES, Ivan and BACK, William, 1964, Dolomite solubility in ground water: *Art. 160* in *U.S. Geol. Survey Prof. Paper* 475-D, D179-180.
GARRELS, R. M. (1960) Mineral Equilibria: Harper and Bros., New York, 254 p.
HANSHAW, B. B., BACK, William, and RUBIN, Meyer, 1965, Radiocarbon determinations for estimating ground-water flow velocities in central Florida: *Science*, Apr., 23, 148, no. 3669, 494-495.
HEALY, H. G., 1962, Piezometric surface of the Florida aquifer in Florida, July 6-17, 1961: *Fla. Geol. Survey*, Map Series No. 1.
HSU, K. J., 1963, Solubility of dolomite and composition of Florida ground waters: *J. Hydrol.*, 1, 288-310.
LEGGRAND, H. E., and STRINGFIELD, V. T., 1965, Development of permeability in the Tertiary limestones of the southeastern states, U.S.A.: (in press) *International Assoc. Sci. Hydrology*.
PRIDE, R. W., MEYER, F. W., and CHERRY, R. N., 1961, Interim report on the hydrologic features of the Green Swamp area in central Florida: *Fla. Geol. Survey Circ.*, no. 26, 968.
STRINGFIELD, V. T., 1936, Artesian water in the Florida peninsula: *U.S. Geol. Survey Water-Supply Paper* 773-C, pp. 115-195.

19

Copyright ©1968 by The Geological Society of America
Reprinted from Geol. Soc. America Bull. **79**:19-45 (1968)

Chemical and Hydrologic Factors in the Excavation of Limestone Caves

JOHN THRAILKILL *Dept. of Geology, University of Kentucky, Lexington, Kentucky 40506*

Abstract: Various hydrologic and chemical processes have been investigated in an attempt to explain the development of caves in the region just beneath the water table (shallow-phreatic zone). Special attention was paid to the differences between the Darcy flow of granular aquifers and the laminar or turbulent flow of limestone aquifers, and to the ways in which water in the limestone aquifer may become undersaturated with respect to calcite.

The pattern of flow in a limestone aquifer, although dependent on the boundary conditions, is similar under Darcy, laminar, or turbulent flow, provided the aquifer is wide relative to its depth and the permeability distribution is homogeneous.

Although the vadose water which supplies an aquifer is often saturated or supersaturated with respect to calcite, such water may become undersaturated by being cooled or by being mixed with water which is in equilibrium with a lower P_{CO_2}. The water of surface streams may be undersaturated with respect to calcite and may enter the aquifer by capture or by backflooding from effluent streams.

The supply of vadose water to a limestone aquifer is commonly irregular. In an area where no water crosses the water table, flow in the aquifer will be nearly horizontal and water from sources adjacent to the area will follow shallow flow-paths. If such water is undersaturated with respect to calcite, caves will be excavated.

Mammoth Cave, Kentucky, was probably formed by undersaturated water spilling from an impermeable bed. After reaching the water table, this water followed shallow-phreatic paths in the area beneath the impermeable bed. Backflooding from the Green River may have aided in the excavation of the cave.

CONTENTS

Introduction	20
Acknowledgments	20
Hypotheses of cave origin	21
The ground-water body in limestone	23
Flow considerations	24
General statement	24
Flow in limestone	25
The pipe network	26
Summary	28
Chemical considerations	29
Vadose seepage	29
Vadose flows	31
Temperature effect	31
Mixing effect	32
Flowrate effect	34
Other effects	36
Summary	37
Flow patterns in the phreatic zone	37
General statement	37
Uniform supply to the water table	38
Single discrete source	38
Multiple discrete sources	39
Irregular supply to the water table	40
Backflooding	40
Underflow	41
Mammoth Cave, Kentucky	41
Conclusions	42
References cited	43

Figure	
1. Patterns of ground-water flow in limestone	22
2. Concept of limestone aquifer used	23
3. Alternate concepts of limestone aquifer	23
4. Reynolds number—friction factor plot	25
5. Square pipe networks	26
6. Elongated pipe networks	27
7. Flow division in loop	28
8. Calcite saturation diagram, log scales	29
9. Calcite saturation diagram, arithmetic scales	33
10. Log concentration—log discharge diagram	34
11. Log A_C—log discharge diagram	35
12. Uniform supply to water table	38
13. Single discrete source	39
14. Multiple discrete sources	39
15. Irregular supply to water table	40

Table	
1. Values of equilibrium constants	30

207

INTRODUCTION

Investigations of limestone caves by various geologists have led to three generalizations regarding cave origin. More or less in the order of their appearance, these are: (1) most limestone caves are the result of solution by cold meteoric waters; (2) many of these solution caves were excavated when the rock was completely filled with water (i.e., below the water table); and (3) some of these subwater-table caves exhibit horizontal surfaces or a horizontal distribution of passages which are unrelated to bedding or other structures of the enclosing rocks.

Although some small caves may have been formed as a result of hydrothermal activity, and others have been formed or significantly enlarged by mechanical erosion, few geologists would question the validity of the first statement.

There is probably less unanimity on the second statement. In the writer's opinion, however, numerous features of caves indicate that extensive solution has occurred while the caves were filled with water. These features and their interpretation have been discussed at length by Davis (1930) and Bretz (1942).

The substance of statement three has been less thoroughly discussed in the literature. In many caves which have been formed beneath the water table, well-defined levels exist. Although the shapes of rooms and the local distribution of passages (especially in plan view) are usually controlled by joints or bedding planes, the cave as a whole will often be confined to one or more well-defined elevations. In more steeply dipping or vertical beds, a general horizontality of passages (for an example see Davies, 1960, p. 8) is obviously due to a control other than bedding. There are, of course, caves which show little or no development of levels.

The most reasonable explanation for levels in phreatic caves is that they have been localized by a water table. It is difficult to conceive of any other control and, to the writer's knowledge, none has been suggested. Caves with levels, therefore, seem to have been excavated just below a water table in the so-called *shallow-phreatic zone*. The principal purpose of this paper is to attempt to identify one or more mechanisms for shallow-phreatic cave excavation.

Before this can be done, however, two subsidiary questions must be explored. The first concerns the nature of flow in limestone aquifers. Due to the size and distribution of their pores, the hydrologic properties of limestone aquifers are qualitatively different from those of granular aquifers. Because our concepts of subwater-table flow are largely based on flow in granular aquifers, it is necessary to examine the differences between the two types of aquifers before discussing cave development based on these concepts.

The second subsidiary question is a chemical one. Water in limestone aquifers is usually thought to be meteoric water which has moved down from the surface through the vadose zone. As all normally accessible caves are now in the vadose zone, this downward-moving water is easily studied. We find, however, that it is usually saturated or supersaturated with respect to calcite and is actively depositing $CaCO_3$. Davis (1930, p. 477) began his paper on cave origin by asking if the change from cave excavation to secondary deposition did not indicate a change in environment, a change that he later decided was due to elevation above the water table. The question remains how water which is depositing $CaCO_3$ above the water table can dissolve $CaCO_3$ below it.

The emphasis in this paper is on hydrologic and chemical factors which are believed to be of general applicability in cave excavation. Undoubtedly, stratigraphic, structural, and geomorphologic influences are of great importance in controlling the development of individual caves. Except during a brief excursion into the origin of Mammoth Cave, Kentucky, however, these geologic influences will not be discussed.

Finally, although this paper deals specifically with the large "pores" in limestone known as caves, many of the arguments and conclusions should also apply to the general development of porosity in limestone by solution.

ACKNOWLEDGMENTS

This paper, which represents a portion of a dissertation presented to Princeton University in partial fulfillment of the degree Doctor of Philosophy, was prepared while the writer was the recipient of a William Libbey Fellowship from Princeton University and a National Science Foundation Graduate Fellowship. Thanks are extended to William Back, J. M. R. de Wiest, Sheldon Judson, G. W. Moore, and, especially, to H. D. Holland for their critical reviews of the manuscript and suggestions for its improvement. All errors of fact or of reasoning are, of course, the sole responsibility of the author. Appreciation is also expressed to the superintendents of Carlsbad Caverns and Mammoth Cave National Parks for permitting access to the caves and other areas under their jurisdiction.

HYPOTHESES OF CAVE ORIGIN

Prior to the appearance of a paper by W. M. Davis in 1930, most of the explanations of cave development advanced in the American literature held that caves were dissolved in limestone by meteoric water which descended from the surface, mainly along joints, and then flowed laterally to the valley of a surface stream. The lateral flow was stated or implied to be on a less permeable (or less soluble) bed or along the "top" of the water table. As it would now be characterized, cave excavation was considered to take place in the vadose zone and at the interface between the vadose and phreatic zones. Explanations of this sort were advanced or assumed in more than a dozen papers in the early part of the 20th century. Of these, the following are more important.

In writing on the caves of southern Indiana, Greene (1909), Beede (1911), Malott (1921), and Addington (1927) all considered the water table unimportant in controlling cave development and, in fact, rarely mentioned it. The cause of horizontal flow and hence the cause of cave levels was usually taken to be an impermeable bed or the base level of a surface stream.

In contrast, Matson, Weller, and Swinnerton, in papers dealing with Kentucky caves, all considered the water table significant in determining the site of lateral flow. According to Matson (1909, p. 44–45), "In all limestone regions there are water passages below the levels of surface streams. In some places the openings are simply small pores in the limestone, but in the Blue Grass region they appear to be channels which closely resemble the caves." Weller (1927, p. 48) considers channels developed below the water table to be of great importance in that they allow the water table to fall so that the channels may be occupied by streams flowing at the water table, which he believes is the major site of cave excavation. A similar interpretation was made by Swinnerton (1929). Although these writers apparently did not think that circulation below the water table actually formed large caves, they were well aware that solution took place there.

Quite different concepts of cave formation were held during this period in Europe (summarized by Zötl, 1961, p. 141–144). Two of the leading workers, Katzer (1909) and Bock (1913), considered that rapidly flowing water in underground conduits is an important cause of cave formation. These writers rejected the idea of an integrated ground-water body in limestone. Because they visualized the water in the conduits as being under pressure, however, the site of cave formation is below the water table (or piezometric surface) and in the phreatic zone according to our terminology. Earlier, Grund (1903) had held that caves were formed in the upper part of a continuous body of ground water, and the principal debate in the European literature during this period (and, for that matter, up to the present) centered on the question of the continuity of ground water in karst areas. This "single aquifer" versus "multiple aquifers" question will be discussed in a later section.

In 1930, William Morris Davis published his classic paper on the origin of limestone caves. He suggested that, because most caves are now the sites of deposition rather than of solution, they may have been formed in an environment different from their present one. Because he considered that various features found in many caves indicated that they were formed when full of water, he proposed that such caves were excavated in the phreatic zone. The decline of the water table since their formation was ascribed to rejuvenation, and thus such caves were considered to have been formed in an earlier geomorphologic cycle. Davis therefore called them *two-cycle caves* and his explanation of their development the *two-cycle theory*. Earlier ideas involving cave excavation "by vadose or water-table streams" he termed the *one-cycle theory* (Davis, 1930, p. 480).

As Davis acknowledged, these ideas were similar to those of Grund. Davis, however, was proposing cave formation in the phreatic zone, while Grund, essentially, was arguing for the existence of a phreatic zone in limestone, For a flow model Davis used that of King (1899), redrawing it to show the water moving along joints and bedding planes (Fig. 1A).

In 1932, Piper concluded that deep circulation must be rapid if the water is not to become saturated with respect to calcite at shallow depths. He felt that the caves of the area in Tennessee which he studied were formed near the water table, especially in the zone through which it fluctuated seasonally (Piper, 1932, p. 71–74).

Also in 1932, Swinnerton amplified the idea of cave formation in the vicinity of the water

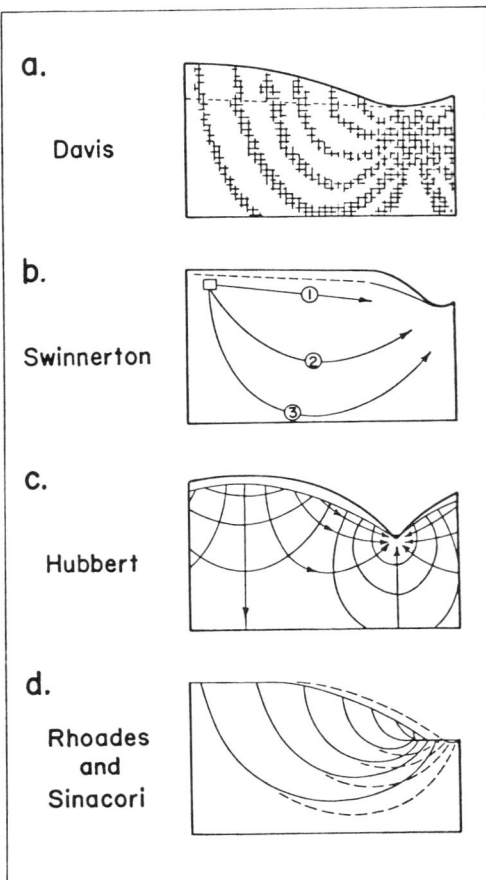

Figure 1. Patterns of ground-water flow in limestone: (a) *after* Davis (1930, fig. 31); (b) *after* Swinnerton (1932, fig. 1); (c) *after* Hubbert (1940, fig. 45); (d) *after* Rhoades and Sinacori (1941, fig. 3).

table. He termed this concept, which was similar to those of Matson (1909) and Weller (1927), the *water-table hypothesis*, and, like Piper, suggested that floor waters flowing at the water table would aid in cave excavation. Swinnerton based his view of ground-water circulation on the ideas of Finch (1904), who held that the ground-water body in nongranular rock consists of an upper zone of mobile water and a lower zone of static water. It is interesting to note that this is the same picture conceived by Grund (1903), who believed that caves were formed in an upper zone of flowing "karst water" above a lower zone of stagnant groundwater.

Swinnerton considered that if water were to flow beneath the water table by all possible paths (Fig. 1B), the nearly horizontal path just below the water table would carry the most water because it is the shortest. He also believed that the water of this path would be less saturated with respect to calcite than that of the deeper paths.

Gardner (1935) and Malott (1938) restated the case for a vadose origin for caves and emphasized the importance of lithology and the local physiographic development.

Hubbert (1940, p. 927–930) analyzed Swinnerton's conception of subwater-table flow and showed that, in the general case, ground water will follow deeply curving paths and that alternate paths are not possible (Fig. 1C). We shall see, however, that for certain types of flow Swinnerton's model was approximately correct.

Rhoades and Sinacori (1941) presented a picture similar to Hubbert's and concluded that solution would be greatest where the flow was most concentrated. Cave development thus began at the point of outflow to a surface stream and progressed horizontally away from the stream, with the flowlines adjusting continuously to an outflow at the end of the cave (Fig. 1D).

The two-cycle theory of Davis received support in a paper by Bretz (1942), in which he described many of the features to be found in caves. He divided these features into those formed when the caves were completely filled with water and those imposed later by free-surface vadose streams.

Although the existence of caves with horizontal passages not directly related to bedding had been noted by several workers, including Bretz (1942, p. 705), Sweeting (1950) was the first to present a clear description of such passages in a paper dealing with the Ingleborough district of England. The larger cave passages of this region are at well-defined levels, which she correlated with earlier higher stands of the water table. The passages appear to have been excavated when completely filled with water.

Conclusions similar to those of Sweeting were reached by Davies (1960) in a paper based on observations made in hundreds of caves in the eastern United States. Davies felt that most of these caves were formed just below the water table at times when it was relatively stable. He also noted that many caves seem to be graded toward surface valleys, that caves are more abundant in the upper parts of drainage basins, and that larger caves tend to be adjacent to major valleys.

Woodward (1961) presented an hypothesis in which caves are formed rapidly during the water-table adjustments accompanying stream piracy.

Recently, Howard (1964) has investigated the development of caves theoretically and suggested a mechanism similar to that of Rhoades and Sinacori (1941) to explain cave excavation in the shallow-phreatic zone.

The accumulating body of observational evidence (see Sweeting, 1950; Davies, 1960; White, 1960; and Wolfe, 1964) seems to show that there are a number of caves, including some large cave systems, which have been formed in the vicinity of the water table, probably just below it. This environment has come to be called the *shallow-phreatic zone*. The principal explanation for cave development in this zone is that of Swinnerton (1932) mentioned above. As we will see, however, this explanation falls short of providing a theoretical basis for considering the shallow-phreatic zone to be the preferred region for cave excavation.

THE GROUND-WATER BODY IN LIMESTONE

The terms and concepts used to describe limestone aquifers have, in general, been transferred from aquifers composed of granular rocks. The two types of aquifers, however, are widely recognized as being qualitatively different, and a brief examination of the way in which limestone aquifers may be described is necessary.

In granular rocks, an *aquifer* may be considered as a body of permeable rock having characteristic hydrologic properties and containing a continuous body of ground water. Neglecting, for the moment, such phenomena as seepage surfaces and leaky aquifers, an aquifer is bounded by *permeability barriers* (across which there is considered to be no flow), which separate it from impermeable rock, or by a *water table* (at which the pressure is atmospheric), which separates it from the *vadose zone*, or by both. Where the aquifer is bounded above by a permeability barrier, the water table is replaced by a *piezometric surface*. Finally, the *phreatic zone* will be taken to be the same as the aquifer, following Meinzer (1923, p. 5). In the above statements, of course, the modifiers *permeable* and *impermeable* should be taken to indicate values of permeability that are greater than or less than a value selected as a threshold in a particular context.

All the above terms refer to features which are much larger than a well, which is the "instrument" used to examine them. In contrast, the actual voids (pores) and intervoid volumes (grains) of the granular aquifer are usually much smaller than a well. This clear-cut difference in scale allows analyses on two conceptual levels, and although the true voids of an aquifer are the pores, the aquifer as a whole may be treated as a void in which special volume and flow relationships hold.

In a limestone aquifer in which the significant porosity and permeability is in large solution channels (caves), the true voids and intervoid volumes are intermediate in scale between the two types of granular aquifer "voids." It must be decided, therefore, whether to consider caves as pores or aquifers. If they are treated as pores (Fig. 2), the aquifer be-

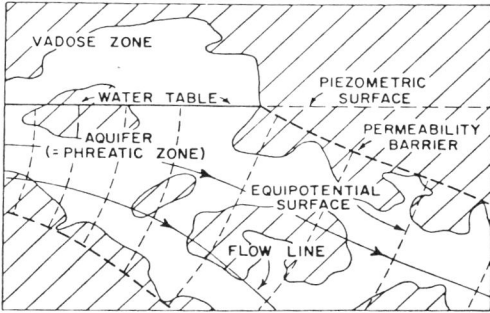

Figure 2. Concept of limestone aquifer (region of cave development = aquifer).

comes the region of cave development and the bounding permeability barrier will be an envelope enclosing this region. The bounding water table will extend through caves and rock alike, in the same way that the water table which bounds a granular aquifer extends through both grains and pores.

Alternatively, caves may be identified with the aquifer (Fig. 3). In this case, the bounding

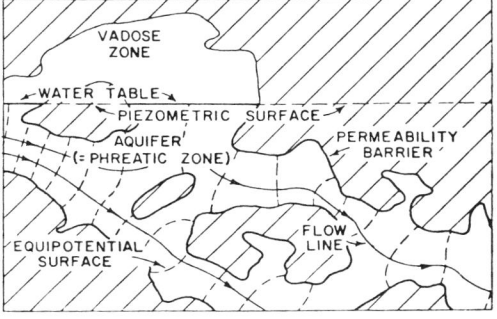

Figure 3. Alternate concept of limestone aquifer (caves = aquifer).

permeability barrier is the cave walls, and only the actual water surface in passages that extend into the vadose zone would be the water table.

In studies in which the emphasis is on flow in specific cave passages, this latter view may be useful. For most purposes, however, it is easier to consider the aquifer to be the entire zone of cave development, and this convention will be followed in this paper. One advantage of this approach is that it allows a generalized flow net (Fig. 2) to be drawn, rather than one restricted to the actual openings (Fig. 3).

Regardless of whether the limestone aquifer is taken to be the entire region of cave development or only the actual caves, two or more aquifers may occur in close horizontal or vertical proximity. Such aquifers may be entirely separate, in the sense that a continuous body of ground water does not exist between them. This would be the case when a perched aquifer overlies a "main" aquifer, with the two separated by surfaces at which the pressure is one atmosphere (water tables or seepage surfaces) and by permeability barriers. Two aquifers may also be separated only by permeability barriers.

Although a continuous body of ground water may exist in a region of rock, variations in hydrologic properties, such as those due to less permeable beds and sediment-filled channels, will often produce steep head gradients over short distances even under low-flow conditions. Such steep gradients will be reflected by apparently discontinuous jumps in the water table or piezometric surface. A body of ground water completely surrounded by such jumps would probably best be considered a separate, although leaky, aquifer. To be strictly consistent with the definition of an aquifer being used here, the zone of low permeability which produces the head jump should also be defined as a separate aquifer.

All of the above relationships should apply to any aquifer under steady-state conditions. The position of the water table, magnitude of the head jumps, and other features of the aquifer, however, will be dependent on the flow rate. When larger than normal amounts of water are delivered to an aquifer, the water table (or piezometric surface) will rise, and such a flooded aquifer may be vastly different from the low-flow aquifer. Although it would be simpler in almost every way to discuss an aquifer in terms of its low-flow configuration, it appears that cave excavation is more directly related to flooded aquifers. This introduces difficulties which will be discussed further in the concluding section.

FLOW CONSIDERATIONS

General Statement

It is reasonable to assume that the flow pattern below the water table is of considerable importance in cave excavation. As we have seen, most explanations of cave origin in the phreatic zone have been based on principles of ground-water flow in granular rocks. When large-solution cavities and caves are present in the aquifer, however, the mode of flow is laminar or turbulent in conduits, not the Darcy flow of water in granular rocks. Also, it is probable that the flow mode changes with time as cavities are enlarged and integrated. We must therefore briefly examine and compare the various modes of flow.

Water flowing slowly in granular rocks obeys Darcy's Law, which may be written

$$Q = Nd^2\rho \frac{g}{\mu} \frac{dh}{ds}$$

where Q is the bulk velocity (flow volume per unit area) and dh/ds is the headloss gradient. The remaining terms are those that pertain to or act upon the fluid (ρ, density; g, gravitational acceleration; and μ, viscosity) and those of the medium (N, a shape factor, and d, a characteristic length). For a given fluid and medium, these latter five terms may be collected into a single constant, and it is seen that the flow varies linearly with the headloss gradient. A variety of mathematical treatments are therefore possible, including construction of a flow net.

In small pipes and channels at moderate velocities, flow is typically laminar, and the headloss and flow in a circular pipe are related by the Hagen-Poiseuille Law, which may be written

$$Q = \frac{d^2}{32} \rho \frac{g}{\mu} \frac{dh}{ds}$$

This expression is seen to be the same as that for Darcy flow (as is widely recognized, of course) except for the term $d^2/32$, where d is now the diameter of the conduit. It would have been possible, therefore, to consider Darcy flow a special case of laminar flow (or vice versa). Because of the different significance of the velocity and other parameters in Darcy and laminar flow, however, it is believed that the treatment is simplified by considering them separately.

The third mode, turbulent flow, occurs in

pipes or channels when the flow is rapid or the conduits are of large diameter. The Darcy-Weisbach expression for turbulent flow in a circular pipe (after Vennard, 1961, p. 280) may be written

$$Q^2 = \frac{2dg}{f}\frac{dh}{ds}$$

It will be noted that in this equation Q is no longer a linear function of the headloss gradient, that ρ and μ are not explicitly stated, and that the (Darcy) *friction factor f* has been introduced (d is again the diameter of the pipe). In order to relate this expression to those for Darcy and laminar flow, the *Reynolds number* $\mathbf{R}$ must be discussed, which will also enable the ranges of applicability of the expressions for Darcy, laminar, and turbulent flow to be determined.

For flow in circular pipes, $\mathbf{R} = Qd\rho/\mu$, with Q interpreted as the average velocity and d the diameter. The relationship between $\mathbf{R}$ and f is somewhat complex. Experimentally, f is a function only of $\mathbf{R}$ at small $\mathbf{R}$, of both $\mathbf{R}$ and e/d at intermediate $\mathbf{R}$, where e is the height of irregularities on the wall of a pipe of diameter d, and only of e/d at high $\mathbf{R}$ (Vennard, 1961, p. 283-284). These relationships are shown in Figure 4. For $\mathbf{R}$ less than about 2100, $f = 64/\mathbf{R}$ and the Darcy-Weisbach expression reduces to the Hagen-Poiseuille equation. This is the region of laminar flow.

For values of $\mathbf{R}$ greater than 2100 but less than about 21,000 for rough pipes (much higher for smooth pipes), f is generally a function of both $\mathbf{R}$ and e/d, and its exact value may be determined most easily from a diagram similar to Figure 4. At still higher values of $\mathbf{R}$

$$\frac{1}{\sqrt{f}} = 1.14 + 2.0 \log \frac{d}{e}$$

(Vennard, 1961, p. 298). Because in this range f is not a function of $\mathbf{R}$, and hence not of Q, the headloss in the Darcy-Weisbach equation is proportional to Q^2, with the proportionality constant made up only of terms which are constant for a given fluid and conduit.

For flow through granular material, $\mathbf{R}^* = Qd\rho/\mu$. Because Q, here the bulk velocity, is lower than the flow velocity, and because d is now the average grain diameter, it is apparent that $\mathbf{R}^*$ and $\mathbf{R}$ are not easily compared. In Darcy flow, the relationship $f = 250/\mathbf{R}^*$ provides a good fit to experimental data (see diagram in Todd, 1959, p. 49). At $\mathbf{R}^*$ greater than about 1, this relationship no longer holds and Darcy's Law becomes invalid. This is probably due to the increased importance of inertial forces (Hubbert, 1940, p. 821).

Flow in Limestone

Before investigating patterns of flow in limestone, some mention should be made of the presumed changes in flow modes as progressively larger conduits are dissolved out.

It is obvious that limestone at the start of the solution process must possess some permeability due to fractures (joints or bedding planes) or intergranular pores. Whatever the nature of such openings, their existence must be invoked here, as in other papers on cave origin, as an initial condition.

If the initial openings are small intergranular pores, the initial flow will be Darcy. Assuming $Q = 10^{-1}$ cm/sec, Darcy flow will persist as the pores are solutionally enlarged until $\mathbf{R}^* = 1$, corresponding to a grain diameter of 1.31 mm. If the flow paths are tortuous, the flow at larger pore diameters cannot easily be described. If, however, the conduit formed by a series of pores is fairly straight, we may consider the flow to be similar to laminar flow in a pipe. This would also be a reasonable assumption if a fracture intersection had been the initial opening. If we specify that there is one

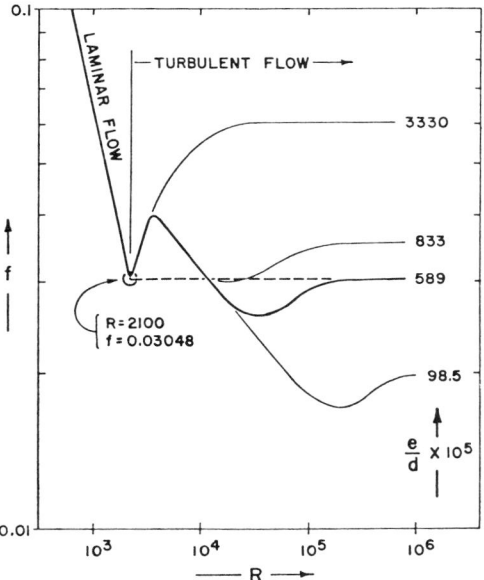

Figure 4. Plot of Reynolds number $\mathbf{R}$ against friction factor f *after* Vennard (1961, p. 283, fig. 119). *See* text for explanation.

213

such pipe per cm² of the rock (in cross section), that its diameter is 1 mm, and the bulk velocity Q remains 10^{-1} cm/sec, then the actual velocity in the pipe will be 12.74 cm/sec and $\mathbf{R} = 97$ (for water at 10° C, $\mathbf{R} = 76.3\ Qd \times \text{sec/cm}^4$ in a circular pipe, where Q is the pipe velocity).

For a given value of the bulk velocity Q, both the actual water velocity and $\mathbf{R}$ will decrease as the pipe is enlarged. If there is one pipe per cm², the actual velocity will equal the bulk velocity when the cross-sectional area of the pipe reaches 1 cm², and $\mathbf{R}$ will here be 8.6. Because this would mean that all the rock had been removed, it is apparent that as certain pipes are enlarged, others will either grow more slowly (as their flow is captured), become amalgamated with the growing pipes, or become plugged.

Hereafter Q will denote the average velocity in a pipe. If it is held constant at 10^{-1} cm/sec, $\mathbf{R}$ will increase as the pipe diameter increases. The transition to turbulent flow ($\mathbf{R} = 2100$) will occur at a pipe diameter of 2.75 m.

The processes by which certain pipes are enlarged and the flow becomes integrated are undoubtedly of importance in cave development. Lack of space precludes their investigation here. It is believed, however, that the conclusions reached in this paper will remain valid regardless of the exact nature of flow integration.

Although there is almost no data on the actual size of subwater-table conduits and on the velocities of flow in them, it is apparent that the flow mode in most limestone aquifers will be laminar or turbulent or both, not Darcy.

The Pipe Network

In order to examine the distribution of flow in a hypothetical limestone aquifer, we will use the simple pipe network shown in Figure 5. All of the segments of the network are specified

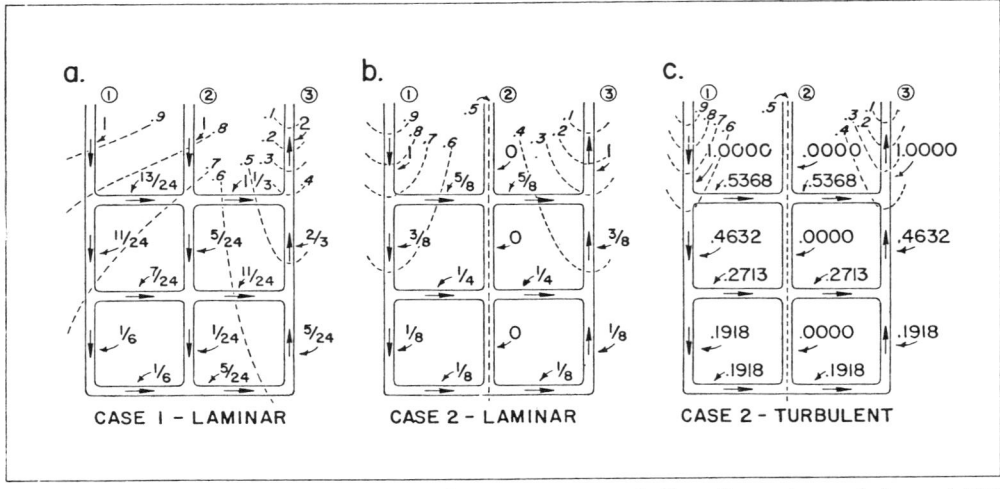

Figure 5. Square pipe networks. Figures adjacent to segments indicate flow volume relative to 1.0000 entering at point 1. Equipotential lines dashed; figures indicate potential relative to 1.0 at point 1 and 0.0 at point 3.

as having the same diameter and Q may be considered the flow volume. At each junction (or elbow), the flow leaving is equal to the flow entering, which yields nine equations. By specifying the flow leaving or entering the network at points 1, 2, and 3 (Fig. 5), three boundary equations are obtained. If the flow is laminar, if all the fluid and environmental parameters are the same, and if all the segments are of equal length, then the headloss along each segment is equal to Q multiplied by a constant. Because the head at any point in the network is independent of path, we may equate the headloss along alternate paths, substitute Q, and eliminate the constant. The simplest method is to sum the headloss about each of the four loops (after adopting a sign convention) and set it equal to zero. This yields 4 additional equations for a total of 16, of which only 15 are independent. The value of Q for each of the 15 segments of the network may therefore be determined.

It is evident that the simple network of

Figure 5 could be extended indefinitely and solutions would still be obtainable (although practical only on a computer). It is also evident that the whole procedure is identical to methods of solving electrical networks, as has been discussed by numerous workers.

It might be noted that Kaye (1957, p. 43) examined a similar problem. As he stated, he did not consider headlosses, and his solution was thus only one of many possible ones.

equal amounts of water introduced at points 1 and 2, with the combined flow discharging at point 3. This may be likened to a "normal" aquifer in which water is uniformly supplied to the water table and discharged at a stream. In the second situation, Case 2, all the water enters the system at point 1 and discharges at point 3, with no flow at point 2. This resembles an aquifer in which water is contributed by an influent stream or sinkhole and flows to an

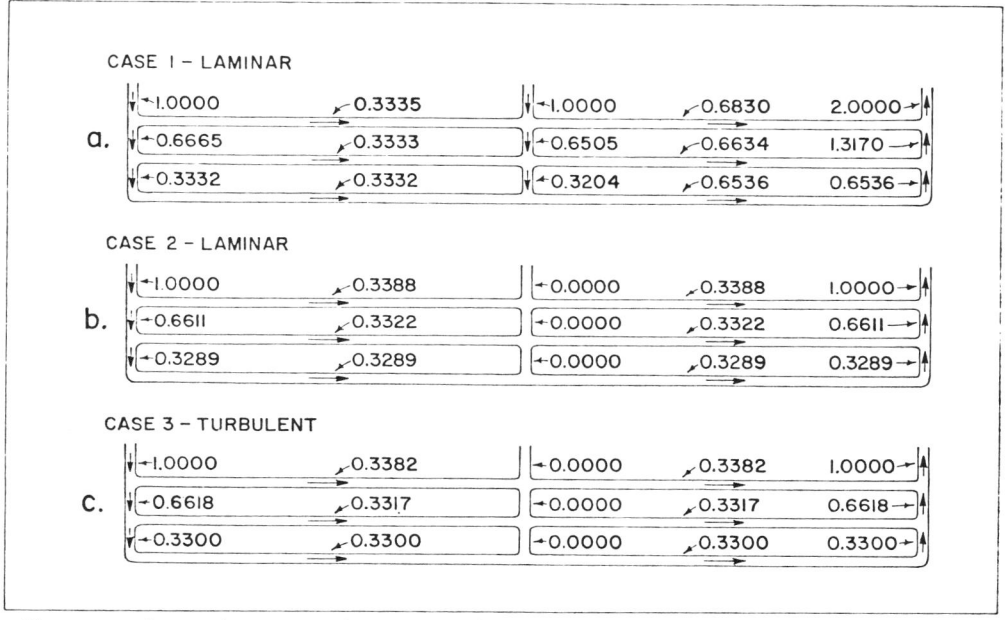

Figure 6. Elongated pipe networks. Figures indicate flow volume when horizontal segments are 100 × vertical segments (note 10 × vertical exaggeration).

Computations for turbulent flow are somewhat more complex. Using the Darcy-Weisbach expression, and assuming constant f (i.e., large R), the four headloss equations will be the same as before with Q^2 substituted for Q.

In addition to the headloss along a pipe given by the Hagen-Poiseuille and Darcy-Weisbach equations, head will also be lost at projections, corners, junctions, and sudden enlargements and contractions. Such so-called "minor losses" are undoubtedly important in real caves, but difficult to evaluate here. It will only be said that if the amount of headloss due to such features is proportional to the length of the pipe, the above analysis will remain valid.

The simple pipe network permits the investigation of two situations of geologic interest. The first of these, Case 1, consists of

effluent stream. The two-dimensional network, of course, only approximates a real, three-dimensional geologic situation.

Because the thickness of an aquifer is commonly measured in tens of meters while its horizontal extent is in kilometers, the network would be more realistic if the horizontal segments were 100 times the length of the vertical segments. Flow in such a network may be calculated by multiplying the appropriate terms in the headloss equations by 100.

Computations of flow distribution were made for both square and elongated networks for Case 1, laminar flow, and Case 2, laminar and turbulent flow. The results are shown in Figures 5 and 6. To facilitate comparisons, potential contours analogous to those of a flow net have been drawn on each of the square networks (Fig. 5). The justification for sketching them in the areas between the pipes was outlined earlier in the discussion of the groundwater body in limestone. Similarly, orthogonal flow lines could also have been drawn.

The equations for Darcy flow and laminar flow are equivalent, and the potential contours for laminar flow (Fig. 5A and B) are essentially the same as those that would be developed in Darcy flow (although the small number of pipes produces distortions). The potential contours for turbulent flow (Fig. 5C), however, are quite different from those of laminar flow (Fig. 5B). The flow in the horizontal segments is more equally divided in turbulent flow than in laminar flow.

Under Case 2 conditions in the elongated network, however, the division of flow in the three horizontal segments is within one per cent of being equal in both laminar (Fig. 6B) and turbulent (Fig. 6C) flow. Furthermore, in the elongated network the flow distribution is virtually the same in laminar and turbulent flow; the flow in corresponding horizontal segments in Figure 6B and C is within one per cent of being identical. Although turbulent flow under Case 1 conditions was not analyzed, it would probably show an equally slight difference from Case 1, laminar (Fig. 6A).

If the head in Case 2, laminar (Figures 5B and 6B) is increased so that more water is forced through the system, **R** will eventually exceed the critical value (here taken to be 2100), and turbulent flow (Figures 5C and 6C) will commence. Because more water is flowing through the upper horizontal segments than the lower ones (although the difference is slight in the elongated network), turbulent flow will occur first in the upper segment. It is of interest, therefore, to examine the flow distribution when both laminar and turbulent flow occur in the same network.

For laminar flow at **R** = 2100, f = 0.03048. By using this value of f at all higher **R**, a jump in f is avoided at **R** = 2100 when the flow becomes turbulent. This corresponds to a value of e/d of 0.00589, which would seem to be a realistic value, being that of a pipe (or cave passage) 1.8 m in diameter with projections on its walls 1 cm high. This value of f is shown by the dashed line in Figure 4. Note that for intermediate values of **R**, the experimental value of f for this e/d varies only from about 0.025 to 0.040.

For simplicity, only the loop shown in Figure 7A was analyzed. The increase in flow

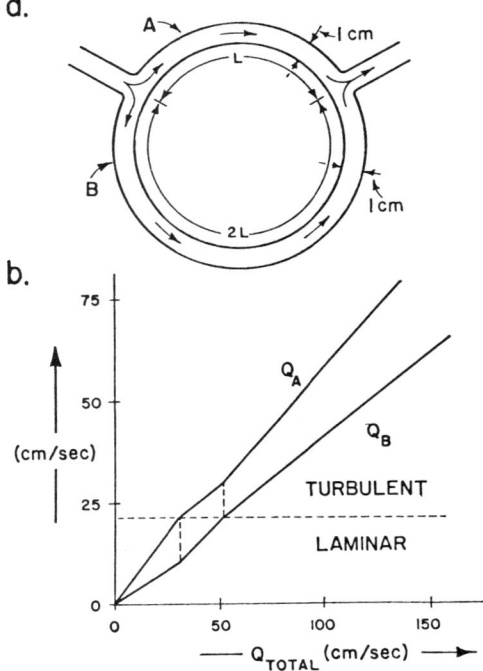

Figure 7. Simple loop showing flow division at various velocities Q. (a) Diagram of loop. (b) Division of flow. See text for discussion.

in each of the pipes with increasing total flow is presented in Figure 7B. It can be seen that, with turbulent flow in the shorter segment and laminar flow in the longer, the division of flow is intermediate between the wholly laminar and wholly turbulent cases. The flow division is similar when the two segments are of different diameters instead of different lengths.

Summary

From the preceding analysis, it would appear that the flow pattern in a limestone aquifer is approximately the same under conditions of Darcy, laminar, turbulent, or mixed laminar and turbulent flow, provided that (1) there are enough pipes in the aquifer to prevent gross distortions of the flow net, (2) the lateral extent of the aquifer is large relative to its thickness, and (3) there are no systematic variations in the size or shape of the conduits.

Furthermore, there seems to be no inherent hydrologic reason for the flow in shallow conduits to be significantly greater than in deeper ones (in aquifers of geologically reasonable proportions). The major cause of shallow phreatic solution apparently lies elsewhere.

CHEMICAL CONSIDERATIONS

Vadose Seepage

Water above the water table and below the soil zone was called *vadose water* by Meinzer (1923, p. 26). It includes water held by surface tension (of no particular interest here), slow seepage, and discrete flows. Although less obvious than the flows, seepage probably

Also, the presence of complexes such as $CaHCO_3^+$ and $CaCO_3^\circ$ (Garrels and Thompson, 1962) was ignored; all Ca^{+2} in solution was considered to be uncombined. The absolute error in the calculated values of A_C ($= a_{Ca}^{+2} \cdot a_{CO_3}^{-2}$) due to these simplifications is on the order of 10 percent or so at values of the pH between about 6 and 9. The following discussion, however, will deal mainly with differences

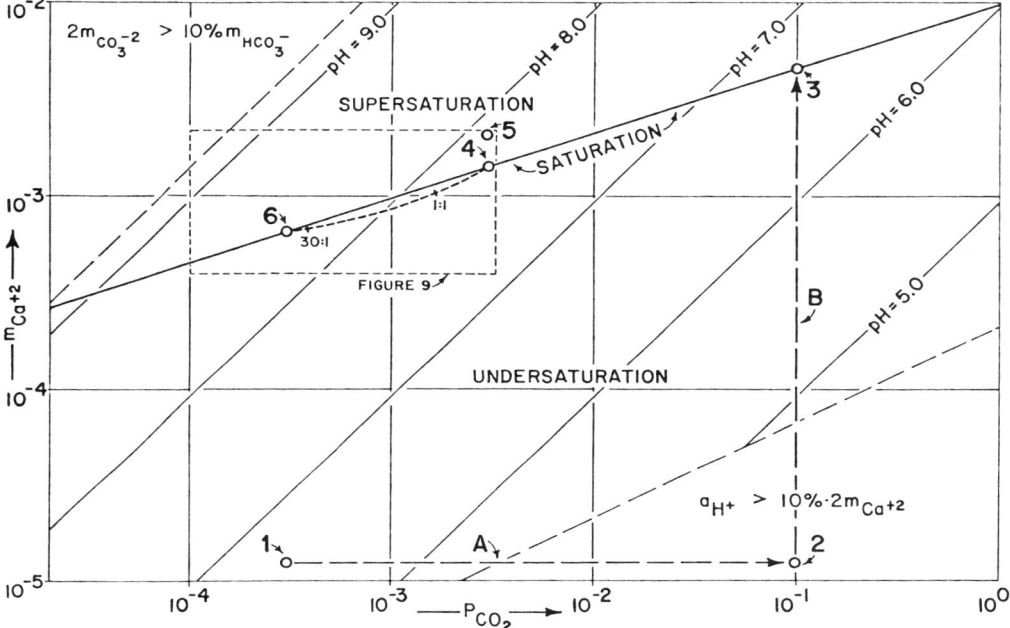

Figure 8. Calcite saturation diagram. Plot of log m_{Ca}^{+2} (moles/liter) against log equilibrium P_{CO_2} (atm). See text for discussion.

comprises the bulk of the water moving through the vadose zone in most areas. Here called *vadose seepage*, it is meteoric water that slowly percolates through the rock.

Because the ground water body of limestone terrains may have been derived from vadose seepage, a review of the chemical changes that may take place before and at the time this water arrives at the water table is pertinent. In the following discussion, a pure limestone terrain is assumed whose waters have compositions which lie within the system $CaCO_3$–CO_2–H_2O. In order to simplify calculations, the small terms in the charge-balance equation $m_{H^+} + 2m_{Ca}^{+2} + m_{CaHCO_3^+} = m_{HCO_3^-} + 2m_{CO_3}^{-2} + m_{OH^-}$ were neglected and the equation reduced to $2m_{Ca}^{+2} = m_{HCO_3^-}$.

in A_C between waters which differ in pH by less than one pH unit. The concentration of the ignored species is generally a function of the pH, and the relative error in A_C will be small. Much of the discussion follows similar treatments by Garrels (1960, p. 43–60) and especially by Holland and others (1964).

The chemical changes discussed below have been plotted on Figure 8. Although all computations (unless otherwise stated) have been made using activity coefficients obtained from solutions of the Debye-Hückel equation, the various activity coefficients were assumed to be unity in the construction of this diagram. The compositions and pH of the various points are therefore only approximate. The temperature was taken to be constant at 10° C. The lines of constant pH were obtained from the equation

$$a_{\mathrm{H}}^+ = \frac{K_1 B P_{CO_2}}{2m_{Ca^{+2}}}$$

derived from expressions for the first dissociation of carbonic acid, the solubility of CO_2, and the charge-balance equation. Values used for the constants are given in Table 1.

Point 1 in Figure 8 represents rainwater

tersected, however, the composition will change.

Variations in the composition of vadose seepage as it passes through a cave have recently been studied by Holland and others (1964). Their best data were from Luray Caverns, Virginia, which is in a mixed limestone and dolomite terrain. Their findings, as outlined below, are probably equally applicable to caves in limestone.

TABLE 1. VALUES OF EQUILIBRIUM CONSTANTS

	10°	15°C	reference
log K_1	− 6.4640	− 6.4187	Harned and Davis (1943)
log K_2	−10.490	−10.430	Harned and Scholes (1941)
log K_C	− 8.150	− 8.215	Garrels (1960)
log B	− 1.2695	− 1.3412	Harned and Davis (1943)

with a $c_{Ca^{+2}}$ of 0.5 ppm, which is seemingly a reasonable value (see Carroll, 1962, p. 6). The rainwater is assumed to be in equilibrium with a P_{CO_2} of 3×10^{-4} atm (that of the normal atmosphere).

As noted by Adams and Swinnerton (1937, p. 506), the partial pressure of CO_2 in the soil air may be considerably higher than that of the normal atmosphere; apparently a value of 0.1 atm is not uncommon. If we allow the rainwater to equilibrate with a vapor phase containing CO_2 at this pressure, its composition will follow path A to point 2. When this water dissolves calcite (while remaining in contact with a vapor phase P_{CO_2} of 0.1 atm), its composition will move along path B to point 3, where it is saturated with respect to calcite. Paths A and B correspond, respectively, to stages one and two of Holland and others (1964, p. 37–41). These authors also discussed the equilibration of water with limestone in the absence of a vapor phase (Holland and others, 1964, p. 40). Water flowing in small channels rapidly becomes saturated with respect to calcite (Weyl, 1958) and saturation with respect to both CO_2 and calcite probably takes place above or at the base of the soil zone. Equilibration of vadose seepage in the absence of a vapor phase, therefore, is probably uncommon.

In the event that no ventilated cave is encountered by the vadose seepage before it reaches the water table, its composition will remain that of point 3. If a ventilated cave is in-

The Luray vadose seepage had an A_C of from 0.80×10^{-3} to 4.5×10^{-3}, and was in equilibrium with a P_{CO_2} of from 3.3×10^{-3} to 7.2×10^{-3} atm (Holland and others, p. 61). Thus the water was supersaturated with respect to calcite ($K_{Calcite}$ at 15°C = 0.61×10^{-8}; Garrels, 1960, p. 58) and in equilibrium with a P_{CO_2} higher than that of the normal atmosphere (0.3×10^{-3} atm). In general, the water in pools had a lower A_C and was in equilibrium with a lower P_{CO_2} than dripping water.

Although the seepage was in equilibrium with a P_{CO_2} considerably less than the 0.1 atm postulated earlier, the supersaturation with respect to calcite indicated that either the equilibrium P_{CO_2} was originally higher or that evaporation had occurred. By demonstrating that the concentration of Mg^{+2} stayed nearly constant as the water moved through the cave, Holland and others (1964, p. 59) showed that evaporation was not important. Also, the activity product for dolomite in the vadose seepage was so high that a soil P_{CO_2} in the neighborhood of 0.1 atm was indicated.

The Luray data show that the change in composition of vadose seepage in a ventilated cave is due both to the escape of CO_2 from solution and to the precipitation of calcium carbonate. The escape of CO_2 is controlled by the P_{CO_2} of the cave atmosphere and the rate with which water equilibrates with that atmosphere. The P_{CO_2} of the Luray atmosphere was not specifically investigated, but there was evidence that it may have been higher than 0.86

× 10^{-3} atm (Holland and others, 1964, p. 62). Even higher values exist in some caves: Roques (1963, p. 382) found a P_{CO_2} of 1.4 × 10^{-3} atm in the Grotte de Moulis, France; and my study in Carlsbad Caverns, New Mexico, indicates values in that cave of about 4 × 10^{-3} atm. The P_{CO_2} of the atmosphere of caves encountered by vadose seepage would normally depend on the degree of ventilation of the cave relative to the amount of CO_2 introduced by the seepage. The considerable range in P_{CO_2} with which the vadose seepage in Luray is in equilibrium shows that equilibration is not extremely rapid. Other factors determining the equilibrium P_{CO_2} of the water leaving the cave are undoubtedly its residence time in the cave and the ease with which equilibration takes place, which is influenced by such things as film thickness, pool depth, and amount of agitation.

Although the equilibration of P_{CO_2} is apparently slow, the supersaturation of vadose seepage with respect to calcite (and aragonite) indicates that the precipitation of calcium carbonate is much slower. Here again the residence time of the water in the cave is an important factor in determining the degree of supersaturation of the water.

To summarize, the presence of ventilated caves in the path of vadose seepage will cause a shift to the left in the composition of the solution at point 3 in Figure 8. If the escape of CO_2 is slow, or if the water remains in the cave long enough for all excess $CaCO_3$ to precipitate following rapid CO_2 loss, the composition of the water will lie along the saturation line. Point 4 represents water that has equilibrated to a P_{CO_2} of 3 × 10^{-3} atm and is saturated with respect to calcite. Much of the vadose seepage in Carlsbad Caverns is approximately saturated with respect to calcite (i.e., $A_C = K_C$), probably because of the long residence time of the water in this large cave.

When CO_2 escape is relatively fast and the residence time of the seepage in the cave is short, however, the composition of the water will move into the field of supersaturation to a point such as 5 (Fig. 8). At point 5, $A_C = 3K_C$, which is probably characteristic of the vadose seepage leaving many caves. The data of Holland and others (1964, p. 61) show the mean supersaturation of vadose seepage in Luray to be $2.9K_C$.

Finally, point 6 represents water that has equilibrated to the P_{CO_2} of the normal atmosphere and is not supersaturated with respect to calcite. This composition may be attained after long residence times in especially well ventilated (or relatively dry) caves.

Vadose Flows

Flows of water larger than what is here called vadose seepage are not uncommon in limestone caves. Any continuous flow ranging from a small trickle to a large underground stream (above the water table) will be termed a *vadose flow*. Although some are merely the accumulation of vadose seepage, more often such flows are a captured surface stream or are from an area of concentrated surface or soil drainage, such as the bottom of a sinkhole.

The chemistry of vadose flows has been little studied, but it appears that they are often undersaturated with respect to calcite. They are responsible for the vadose modifications of caves which have been discussed by Bretz (1942). Their undersaturation, as contrasted with the saturation or supersaturation of vadose seepage, is probably due to their different origin, the volume and speed of their flow, and to other factors which will be discussed later.

Temperature Effect

Although vadose seepage is generally saturated with respect to calcite when it arrives at the water table, there are several effects which may induce undersaturation when it joins the ground water. One of these is the inverse relationship between temperature and calcite solubility. In a geologic situation in which the incoming vadose seepage is cooled at the water table, cave excavation might be expected.

Before discussing the magnitude of the undersaturation produced by the cooling and other effects, however, it is desirable to make some estimate of the amount of solution (and hence undersaturation) necessary to produce a cave. We will postulate a "minimum" cave having a length of 500 m and a mean cross-sectional area of 1 m². Its volume is therefore 5 × 10^8 cm³. We specify that it receives the infiltration of an area of 1 km² (if necessary we may lengthen the cave while reducing the mean cross-sectional area to retain the specified volume), and that this infiltration is one-half an annual precipitation of 100 cm. The flow through the cave will therefore be 5 × 10^{11} cm³/yr. If we assume the rock is solid calcite, the undersaturation necessary to dissolve the cave in 100,000 years is 0.0108 ppm Ca^{+2}. This will be referred to as the *standard minimum undersaturation*.

Using the standard equations for carbonate

equilibria (Garrels, 1960, p. 45), and assuming $\gamma_{H_2CO_3} = 1$ (Garrels, 1960, p. 53),

$$m_{Ca}^{+2} = \left(\frac{K_1 K_C B P_{CO_2}}{\gamma Ca^{+2} \cdot \gamma^2 HCO_3^{-} \cdot 4K_2} \right)^{1/3}$$

Values of K_1, K_2, K_C, and B for 10 and 15°C are given in Table 1, and values for 11° were interpolated linearly from a logarithmic plot. The activity coefficients γ_{Ca}^{+2} and $\gamma_{HCO_3}^{-}$ were calculated from the Debye-Hückel equation using values for the appropriate constants at 10°.

Water at 11°C in equilibrium with a P_{CO_2} of 3×10^{-4} atm (that of the normal atmosphere) contains 29.3 ppm Ca^{+2} when saturated with calcite. When this water is cooled to 10°C in the absence of a vapor phase (closed to CO_2), very little additional calcite will be dissolved (c_{Ca}^{+2} remains 29.3 ppm). If the solution is then opened, or was always open, to an atmosphere with the above P_{CO_2}, however, the c_{Ca}^{+2} of the saturated solution increases to 29.8 ppm. Thus an additional 0.5 ppm Ca^{+2} may be dissolved, which is equivalent to 50 times the standard minimum undersaturation. At higher values of the equilibrium P_{CO_2}, the increase in c_{Ca}^{+2} is even greater (1.3 ppm, or 130 times the standard minimum undersaturation, at $P_{CO_2} = 3 \times 10^{-3}$ atm).

It would appear, therefore, that cooling is a likely mechanism for cave excavation, assuming the system is open to a P_{CO_2} at least equal to that of the normal atmosphere, which is probably the case, and that cooling actually occurs. Although the temperature of ground water at moderate depths usually approximates the mean temperature of the region, temperature fluctuations of one degree or more are not uncommon. During the summer, infiltrating water may be several degrees warmer than the ground-water body. Usually, cooling of this water takes place in or just below the soil zone and, unless the water table is very shallow, the infiltrating water is probably close to the temperature of the ground water when it reaches the water table. It has been shown, however, that very slight temperature differences will be sufficient to create the standard minimum undersaturation. Further, although seasonal fluctuations might average out to a net warming of the infiltrating water at the water table (and hence supersaturation), the kinetics are apparently much more favorable for solution than for deposition, and more calcite might be dissolved during short periods of slight undersaturation than would be deposited during longer periods of considerable supersaturation.

As was discussed earlier, vadose seepage that has transected caves in the vadose zone will tend to be supersaturated with respect to calcite. This will, of course, tend to counteract any undersaturation induced by cooling. This point will be explored further in the following section.

Although the preceding discussion was concerned with vadose seepage, cooling at the water table is probably even more likely in the case of vadose flows. Because vadose flows generally traverse the vadose zone more rapidly than vadose seepage (and are in less intimate contact with the rock), they are probably less likely to have cooled before reaching the water table.

Mixing Effect

Undersaturation with respect to calcite will also result when two solutions, both saturated with respect to calcite but in equilibrium with vapor phases at different values of P_{CO_2} are mixed. The basic phenomenon, which was recently pointed out by Bögli (1963) and by Arntson (1964), is illustrated in Figure 8. The composition of mixtures of waters with compositions represented by points 4 and 6 will lie along the curve (1:1 and 30:1 mixtures are shown). When a given mixture is allowed to react with calcite with the system closed to CO_2, its composition will move up and to the left until the saturation line is reached by following a curve similar to the ones shown by Holland and others (1964, p. 39) for solution in the absence of a vapor phase. The configuration of these solution curves (not shown) determines the amount of additional Ca^{+2} that may be dissolved and hence the proportions of the mixture which yields maximum undersaturation.

If it is initially assumed that all activity coefficients are equal to one, when saturated water with an equilibrium P_{CO_2} of 3×10^{-3} (point 4 in Figure 8) is mixed with an equal volume of saturated water with an equilibrium P_{CO_2} of 3×10^{-4} (point 6), the resulting c_{Ca}^{+2} is computed to be 42.4 ppm. After the solution reacts with calcite with the system closed to CO_2, $c_{Ca}^{+2} = 43.0$ ppm. Thus the amount of calcite dissolved is calculated to be about 60 times the standard minimum undersaturation.

When activity coefficients derived from the Debye-Hückel equation are used, a 1:1 mixture of the same two solutions will have a calculated

c_{Ca}^{+2} of 48.4 ppm. After saturation with calcite (while closed to CO_2), the calculated c_{Ca}^{+2} = 49.7 ppm. The amount of calcite dissolved will be 130 times the standard minimum undersaturation. All of the above calculations are at 10°C.

Although the maximum undersaturation per unit volume of mixture will result when about equal parts of the two solutions are mixed (Bögli, 1964, p. 86), the *total* undersaturation (undersaturation per unit volume times total water), relative to the high P_{CO_2} water, will be greater when larger amounts of low P_{CO_2} water are in the mixture. The exact amount of total undersaturation for specific proportions of the two end members is difficult to determine. With increasing amounts of low P_{CO_2} water, however, the total undersaturation will approach a limiting value. This value is determined by the slope of the saturation curve at the composition of the low P_{CO_2} water (point 6 in Fig. 9) and the composition of the waters

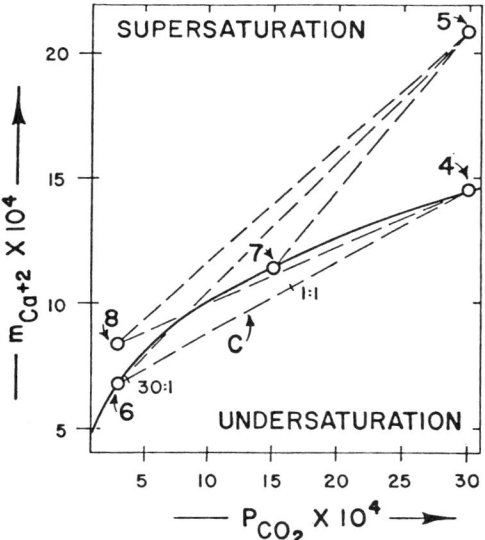

Figure 9. Calcite saturation diagram. Plot of m_{Ca}^{+2} (moles/liter) against equilibrium P_{CO_2} (atm). See text for discussion.

being mixed. When these waters have the compositions indicated by points 4 and 6, this limiting value of the undersaturation is 3.2 "ppm" relative to the high P_{CO_2} water (if all the undersaturation is assumed to be "concentrated" in the high P_{CO_2} water). This value is probably very nearly attained when the proportion of low P_{CO_2} to high P_{CO_2} water is 30:1 or less (possibly much less).

If we identify the high P_{CO_2} water with vadose water and the low P_{CO_2} water with phreatic water (i.e., ground water), the mixing of 1 part vadose water with 30 or more parts of phreatic water will result in an undersaturation of 3.2 ppm Ca^{+2} (relative to the vadose water), which is more than 300 times the standard minimum undersaturation. This figure, however, probably represents a maximum. Undersaturation due to mixing may be partially or completely defeated by at least four circumstances, all of which may commonly occur.

First, the vadose water may be in equilibrium with a low P_{CO_2}. Although vadose seepage is usually in equilibrium with a P_{CO_2} considerably higher than that of the normal atmosphere, the equilibrium P_{CO_2} of vadose flows may often be low. Chemical data on vadose flows is lacking, but it seems likely that flows representing captured surface streams which descend rapidly to the water table may be in equilibrium with a P_{CO_2} which is little, if any, higher than that of the normal atmosphere.

Second, phreatic water may be in equilibrium with a high P_{CO_2}. Accurate determination of the P_{CO_2} of water involves either a careful pH measurement combined with a chemical analysis or direct measurement of the P_{CO_2}. Unfortunately, the pH data available for limestone ground waters are almost always laboratory determinations made some time after collection. These values are, in general, not representative of the *in situ* pH (Back, 1963, p. 46). As discussed in the next section, a similar situation exists with regard to surface streams in limestone areas.

The data of G. W. Moore (*in* Hostetler, 1964, Table 1) suggests that a higher equilibrium P_{CO_2} than that of the normal atmosphere may exist in some ground-water bodies. Moore (written communication, 1962) determined the P_{CO_2}, pH, and $m_{HCO_3^-}$ of water from a deep lake in Black Chasm Cave, California, within a few hours of collection. The water was found to be approximately saturated with respect to calcite and to be in equilibrium with a P_{CO_2} in the neighborhood of 4×10^{-3} atm.

Figure 9 shows a portion of Figure 8 replotted with arithmetic coordinates (and retaining the simplifying assumption that all $\gamma = 1$). The mixing curves are now straight lines joining the compositions of the two waters being mixed. If the ground water is in equilibrium with a P_{CO_2} of 1.5×10^{-3} (point 7), which is still much lower than that of the Black Chasm Cave water, its mixtures with vadose seepage (point 4), whose compositions lie along a line (not shown) joining the two points, will be only very slightly undersaturated.

A third phenomenon which will tend to counteract the mixing effect is supersaturation of the ground water with respect to calcite. As much vadose seepage. If lines representing the compositions of mixtures are drawn from point 5 to the various ground-water compositions (Fig. 9), little or no undersaturation is indicated.

In summary, although the mixing effect will result in significant undersaturation under optimum conditions, it tends to be ineffective when vadose seepage is supersaturated with respect to calcite or when the ground water,

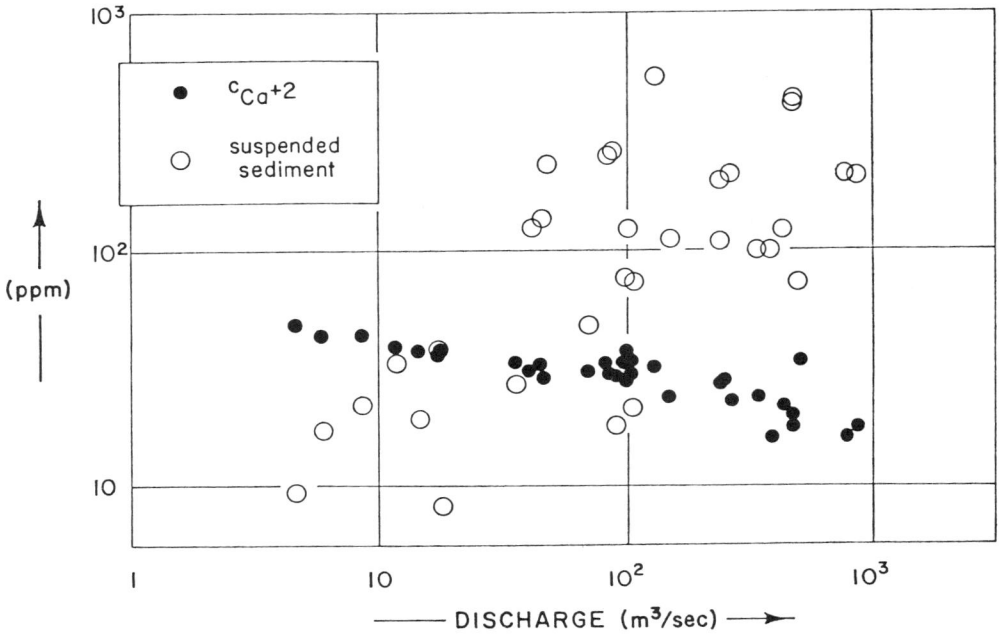

Figure 10. Log concentration—log discharge D diagram for Green River at Munfordville, Kentucky, for 1957. Data from U. S. Geol. Survey (1960, p. 497–501; 1962, p. 612–615).

mentioned earlier, the deposition of calcite is often not as rapid as the outgassing of CO_2, and ground water derived from vadose seepage would be expected to reflect this. Subwatertable deposition is not common in caves (except in "crystal caves"), and it would seem that ground water whose equilibrium P_{CO_2} is lowered by ventilation may often remain supersaturated until it leaves the aquifer. Such supersaturated ground water ($A_C = 2 K_C$) is represented by point 8 in Figure 9, and it is qualitatively apparent that little or no undersaturation results when this water is mixed with saturated vadose seepage (point 4).

Finally, the supersaturation of vadose seepage will also reduce or eliminate undersaturation due to mixing. It was stated earlier that point 5 (Figs. 8 and 9) was probably representative of having been derived from vadose seepage, is in equilibrium with a high P_{CO_2} or is supersaturated with respect to calcite. If the ground water were not derived from vadose seepage, the mixing effect might be a more significant mechanism in cave excavation. Such a case, in which the ground water is derived at least partly from surface streams and vadose flows, will be discussed in the next section.

Flowrate Effect

A third effect which may operate to yield water undersaturated with respect to calcite is the decrease in the Ca^{+2} concentration in surface streams during times of high flow. Figure 10, based on published data (U. S. Geol. Surv., 1960, p. 497–501; 1962, p. 612–615) shows this effect in the Green River near Mammoth

Cave, Kentucky. Also shown in Figure 10 is the increase in suspended sediment with increasing flowrate.

The decrease in $c_{Ca^{+2}}$ with increasing flow suggests a similar relationship between A_C ($= a_{Ca^{+2}} \cdot a_{C_{3}{}^{-2}}$) and flowrate. In natural waters with $pH < 8$, $a_{CO_3^{-2}}$ is small and is best determined from $m_{HCO_3^-}$ using measured pH and ionic strength. Calculations for the Green River based on the reported pH indicated $A_C < K_C$ (25°C) at all flowrates (Fig. 11).

It seems odd that a slow-flowing stream traversing a limestone terrain should be consistently undersaturated with respect to calcite. Also, calculations using the reported pH values show the river water at all flowrates to be in equilibrium with a P_{CO_2} between 1.3×10^{-3} and 8×10^{-3} atm, which is considerably higher than that of the normal atmosphere. Although a large part of the river water at low flowrates is probably derived from seepage which has a high equilibrium P_{CO_2}, this water should be nearly saturated with respect to calcite. During floods, much of the water is surface runoff in equilibrium with a P_{CO_2} near that of the normal atmosphere. The calculated P_{CO_2} values show little correlation with flowrate.

The published pH values were measured in the laboratory and the departure of such determinations from field measurements has already been discussed. According to G. D. Whetstone (written communication, 1964), comparisons between field and laboratory pH measurements of water samples from a stream

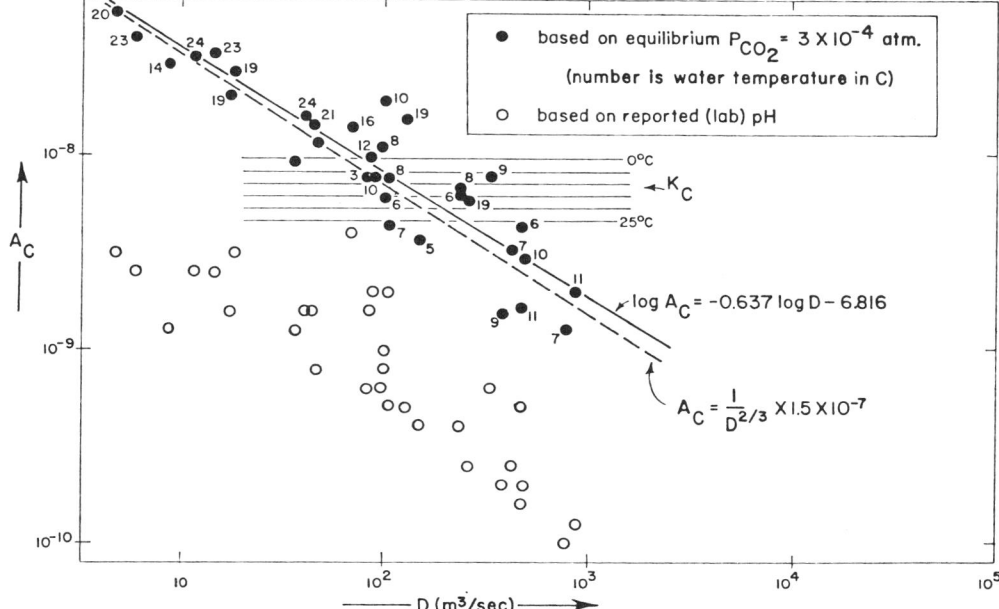

Figure 11. Log A_C ($= a_{Ca^{+2}} \cdot a_{CO_3^{-2}}$)—log discharge D diagram of Green River for year 1957. (U. S. Geol. Survey, 1960, 1962).

geologically similar to the Green River showed large discrepancies. Values of A_C were therefore recalculated using the assumption that the Green River water was in equilibrium with the P_{CO_2} of the normal atmosphere. Although the equilibrium P_{CO_2} may actually be somewhat higher (especially at low flowrates), this assumption yields maximum values for A_C and hence minimum undersaturation at high flowrates. The results of these calculations are plotted in Figure 11.

A straight line was fitted to the data by a regression of log A_C on log discharge (Fig. 11, solid line). A_C appears to vary approximately with the 2/3 power of the discharge (dashed line), and this relationship also seems to hold if the published pH data are used to compute A_C.

Assuming the water is in equilibrium with the P_{CO_2} of the normal atmosphere (3×10^{-4} atm), the river is supersaturated with respect to calcite at low flowrates and undersaturated at high flowrates. The exact amount of departure from saturation depends on the temperature. As with the temperature and mixing effects discussed earlier, the ionic strength plays a significant role in the value of A_C. Here, however, the decreasing ionic strength at high flowrates tends to counteract the effect of decreasing $m_{Ca^{+2}}$.

The concentration of Ca^{+2} in the Green River during the greatest 1957 flow was 18 ppm. Due to the low P_{CO_2} with which the water is assumed to be in equilibrium (3×10^{-4} atm), very little calcite will be dissolved when the water reacts with limestone in the absence of a vapor phase. In the presence of an atmosphere with the above P_{CO_2}, however, 11.4 ppm Ca^{+2} may be dissolved to a saturation $c_{Ca^{+2}}$ of 29.4 ppm (assuming no change in ionic strength other than that due to CO_2 and $CaCO_3$ solution). This is more than 1000 times the standard minimum undersaturation.

Although the Green River is a large perennial stream, undersaturation, at least at high flowrates, probably also occurs in the smaller and intermittent streams in limestone areas. Such streams often sink into the limestone and their water may join that beneath the water table. These sinking streams make up a large part of what was earlier termed vadose flow. If the P_{CO_2} of the caves in the vadose zone through which they flow is at least as high as that of the normal atmosphere (usually it will be higher), the water of these streams should be able to dissolve large amounts of calcite, especially during floods. This solution, possibly aided by the temperature effect, would be expected both in the vadose zone and the upper part of the phreatic zone. Also, the mixing effect may allow further solution in the phreatic zone. If the ground water is in equilibrium with a P_{CO_2} higher than that of the normal atmosphere and the incoming vadose flow is in equilibrium with a lower P_{CO_2}, the situation would be the reverse of that considered earlier, and the mixing effect would be somewhat less pronounced.

The water of surface streams may also be introduced into the aquifer in another way. When the level of an effluent stream rises during floods, water will move directly into the aquifer from the stream. This backflooding process (which will be discussed further) may greatly influence the composition of the ground-water body. Even if ventilation is lacking and direct solution is slight, the low P_{CO_2} of the river water in the aquifer will enhance the mixing effect.

Other Effects

In addition to the three effects which have been discussed, there are, of course, other ways in which the ground water in limestone may become undersaturated. Due to the geothermal gradient, the temperature of ground water may rise with depth with a corresponding decrease in calcite solubility. There is, however, little evidence for a consistent temperature rise in the upper few hundred meters of limestone aquifers (Schneider, 1964, p. 213). The large and rapid flows of water in this region are apparently sufficient to swamp out the geothermal gradient.

Another effect examined was the increase in solubility with depth due to the rise in hydrostatic pressure. Recently, Sipple and Glover (1964) determined the variations in calcite solubility for various combinations of hydrostatic pressure, temperature, and initial equilibrium P_{CO_2}. Calculations using their data for 30°C and 0.1 atm initial equilibrium P_{CO_2} yield a solubility increase of 2×10^{-4} ppm Ca^{+2} per meter of water in the pressure range from 35 to 570 m of water. Assuming this gradient extends to one atmosphere total pressure, the increase in saturation $c_{Ca^{+2}}$ at 100 m depth is 0.02 ppm. The corresponding increase with an initial equilibrium P_{CO_2} of 0.3 atm is 0.12 ppm. Although these increases in solubility are in excess of the standard minimum undersaturation, the slight solution of calcite (especially in the absence of a vapor phase) which might result is negligible relative to the other effects which have been discussed.

There are other possible mechanisms for producing water undersaturated with respect to calcite in the aquifer. In some geologic situations phreatic water may contain appreciable quantities of other ions (especially Mg^{+2} and SO_4^{-2}). This will, in general, result in a lowering of the activities of Ca^{+2} and CO_3^{-2} due to the increase in ionic strength and the formation of complex ions, which will tend to cause even greater undersaturation during mixing with vadose or other waters. Oxidation-reduction reactions, possibly involving living organisms, have been suggested as a cause of undersaturation (Moore, 1960). Although such reactions should be investigated further, there is, at present, little or no evidence that they are quantitatively significant.

Summary

In the preceding discussion it has been shown that, in all probability, the high P_{CO_2} with which vadose seepage is in equilibrium is, in general, not directly responsible for cave excavation below the water table. Because the considerable solution made possible by the uptake of CO_2 in the soil zone occurs before the water penetrates very far into the limestone, the consequent reduction of the landscape is essentially a surface process. Only under special conditions, such as in the absence of calcium carbonate near the surface, will caves be formed by this type of solution. Because sinkholes are largely created by this surface solution, the formation of karst and the excavation of caves appear to be only indirectly related.

The temperature, mixing, and flowrate effects which have been discussed seem to be both theoretically likely and quantitatively sufficient to excavate caves in geologically reasonable lengths of time. Due to the paucity of field data, the extent to which each of these actually operates is not known, and no attempt will be made to identify any one of these effects as the principal chemical mechanism in cave excavation. It is sufficient to note that each of the three effects tends to produce more undersaturation in the upper part of the groundwater body (shallow-phreatic zone) than in the deeper parts.

FLOW PATTERNS IN THE PHREATIC ZONE

General Statement

Subject to certain conditions, the patterns of laminar or of turbulent flow or of both in limestone aquifers are similar to those of Darcy flow in granular rocks. It should be possible, therefore, to obtain some insight into the probable patterns of flow in limestone aquifers by an examination of Darcy flow nets based on various boundary conditions. Also, estimates may be made of the amount and site of solution using our analysis of the chemistry of limestone waters.

Before proceeding to a discussion of these patterns, however, it is necessary to examine two additional items: the slope of the water table in limestones and the integration of vadose seepage into vadose flows.

Although the exact configuration of a water table is difficult to determine, its slope is generally a function of the amount of water delivered to it and of the permeability of the rocks beneath it. When subwater-table caves are present, the permeability of the limestone will be very high and, unless very large amounts of water are flowing, the water table will be nearly horizontal. This may be illustrated by using the "minimum" cave and yearly infiltration discussed earlier. The cave will be assumed to have a circular cross section 1 m² in area, and an e/d ratio of 0.00589. If the yearly infiltration of 5×10^{11} is assumed to flow through the entire length of the cave, the Darcy-Weisbach expression yields a headloss gradient (and hence a water-table slope) of 3.5×10^{-5}, or 3.5 cm/km, when the infiltration is evenly distributed in time.

It was suggested by O. Lehman (*in* Scheidegger, 1961, p. 315) that different velocities in different parts of the aquifer would produce variations in the elevation of the water table. It should be noted that this will generally be negligibly small. The velocity head (equal to $V^2/2g$) for a flow velocity of one meter per second is only 5.1 cm. The flow velocity in the case described above will be 1.6 cm/sec, and the velocity head will be 1.3×10^{-3} cm.

The rate of infiltration, of course, will be greater during some parts of the year than others. Even if the flowrate is ten times the above, however, the water table will fall only 3.5 m per km. The flow velocity will be 16 cm/sec and the velocity head, 1.3×10^{-1} cm.

Thus if caves are present in an aquifer the water table will be almost level and is best shown on cross sections as a simple horizontal line.

It was stated earlier that flowing water in the vadose zone may be conveniently divided into vadose flows and vadose seepage, and that these two types of vadose water seem to have somewhat different chemical characteristics. Some vadose flows, however, may be derived largely or entirely from collected vadose seepage. Such integration of vadose seepage may be caused by its being intercepted by an impermeable horizon in the vadose zone. Water will flow to breaks in such a horizon or to low spots along its margin and descend to the water table as vadose flows. If the impermeable horizon is overlain by noncarbonate rocks and a noncarbonate soil, the spilled vadose flows will be in equilibrium with a high P_{CO_2} from the soil but will be undersaturated with respect to calcite.

225

In the discussion of flow patterns which follows, the reader should keep in mind the proportions of the aquifer. Although limestone aquifers, of course, vary considerably in their dimensions, they are almost always much wider than they are thick. A typical aquifer might be 10 km wide by 100 m thick, and a small aquifer might be only 1 km wide. It is impractical to draw a true-scale cross section of such aquifers. As in most geologic cross sections the vertical dimension must be exaggerated if internal detail is to be shown. Special procedures must be followed, however, in constructing a flow net on such an exaggerated section (e.g., the net will not be orthogonal).

Although the flow patterns which have been sketched (Figs. 12–15) are therefore only approximations of the real patterns, it is believed that they show the essential features of the various cases. In particular, they indicate the differences in flow between the shallow-phreatic zone and the rest of the aquifer, where the shallow-phreatic zone is on the order of 10 m deep and represents the upper tenth of an aquifer which is about 100 m deep.

Uniform Supply to the Water Table

The first flow pattern that will be examined is the one used by Davis (1930) and by Rhoades and Sinacori (1941) as a basis for their hypotheses of cave origin. This pattern (Fig. 12) results from a spatially uniform supply to the water table and an outflow at an effluent stream. The water table approximates an equipotential surface and the flow lines just beneath the water table are nearly vertical. It may be compared with Case 1 of the pipe network (Figures 5A and 6A), but the flow in Case 1 is distorted by the small number of pipes.

The uniform supply to the water table required for this flow pattern to develop is probably not common in limestones, since their vadose-zone permeability is usually rather unequally distributed. Where uniform supply occurs, however, the water will be largely or entirely vadose seepage, which is usually saturated or supersaturated with respect to calcite. Some undersaturation may be produced at the top of the phreatic zone by the temperature and mixing effects (shaded area, Fig. 12), but this is probably a minor effect. Because the flow in the shallow-phreatic zone is nearly vertical, it is difficult to see how long horizontal passages could be produced.

Davis (1930) availed himself of this flow pattern only to show that deep phreatic flow was possible (Figure 1A). Rhoades and Sinacori (1941) suggested that cave excavation should occur near the outflow because flow is concentrated there (Figure 1D). Water near the outflow, however, is more likely to be saturated with respect to calcite than that nearer the source.

Single Discrete Source

A rather different flow pattern results when all of the water which reaches the water table between outflows is concentrated at a point

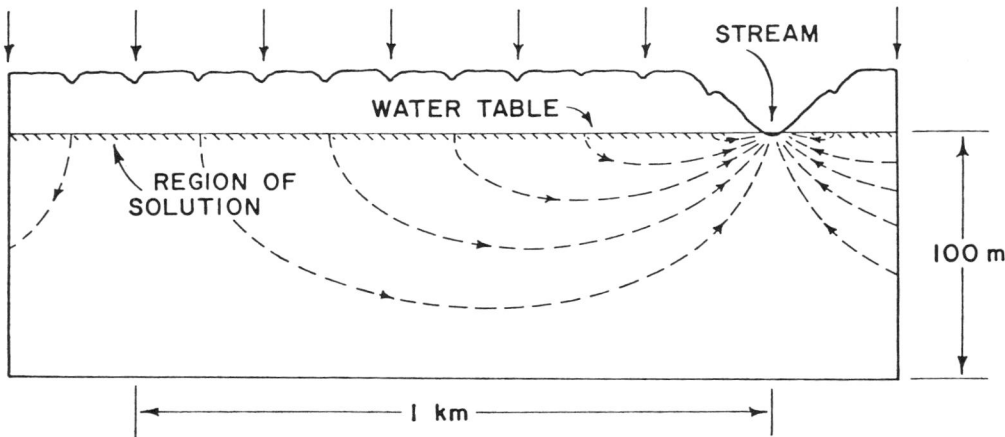

Figure 12. Section through a small limestone aquifer showing approximate flow lines with uniform supply to water table. Vertical exaggeration 4 ×. *See* text for discussion.

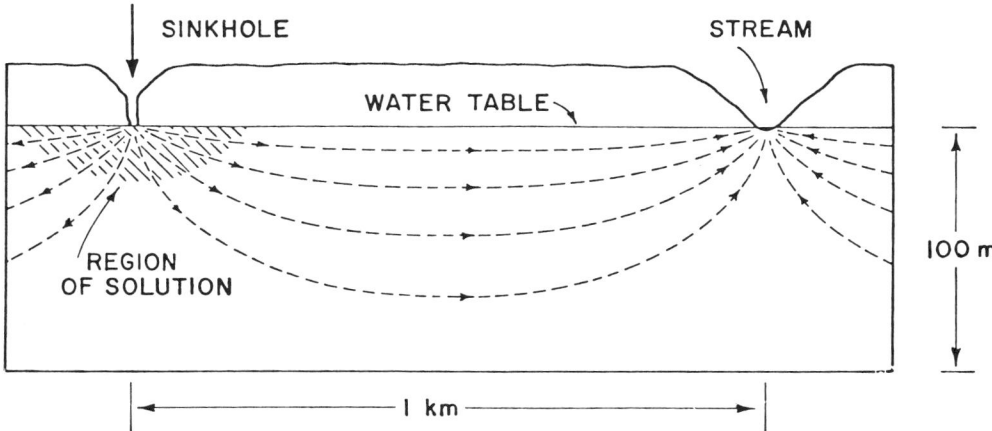

Figure 13. Section through a small limestone aquifer showing approximate flow lines with a single discrete source. Vertical exaggeration 4 ×. *See* text for discussion.

(Fig. 13). Here the equipotential surfaces descend vertically just below the water table, and the water table, in section, is a flowline. This situation corresponds to Case 2 described earlier.

It is of interest to note that the water at the source follows "alternate paths" as described by Swinnerton (1932) and hence this model is apparently the one he had in mind (Figure 1B). Swinnerton believed that the shallow paths, being shorter, would carry more water and hence dissolve more actively (Swinnerton, 1932, p. 674). It has been shown, however, that in an aquifer of normal proportions this difference is negligible.

Because the inflowing water is concentrated at a point, it would be a vadose flow. As discussed earlier, such a flow may be undersaturated with respect to calcite or undersaturation may be produced at the water table. Subwatertable solution might therefore be expected in the vicinity of the source (shaded area, Fig. 13), but not elsewhere in the shallow-phreatic zone.

Multiple Discrete Sources

A third type of flow pattern will develop when several discrete sources are located between points of outflow. Flow from sources near an outflow will be unidirectional toward the outflow and lie close to the water table, while water from more distant sources will be forced to follow deeper flow paths, as illustrated in Figure 14. It is evident that as the density of sources is increased, the resulting flow pattern would approach that shown in Figure 12.

It will be noted (Fig. 14) that water just "downstream" from a source will tend to follow nearly horizontal paths beneath the water table, which will be a flowline in these areas. The sources will be vadose flows, and the undersaturation associated with the sources would tend to cause solution by more or less horizontally flowing water just beneath the water table (shaded areas, Fig. 14). The actual path of this flow in three dimensions (and hence the regions

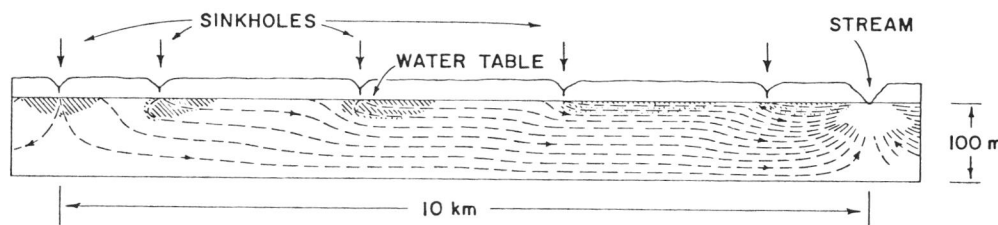

Figure 14. Section through a limestone aquifer showing approximate flow lines with multiple discrete sources. Vertical exaggeration 10 ×. Shaded areas indicate zones of solution. See text for discussion.

227

of solution) will, of course, be much more complex than that shown.

Irregular Supply to the Water Table

The most general flow pattern in limestone results from supply to the aquifer by both vadose seepage and vadose flows (Fig. 15). Water from the larger sources (vadose flows) will gradually be forced down in the aquifer by water which arrives at the water table downstream from the source. In areas where such downstream inflows are absent, the pattern of flow from the larger sources will be the same as in the preceding case. Solution will be expected in the shaded areas of Figure 15.

Here again the two-dimensional flow pattern

When subwater-table cave development has been caused by flow in one of these patterns, of course, the flow during backflooding will tend to follow that pattern.

A situation akin to backflooding will occur when large amounts of water are delivered to one area of the water table, and adjacent areas serve as temporary discharge points.

The backflooding process is, of course, an aspect of bank storage. A search of the literature has failed to disclose any discussion of bank storage or backflooding relative to cave excavation, although E. R. Pohl apparently considered it at one time (*in* Muma and Muma, 1944, p. 2). It has been briefly discussed by the writer earlier (Thrailkill, 1964).

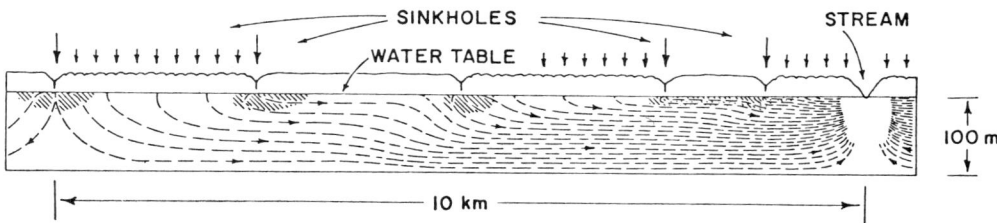

Figure 15. Section through a limestone aquifer showing approximate flow lines with spatially irregular supply to water table. Vertical exaggeration 10 ×. Shaded areas indicate zones of solution. *See* text for discussion.

is only an approximation of an actual three-dimensional pattern. Not only will the water from a vadose flow be forced into deeper or shallower paths by the effects of water from other sources, but it will also be deflected laterally. Thus such flow, and the accompanying solution, will tend to be concentrated in areas where little water is reaching the water table.

Back flooding

As has been shown, the water table in limestones in which caves are developed will usually be nearly horizontal. When the level of the stream at the outflow rises during a flood, therefore, the direction of flow in the aquifer will be easily reversed and water will move into the aquifer from the stream. The flow pattern will be determined largely by the distribution of cavities in the limestone. Even in the unlikely case that such cavities are uniformly distributed, the pattern of flow will depend on the distribution of "discharge points," that is, on the distribution of openings above the low water table to which the water must flow. Thus the flow pattern during backflooding may resemble any of the patterns shown in Figures 12, 13, 14, and 15.

The undersaturation of surface streams during floods (flowrate effect) has already been discussed. It was shown that, although such water may be greatly undersaturated with respect to calcite, it will be relatively ineffective in dissolving limestone due to its low CO_2 content. When the water comes into contact with an atmosphere whose P_{CO_2} is that of the normal atmosphere or greater, however, CO_2 may be dissolved and solution of limestone take place. This is most likely to occur at the water table where the backflooded water is open to the atmosphere of vadose zone caves. Also, mixing of the backflooded water with higher P_{CO_2} water already in the aquifer may result in undersaturation due to the mixing effect.

Even if direct solution is unimportant, backflooding will furnish the aquifer with water unsaturated with respect to calcite and in equilibrium with a low P_{CO_2}. It was earlier pointed out that undersaturation due to mixing will be greatest when the phreatic water has these characteristics, and the mixing of vadose water (especially vadose flows) with backflooded water should result in undersaturation and solution at the water table.

Finally, backflooding may explain some aspects of cave distribution, such as the greater abundance of large caves near major streams (Davies, 1960, p. 7) and the occurrence of caves in areas where surface solution features (karst) are rare or absent. Also, although the concentration of dissolved species in streams decreases during a flood, the content of suspended material increases (Fig. 10). Some of the diverse sediment types found in caves may therefore have been introduced by backflooding. This explanation has been advanced by Pohl (1936) and by W. E. Davies and E. C. T. Chao (unpublished report) for deposits in Mammoth and nearby caves in Kentucky.

Underflow

The final flow pattern of significance in cave excavation which will be discussed here is the underflow of an effluent stream. In response to the downstream decrease in head, water will flow approximately parallel to the stream beneath its bed and banks. Similar flow will occur across sharp bends in the stream and from a lake or reservoir to discharge points downstream (see for example, Hendrickson and Jones, 1952, Fig. 8).

Although the writer knows of no large cave systems that have been formed by underflow, the subwater-table cavities in the beds of many streams (see, for example, Moneymaker and Rhoades, 1945) have probably been dissolved out in this way. As might be expected, solution by underflow seems to be limited to a depth of about 50 m below the stream bed (or water table).

MAMMOTH CAVE, KENTUCKY

This paper has attempted to identify hydrologic and chemical processes which may be of importance in cave excavation. If the ideas which have been expressed have any validity, of course, they must furnish an explanation of the origin of individual caves and cave systems. The conclusions contained in the following discussion of Mammoth Cave should, however, be considered tentative (or even speculative) until more data are available.

Although a complete description or a satisfactory map of Mammoth Cave have never been published, some descriptive material and information on the geologic setting will be found in Bretz (1942), Pohl (1955), and Haynes (1964).

The nearly horizontal limestone in which Mammoth Cave has been excavated is overlain by a predominantly sandstone unit with a thin basal shale which forms an impermeable horizon. The sandstone crops out at the surface, and vadose seepage collected by the impermeable horizon spills from it at the edge of the plateau which contains the cave. Virtually no seepage enters the cave beneath the sandstone, and the flow pattern is probably similar to Figure 14.

The spilling water forms vadose flows which, because they are derived from a nonlimestone soil, are undersaturated with respect to calcite. These flows are actively dissolving vertical shafts, called *domepits*. As the edge of the plateau retreats, the domepits first tend to fill with stalactitic deposits (because they are then overlain by a limestone soil) and later become incorporated into the widening valley or sink. Concurrently, the spilling water excavates new domepits beneath the retreating edge of the plateau.

Mammoth Cave, with its long, nearly horizontal passages, appears to have been excavated mainly in the shallow-phreatic zone. The spilling vadose flows would seem to have been a likely source of undersaturated water. Most of the domepits now active, however, do not extend down to the water table; their vadose flows apparently become saturated with respect to calcite before they reach the phreatic zone. This saturation is probably due as much to the escape of CO_2 from the water as to actual solution of the limestone.

It seems reasonable to suppose that in the past, undersaturated vadose flows and hence domepits did reach the water table, both because the cave was less well ventilated and because the water table was higher. These earlier domepits, which would have since been destroyed by the retreat of the plateau margin, would have had their bases at the ends of shallow phreatic passages if this interpretation is correct.

After being delivered to the water table by vadose flows down domepits, the undersaturated water would tend (for reasons discussed earlier) to follow shallow-phreatic flows paths, at least initially. In plan, these flow paths would be into areas of minimum inflow to the water table (under the plateau). This would explain the situation at Mammoth Cave, where more than 70 km of passages are located beneath a ridge which is only 7 km long and seldom more than 1500 m wide.

In addition, backflooding occurs at Mammoth Cave. The cave is situated near the Green River, whose undersaturation at high flowrates has already been discussed, and river water is known to move into the cave during floods

(Hendrickson, 1961). This water, if it acquires CO_2 from air-filled caves in the vadose zone, should be effective in dissolving limestone just beneath the water table. Mixing of this low P_{CO_2} water with high P_{CO_2} vadose flows (even if such flows are saturated with respect to calcite) will result in further undersaturation.

Davies (1960, p. 7) states that each of the two levels (see below) of Mammoth Cave has a two percent slope toward the Green River. If this represents the water-table slope at the time of their excavation, its steepness could be due to a lack of open subwater-table caves in the aquifer (which does not seem likely), or to high flow rates in the aquifer. It might also be due to a pushing down of the dissolving water beneath the water table by other sources, in the manner described earlier. It does not appear to represent control by bedding which, although dipping toward the river, has an inclination of only about one percent (Haynes, 1964). It might be noted here that shallow-phreatic cave passages in southern Pocohantis County, West Virginia, have a slope of about one percent (which is still rather steep) in limestone which dips four degrees in the opposite direction (Wolfe, 1964, p. 51 and 56).

Opinions differ as to the number and even the existence of "levels" in Mammoth Cave (see, for example, Davis, 1930, p. 598–600; Bretz, 1942, p. 789; and Davies, 1960, p. 7). If the primary cause of passage development has been the vadose flow process outlined above, the elevation of any one shallow-phreatic conduit would depend on the elevation of the water table at the time a particular domepit or group of domepits was active. If the decline of the water table was intermittent, elevation accordancy or near-accordancy of several shallow-phreatic passages would be expected.

Much of the underground drainage from the Pennyroyal sinkhole plain south of the cave area flows north (in the aquifer) and discharges into the Green River (Bretz, 1942, p. 762–764). Although this water is probably saturated with respect to calcite, it may have influenced the cave-forming processes described above. Because the Pennyroyal drainage enters the aquifer farther from the outflow than that in the cave area, it will follow deeper paths (Fig. 15) and thus cause water from the domepits to follow even shallower and more nearly horizontal paths than it otherwise would.

CONCLUSIONS

As set forth in the introduction, the purpose of this paper was to explore and attempt to answer three questions regarding the development of limestone caves. The first of these concerned the validity of applying to limestone aquifers the flow concepts that have been derived for granular aquifers. It was shown that, where solutional development is homogeneous and the limestone aquifer is much wider than it is thick there is little difference in the flow distribution in the two types of aquifers. Although the paucity of data usually makes necessary the assumption that a limestone aquifer is homogeneous, it is obvious that this is seldom actually the case. When information on the actual distribution of cavities is known, it would probably be best to divide the general aquifer into smaller aquifers, within each of which the flow parameters are approximately uniform.

A second question had to do with the source of the undersaturation which permits caves to be dissolved. It appears that, although the uptake of CO_2 in the soil bears little direct relationship to cave excavation in the phreatic zone, there are ways in which phreatic water may become undersaturated. These include the effects of temperature change, mixing of dissimilar waters, and floods in surface streams. It is not possible at present to assess the relative importance of these mechanisms. Undoubtedly, local conditions will strongly influence their relative effectiveness. Because of the very slight undersaturation needed to excavate a cave, the only certain conclusion that can be drawn is that none of these mechanisms have acted universally or with great effectiveness, since even the largest caves are volumetrically insignificant as landforms.

The third and principal question pertained to the reasons for the localization of cave development in the shallow-phreatic zone. Several hydrologic and chemical factors have been suggested which would explain the occurrence of shallow-phreatic caves. The pattern of flow in the aquifer which develops when water is supplied to the water table at a number of discrete points or, to a lesser extent, irregularly, contains regions in which the flow is in nearly horizontal paths just beneath the water table. This horizontally flowing water probably reaches the water table as a vadose flow. If such a vadose flow is undersaturated with respect to calcite, either because of the flowrate effect or because it is derived from a noncarbonate soil, or if undersaturation is produced at the water table by the mixing or temperature effects, shallow-phreatic solution would be expected. In addition, backflooded water may follow a similar flow pattern, and such water in the shallow-phreatic zone may dissolve limestone as CO_2 is acquired. Even if backflooded water does not dissolve directly, its presence in the aquifer may enhance the mixing and temperature effects. Although vadose seepage is usually saturated or supersaturated with respect to calcite, undersaturation due to mixing may occur when the seepage mixes with the ground water. This would result in solution just beneath the water table (see Fig. 12), and adjustment of flow in the aquifer to the zone of higher permeability thus created would take place, resulting in increased shallow-phreatic flow.

Cave excavation in the shallow-phreatic zone may be favored by factors other than those which have been discussed in this paper. These include the consideration that, with a falling water table, the higher passages in the aquifer would probably be older, larger, and hence carry more flow than the lower passages; the sealing of lower cavities by the fallout of insoluble detritus from the higher ones; the occurrence of highly mineralized, and hence heavier, water in the deeper parts of an aquifer which would tend to force the fresh water into shallower paths; and the fact that, in horizontal beds, the vertical permeability will tend to be less than the horizontal permeability. Except for the first (which would probably be easily swamped out by other influences), these are all factors which would depend greatly on the local geologic environment. Although they may be quite important in many areas, they probably serve more to modify than to dominate the hydrologic and chemical processes which have been discussed.

It will be noted that many of the processes thought to be of importance in cave excavation will operate most effectively during floods. Thus it would appear that explanations of cave origin should be made with reference to flooded, rather than to low-water aquifers. As was mentioned earlier, this tends to cause conceptual difficulties. Although the earlier discussion of the nature of limestone aquifers will apply to any steady-state aquifer, a flooded aquifer probably seldom attains a steady state. Under nonsteady-state conditions, the water table will exhibit marked elevation differences whose magnitude and position will vary rapidly with time (flood waves), as will the head and the flow patterns of the underlying aquifer. Also, there will be almost as many different flooded aquifers as floods.

It is, of course, possible to consider the region between the low water table and the flooded water table as a zone of flooding in the vadose zone. This suggests that there is a separation, or at least a qualitative difference, between the flow in the upper and lower parts of the flooded aquifer. Such a difference may, in fact, exist. The largest room of Carlsbad Caverns, New Mexico, has a nearly horizontal bedrock floor which is unrelated to bedding. The room itself was apparently excavated when full of water. If the position of the floor was determined by a water table (although there may be other explanations), it would suggest that solution was localized between a low and a high water table. Davies (1960, p. 8) reports a similar flat bedrock floor in a West Virginia cave. Although the bedrock floor of most shallow-phreatic caves is covered with sediment and collapsed blocks, the writer suspects that many, if not most, shallow-phreatic caves have flat bedrock floors. It is not known why cave excavation should be so sharply limited downward, but if the floor of a cave is determined by a low water table, it may eventually be convenient to separate the zone of flooding (shallow-phreatic zone?) from the rest of the aquifer. In any case, it seems best to consider the upper limit of the shallow-phreatic zone as an "average, steady-state, flood water table."

REFERENCES CITED

Adams, C. S., and Swinnerton, A. C., 1937, Solubility of limestone: Am. Geophys. Union Trans., pt. 2, p. 504–508.

Addington, A. R., 1927, Porter's Cave and recent drainage adjustments in its vicinity: Ind. Acad. Sci. Proc., v. 36, p. 107–116.

Arnston, R. H., 1964, Effect of temperature and confining pressure on the solubility of calcite at constant CO_2 concentrations [abs.]: Geol. Soc. America Spec. Paper 76, p. 6.

Back, William, 1963, Preliminary results of a study of calcium carbonate saturation of ground water in central Florida: Internat. Assoc. Sci. Hydrology, Année 8, no. 3, p. 43–51.

Beede, J. W., 1911, The cycle of subterranean drainage as illustrated in the Bloomington, Indiana, quadrangle: Ind. Acad. Sci. Proc., v. 20, p. 81–111.

Bock, Herman, 1913, Der Karst und seine Gewässer: Mitt. Höhlenkunde, v. 6, no. 3.

Bögli, Alfred, 1963, Beitrag zur Entstehung von Karsthöhlen: Die Höhle, v. 14, p. 63–68.

—— 1964, Mischungskorrosion—ein Beitrag zum Verkarstungs-problem: Erdkunde, v. 18, p. 83–92.

Bretz, J. H., 1942, Vadose and phreatic features of limestone caverns: Jour. Geology, v. 50, p. 675–811.

Carroll, Dorothy, 1962, Rainwater as a chemical agent of geologic processes—a review: U. S. Geol. Survey Water-Supply Paper 1535–G, 18 p.

Davies, W. E., 1960, Origin of caves in folded limestones: Natl. Speleol. Soc. Bull., v. 22, p. 5–18.

Davis, W. M., 1930, Origin of limestone caverns: Geol. Soc. America Bull., v. 41, p. 475–628.

Finch, J. W., 1904, The circulation of underground aqueous solutions: Colo. Sci. Soc. Proc., v. 7, p. 193–252.

Gardner, J. H., 1935, Origin and development of limestone caverns: Geol. Soc. America Bull., v. 46, p. 1255–1274.

Garrels, R. M., 1960, Mineral equilibria: New York, Harper & Brothers, 254 p.

Garrels, R. M., and Thompson, M. E., 1962, A chemical model for sea water at 25°C and one atmosphere total pressure: Am. Jour. Sci., v. 260, p. 57–66.

Greene, F. C., 1909, Caves and cave formation of the Mitchell limestone: Ind. Acad. Sci. Proc. for 1908, p. 175–184.

Grund, Alfred, 1903, Die Karsthydrographie: Geog. Abh. A. Penck, v. 7, no. 3, 200 p.

Harned, H. S., and Davis, Raymond, 1943, The ionization constant of H_2CO_3 in water and the solubility of CO_2 in water and aqueous solutions from 0° to 50°: Am. Chem. Soc. Jour., v. 65, p. 2030–2037.

Harned, H. S., Davis, Raymond, and Scholes, S. R., 1941, The ionization constant of HCO_3—from 0° to 50°: Am. Chem. Soc. Jour., v. 63, p. 1706–1709.

Haynes, D. D., 1964, Geology of the Mammoth Cave quadrangle, Kentucky: U. S. Geol. Survey Map GQ–351.

Hendrickson, G. E., 1961, Sources of water in Styx and Echo Rivers, Mammoth Cave, Kentucky: U. S. Geol. Survey Prof. Paper 424–D, p. 41–42.

Hendrickson, G. E., and Jones, R. S., 1952, Geology and ground-water resources of Eddy County, New Mexico: N. Mex. Bur. Mines Mineral Res. Ground-Water Rept. 3, p. 1–169.

Holland, H. D., Kirsipu, T. V., Heubner, J. S., and Oxburgh, U. M., 1964, On some aspects of the chemical evolution of cave waters: Jour. Geology, v. 72, p. 36–67.

Hostetler, P. B., 1964, The degree of saturation of magnesium and calcium carbonate minerals in natural waters: Internat. Assoc. Sci. Hydrology Pub., no. 64, p. 34–49.

Howard, A. D., 1964, Processes of limestone cave development: Internat. Jour. Speleol., v. 1, p. 47–60.

Hubbert, M. K., 1940, The theory of ground water motion: Jour. Geology, v. 48, p. 785–944.

Katzef, Friedrich, 1909, Karst und Karsthydrographie: Zur Kunde der Balkanhalbinsel, no. 8, 94 p.

Kaye, C. A., 1957, The effect of solvent motion on limestone solution: Jour. Geology, v. 65, p. 35–46.

King, F. H., 1899, Principles and conditions of the movements of ground water: U. S. Geol. Survey 19th Ann. Rept., v. 2, p. 59–294.

Malott, C. A., 1921, A subterranean cut-off and other subterranean phenomena along Indian Creek, Lawrence County, Indiana: Ind. Acad. Sci. Proc., v. 31, p. 203–210.

—— 1938, Invasion theory of cavern development (abs.): Geol. Soc. America Proc. for 1937, p. 323.

Matson, G. C., 1909, Water resources of the Bluegrass region, Kentucky: U. S. Geol. Survey Water-Supply Paper 233, p. 1–223.

Meinzer, O. E., 1923, Outline of ground-water hyhrology, with definitions: U. S. Geol. Surv. Water-Supply Paper 494, p. 1–71.

Moneymaker, B. C., and Rhoades, R. F., 1947, Deep solution channel in western Kentucky: Geol. Soc. America Bull., v. 56, p. 39–44.

Moore, G. W., 1960, Introduction to the origin of limestone caves: Natl. Speleol. Soc. Bull., v. 22, p. 3–4.

Muma, M. H., and Muma, K. E., 1944, A glossary of speleology: Natl. Speleol. Soc. Bull., v. 6, p. 1–10.

Piper, A. M., 1932, Ground water in north-central Tennessee: U. S. Geol. Survey Water-Supply Paper 640, p. 1–238.

Pohl, E. R., 1936, Geologic investigations at Mammoth Cave, Kentucky: American Geophys. Union Trans. for 1936, p. 332–334.

—— 1955, Vertical shafts in limestone caves: Natl. Speleol. Soc. Occasional Papers, no. 2, 24 p.

Rhoades, R. F., and Sinacori, M. N., 1941, Patterns of ground water flow and solution: Jour. Geology, v. 49, p. 785-794.

Roques, H., 1963, Observations physico-chimiques sur les eaux d'alimentation de quelques concretions: Ann. Spéléologie, v. 18, p. 377-404.

Scheidegger, A. E., 1962, Theoretical geomorphology: Berlin, Springer-Verlag; Englewood Cliffs, N. J., Prentice-Hall, Inc., 333 p.

Schneider, Robert, 1964, Relation of temperature distribution to ground-water movement in carbonate rocks of central Israel: Geol. Soc. America Bull., v. 75, p. 209-216.

Sipple, R. F., and Glover, E. D., 1964, Solution alteration of carbonate rocks: the effects of temperature and pressure: Geochim. et Cosmochim. Acta, v. 28, p. 1401-1417.

Sweeting, M. M., 1950, Erosion cycles and limestone caverns in the Ingleborough district (England): Geog. Jour., v. 115, p. 63-78.

Swinnerton, A. C., 1929, Changes in base-level indicated by caves in Kentucky and Bermuda [abs.]: Geol. Soc. America Bull., v. 40, p. 194.

—— 1932, Origin of limestone caves: Geol. Soc. America Bull., v. 43, p. 663-693.

Thrailkill, J. V., 1964, Cave solution by backflooding from surface streams [abs.]: Geol. Soc. America Spec. Paper 76, p. 165-166.

Todd, D. K., 1959, Ground water hydrology: New York, John Wiley & Sons, Inc., 336 p.

U. S. Geol. Survey, 1960, Quality of surface waters of the United States, parts 1-4: U. S. Geol. Survey Water-Supply Paper 1520, p. 1-641.

—— 1962, Quality of surface waters of the United States, parts 1-4: U. S. Geol. Survey Water-Supply Paper 1571, p. 1-773.

Vennard, J. K., 1961, Elementary fluid mechanics: New York, John Wiley & Sons, Inc., 570 p.

Weller, J. M., 1927, The geology of Edmonson County: Kentucky Geol. Survey, ser. 6, v. 28, p. 1-246.

Weyl, P. K., 1958, The solution kinetics of calcite: Jour. Geology, v. 66, p. 163-176.

White, W. B., 1960, Terminations of passages in Appalachian caves as evidence for a shallow phreatic origin: Natl. Speleol. Soc. Bull., v. 22, p. 43-53.

Wolfe, T. E., 1964, Cavern development in the Greenbrier Series, West Virginia: Natl. Speleol. Soc. Bull., v. 26, p. 37-60.

Woodward, H. P., 1961, A stream piracy theory of cave formation: Natl. Speleol. Soc. Bull., v. 23, p. 39-58.

Zötl, Josef, 1961, Die Hydrographie des nordostalpinen Karstes: Steirische Beitr. Hydrogeologie for 1960/61, no. 2, p. 54-183.

Manuscript Received by the Society August 5, 1966

The geochemistry of some carbonate ground waters in central Pennsylvania

Donald Langmuir

Abstract—The field pH and content of Ca^{2+}, Mg^{2+}, HCO_3^-, dissolved oxygen and other aqueous species have been measured in 29 spring waters and 29 well waters in folded and faulted Paleozoic carbonate rocks near State College, Pennsylvania. Most of the springs issue from limestone; most of the well waters are pumped from dolomite. Average specific conductances of the spring and well waters were 347 and 499 $\mu\Omega^{-1}$, respectively, with polluted well waters having conductances as high as 945 $\mu\Omega^{-1}$. Molar Ca^{2+}/Mg^{2+} ratios as low as 0·6 in the well waters have resulted from incongruent solution of dolomite. Theoretical treatment of relationships between P_{CO_2}, pH and HCO_3^- content shows that solution of calcite or dolomite by undersaturated waters leads to increases in both pH and HCO_3^- content. This relation is approximately obeyed by the spring waters at an average CO_2 pressure of $10^{-2 \cdot 2}$ atm. In contrast, solution or precipitation of carbonates by waters near saturation with respect to them, results in inverse relationships of pH and HCO_3^-; a behavior closely obeyed by the well waters which have CO_2 pressures up to $10^{-1 \cdot 6}$ atm. Computer calculation of ground water saturation with calcite and dolomite at ground water temperatures (8–14°C) was made using remeasured solubility product data for calcite, and considering the effect of $CaSO_4^0$, $MgSO_4^0$ and $MgHCO_3^+$ ion pairs. Within the uncertainty of the chemical analyses and thermochemical data, 12 well waters and 3 spring waters were saturated with calcite, whereas 7 well waters and 3 spring waters were saturated with dolomite. None of the ground waters significantly exceeded saturation with respect to either carbonate.

INTRODUCTION

STUDIES of the geochemistry of ground waters in carbonate rocks (BACK, 1963; BARNES and BACK, 1964; HOSTETLER, 1964; HOLLAND *et al.*, 1964; BARNES, 1965; BACK and HANSHAW, 1970) have indicated that a large percentage of these waters are highly supersaturated with respect to calcite and dolomite. The conclusion thus emerges that the theoretical solubilities of calcite and dolomite do not constitute real limits to the solubilities of these minerals in ground water. There is little doubt that some carbonate ground waters become supersaturated with calcite and or dolomite by CO_2 exsolution (particularly in caves), by mixing with other ground or surface waters, by increases in temperature or by solution of more soluble minerals containing calcium, magnesium or carbonate species. Nevertheless, some results of previous work are questionable, particularly when calculations of the degree of saturation with calcite or dolomite were based on laboratory rather than field pH measurements, when field temperatures were not measured or were not used in the calculations, or when the presence of ion pairs or complexes was not taken into account. In this study these sources of error have been avoided, and corrected solubility data for calcite and dolomite employed. The results suggest that fewer ground waters in carbonate rocks are supersaturated with calcite and dolomite than has been previously supposed.

GEOHYDROLOGY OF THE STUDY AREA

The study area is in the Appalachian Mountain Section of the Valley and Ridge Province in Centre County, Pennsylvania. A geologic map (Fig. 1) and

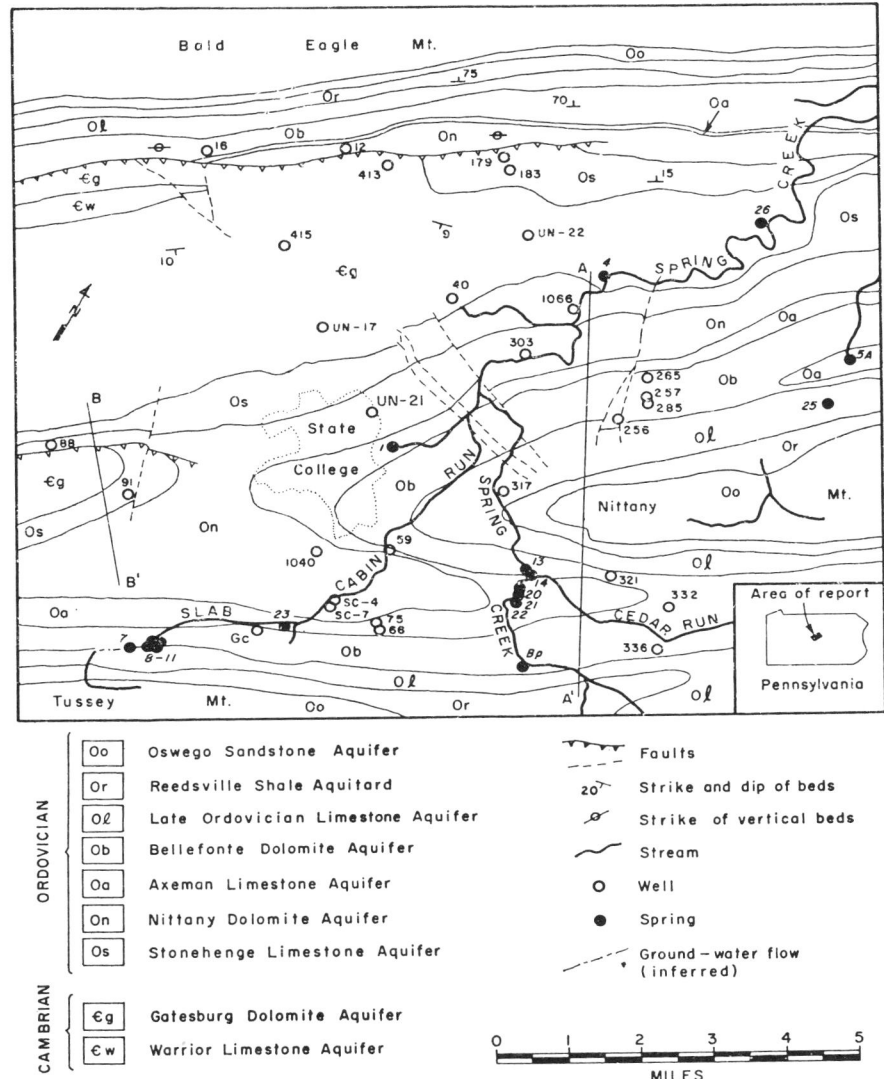

Fig. 1. Geology of the study area modified after BUTTS and MOORE (1936), CARUCCIO (1963), CLARK (1965) and LANDON (1963), showing locations of major streams, 17 of the springs, and all the wells sampled. Also shown are locations of geologic structure sections A–A' and B–B' which are given in Fig. 2.

sections of the area (Fig. 2) show northeast trending ridges of Ordovician sandstone and shale rising approximately 1000 ft above valleys underlain by alternating beds of Cambrian and Ordovician limestone and dolomite. Bald Eagle and Tussey Mountains form the flanks of a regional anticlinorium which strikes N 40°E. Nittany Mountain is a major synclinal ridge within the anticlinorium, which also contains two thrust faults and a number of normal and cross faults. The three

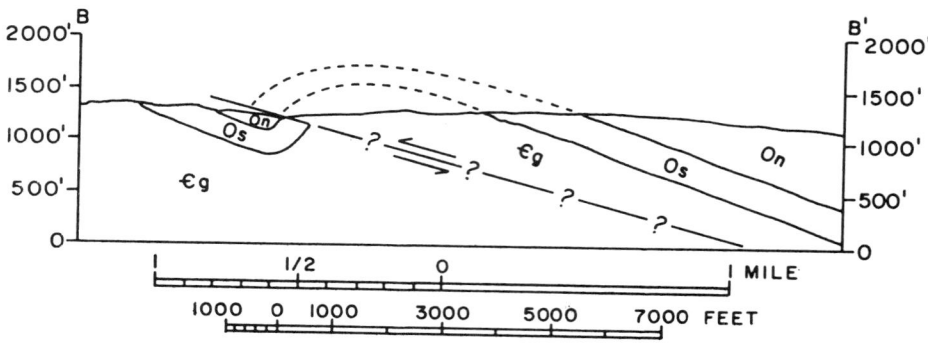

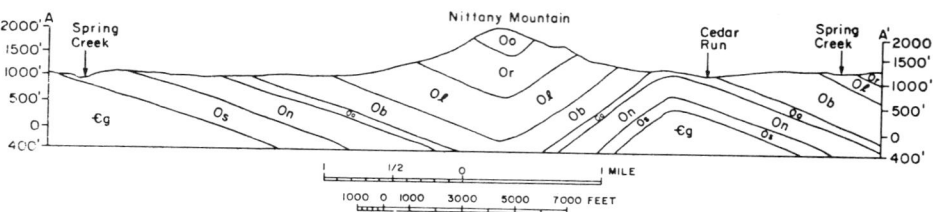

Fig. 2. Geologic structure sections located in Fig. 1. Vertical exaggeration 2 times.

mountain ridges attain elevations between 1800 and 2300 ft, and are capped by non-carbonate rocks including the Oswego Sandstone and Reedsville Shale. The carbonate rocks form a gently rolling surface with elevations which range from about 1400 ft adjacent to the ridges and in northwestern parts of the area underlain by the Gatesburg Formation, to 800 ft along the lower reaches of Spring Creek.

Among the carbonate rocks, the Axeman and Stonehenge Limestones are relatively pure, massive limestone with occasional thin beds or layers of dolomite. The Bellefonte Dolomite is a nearly pure, massive dolomite. The Nittany dolomite is also relatively pure and massive, but locally contains thin limestone beds. The Gatesburg Dolomite is mostly a sandy dolomite interbedded with orthoquartzite, and contains negligible limestone.

Annual precipitation in the study area averages 39 in. as rain, and is distributed with a slight excess during summer months. Streams have incised youthful valleys in the carbonates to depths 50–300 ft below the surrounding land surface (LATTMAN and PARIZEK, 1964). West of State College in upland areas which are underlain by the Gatesburg Formation and Stonehenge Limestone, the water table ranges from 50 to 400 ft below the land surface. Most tributary valleys are dry with the water table from 10 to 75 ft below the valley bottoms (LATTMAN and PARIZEK, 1964). Slab Cabin Run, Spring Creek, and Cedar Run tend to be losing streams during summer months. North of State College at lower elevations, Spring Creek is usually a gaining stream and acts as a regional drain for both surface and ground waters in the mapped area.

Ground water flow in carbonate rocks other than the Gatesburg Formation usually parallels strike and is through secondary openings such as conduits and fractures enlarged by solution. Similar features are present in the Gatesburg, however dolomitic sandstone beds within this formation have substantial primary as well as secondary porosity.

This study deals with water from 29 springs and 29 wells in the carbonate rocks. All the wells and 17 of the springs are located in Fig. 1. The other 12 springs are in similar hydrogeologic settings within 20 miles of the mapped area. Of the 29 springs, 22 issue from limestone and 7 from dolomite. Of the 29 wells, 20 are in dolomite, and 9 in limestone. These proportions reflect the greater areal extent of dolomite, and the location of most populated areas on dolomite rocks.

Most of the springs (those of Group 1) issue from limestone near the foot of the mountain slopes. Included in this category are the five springs near the headwaters of Slab Cabin Run. These springs are fed by waters which have moved short distances downslope as streams or shallow ground waters, and which enter the carbonates in the subsurface through sinkholes, stream bottom sediments or zones of rock fracture. Springs of this type have discharges which range from a few gallons to several thousand gallons a minute, and which usually increase substantially in flow and turbidity within a few days after a heavy rainfall. Group 1 springs have subsurface residence times which often exceed 2–6 days based on fluorescein dye tracing (JACOBSON and LANGMUIR, 1970). Although all the springs sampled in this category are perennial, their flow tends to be greatly reduced during dry, summer months.

Springs of a second category (Group 2 springs) located further downvalley, act as regional ground water discharge points. Their flows range from several, up to thousands of gallons a minute, and tend to be more constant than flows of Group 1 springs. Springs 1 and 4 which discharge into Spring Creek northeast of State College are of this type. E. T. SHUSTER (oral communication) followed chemical changes with time in Group 2 springs 1 and 17. Neither spring varied more than a few percent in specific conductance during a year of observation. However, discharge rates of spring 1 (R. R. KOUNTZ, written communication) and spring 17 (E. T. SHUSTER, written communication) both indicate a lag of several months in response to periods of maximum ground water recharge in early spring and late fall. Flow data for spring 1 are most complete and show a secondary, rapid response to individual rains of a few days or less superimposed on the long term flow cycle. Assuming spring 1 is typical of other Group 2 springs, most of the water feeding these springs has a ground-water residence time measured in months, and presumably is derived from infiltration through soils or from relatively diffuse or long-distance ground water flow. A smaller fraction of the spring discharge is provided by storm runoff which reaches the ground water table via sinkholes, zones of rock fracture, or thin soil cover.

Wells in the carbonates tap ground water under water table or in a few cases semi-confined conditions. Well yields range from a few gallons a minute (gpm) in some upland or mountain slope areas, to 1000–10,000 gpm in wells such as SC-7 and

SC-11 which tap fractured and or cavernous rocks adjacent to perennial streams.

Detailed water level information is not available for the 29 wells in this study. However, a review of yearly water level variations in 13 observation wells tapping carbonate ground waters in a variety of lithologic and topographic settings within the study area (T. GIDDINGS, written communication) permits several generalizations. First, water levels in most wells rise in response to rainfall within 5–10 days after precipitation. Exceptions to this are wells which tap the Gatesburg and Stonehenge Formations in upland areas north and west of State College beneath a hundred or more feet of residual soil. Water level changes in these wells are gradual and may show little if any response to rainfall.

Except for a few wells adjacent to perennial streams, the wells receive much of their recharge by infiltration of precipitation through soils. Thus, because of evapotranspiration most wells obtain little if any ground water recharge between June and October as reflected in a gradual water level decline during that period. Recharge to most wells is also reduced, but to a less extent, by frozen ground and snow cover from December through March.

The chemical character of ground waters in the study area is controlled by both hydrogeology and land use practices. Approximately 40,000 people live in the mapped area. Most habitations close to State College receive public water supply and are sewered. However, 23 of the wells sampled in this study, supply water to private homes or farms and are vulnerable to contamination by discharges from adjacent septic tanks, and by infiltrating waters which have contacted animal wastes and inorganic fertilizers. Other pollution sources which also affect both spring and well waters include road salt, sewage plant discharges and open dumping. The important ionic species in unpolluted carbonate ground waters are Ca^{2+}, Mg^{2+} and HCO_3^- Concentrations of Na^+, Cl^-, SO_4^{2-} or NO_3^- in excess of about 10 ppm have been introduced in pollution. Fifty local wells and springs including several sampled in this study, were sampled five or more times and all found to be bacterially polluted at least once by the Pennsylvania Department of Health (PARIZEK et al., 1967). Forty-one of the same wells sampled 10 to 22 times by PARIZEK et al., were found to be bacteriologically polluted at least five times.

THERMOCHEMICAL DATA

· Values of K_1, the first dissociation constant of H_2CO_3, and K_{CO_2} used in this paper are from HARNED and DAVIS (1943). By definition, $K_{CO_2} = [H_2CO_3]/P_{CO_2}$. The brackets in this expression denote the activity of H_2CO_3, which by convention equals the sum of dissolved H_2CO_3 and CO_2. Values of K_2, the second dissociation constant of H_2CO_3, are from HARNED and SCHOLES (1941).

Values of K_c, the activity product of calcite are based on reversed laboratory solubility measurements at 5, 12, 15 and 20°C (JACOBSON and LANGMUIR, 1971), and on 25°C solubility measurements described by the author (1968). Resultant smoothed values of $-\log K_c$ at five degree intervals from 0 to 25°C are, respectively, 8·340, 8·345, 8·355, 8·370, 8·385 and 8·400, and are probably accurate to ±0·02 units. These values are much smaller than K_c values based on LARSON and BUSWELL (1942) which are in general use among geochemists. At 0 and 25°C the constants calculated by LARSON and BUSWELL (see GARRELS and CHRIST, 1965) are 207 per cent and 11·4 per cent smaller than the measured values given above. Relative to the constants of LARSON and BUSWELL, the constants given above markedly increase the degree of saturation of most ground waters with respect to calcite.

K_d, the activity product of dolomite, is taken as $10^{-17.0}$ at 25°C based on solubility measurements in distilled water and dilute $MgCl_2$ solutions (LANGMUIR, 1964). This is in excellent agreement with $K_d = 10^{-17.05 \pm 0.3}$ at 25°C based on heat of solution and third law entropy measurements (R. A. ROBIE, verbal communication). Values of K_d below 25°C were calculated from the equation

$$-\log K_d = 17\cdot 0 + \frac{\Delta H^\circ_{298}}{4\cdot 576}\left(\frac{1}{T} - \frac{1}{298}\right) - \frac{\Delta C^\circ_{p_{298}}}{1\cdot 987}\left[\frac{1}{2\cdot 303}\left(\frac{298}{T} - 1\right) - \log\frac{298}{T}\right] \quad (1)$$

which assumes the heat capacity, $\Delta C^\circ_{p_{298}}$, of the reaction

$$CaMg(CO_3)_2 = Ca^{2+} + Mg^{2+} + 2CO_3^{2-} \quad (2)$$

is constant below 25°C. In equation (1), T is temperature in degrees Kelvin, ΔH°_{298}, the enthalpy of reaction (2), equals -9280 cal/mol (LANGMUIR, 1964). $\Delta C^\circ_{p_{298}} = 224$ cal/deg mol is calculated from $C^\circ_{p_{298}} = 37\cdot 7$ cal/deg mol for dolomite (STOUT and ROBIE, 1963), and heat capacities of the ions at 25°C from CRISS and COBBLE (1964). Values of $-\log K_d$ based on equation (1) at 0, 5, 10, 15 and 20°C, are 16·56, 16·63, 16·71, 16·79 and 16·89, respectively, and are probably accurate to $\pm 0\cdot 2$ log units.

CALCULATIONS OF CARBONATE EQUILIBRIA

All calculations of carbonate equilibria were made using individual ion activities calculated from the definition $a_i = \gamma_i m_i$ where a_i, γ_i and m_i are the activity, activity coefficient and molality of species i, respectively. Ion activity coefficients were computed from the ionic strength (I) using the Debye–Huckel equation where I is defined as

$$I = \tfrac{1}{2}\sum m_i z_i^2 \quad (3)$$

and z_i is the valence of ionic species i (GARRELS and CHRIST, 1965). When a complete ground water chemical analysis was available, I was calculated with expression (3). When only a partial analysis was made, I was estimated from the specific conductance (μ) in $\mu\Omega^{-1}$, and the empirical relation

$$I = 1\cdot 88 \times 10^{-5}\,(\mu). \quad (4)$$

This relation is based on specific conductance measurements and total chemical analyses of 56 ground waters from the study area. For 51 of the 56 waters, equation (4) gave an estimate of I within ± 10 per cent of the true value. A possible error of ± 10 per cent in I leads to an error of ± 2 per cent in the ion activity products of calcite or dolomite for the ionic strength range of $3\cdot 3 \times 10^{-3}$ to $1\cdot 8 \times 10^{-2}$ observed in this study. This in turn is equivalent to a maximum uncertainty of about $\pm 0\cdot 01$ units in SI_c or SI_d (see below).

The saturation index of calcite (SI_c) equals $\log(IAP_c/K_c)$, where IAP_c is the ion activity product $[Ca^{2+}][CO_3^{2-}]$ in solution. Similarly SI_d, the saturation index of dolomite equals $\log(IAP_d/K_d)^{1/2}$, where IAP_d is the ion activity product $[Ca^{2+}][Mg^{2+}][CO_3^{2-}]^2$ in solution. When SI_c or SI_d equals zero the ground water is saturated with respect to the carbonate in question. Negative saturation indices denote undersaturation; positive indices supersaturation. The saturation indices are related to the Gibbs free energies of solution in calories, of one mole of calcite and one-half mole of dolomite, by the expression $SI = \Delta G°/4\cdot576T$.

Ion pairs which are sometimes present in local carbonate ground waters in significant amounts are $CaSO_4^0$, $MgSO_4^0$ and $MgHCO_3^+$. The $CaHCO_3^+$ ion pair is unimportant and may be ignored (LANGMUIR, 1968). Values of $K_{CaSO_4^0}$, the dissociation constant for $CaSO_4^0$ may be calculated with the equation

$$\log K_{CaSO_4^0} = -2\cdot20 - 0\cdot0044t \tag{5}$$

in which t is the temperature in degrees Celsius. Equation (5) is based on measurements at 0 and 25°C by BELL and GEORGE (1953). Values for $K_{MgSO_4^0}$ may be computed with the expression

$$\log K_{MgSO_4^0} = -2\cdot03 - 0\cdot0132t \tag{6}$$

based on measurements from 20 to 35°C by JONES and MONK (1952). HOSTETLER (1963) gives $K_{MgHCO_3^+} = 10^{-0\cdot95}$ at 25°C. In the absence of data at other temperatures this value may be assumed constant below 25°C. For well waters from this study highest in Ca^{2+}, Mg^{2+}, HCO_3^- and SO_4^{2-} (Table 1), the maximum effect of ion pairs on SI_c and SI_d is about $-0\cdot02$ units. For spring waters in Table 1 having SO_4^{2-} analyses, the maximum effect of ion pairs on SI_c and SI_d is $-0\cdot01$ units. Thus, although SO_4^{2-} analyses were not made for all the spring waters, ignoring sulfate ion pairs in these samples should have introduced a negligible error in SI_c and SI_d values.

The most important uncertainty among measured values is in the pH. A possible error of $\pm0\cdot05$ pH units leads to an uncertainty of $\pm0\cdot05$ units in SI_c or SI_d. Double buffer checks in the field were generally well within $\pm0\cdot05$ pH units, however, in view of uncertainties in Ca^{2+}, Mg^{2+} and HCO_3^- analyses, in calculations of carbonate ion activity products, and in values of K_c and K_d, the total uncertainty is probably about $\pm0\cdot1$ units in SI_c, and may be slightly larger in SI_d. We will therefore assume that all values of the indices within $\pm0\cdot1$ units of zero are saturated with respect to the carbonate in question. Ion activities, P_{CO_2}, and SI_c and SI_d values were computed for each of the 29 spring waters and 29 well waters based on the chemical analytical data in Table 1. Ionic strengths were corrected for ion pairing. Methods of calculation are described by GARRELS and THOMPSON (1962). Computations were programmed in Fortran IV and run on an IBM 360-67 computer.

GROUND WATER CHEMICAL ANALYSES

Chemical analyses were performed on water samples from each of the 29 springs and 29 wells. The springs were sampled in August 1968, or May and September 1969; the wells in late August, 1970. All samples were clear at the time of collection. However, to avoid the risk that suspended $CaCO_3$ might be present in some saturated well waters, a number of these were passed through Whatman 40 (medium grade) filter paper and acidified immediately upon collection. The Ca^{2+} and Mg^{2+} content of these samples and unfiltered acidified samples was identical, proving the absence of suspended carbonates. Analyses for cations were done on field-acidified samples by atomic absorption spectroscopy. The alkaline earths in some well waters were also analyzed by EDTA titration. Bicarbonate was determined within 24 hr of sampling by titration with standard H_2SO_4 or $KH(IO_3)_2$ to the inflection point between pH 4·1 and 4·5 (BARNES, 1964). Samples for HCO_3^- analysis were packed in ice in the field, and brought to laboratory temperatures just prior to titration. Specific ion electrodes were used to measure Cl^- and NO_3^- concentrations (LANGMUIR and JACOBSON, 1970). Sulfate values in the spring waters were measured by the visual thorin method, and in the well waters with a Helig turbidimeter (RAINWATER and THATCHER, 1960). Dissolved oxygen (DO) was determined in the field with a thermally compensated model 51 Yellow Springs portable oxygen meter. The meter was calibrated against the oxygen content of air at ground water temperature. DO values reported in percent of saturation, have been corrected for elevation, and are probably accurate to ± 10 per cent. The pH of all spring and well waters was

Table 1. Chemical data for the spring and well waters analyzed in this study. complexes. Specific conductance (μ) is in micromhos at 25°C. A dash means the (SI_c) and dolomite (SI_d)

Springs

No.	$T(°C)$	Ca^{2+}	Mg^{2+}	Na^+	K^+	HCO_3^-	SO_4^{2-}	Cl^-
1	10·0	47	24	3·8	—	234	14	10
4	10·0	38	21	3·5	1·4	191	—	—
5A	10·5	60	9	6	2·1	167	—	9·5
7	12·0	41	4·4	6·5	2·1	134	—	12
8	12·0	37	7·0	6·0	2·4	128	—	12
9	11·0	50	6	—	—	177	—	—
10	10·0	51	3	—	—	185	—	—
11	10·4	47	3	—	—	168	—	—
13	11·0	55	10	1·5	1·2	199	19	—
14	10·0	58	37	2·5	1·2	270	22	—
15	12·0	39	11	—	—	159	—	—
16	12·3	39	11	—	—	156	—	—
17	9·0	33	25	—	—	199	—	—
18	10·1	39	26	—	—	217	—	—
19	9·3	44	12	—	—	217	—	—
20	10·0	46	31	2·5	1·1	238	28	—
20A	10·2	56	5	—	—	212	—	—
21	9·8	42	23	3·0	2·6	196	—	—
22	9·9	50	26	3·5	1·1	230	24	—
23	12·2	50	33	5·0	1·8	220	—	—
24	10·2	36	6·2	2·0	1·5	107	—	—
25	10·5	70	4·4	2·5	0·7	182	28	—
26	14·2	52	28	7·0	2·2	214	—	—
30	9·5	55	7·0	—	—	175	—	3·5
31	9·5	57	9·0	—	—	179	—	5·0
32	10·0	63	9·6	—	—	190	—	5·2
33	9·0	65	3·5	—	—	178	19	—
BP	12·5	35	4·6	2·0	1·2	101	—	—
MC	8·0	24	3·9	—	—	81	—	—

measured in the field after electrode and meter calibration in nominal pH 4, and 6·86 or 7 buffers brought to within ±1°C of ground-water temperature. Measurements were made with a combination pH electrode and an Orion Model 407 battery powered millivolt-pH meter. Double buffer checks were always within ±0·05, and usually within ±0·03 pH units. Hydrogeologic data, chemical analyses and computed equilibrium CO_2 pressures and SI_c and SI_d values for the spring and well waters are given in Table 1.

Ground Water Geochemistry

General description of spring and well waters

Some chemical quality parameters for the spring and well waters are generalized in Table 2. Recognizable differences in chemistry exist between spring waters which

Ion concentrations are total values in parts per million, and are uncorrected for particular ion or parameter was not analyzed. P_{CO_2} and saturation indices of calcite are defined in the text.

				Springs				
NO_3^-	μ	pH	DO (% sat)	Rock type	Log P_{CO_2} (atm)	Ca^{2+}/Mg^{2+} (molar)	SI_c	SI_d
14	400	7·63	—	ls	−2·35	1·2	−0·04	−0·09
—	355	7·57	68	dol	−2·38	1·1	−0·27	−0·30
13	387	7·32	53	ls	−2·18	4·0	−0·38	−0·68
10	280	7·13	72	ls	−2·08	5·6	−0·78	−1·16
12	280	7·25	74	ls	−2·22	3·2	−0·73	−0·97
—	373	7·39	—	ls	−2·23	5·1	−0·35	−0·70
—	328	7·31	—	ls	−2·13	10	−0·41	−0·92
—	328	7·28	68	ls	−2·14	9·5	−0·51	−1·00
—	390	7·20	100	ls in dol	−1·99	3·3	−0·46	−0·72
—	475	7·36	68	ls	−2·02	0·95	−0·18	−0·17
—	290	7·42	83	ls	−2·29	2·2	−0·44	−0·60
—	270	7·32	79	ls	−2·20	2·2	−0·54	−0·70
—	338	7·80	84	dol	−2·60	0·80	−0·10	−0·06
—	381	7·59	91	dol	−2·34	0·91	−0·19	−0·17
—	392	7·33	88	ls	−2·09	2·2	−0·41	−0·59
—	460	7·46	68	ls	−2·18	0·90	−0·23	−0·22
—	316	7·49	83	ls	−2·25	6·8	−0·12	−0·55
—	340	7·49	64	ls	−2·29	1·1	−0·30	−0·33
—	430	7·45	64	ls	−2·18	1·2	−0·22	−0·26
—	450	7·31	64	ls	−2·05	0·92	−0·34	−0·31
—	238	7·25	82	ls	−2·30	3·5	−0·83	−1·11
—	350	7·11	42	ls	−1·94	9·7	−0·49	−0·98
—	450	7·52	45	dol	−2·26	1·1	−0·09	−0·10
14	320	7·35	—	ls	−2·20	4·8	−0·37	−0·71
16	344	7·42	—	ls	−2·26	3·9	−0·28	−0·58
16	368	7·11	—	ls	−1·92	4·0	−0·51	−0·82
—	324	7·28	58	ls	−2·12	11	−0·37	−0·91
—	220	7·30	53	dol	−2·36	4·6	−0·78	−1·10
—	180	7·22	—	ls	−2·40	3·7	−1·17	−1·48

Table 1.

Wells

No.	$T(°C)$	Ca^{2+}	Mg^{2+}	Na^+	K^+	HCO_3^-	SO_4^{2-}	Cl^-	NO_3^-
12	10·2	50	25	1·0	0·9	237	14	2·2	22
16	11·1	43	18	1·0	1·0	189	16	2·0	8·3
40	10·1	40	25	1·6	1·9	231	8	3·0	8·7
59	14·4	91	35	3·1	0·3	321	36	37	33
66	10·5	55	15	2·5	1·8	195	12	15	24
75	9·9	92	19	5·7	1·9	357	7	7·4	23
88	9·1	63	25	1·2	1·1	269	12	8·1	37
91	11·2	92	15	13	6·3	287	20	23	56
179	10·9	56	25	3·3	1·1	250	14	7·9	16
182	10·0	83	17	8·5	6·3	279	27	17	38
256	12·0	71	61	9·9	1·2	292	87	110	49
257	11·5	54	54	4·3	0·8	373	29	21	20
265	10·6	50	37	1·2	0·4	276	18	17	26
285	13·0	56	55	0·3	1·4	387	43	5·4	10
303	11·8	75	35	5·7	1·6	352	26	32	23
317	10·8	97	21	6·5	1·1	311	74	7·5	12
321	11·7	38	19	2·1	0·4	180	12	4·2	15
332	13·0	56	29	1·0	0·8	263	17	4·7	27
336	9·8	68	56	1·1	0·9	438	29	3·2	7·7
413	10·5	41	21	2·2	0·9	193	18	6·1	13
415	9·3	17	11	0·6	0·7	100	3	1·8	7·5
1040	10·6	56	59	5·3	1·3	396	12	12	33
1066	12·2	41	31	2·5	1·2	303	30	3·6	11
SC-7	10·0	41	28	2·6	2·3	234	13	5·2	21
SC-11	10·6	41	26	1·4	1·3	228	12	5·1	16
UN-17	9·5	22	15	0·5	0·7	128	8	1·7	6·5
UN-21	10·0	36	34	3·7	1·2	224	28	9·8	14
UN-22	10·0	24	16	0·7	1·3	143	1	2·0	11
GC	10·4	47	33	4·1	1·3	275	11	14	26

issue at the foot of mountain slopes (Group 1), and those which discharge further downvalley (Group 2). Group 1 springs are generally lower in μ (specific conductance), pH, and in SI_c and SI_d values than springs of Group 2. However, no systematic differences exist between the DO (dissolved oxygen) and equilibrium P_{CO_2} contents of the two groups. At 1000 ft elevation and 10°C, the average spring water DO content in Table 2 is equivalent to $P_{O_2} = 10^{-0.84}$ atm. Because the ground waters which feed springs have flowed mostly under water table conditions, the average O_2 and CO_2 content of the spring waters may equal the average contents of these gases in the gas phase at the water table.

Ground water from the Gatesburg Dolomite Aquifer in most cases differs from water in other carbonate rocks. The five well waters in the Gatesburg have among them the lowest μ, highest pH and lowest P_{CO_2} values of all the spring and well waters. These pH and P_{CO_2} values reflect the paucity of humus material, and well drained sandy character of soils on the Gatesburg Formation. The low μ values are a consequence of the abundance of sandy beds in the Gatesburg, the reduced

(continued)

μ	pH	DO (% sat)	Wells Intake interval or well depth (ft)	Rock type	Log P_{CO_2} (atm)	Ca^{2+}/Mg^{2+} (molar)	SI_c	SI_d
418	7.54	104	78	dol	−2.25	1.2	−0.09	−0.14
331	7.74	100	45–50	dol	−2.54	1.5	−0.03	−0.11
368	7.75	113	108	dol	−2.47	0.97	+0.02	+0.02
694	7.19	73	128	ls	−1.76	1.6	−0.04	−0.12
396	7.45	85	101	ls	−2.25	2.3	−0.21	−0.39
597	7.18	56	112	ls	−1.72	3.0	−0.04	−0.28
522	7.45	96	127	dol	−2.12	1.5	−0.05	−0.16
648	7.25	117	189	ls	−1.88	3.9	−0.05	−0.34
450	7.41	92	67	ls	−2.10	1.4	−0.14	−0.21
600	7.36	97	172	ls	−2.01	2.9	−0.01	−0.25
945	7.06	38	257	dol	−1.69	0.71	−0.38	−0.30
644	7.16	77	50	dol	−1.68	0.61	−0.27	−0.16
490	7.41	76	75	dol	−2.06	0.82	−0.17	−0.13
635	7.20	31	320	dol	−1.70	0.62	−0.18	−0.06
675	7.16	30	44	dol	−1.70	1.3	−0.14	−0.19
605	7.30	60	185	ls	−1.90	2.8	+0.03	−0.19
321	7.77	88	139	dol	−2.59	1.2	−0.06	−0.10
467	7.55	102	88–133	dol	−2.21	1.2	+0.04	+0.02
720	7.17	65	175	dol	−1.63	0.74	−0.12	−0.06
344	7.66	87	114–172	dol	−2.46	1.2	−0.14	−0.17
173	8.15	100	284–298	dol	−3.23	0.95	−0.27	−0.27
682	7.15	83	198	dol	−1.65	0.58	−0.25	−0.13
536	7.23	88	30	ls	−1.83	0.80	−0.37	−0.31
422	7.40	79	71–165	dol	−2.12	0.89	−0.32	−0.31
402	7.41	77	83–155	dol	−2.13	0.96	−0.31	−0.31
215	8.04	93	31–188	dol	−3.02	0.90	−0.19	−0.17
427	7.65	99	27–375	dol	−2.39	0.63	−0.17	−0.07
239	8.03	93	97–343	dol	−2.95	0.94	−0.10	−0.09
518	7.39	85	120	dol	−2.04	0.86	−0.21	−0.19

solubility of dolomite at low CO_2 pressures, and the absence of pollution in the sparsely populated areas underlain by this formation. Excluding the Gatesburg well waters, for the remaining well waters specific conductances range from 321

Table 2. Ranges and averages of some chemical quality parameters for the spring and well waters

Parameter	Spring waters		Well waters	
	Range	Average	Range	Average
μ	180–475	347	173–945	499
DO (% sat)	42–100	71	30–117	82
Ca^{2+}/Mg^{2+} (molar)	0.8–11	3.4	0.6–3.9	1.3
pH	7.11–7.80	7.37	7.06–8.15	7.46
P_{CO_2} (atm)	$10^{-1.92}$–$10^{-2.60}$	$10^{-2.18}$	$10^{-1.63}$–$10^{-3.23}$	$10^{-1.99}$
SI_c	−1.17 to −0.04	−0.41	−0.38 to +0.04	−0.15
SI_d	−1.48 to −0.06	−0.63	−0.39 to +0.02	−0.18

to 945 $\mu\Omega^{-1}$, and average 548 $\mu\Omega^{-1}$, pH ranges from 7·06 to 7·77 and averages 7·34, and CO_2 pressures range from $10^{-2·59}$ to $10^{-1·63}$ atm and average $10^{-1·93}$ atm. Non-Gatesburg well waters, then, are usually higher than the spring waters in specific conductance and P_{CO_2} content.

As is evident from Tables 1 and 2, and Fig. 3, the well waters are generally nearer to saturation with calcite and dolomite than are the spring waters. Most spring waters are closer to saturation with calcite than with dolomite, consistent with the occurrence of 24 springs in limestone rocks. Of 21 well waters from dolomite, 15 are closer to saturation with dolomite than with calcite. The remaining 6 are more nearly saturated with calcite than with dolomite. Molar Ca^{2+}/Mg^{2+} ratios in these 6, range from 1·2 to 1·5, and indicate contact with calcite-bearing rocks.

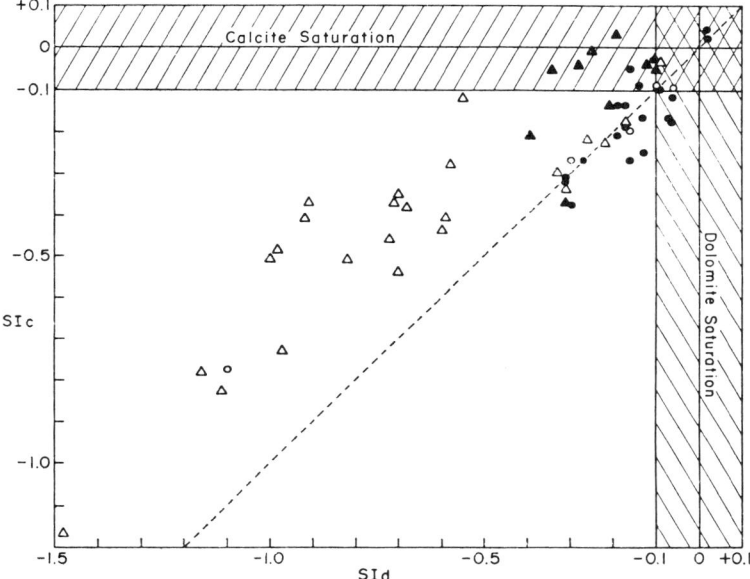

Fig. 3. SI_c vs. SI_d for the spring and well waters. Spring waters are denoted by open symbols; well waters by solid symbols. Triangles denote a limestone source rock; circles a dolomite source rock. $SI_c = SI_d$ along the dashed line. Crosshatched area shows the limits of uncertainty in SI_c and SI_d.

None of the spring or well waters are significantly supersaturated with calcite or dolomite. Based on the uncertainty of $\pm 0·1$ units in SI_c and SI_d, 3 spring waters are saturated with both minerals. Similarly, 12 well waters are just saturated with calcite, and 7 just saturated with dolomite. These results show that the theoretical solubilities of stoichiometric calcite and dolomite represent real limits on the relative concentrations of Ca^{2+}, Mg^{2+}, HCO_3^- and H^+ ions in the ground water. This is true even in well waters which are polluted, having specific conductances above 600–700 $\mu\Omega^{-1}$, and SO_4^{2-}, Cl^- and or NO_3^- ion concentrations in excess of 10–15 ppm.

Ca^{2+}/Mg^{2+} ratios in the ground water have both hydrologic and geochemical implications. Ground water flow from spring 7 towards spring 8 is along a limestone–dolomite contact so that the molar Ca^{2+}/Mg^{2+} ratio decreases with flow from 5·6 to 3·2. In general, Ca^{2+}/Mg^{2+} ratios show that ground water has flowed largely within the formation in which it is located. Ratios are consistent with the fact that most springs issue from limestone and most well waters are pumped from dolomite rocks. With the exception of springs 1, 14 and 20–23, and wells 59, 179 and 1066, Ca^{2+}/Mg^{2+} ratios equal or exceed 2·2 for ground waters in mapped limestone rocks. Lower ratios reflect the presence of dolomitic beds in the Axeman and Stonehenge Limestones, or flow originating in adjacent dolomite rocks. Such cross-formational ground water flow may occur along zones of rock fracture enlarged by solutioning (LATTMAN and PARIZEK, 1964). Excepting spring Bp, Ca^2/Mg^{2+} ratios in mapped dolomite areas do not exceed 1·5. Waters feeding spring Bp flow from the Late Ordovician Limestone and issue from alluvial materials overlying the Bellefonte Dolomite near the contact of the two formations, apparently without reaching the dolomite rock.

Fourteen well waters and 5 spring waters have molar Ca^{2+}/Mg^{2+} ratios ranging from 0·58 to 0·95. Such ratios are not possible unless waters with $Ca^{2+} < Mg^{2+}$ are mixed with ground waters in dolomite, or Ca^{2+} is precipitated from these ground waters. No minerals with $Ca^{2+} < Mg^{2+}$ are present in local carbonate rocks (RAUCH and WHITE, 1970). With the exception of small amounts of MgO added to a few magnesium-deficient local topsoils (nearly all such soils are treated instead with crushed dolomite-limestone rock), there are no culturally introduced substances with molar Ca^{2+}/Mg^{2+} values less than unity. The possibility also exists that because calcium phosphate compounds are slightly less soluble than magnesium phosphate compounds (McCARTY et al., 1970), phosphorus in ground water recharge accounts for the low Ca^{2+}/Mg^{2+} values. Local sources of phosphorus include phosphate fertilizers and farm animal wastes, and phosphate detergents and human wastes in septic tank and sewage effluent discharges. However, most of the phosphate fertilizer used locally is superphosphate and so contains SO_4^{2-} (W. W. HINISH, verbal communication). Also, detergent phosphate is present as sodium pyrophosphate ($Na_4P_2O_7$) or sodium triphosphate ($Na_5P_3O_{10}$), and animal and human wastes also contain high sodium concentrations. Examination of the analyses in Table 1 shows no significant correlation between Ca^{2+}/Mg^{2+} ratios and either Na^+ or SO_4^{2-} content, so that a different explanation for the low Ca^{2+}/Mg^{2+} ratios must be sought.

Seventeen of the 29 wells sampled for this study in late August 1970, were also sampled and analyzed in December 1967, or January, 1968. At the earlier time in 13 observation wells already discussed (and so probably also in the sampled wells), water levels were generally 10 ft or more above their levels in late August, 1970. These higher levels were a consequence of several rainfalls each of 1–3 in. which occurred after mid-September, 1967. A comparison between the chemical analytical data collected in the winter of 1967–68, and summer of 1970, shows μ and Mg^{2+} values averaging 17 and 20 per cent higher respectively, but Ca^{2+} values 9 per cent lower in the 1970 than in the 1967–68 samples. A similar comparison among 7 of the 17 wells (256, 257, 265, 336, SC-7, SC-11 and 1040) in which molar Ca^{2+}/Mg^{2+} ratios were less than unity in August 1970, showed that μ and Mg^{2+} values had increased by 16 and 24 per cent, respectively, but Ca^{2+} had decreased by 23 per cent between the time of high water levels in 1967–68, and lower water levels in August 1970. These data indicate that precipitation of calcite from most of the 17 well waters took place at some time between the two sampling dates. This precipitation must have occurred concurrently with solution of dolomite.

The $[Ca^{2+}]/[Mg^{2+}]$ ratio in a ground water in equilibrium with both calcite and dolomite is defined by the expression

$$K_c^2/K_d = [Ca^{2+}]/[Mg^{2+}] \quad (7)$$

(HOLLAND et al., 1964). Table 3 lists $[Ca^{2+}]/[Mg^{2+}]$ ratios for calcite–dolomite equilibrium between 0 and 25°C computed from the K_c and K_d values used in this report. The $[Ca^{2+}]/[Mg^{2+}]$ ratios in local ground waters are usually about 2–3 per cent larger than corresponding molar Ca^{2+}/Mg^{2+} ratios in Table 1, because $\gamma Mg^{2+} > \gamma Ca^{2+}$, and because of Mg^{2+} and Ca^{2+} ion pairing. Four of the well waters (temperatures 10–13°C) have $[Ca^{2+}]/[Mg^{2+}]$ ratios ranging from 0·60 to 0·65. If 0·60 is taken as the ratio for calcite–dolomite equilibrium at 10°C, then substituting this value and the value for K_c at 10°C into equation (7) yields $K_d = 10^{-16.49}$. With equation (1), this K_d value, and the same heat capacity and enthalpy values for dolomite solution employed previously, we may compute $K_d = 10^{-16.85}$ at 25°C for the natural

Table 3. $[Ca^{2+}]/[Mg^{2+}]$ ratios for equilibrium between calcite and dolomite from 0 to 25°C. Column I lists ratios computed from equation (7) using K_c and K_d values adapted in this report. Column II lists ratios computed assuming a ratio of 0·60 at 10°C (see text)

$T(°C)$	I	II
0	0·76	0·36
5	0·87	0·47
10	0·99	0·60
15	1·12	0·76
20	1·32	0·93
25	1·58	1·12

dolomite rocks in the study area. With $K_d = 10^{-16.49}$ at 10°C, the SI_d values in Table 1 would be reduced by about 0·1 units, so that all the spring waters and all but two of the well waters would be undersaturated with respect to the natural dolomite. Although an error in the adapted K_d values for ideal dolomite could account for the disagreement between computed and measured $[Ca^{2+}]/[Mg^{2+}]$ ratios at 10°C in Table 3, more likely the measured ratio reflects the greater solubility of the natural dolomite than the ideal dolomite used to derive K_d at 25°C. Mixed layering, or perhaps more important several mole percent excess $CaCO_3$ in the natural dolomite may account for its increased solubility.

GOLDSMITH and GRAF (1958) have shown that Paleozoic dolomites often contain 4 mole percent excess $CaCO_3$ and so have a composition which may be written $Ca_{0.54}Mg_{0.46}CO_3$. The free energy of formation of such a solid solution (G_s^0) is related to that of pure calcite (G_c^0) and pure dolomite (G_d^0) by the expression

$$G_s^0 = N_c G_c^0 + N_d G_d^0 + G_m^0 \quad (8)$$

where N_c and N_d are the mole fraction of calcite (as $CaCO_3$) and dolomite (as $Ca_{0.5}Mg_{0.5}CO_3$) in the solid solution respectively, and G_m^0 is the free energy of mixing of calcite and dolomite. Assuming as an approximation that dolomite is an asymmetrical regular solution, and with solution constants from LERMAN (1965) we may compute $G_m^0 = 90$ cal/mole at 25°C for dolomite with 4 mole percent excess $CaCO_3$. $G_c^0 = -269,980$ cal/mole (LANGMUIR, 1968). With free energies of formation for Ca^{2+}, Mg^{2+} and CO_3^{2-} of $-132,350$ cal/mole (LANGMUIR, 1968), $-108,760$ cal/mole, and $126,200$ cal/mole (LANGMUIR, 1965), respectively and $K_d = 10^{-17.0}$, we may compute $G_d^0 = -258,350$ cal/mole, and with equation (8), $G_s^0 = -259,191$ cal/mole. Equilibrium between the Ca^{2+}-rich dolomite and calcite may be written

$$CaCO_3 + 0.46\, Mg^{2+} = Ca_{0.54}Mg_{0.46}CO_3 + 0.46\, Ca^{2+}. \tag{9}$$

Based on the free energy data just given the reaction free energy of (9) is -62 cal/mole, $[Ca^{2+}]/[Mg^{2+}] = 1.26$, and through expression (7), $K_d = 10^{-16.90}$ at 25°C. This last value is in good agreement with $K_d = 10^{-16.85}$ at 25°C based on $[Ca^{2+}]/[Mg^{2+}] = 0.6$ at 10°C in study area ground waters and suggests that this minimal ratio may be controlled by the solubility of dolomite containing several percent excess $CaCO_3$.

The Na^+ and K^+ deficit

The Na^+ and K^+ content of ground waters in Table 1 never exceeds 13 and 6.3 ppm, respectively. In contrast SO_4^{2-}, Cl^- and NO_3^- values all commonly exceed 10 ppm, and are as high as 87, 110 and 56 ppm, respectively. The relatively low Na^+ and K^+ concentrations indicate that the bulk of these ions are being replaced by Ca^{2+} and Mg^{2+} in soils through ion exchange, or that they are not present in ground water recharge to begin with. Probably both explanations are locally correct. Na^+ and K^+ values comparable to observed concentrations of SO_4^{2-}, Cl^- and NO_3^- could be expected in many of the well waters, which are vulnerable to contamination from adjacent septic tanks. Rock salt and smaller amounts of calcium chloride are added to paved surfaces in winter. The rock salt has been shown to contaminate some of the shallow spring waters (JACOBSON and LANGMUIR, 1970). Leaching of KCl fertilizers, the predominant source of added potassium in soils must also occur. These sources of the alkali metals are probably also the chief sources of Cl^- in the ground water. In a few well waters the total molarity of alkali metals exceeds that of chloride, but in most cases the opposite is true, suggesting that some Na^+ and K^+ have been removed by ion exchange in soils.

The deficit in alkali metals also suggests that most of the SO_4^{2-} and NO_3^- in the ground water are derived from sources free of or low in these metals. Sulfate may arise from the oxidation of pyrite or marcasite which comprise up to 1.7 per cent by volume of local carbonate rocks (RAUCH and WHITE, 1970). Other likely sulfate sources are leaching of superphosphate fertilizer which contains SO_4^{2-} in the Ca^{2+} or H^+-ion form, or leaching of ammonium sulfate fertilizer. Major nitrate sources in the ground water are probably nitrogen fertilizers, which are chiefly urea and ammonium nitrate, but also include anhydrous ammonia and ammonium sulfate (W. W. HINISH, verbal communication). Oxidation of the reduced sulfur in the ferrous disulfides to SO_4^{2-}, and of the reduced nitrogen in fertilizers to NO_3^- produces

H^+ ions. The H^+ ions leached from superphosphate fertilizer and H^+ ions from the oxidation of sulfur and nitrogen species in turn must attack the carbonate rocks to form carbonic acid, which may account for the relatively high CO_2 pressures in some of the well waters.

Carbonate solution and precipitation models

The greater P_{CO_2} and HCO_3^- values and saturation levels of most well waters than spring waters (Table 2) suggest basic differences in their chemical evolution. In water which contains only alkaline earth cations and carbonate species, changes in chemistry may be described in terms of the pH and HCO_3^- content. Although not exact for study area ground waters which contain other ionic species, this simplification permits graphic comparison of the chemical behavior of spring and well waters. There are three general ways that the water chemistry can change with time. First, in a ground water unsaturated with carbonate minerals which becomes isolated from fresh sources of CO_2, the decrease in CO_2 content with time is accompanied by an increase in HCO_3^- as is evident from the solution reaction for calcite

$$CaCO_3 + CO_2 + H_2O = Ca^{2+} + 2HCO_3^- \qquad (10)$$

and the similar reaction for dolomite. Ignoring the difference between activities and concentrations, when mCO_3^{2-} is negligible (pH < 8·3)

$$d(P_{CO_2}) = -\tfrac{1}{2} d(mHCO_3^-). \qquad (11)$$

Because $P_{CO_2} = [H_2CO_3]/K_{CO_2}$, expression (11) is equivalent to

$$d(mH_2CO_3) = -\tfrac{1}{2} d(mHCO_3^-). \qquad (12)$$

Through K_1, $d(mH_2CO_3)$ may be replaced to give

$$d(mHCO_3^-)/d(pH) = 2 \cdot 30[(mHCO_3^-)/(2 + K_1 \times 10^{pH})] \qquad (13)$$

Integration of (13) yields

$$(mHCO_3^-)(2 \times 10^{-pH} + K_1) = C \qquad (14)$$

where C is constant for a given water. Knowing the initial pH and $mHCO_3^-$ content of a water, we can solve for C. Curves in Fig. 4, labeled 'no CO_2 added' have been calculated with expression (14). This model, however, does not generally apply to study area ground waters because of the unconfined and open nature of the carbonate aquifers. This openness is evidenced by the rapid response to rain of spring discharges and water levels in wells, and the high DO content, and usual presence of significant amounts of bacteria and pollutant ions such as SO_4^{2-}, Cl^- or NO_3^- in the ground water.

In the second model, if a ground water unsaturated with calcite and dolomite is continuously resupplied with CO_2 as the gas is depleted by solution of the carbonates, both the pH and HCO_3^- content will increase with time. In Fig. 4, the water chemistry will change diagonally upwards along lines of constant P_{CO_2}. These lines are calculable with the equation

$$mHCO_3^- = K_{CO_2} \cdot K_1 \cdot P_{CO_2} \cdot 10^{pH}/\gamma HCO_3^-. \qquad (15)$$

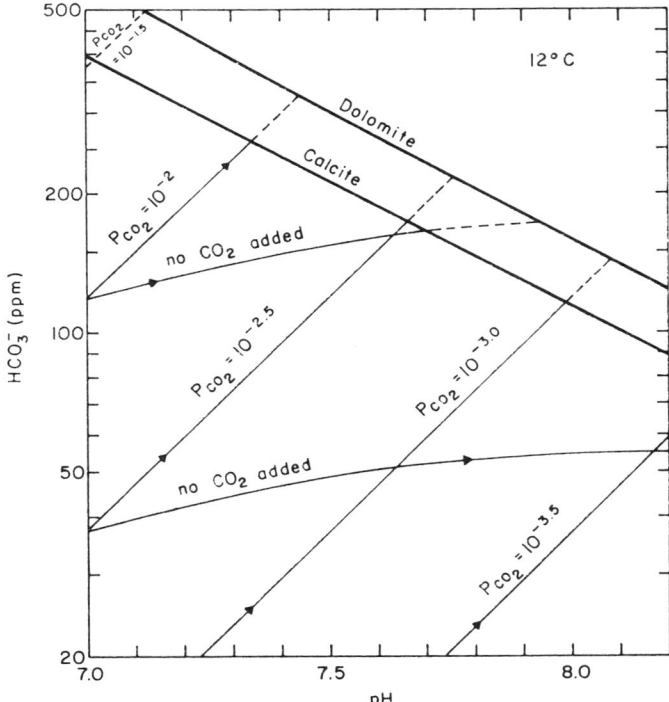

Fig. 4. Possible approaches to equilibrium of ground water in contact with calcite and (or) dolomite at 12°C. The solubility curves of dolomite, and calcite plus dolomite are the same at this temperature.

A plot of the pH and HCO_3^- content of the 29 spring waters is shown in Fig. 5. The figure also shows regression lines calculated from the data assuming pH or HCO_3^- the independent variable, and a line drawn at the average spring water CO_2 pressure of $10^{-2.2}$ atm. The pH and HCO_3^- content of Group 2 spring waters generally exceeds that of Group 1 spring waters. In other words the pH and HCO_3^- values of spring waters tend to increase with their subsurface residence times. The increase occurs at an average CO_2 pressure of $10^{-2.2}$ atm, so the springs roughly obey the second model.

In the third model, a ground water which is at or near saturation with the carbonates may change in chemistry up or down a saturation curve in Fig. 4. Equations for calcite saturation, dolomite saturation and calcite plus dolomite saturation curves, are respectively

$$-\log (mHCO_3^-) = \tfrac{1}{2}[pH - \log (2K_c/K_2) + \log (\gamma_{Ca}\gamma_{HCO_3})] \tag{16}$$

$$-\log (mHCO_3^-) = \tfrac{1}{2}[pH - \log (4K_d^{1/2}/K_2) + \tfrac{1}{2} \log (\gamma_{Ca}\gamma_{Mg}\gamma_{HCO_3}^2)] \tag{17}$$

and

$$-\log (mHCO_3^-) = \tfrac{1}{2}\left[pH - \log \frac{2}{K_2}\left(\frac{K_c}{\gamma_{Ca}\gamma_{HCO_3}} + \frac{K_d/K_c}{\gamma_{Mg}\gamma_{HCO_3}}\right)\right] \tag{18}$$

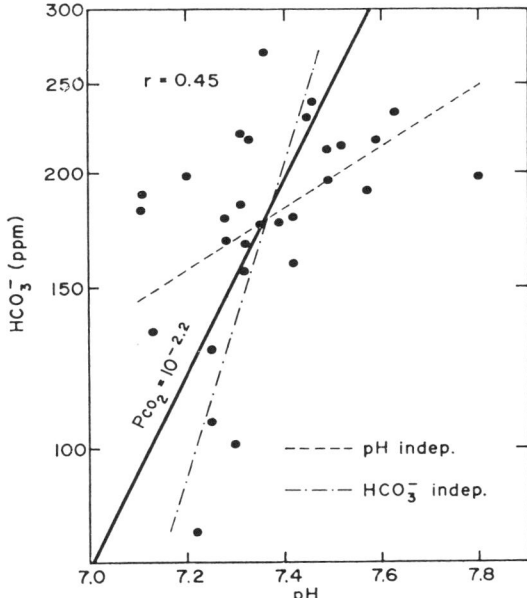

Fig. 5. Plot of pH vs. HCO_3^- (ppm) for the spring waters. Regression lines are drawn assuming HCO_3^- or pH the independent variable. Also shown is the line for $P_{CO_2} = 10^{-2.2}$ atm at 12°C.

These equations have been derived using the law of electroneutrality and equilibrium constant expressions for carbonic acid species, calcite, and dolomite. At pH's below 8·3, mCO_3^{2-} is negligible and ionic strength equals 1·5 ($mHCO_3^-$) from which we may evaluate the ion activity coefficients in expressions (16), (17) and (18). The last two expressions are practically identical at 12°C, so that the solubility of dolomite and of dolomite plus calcite are the same in terms of $mHCO_3^-$ at this temperature.

In this third model changes at saturation can be related to changes in CO_2 pressure or to changes in ground water chemistry which affect P_{CO_2}. When CO_2 is exsolved as in many caves, carbonate minerals may precipitate and water composition move diagonally down saturation curves to the right. When ground water CO_2 content increases, more limestone or dolomite can dissolve, and water composition tends to move up the saturation curves to higher HCO_3^- values and lower pH's. Similarly an increase in pH, Ca^{2+}, HCO_3^- or CO_3^{2-}, can cause precipitation of calcite and thus an increase in P_{CO_2}. The pH and HCO_3^- content of the 29 well waters is plotted in Fig. 6 along with regression lines drawn assuming pH or HCO_3^- the independent variable. Saturation curves for calcite and dolomite are also shown. The regression coefficient proves a strong negative correlation between pH and HCO_3^- content. The plot shows that changes in well water chemistry generally obey the third model. Exsolution of CO_2 from the well waters (or from spring waters) is probably uncommon in the subsurface, in that none of the ground waters were supersaturated with calcite or dolomite. Chief controls on the chemistry of the well waters are probably incongruent solution of dolomite at times of low water levels, and

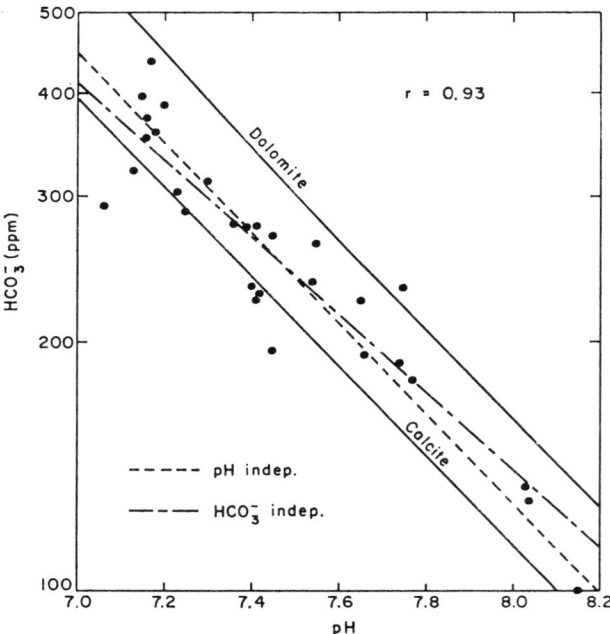

Fig. 6. Plot of pH vs. HCO_3^- (ppm) for the well waters. Regression lines are drawn assuming HCO_3^- or pH the independent variable. Also shown are lines denoting the theoretical solubilities of calcite and dolomite in pure water at 12°C.

at times of high water levels, dilution by ground water or ground water recharge lower in P_{CO_2}, Mg^{2+}, HCO_3^-, μ, and SI_c and SI_d values.

Comparison with published results

Results of this study may be compared with those of other workers including HOSTETLER (1964), HOLLAND et al. (1964), BARNES (1965) and BACK and HANSHAW (1970), who have all reported examples of carbonate ground waters substantially supersaturated with respect to calcite and or dolomite. Most of these workers accepted the K_c values of LARSON and BUSWELL (1942), did not consider changes in K_d with temperature, and except for HOSTETLER (1964) ignored the effect of ion pairs or complexes on saturation. Comparison between their results and those of this study is made after recalculating SI_c and SI_d from their basic chemical analytical measurements, considering ion pairs, and using the equilibrium constants and computational methods described in this report.

HOSTETLER'S (1964) saturation results will not be examined in detail because they mostly depend on laboratory pH measurements, which can differ from the field-measured pH by $+0.5$ units or more even when precautions are taken to avoid CO_2 loss. Data given by HOSTETLER for some dolomite ground waters permit computation of $[Ca^{2+}]/[Mg^{2+}]$ values. The lowest value of 0·63 in a well water at 10·3°C ($\mu = 511\ \mu\Omega^{-1}$) is consistent with the lowest values found in this study at about the same temperature.

In water samples from caves in Pennsylvania and Virginia, HOLLAND and his coworkers (1964) observed maximum ion activity products of calcite and dolomite equivalent to SI_c and SI_d values of 3·0 and 1·1, respectively. The substantial supersaturation found in these and other caves is maintained by CO_2 exsolution to the cave atmosphere at a more rapid rate than calcite can precipitate, and the inability of dolomite to precipitate under such conditions. Molar Ca^{2+}/Mg^{2+} ratios in the cave waters were as low as 0·28, and reflect both the incongruent solution of dolomite with precipitation of calcite, and exsolution of CO_2.

Clear Seep and Iron-Rich Seep sampled by BARNES (1965) had supersaturation levels equivalent to $SI_c = 0·83$ and $SI_d = 0·18$, respectively. BARNES recognized that the supersaturation in his waters was maintained by kinetic factors, but could not explain these factors.

Some of the supersaturation levels reported by BACK and HANSHAW (1970) for carbonate ground waters from Yucatan and Florida are also difficult to explain. For 11 well waters from the Yucatan Peninsula which contain up to 4300 ppm dissolved solids, the authors obtained maximum SI_c and SI_d values of 0·49 (well at Muna) and 0·64 (well at Isla Mujeres), respectively. BACK and HANSHAW note that these unconfined ground waters are frequently mixed with underlying shallow saline ground waters, and with fresh rainfall which reaches the water table directly through open cavities and sinkholes. Supersaturation with calcite and dolomite is apparently caused by mixing of the fresh and saline ground waters. RUNNELLS (1969) has examined possible explanations for such mixing behavior. Rates of mixing must be rapid enough to prevent equilibration of the ground water by precipitation of either calcite or dolomite.

Reasons for substantial supersaturation with respect to calcite and dolomite of some artesian ground waters in central Florida are also not obvious. BACK and HANSHAW (1970) point out that the Floridan ground waters increase markedly in SO_4^{2-} content with flow (and age) by solution of traces of gypsum. The less than equivalent increase in Ca^{2+} with flow probably reflects the precipitation of aragonite, which has an activity product of $10^{-8·23}$ at 25°C (LANGMUIR, 1964), equal or nearly equal to the ion activity product $[Ca^{2+}][CO_3^{2-}]$ in many of the Floridan ground waters (HANSHAW et al., 1964). Equilibration of these ground waters with aragonite rather than calcite is probably a consequence of the solubilities of typical aquifer calcites with 2–4 mole % $MgCO_3$, and dolomites with 2–6 mole % excess $CaCO_3$ (HANSHAW et al., 1970). However, maximal saturation values of $SI_c = 0·30$ (well at St. Cloud), and $SI_d = 0·35$ (well at Ellenton) which occur in waters having radio-carbon ages of 12,900 and >25,000 yr, respectively, are probably too high to be explained by the solubilities of nonstoichiometric calcites (see CHAVE et al., 1962) or dolomites. HANSHAW et al. (1970) point out that dolomite precipitation only occurs from waters in which the molar Ca^{2+}/Mg^{2+} ratio is about 0·3 or less. In none of the Floridan ground waters described by BACK and HANSHAW (1970) is this ratio less than unity. Evidently, where it exists, supersaturation with ideal calcite and dolomite is maintained by the imbalance in rates of solution of gypsum and precipitation of calcite, aragonite or dolomite.

SUMMARY

Fifty-eight ground waters from folded and faulted Paleozoic carbonate rocks near State College, Pennsylvania were analyzed for their field pH, and content of

Ca^{2+}, Mg^{2+}, HCO_3^- and other species. Of 29 spring waters sampled, 22 were in limestone, and 7 in dolomite. Twenty of 29 well waters sampled were in dolomite and 9 in limestone. Average specific conductances for the spring and well waters were 347 and 499 $\mu\Omega^{-1}$, respectively. Conductances ranged from 173 $\mu\Omega^{-1}$ in an unpolluted well water from sandy dolomite, to 945 $\mu\Omega^{-1}$ in a polluted well water from massive dolomite. Low ground water concentrations of Na^+ and K^+ ions (both generally <10 ppm) relative to SO_4^{2-}, Cl^- and NO_3^-, which were as high as 87, 110 and 56 ppm, respectively, indicate that some Na^+ and K^+ is removed by ion exchange in soils, or that major sources of these anions are low in alkali metals. Such sulfate sources include oxidation of ferrous disulfides which represent up to 1·7 per cent of the carbonate rocks, and leaching of superphosphate or ammonium sulfate fertilizers. Calcium chloride applied to highways may contribute significant amounts of Cl^- to the ground water in a few areas. Nitrogenous fertilizers are Na^+ and K^+ free and may be the principle source of nitrate.

Equilibrium CO_2 pressures for the spring and well waters ranged from $10^{-3.2}$ to $10^{-1.6}$ atm, with the highest values measured in the well waters. Mean molar Ca^{2+}/Mg^{2+} ratios for the spring and well waters were 3·4 and 1·3, respectively. Ratios as low as 0·6 in several dolomite well waters with temperatures near 10°C were caused by incongruent solution of dolomite with precipitation of calcite, and indicate that the rocks are more soluble than ideal, stoichiometric dolomite, perhaps because they contain several mole percent excess $CaCO_3$.

Calculations of the degree of saturation of the ground waters with calcite and dolomite were made taking into account ion pair formation and employing remeasured solubility product data for calcite. Twelve well waters and 3 spring waters were saturated with calcite, while 7 well waters and 3 spring waters were saturated with dolomite. None of the ground waters was significantly supersaturated with respect to either carbonate mineral.

The pH and HCO_3^- content and degree of saturation with calcite and dolomite of most spring waters increased with their ground water residence time at an average CO_2 pressure of $10^{-2.2}$ atm. The well waters exhibited inverse relationships between pH and HCO_3^- content. Their composition was controlled by adjustments in chemistry under conditions near saturation with calcite and dolomite. These adjustments include incongruent solution of dolomite at times of low water levels, and at times of high water levels, dilution with recharge undersaturated with the carbonates, and lower in P_{CO_2}, Mg^{2+} and HCO_3^- content. Results of this study and an examination of the literature suggest that the chemistry of ground water in carbonate rocks may be more effectively controlled by the theoretical solubilities of pure calcite and dolomite than has been previously supposed.

Acknowledgements—In various stages of preparation, this article has benefitted from helpful critiques penned by IVAN BARNES, BRUCE B. HANSHAW and ARTHUR W. ROSE. ROGER L. JACOBSON formulated the computer program for calculation of P_{CO_2} and carbonate saturation values, and contributed much of the chemical analytical data on the springs. Support for this research was provided by the Mineral Conservation Section, and the Institute for Research on Land and Water Resources, both of the Pennsylvania State University, University Park, Pennsylvania.

REFERENCES

BACK W. (1963) Preliminary results of a study of calcium carbonate saturation of ground water in central Florida. *Int. Assoc. Sci. Hydrology* **8**, No. 3, 43–51.

BACK W. and HANSHAW B. B. (1970) Comparison of chemical hydrogeology of the carbonate peninsulas of Florida and Yucatan. *J. Hydrol.* **10**, 330–368.

BARNES I. (1964) Field determinations of pH and alkalinity. *U.S. Geol. Surv. Water Supply Paper* 1535-H.

BARNES I. (1965) Geochemistry of Birch Creek, Inyo County, California a travertime depositing creek in an arid climate. *Geochim. Cosmochim. Acta* **29**, 85–112.

BARNES I. and BACK W. (1964) Dolomite solubility in ground water. *U.S. Geol. Surv. Prof. Paper* **475-D**, 179–180.

BELL R. P. and GEORGE J. H. B. (1953) Incomplete dissociation of some thallous and calcium salts at different temperatures. *Trans. Farad. Soc.* **49**, 619–627.

BUTTS C. and MOORE E. S. (1936) Geology and mineral resources of the Bellefonte Quadrangle, Pennsylvania. *U.S. Geol. Surv. Bull.* **855**, 111 p.

CARUCCIO F. T. (1963) The hydrogeology of the sewage disposal, experiment area northwest of State College, Pennsylvania. M.Sc. Thesis, The Pennsylvania State University.

CLARK J. H. (1965) Geology of the Ordovician carbonate formations in the State College, Pennsylvania area and their relationships to the general occurrence and movement of ground water. M.Sc. Thesis, The Pennsylvania State University.

CRISS C. M. and COBBLE J. W. (1964) The thermodynamic properties of high temperature aqueous solutions. V. The calculation of ionic heat capacities up to 200°. Entropies and heat capacities above 200°. *J. Amer. Chem. Soc.* **86**, 5390–5393.

GARRELS R. M. and CHRIST C. L. (1965) *Solutions, Minerals, and Equilibria*, 450 pp. Harper & Row.

GARRELS R. M. and THOMPSON M. E. (1962) A chemical model for sea water at 25°C and one atmospheric total pressure. *Amer. J. Sci.* **260**, 57–66.

GIDDINGS M. T. (1971) Verbal communication. Department of Geology and Geophysics, The Pennsylvania State University, University Park, Pennsylvania.

GOLDSMITH J. R. and GRAF D. L. (1958) Structural and compositional variations in some natural dolomites. *J. Geol.* **66**, 678–693.

HANSHAW B. B., BACK W. and DEIKE R. G. (1971) A geochemical hypothesis for dolomitization by ground water. *Econ. Geol.* in press.

HANSHAW B. B., BACK W. and RUBIN M. (1965) Carbonate equilibria and radiocarbon distribution related to groundwater flow in the Florida limestone aquifer, U.S.A. *Int. Assoc. Sci. Hydrology, Symposium of Dubrovnik*, 601–614.

HARNED H. S. and DAVIS R., JR. (1943) The ionization constant of carbonic acid in water and the solubility of carbon dioxide in water and aqueous salt solutions from 0 to 50°C. *J. Amer. Chem. Soc.* **65**, 2030–2037.

HARNED H. S. and SCHOLES S. R., JR. (1941) The ionization constant of HCO_3^- from 0 to 50°C. *J. Amer. Chem. Soc.* **63**, 1706–1709.

HINISH W. W. (1971) Verbal communication. Agronomy Extension, The Pennsylvania State University, University Park, Pennsylvania.

HOLLAND H. D., KIRSSIPU T. V., HUEBNER J. S. and OXBURG U. M. (1964) On some aspects of the chemical evolution of cave waters. *J. Geol.* **72**, 36–67.

HOSTETLER P. B. (1963) Complexing of magnesium with bicarbonate. *J. Phys. Chem.* **67**, 720–721.

HOSTETLER P. B. (1964) The degree of saturation of magnesium and calcium carbonate minerals in natural waters. *Int. Assoc. Sci. Hydrology*, Commission of Subterranean Waters **64**, 34–49.

JACOBSON R. L. and LANGMUIR D. (1970) The chemical history of some spring waters in carbonate rocks. *Ground Water* **8**, No. 3, 5–9.

JACOBSON R. L. and LANGMUIR D. (1971) The activity product of calcite from 0 to 50°C. In preparation.

JONES H. W. and MONK C. B. (1952) Emf. studies of electrolyte dissociation. II. Magnesium and lanthanum sulfates in water. *Trans. Farad. Soc.* **48**, 929–933.

KOUNTZ R. R. (1971) Verbal communication. Department of Civil Engineering, The Pennsylvania State University, University Park, Pennsylvania.

LANDON R. A. (1963) The geology of the Gatesburg Formation in the Bellefonte quadrangle, Pennsylvania, and its relationship to general occurrence and movement of ground water. M.Sc. Thesis, The Pennsylvania State University.

LANGMUIR D. (1964) Stability of carbonates in the system $CaO-MgO-CO_2-H_2O$, Ph.D. Thesis, Harvard University.

LANGMUIR D. (1965) Stability of carbonates in the system $MgO-CO_2-H_2O$. *J. Geol.* **73**, 755–780.

LANGMUIR D. (1968) Stability of calcite based on aqueous solubility measurements. *Geochim. Cosmochim. Acta* **32**, 835–851.

LANGMUIR D. and JACOBSON R. L. (1970) Specific-ion electrode determination of nitrate in some freshwaters and sewage effluents. *Environmental Science and Technol.* 834–838.

LARSON T. E. and BUSWELL A. M. (1942) Calcium carbonate saturation index and alkalinity interpretations. *J. Amer. Water Works Assoc.* **34**, 1667–1684.

LATTMAN L. H. and PARIZEK R. R. (1964) Relationship between fracture traces and the occurrence of ground water in carbonate rocks. *J. Hydrol.* **2**, 73–91.

LERMAN A. (1965) Paleoecological problems of Mg and Sr in biogenic calcites in light of recent thermodynamic data. *Geochim. Cosmochim. Acta* **29**, 977–1002.

MCCARTY P. L., ECHELBERGER W. F. and HEM J. D. (1970) Chemistry of nitrogen and phosphorus in water. *J. Amer. Water Works Assoc.* **62**, 127–140.

PARIZEK R. R., KARDOS L. T., SOPPER W. E., MYERS E. A., DAVIS D. E., FARREL M. A. and NESBITT J. B. (1967) Waste water renovation and conservation. *The Pennsylvania State University Studies*, No. 23, University Park, Pennsylvania.

RAINWATER F. H. and THATCHER L. L. (1960) Methods for collection and analysis of water samples. *U.S. Geol. Surv. Water-Supply Paper* **1454**, 132.

RAUCH H. W. and WHITE W. B. (1970) Lithological controls on the development of solution porosity in carbonate aquifers. *Water Resour. Res.* **6**, 1175–1192.

ROBIE R. A. (1971) Verbal communication. U.S. Dept. of Interior, Geological Survey, Washington 25, D.C.

RUNNELLS D. D. (1969) Diagenesis, chemical sediments, and the mixing of natural waters. *J. Sediment. Petrol.* **39**, 1188–1201.

SHUSTER E. T. (1971) Verbal communication. Department of Geology and Geophysics, The Pennsylvania State University, University Park, Pennsylvania.

STOUT J. W. and ROBIE R. A. (1963) Heat capacity from 11 to 300°K, entropy, and heat of formation of dolomite. *J. Phys. Chem.* **67**, 2248–2252.

MINERALOGIC FACTORS IN NATURAL WATER EQUILIBRIA

Owen P. Bricker
Johns Hopkins University

R. M. Garrels
Northwestern University

Although the composition of surface and sub-surface waters is partly inherited from rain, and is locally influenced strongly by the activities of man, the dominating influence is that of the solid phases with which they come in contact. This paper has been divided into two parts: in the first we attempt to show the influence of the details of carbonate mineralogy on water compositions which we consider an important amplification of classical studies; in the second we have made a first attempt to consider the importance of silicate phases and conclude that several important constituents of natural waters are actually controlled by equilibrium between stable or metastable silicate minerals and the waters that bathe them.

CARBONATE EQUILIBRIA

The $CaCO_3$-H_2O-CO_2 system has received a great deal of attention from geologists, oceanographers, sanitary engineers, chemists, and medical researchers because of its important buffering effect and relation to water hardness. Equations describing $CaCO_3$ solubility equilibria were derived as early as 1890 by van't Hoff. Since then the system has been the object of extensive investigation; a recent systematic treatment summarizing the pertinent data has been given by Weber and Stumm (1).

In many cases, workers in the field of carbonate equilibria have dealt primarily with the chemistry of the aqueous phase and the gas-liquid equilibrium, attaching little significance to the solid phases in the system. Let us focus attention on the solid phases in the $CaCO_3$-H_2O-CO_2 system and investigate what effect, if any, the solid phase has on the equilibrium.

Polymorphism

Table I shows that there are three polymorphic forms of $CaCO_3$ and three hydrates. (Two high pressure polymorphs of $CaCO_3$ also have been described by Bridgman, 36). All occur in nature with the exception of the monohydrate. Each of these compounds has a different solubility, making the carbonate equilibrium relations somewhat more complicated than the simple system usually treated. Which of the solid phases should be considered in making equilibrium calculations pertaining to the natural water system? The basic laws of thermodynamics tell us that only one of the $CaCO_3$ compounds can be in equilibrium with nearly pure water at any chosen temperature and pressure. Free energy calculations based on data from Latimer indicate that at 25°C and one atm total pressure the stable phase is calcite (2). Experimental confirmation is provided by the work of Jamieson who determined the equilibrium P-T curve for the calcite-aragonite inversion (3), by Brooks, et al, who investigated the hydrates of $CaCO_3$ (4), and by Faivre who investigated the precipitation of vaterite and other forms of $CaCO_3$ (5).

At first glance, it would appear that the original simple picture with calcite being the only important phase in natural water systems is justified.

TABLE I   CALCIUM CARBONATE POLYMORPHS AND HYDRATES

Calcite	$CaCO_3$
Aragonite	$CaCO_3$
Vaterite	$CaCO_3$
Synthetic	$CaCO_3 \cdot H_2O$
Trihydrocalcite	$CaCO_3 \cdot 3H_2O$
Ikaite	$CaCO_3 \cdot 6H_2O$

A cursory glance at recent carbonate sediments, which are composed to a large extent of aragonite, is enough to dispel that illusion. Aragonite is being deposited in vast amounts under conditions of temperature and pressure in which calcite is the thermodynamically stable phase (6). Free energy data indicate a difference of only about 230 cal between the stabilities of calcite and aragonite under earth-surface conditions (22). The precipitation and metastable existence of aragonite under these circumstances is not too surprising.

Kinetics of Crystallization

A number of natural waters, fresh and saline, are supersaturated with respect to both calcite and aragonite. Barnes found springs in Inyo County, California that were supersaturated by over 300% with respect to calcite (7). Back investigated groundwaters in central Florida and found that, over extensive areas, the waters were supersaturated with respect to calcite, in some cases by over 250% (8).

A device for determining the degree of saturation has been devised by Weyl, who found that the surface water samples he examined from the Atlantic Ocean were supersaturated with respect to calcite (9). Siever, et al, found that the surface waters of the Atlantic Ocean and the Gulf of California were uniformly supersaturated with respect to calcite, but interstitial waters squeezed from cores of the bottom sediment were undersaturated or saturated (10). Schmalz and Chave observed that the sea waters off Bermuda are supersaturated with respect to calcite (11). They believe the supersaturation is due to metastable equilibrium of sea water with magnesian calcites and fine-grained carbonates with large surface energies. The solid phase that precipitates first under these conditions will be governed by kinetic and biologic factors.

Role of Organisms

Organisms that secrete shells and other hard parts composed of $CaCO_3$ are abundant in the oceans and in may fresh waters. Lowenstam observed a correlation between temperature and the $CaCO_3$ polymorph produced by marine organisms (12). Aragonite appears to be the dominant form of $CaCO_3$ secreted by marine organisms, particularly in warm waters. In cases where aragonite is already present in an environment from breakup of shell material, it will be available to nucleate inorganic aragonite if concentration factors become favorable for precipitation. Under these conditions calcite is unlikely to precipitate even though it is the stable phase.

Organisms apparently can produce local (often, of course, internal) conditions favorable to the precipitation of $CaCO_3$ compounds even where the bulk environment is undersaturated with respect to these materials. Upon death, in many cases, the hard parts produced by organisms are attacked and the more delicate forms completely dissolved before burial, attesting to the undersaturation of the bulk environment. Organisms also can precipitate calcite at 25°C or lower, containing an amount of Mg that would be thermodynamically stable only above 800°C. The highly magnesian calcites precipitated by many organisms are clearly metastable with respect to the bulk environment in which they formed. More will be said about this in a following section.

Stabilization by Trace Elements

The observation that aragonite invariably has a higher Sr content than calcite has led to the speculation that trace amounts of Sr might determine whether calcite or aragonite will precipitate (13). The structure of aragonite can accommodate the relatively large Sr^{+2} ion but the calcite structure cannot. It has been suggested also that trace amounts of Sr might have a stabilizing effect on aragonite (14), but the degree to which aragonite is stabilized by trace amounts of Sr has not been well demonstrated. MacDonald found, on the basis of thermodynamic calculations, that a content of at least 30% of components other than $CaCO_3$ would be required to make aragonite stable with respect to calcite at 25°C (15). It has been observed, however, that aragonite from the Bahama Banks containing a trace of Sr does not invert to calcite when dried and stored in the laboratory, whereas chemically pure aragonite of the same particle size will partially invert to calcite in less than a year under the same conditions (6). Holland, et al, found that it was impossible to precipitate only calcite from solutions of $CaCl_2$ containing $SrCl_2$ at 25°C (16). Some aragonite always was precipitated as well. Only by resorting to high pressures of CO_2 and temperatures above 25°C was it possible to eliminate aragonite from the precipitate.

The stabilizing effect of trace elements on a compound is a factor that must be considered, particularly when dealing with naturally occurring compounds which are seldom pure. The effect of other components in the solution from which the compound precipitates is another factor that may influence the polymorphic form of the precipitating compound.

Metastable Equilibrium

In order to predict the equilibrium composition of the aqueous phase in a system containing $CaCO_3$ or to determine whether a natural water is in equilibrium with the solid carbonate phase, not only the composition, but the structure of the solid phase must be known as well. It should be apparent that one cannot make the assumption that the solid phase in the system will necessarily be the thermodynamically stable phase for the prevailing conditions. In cases such as the calcite-aragonite system the metastable precipitation of aragonite and the sluggish kinetics of its inversion to calcite at low temperature permits the metastable existence of aragonite for geologically important periods of time. In this type of system equilibrium between natural waters and a metastable phase may be established. Figure 1 compares the stability fields of calcite and aragonite at $25^\circ C$, one atm total pressure, and a total dissolved carbonate activity of 10^{-3}. In the pH range of natural waters aragonite will support nearly two times more calcium in solution than will calcite.

Solid Solution

This picture is further complicated by Mg, a common constituent of natural waters. Magnesium can substitute for Ca in the calcite structure, but the size difference between Mg^{+2} and Ca^{+2} ions precludes a complete solid-solution series at earth-surface temperatures. Subsolidus investigations by Graf and Goldsmith show that at $500^\circ C$ the extent of solid solution of $MgCO_3$ is about 5 mole % and at $900^\circ C$ about 27 mole % (17). Reactions in this system are too slow at lower temperatures to permit attainment of equilibrium in the laboratory, but extrapolation of the curve at $25^\circ C$ suggests that very little Mg should be present in calcite at low temperatures.

Perusal of analyses of natural calcites known to have formed at low temperatures shows $MgCO_3$ contents as high as 18 mole %. Chave found that many marine organisms secrete hard parts of magnesian calcite, the Mg content being a function of the particular organism (19). The alga lithophyllum secretes calcite containing about 20 mole % $MgCO_3$. Calcite containing that amount of Mg is metastable below a temperature of about $850^\circ C$. Alderman and Skinner found magnesian calcites precipitating in the Coorong and in other saline lakes in Australia (20). Graf and Goldsmith precipitated highly magnesian calcites from aqueous solution at room temperature (21).

Magnesian calcites are metastable under earth-surface

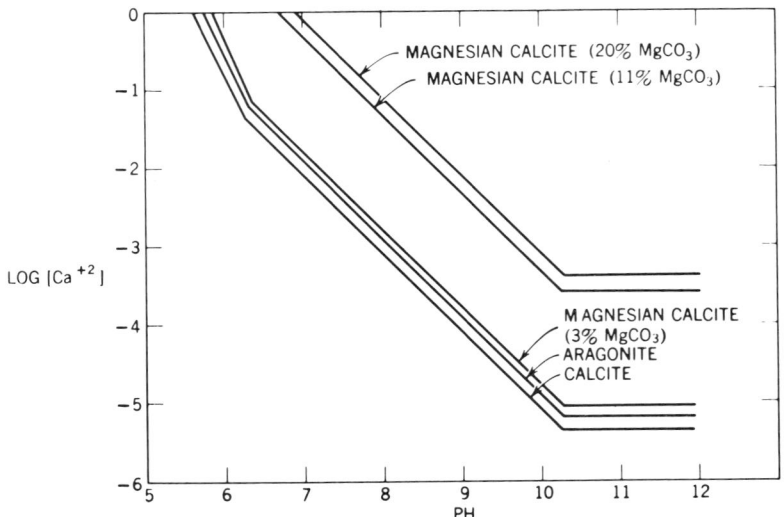

FIGURE 1   SOLUBILITY OF CALCITE, ARAGONITE, AND THREE MAGNESIAN CALCITES AS A FUNCTION OF pH AT 25°C, ONE ATM TOTAL PRESSURE, AND A TOTAL DISSOLVED CARBONATE ACTIVITY OF 10^{-3}.

conditions and lose Mg over geologically short time spans. Calcites from Mesozoic and older rocks are very low in Mg except in unusual circumstances (23). The solubility of magnesian calcites is larger than that of pure calcite (Figure 1) and varies as a function of Mg content (24). Rapid loss of Mg from these calcites, in some cases without disrupting delicate fossil structures, suggests incongruent solution. As the magnesian calcite is dissolved, a purer calcite is reprecipitated, the Mg being lost to the solution. Groundwaters from magnesian carbonate bedrock will have a Mg/Ca ratio higher than that of the rock because of selective loss of Mg^{+2} and may contain Ca concentrations greater than predicted from the solubility of pure calcite.

In addition to magnesian calcite solid solutions, two mixed carbonates of Ca and Mg; dolomite, $CaMg(CO_3)_2$, and huntite, $CaMg_3(CO_3)_4$, occur in nature. Dolomite is a common constituent of carbonate rocks and occurs in beds hundreds of feet thick and of large areal extent. Huntite is metastable with respect to dolomite under earth-surface conditions, and of little importance with respect to the total volume of carbonate sediments.

Huntite precipitates directly from Mg rich groundwaters and may be important in determining water equilibria in certain local environments (24, 26). The precipitation of dolomite, however, is somewhat puzzling. Dolomite constitutes a large percent of the total volume of carbonate rocks and yet only in a few cases has the primary precipitation of this compound been observed. Alderman and Skinner found dolomite precipitating in some saline lakes in Australia (20). The temperature of these lakes has an annual variation between $5^{\circ}C$ and $27^{\circ}C$ whereas salinity and pH vary through a wide range. Maximum precipitation occurs in conjunction with maximum photosynthetic activity during daylight hours, when the pH value may rise as high as 9.3. The precipitate consists of magnesian calcite and dolomite in roughly equal proportions, and amounts to as much as .05 g/100-ml of water. Graf, et al, have described a thin layer of dolomite occurring about a foot below the surface of the Lake Bonneville sediments in Utah which may have been deposited under conditions similar to those presently observed in the Australian lakes (27). Recent investigations by Deffeyes, et al, suggest that dolomite may be formed by the reaction of Mg-rich waters with carbonate sediments (28). On the island of Bonaire, Netherlands Antilles, waters with a molar Mg/Ca ratio of 30:1 commonly have been observed in saline lakes and in pits dug in the carbonate sediment. The high Mg/Ca ratio arises from evaporation of sea water in basins with limited access to the open ocean, resulting in precipitation of gypsum. The Mg-rich waters leave the basins by flowing downward through the underlying carbonate sediments because of their high density. By this mechanism about 1/2 cubic kilometer of carbonate sediments has been dolomitized in less than 10^6 years on Bonaire.

Although precipitation of dolomite from most natural waters would appear to be unimportant as a controlling factor in carbonate equilibria, the extensive areas underlain by dolomite bedrock suggests that the solubility characteristics are not. Water in equilibrium with dolomitic rocks may have a composition quite different from that of waters in equilibrium with calcite or aragonite. Barnes and Back have investigated dolomite solubility in groundwater from carbonate bedrock (29). Many of these waters are supersaturated with respect to calcite but give a solubility product corresponding to dolomite. The kinetics of calcite precipitation apparently are slow enough under these conditions to allow supersaturation of Ca^{+2} ion with respect to calcite and equilibrium with dolomite. Other waters investigated by Barnes and Back are undersaturated or saturated with respect to calcite and apparently in equilibrium with dolomite. Hsu found that most waters he examined from aquifers in Florida were in equilibrium with both dolomite and calcite and had a molar Ca^{+2}/Mg^{+2} ratio of 1 (30). Figure 2 shows the variation of Ca activity in equilibrium with dolomite as a function of pH for certain fixed values of Mg activity and a total dissolved carbonate activity of 10^{-3}. The equilibrium curve for calcite is included for comparison. For groundwaters with pH below 7 and $[Mg^{+2}] \leq 10^{-2}$, dolomite will be more soluble than calcite.

Both calcite and dolomite can accommodate reasonably large amounts of cations other than Ca and Mg in their structures. Iron and manganese are commonly found in calcite and dolomite; more rarely, Zn, Pb, Co, Ni, Cd, Ba, and Sr. Little is known about the extent of solid solution of the latter elements in calcite and dolomite or their effect on the stability of these compounds.

Structural Disorder and Compositional Variation

Many of the primary dolomites are non-stoichiometric to some degree and show varying amounts of disorder in their structure. The term "protodolomite" has been applied to these compounds to distinguish them from stoichiometric, order dolomite (21). The stability of protodolomites would be predicted to vary with degree of non-stoichiometry and degree of disorder, their stability being less the greater the deviation from ideal dolomite. Water in equilibrium with protodolomite thus would vary in dissolved constituents depending upon the composition and degree of order of the solid phase. Detailed information on the variation of stability of protodolomite is not available. Preliminary studies, however, suggest it is small (22).

Figure 3, after Langmuir (22), shows some stability relations in the system $MgO-CaO-H_2O-CO_2$ at $25^\circ C$ and one atm total pressure. The dashed lines represent the change in stability fields resulting from the use of $K_{dol} = 10^{-15}$. The composition of surface sea water is indicated on the diagram. Dolomite is stable in surface sea water although poorly crystalline disordered dolomite, or protodolomite, with a solubility product as large as 10^{-15} is not.

Natural waters may be depleted by Ca both by chemical precipitation of Ca compounds and by utilization of Ca by organisms.

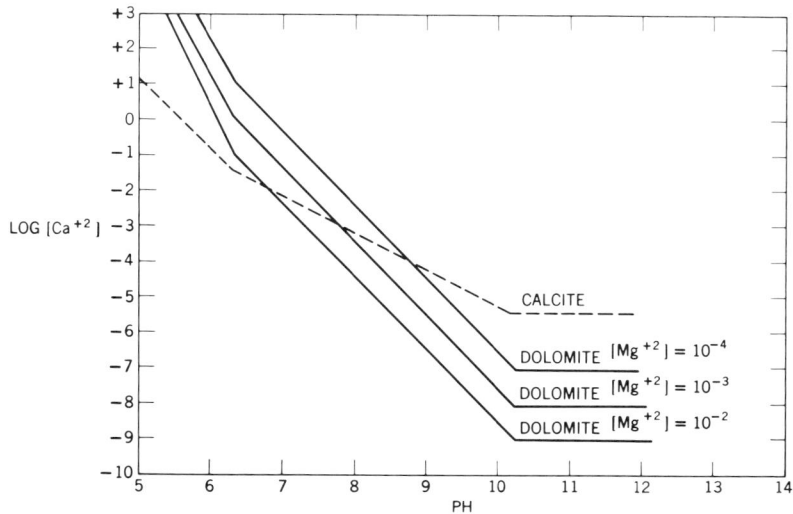

FIGURE 2 ACTIVITY OF Ca^{+2} ION IN EQUILIBRIUM WITH DOLOMITE AS A FUNCTION OF pH FOR FIXED VALUES OF MAGNESIUM ION ACTIVITY AT 25°C, ONE ATM TOTAL PRESSURE, AND A TOTAL DISSOLVED CARBONATE ACTIVITY OF 10^{-3}. SOLUBILITY CURVE OF CALCITE INCLUDED FOR COMPARISON.

Under certain circumstances this may cause a high Mg/Ca ratio. In such cases, with evaporation, Mg concentration can increase to a level at which $MgCO_3$ compounds begin to precipitate. As in the CaCO system, there are a number of $MgCO_3$ compounds (Table II). The stability of each of these compounds is different and will provide a water of unique composition under equilibrium conditions. Again, only one phase (or under certain circumstances two phases) in this system is thermodynamically stable at any chosen temperature and pressure. Kinetic factors, however, make possible the metastable precipitation and persistence of these compounds for relatively long periods of time.

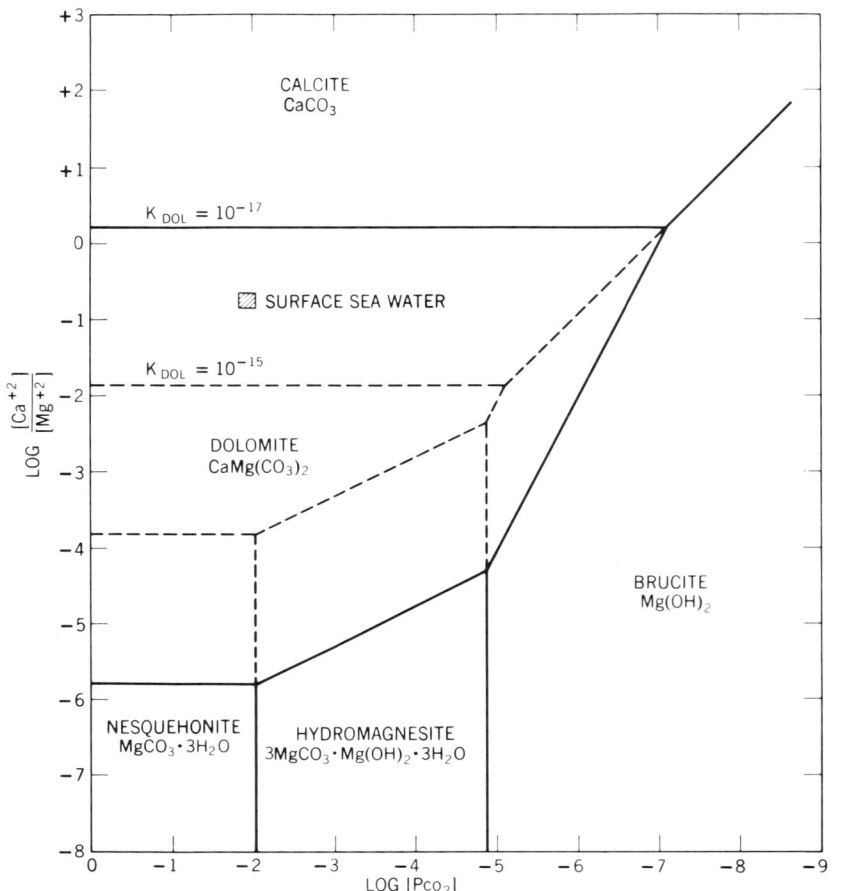

FIGURE 3 STABILITY RELATIONS IN THE SYSTEM MgO-CaO-H_2O-CO_2 AT 25°C AND ONE ATM TOTAL PRESSURE. Dashed lines represent change in stability fields resulting from a hundredfold increase in the solubility of dolomite. (After Langmuir, 22).

Summary

In summary, the composition of surface and near-surface waters is strongly influenced by the carbonate phases present, and especially by the presence of magnesian calcite and dolomite. Furthermore, the processes of solution and precipitation are not symmetric; magnesian calcites exposed to rain water yield solutions with a Mg/Ca ratio equal to that of the solid; the same solution will only precipitate nearly pure calcite, so that Mg becomes enriched relative to Ca. Interpretation of chemical equilibria in natural water systems containing Ca and Mg carbonate compounds can be meaningful only if the detailed chemical and structural variability of the solid phases are considered.

TABLE II   MAGNESIUM CARBONATE COMPOUNDS

Magnesite	$MgCO_3$
Nesquehonite	$MgCO_3 \cdot 3H_2O$
Lansfordite	$MgCO_3 \cdot 5H_2O$
Hydromagnesite	$3MgCO_3 \cdot Mg(OH)_2 \cdot 3H_2O$
Artinite	$MgCO_3 \cdot Mg(OH)_2 \cdot 3H_2O$

SILICATE SYSTEMS

In the preceding section, attention was called to the importance of detailed mineralogy of the carbonate minerals in controlling water compositions. The high rate of reaction of water with carbonate minerals generally is accepted, and it does not seem unreasonable to make calculations based on equilibrium of natural waters with stable or metastable carbonate phases. But the carbonates can be important controls only of Ca, Mg, HCO_3^-, and H^+ ions. What of the Na, K, and silica that are also among the major species in natural waters?

The most obvious source of these constituents is the great variety of silicate minerals that make up 70% or more of the rocks in contact with underground waters and streams. In general, except for their ion exchange properties, these minerals have been ignored or regarded as unimportant in controlling water compositions. Here the bold assumption will be made that silicate minerals, considered as bulk phases, are rapidly reactive and that many water constituents are controlled by equilibrium with one or more silicate phases. Eventually we will retreat somewhat from this extreme view, but think we can document the usefulness of this assumption.

Dissolved Silica and Quartz

The major clue to reactivity of silicate minerals is the concentration of silica in natural waters. The solubility of quartz is about 8 mg/l, and that amorphous silica is about 115 mg/l. The world average for streams (sampled chiefly at the mouths of master streams) is about 13 mg/l, and a plot of the distribution of silica values in streams and in subsurface waters from sandstones and shales shows an abrupt minimum at about 1-2 mg/l and a fairly sharply defined maximum at about 60 mg/l. Furthermore, as shown by Feth, et al, (31) and by Davis (32), silica rises abruptly when rainwater or snow melt comes in contact with silicate rocks, and may be unique in showing little variation of concentration with stream discharge. The picture that emerges is that neither quartz nor amorphous silica exert a control on the silica content of streams or shallow groundwaters, and that the silica is derived from the weathering of silicates. This conclusion has been documented for spring waters from the Sierra Nevada by Feth, et al, who showed that the water compositions are exactly those predicted from alteration of the rocks by soil waters high in CO_2 (31). Calculations based on their data show, in fact, that almost all the dissolved silica must come from silicate minerals, chiefly feldspars. Thus, less than a few percent can be attributed to direct solution of quartz despite the fact that quartz is an abundant constituent of the rocks. More studies like that on the Sierras are needed before this conclusion on the natural chemical inertness of quartz at room temperature can be put to general use, but it is in accord with experimental work in quartz solubility and growth.

Weathering of Silicates

The probable derivation of dissolved silica from silicates indicates structural breakdown of silicate minerals, with concomitant release of cations. The Al in most silicate phases, however, is not at all mobile, so that reactions with surface waters must produce solid products higher in Al than the original silicates. Aluminum is characteristically so low in natural waters (0.1 mg/l) that it is convenient to assume that Al is conserved among the solid phases during reactions. With this stipulation it is possible to make stability diagrams for many of the common silicates on which the phases are described in terms of dissolved constituents of the coexisting solution. No attempt will be made here to utilize such diagrams extensively but one example will show how plotting of natural water compositions on these diagrams almost forces one to conclusions regarding the chemical genesis of the water.

Figure 4, after Hess, (37), is a stability diagram showing relations among some silicate minerals and the aluminum oxide hydrate gibbsite, as best they can be calculated from existing free energy data and inferred from natural occurrences.

Details of the construction of silicate diagrams are given by Garrels and Christ (33). Figure 5 is a plot of the analyses of waters from various rock types showing the ratio of Na^+, H^+, and SiO_2 of a given water are plotted on the silicate diagram, and the resulting point falls into a field containing the point. On the other hand, it definitely is not in equilibrium with the other phases depicted. Figure 6 was obtained by super-imposing Figure 5 on Figure 4. Two obvious relations and their possible explanation are:

a. There is a silica cut-off at about 2 mg/l. This may, of course, be related to the minimum reported analytically, but reasonably can be attributed to a silica control by the two-phase equilibrium gibbsite-kaolinite. Waters draining the Jamaican bauxitic soils, that are dominantly gibbsite-kaolinite mixtures, range from 3 to 6 mg/l SiO_2 close to the theoretical 1 to 2 mg/l calculated from free energy values for equilibrium between the crystalline phases (35).

b. Most of the analyses fall within the field of stability of kaolinite, and seem to be limited more or less by the phase boundaries of kaolinite with the other phases depicted. The sharply defined silica cut-off in the right, at about 60 mg/l SiO_2, may well be the boundary between kaolinite and montmorillonite if other constituents of the waters are taken into consideration.

It seems to us that the picture that emerges is reasonably clear cut. When "aggressive" waters, high in CO_2 and low in dissolved solids, encounter silicates high in cations and silica such as feldspars, they leach silica and cations and leave an aluminosilicate residue with an increased Al-Si ratio (i.e., kaolinite). Initial water attack yields a gibbsite residue but reaction is so rapid that it is only under exceptional conditions that the silica in solution can be kept low enough to prevent the gibbsite from being converted to kaolinite, or to prevent kaolinite from forming in addition to the initial small amount of gibbsite. As the waters continue to attack feldspar, the pH rises, cations and silica increase in concentration, and kaolinite forms until the cations and silica content rise high enough so that montmorillonite begins to form. At that stage, kaolinite apparently tends

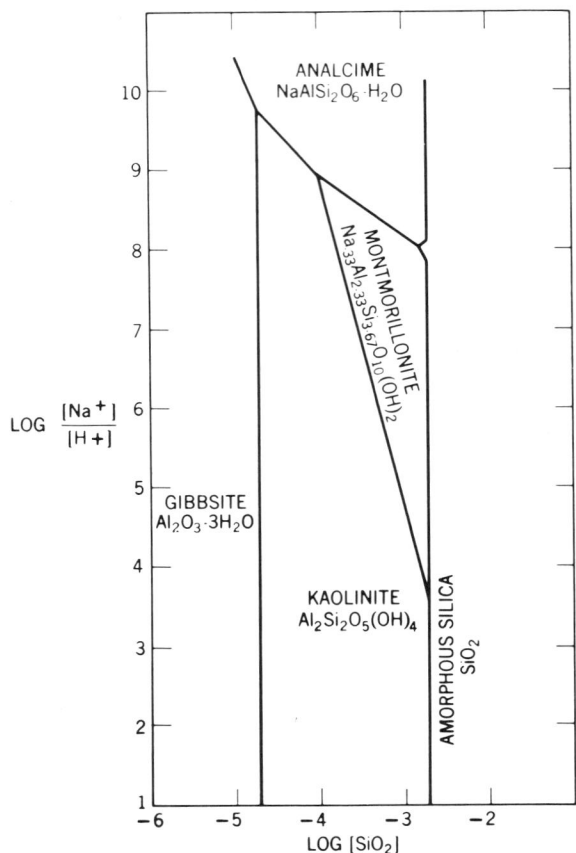

FIGURE 4 STABILITY RELATIONS AMONG SOME SILICATE MINERALS AND GIBBSITE AT 25°C AND ONE ATM TOTAL PRESSURE. (After Hess, 37).

to be converted to montmorillonite, accounting for the limitation of silica content to abut 60 mg/l at the two phase boundary kaolinite-montmorillonite.

The chemical reactions involved in weathering are typified by the alteration of Na feldspar to kaolinite:

$$2\ NaAlSi_3O_8 + 2CO_2 + 3\ H_2O = Al_2Si_2O_5(OH)_4 + 2HCO_3^- + 2Na^+ + 4SiO_2 \quad (17\text{-}1)$$

Na-feldspar Kaolinite

MINERAL EQUILIBRIA   463

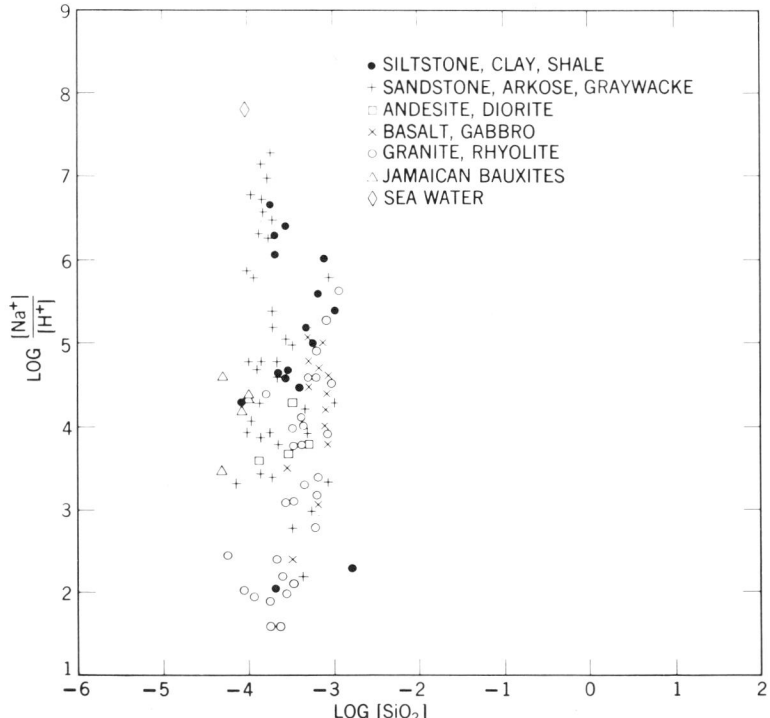

FIGURE 5   SODIUM ION: HYDROGEN ION RATIO OF
           WATERS FROM VARIOUS ROCK TYPES
           PLOTTED AS A FUNCTION OF SiO_2
           CONCENTRATION.

The picture drawn here is that at any stage of the reaction the solution is in equilibrium with kaolinite, but not with Na-feldspar. Note that for a given initial dissolved CO_2, it is possible to calculate pH, Na^+, SiO_2, HCO_3^-, and the amount of kaolinite formed as a function of the amount of CO_2 consumed. In a system isolated from the atmosphere, CO_2 would obviously drop as the reaction proceeds. The parallelism with carbonate equilibria is apparent. Most silicate phase changes can be written similarly if, as postulated, reactions of this type occur rapidly in natural waters, they must be considered in any analysis of pH-controlling factors.

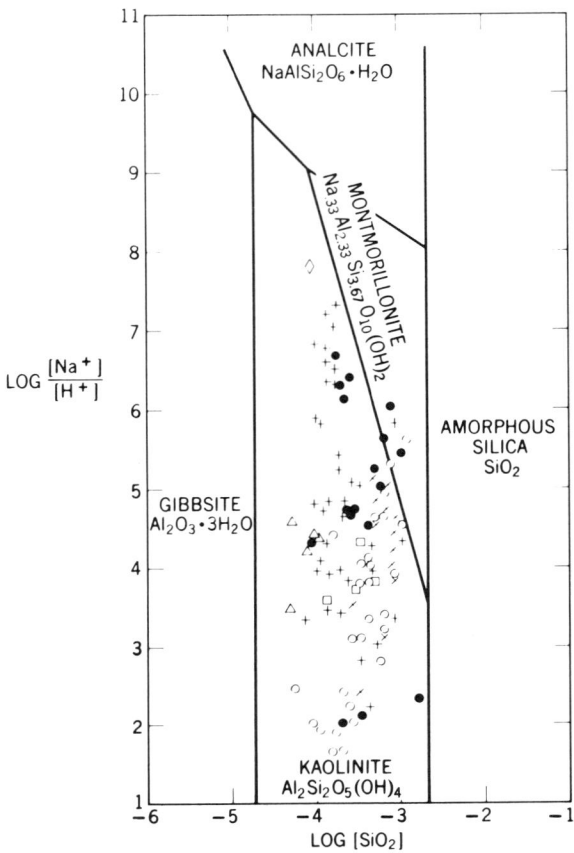

FIGURE 6 SILICATE STABILITY DIAGRAM WITH WATER ANALYSES SUPERIMPOSED.

In Figure 7, an idealized diagram of a part of the Na_2O-Al_2O_3-SiO_2-H_2O system, the shaded area shows predicted changes in water composition of an initially pure water charged with CO_2, on the assumption that it reacts with Na-Ca feldspar (oligoclase) to produce the appropriate stable phases instantaneously. Such a water should form gibbsite until it reaches the gibbsite-kaolinite phase boundary, then convert the gibbsite to kaolinite, and continue to react, producing kaolinite, until the two-phase boundary kaolinite-montmorillonite is reached, after which it should follow the boundary. The points represent water compositions from the

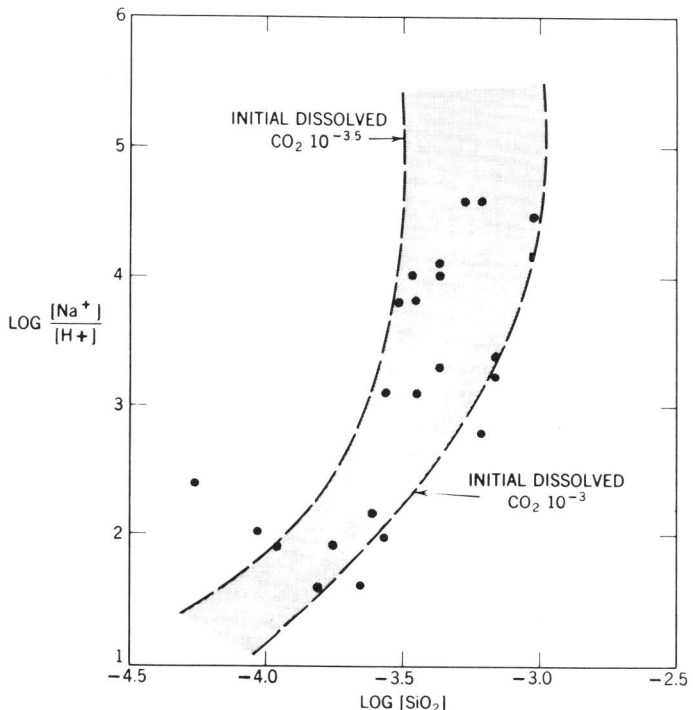

FIGURE 7   COMPOSITION OF EPHEMERAL SPRINGS FROM THE SIERRA NEVADA IN RELATION TO COMPOSITION CHANGES CALCULATED FROM ASSUMPTION OF ATTACK ON Na-Ca FELDSPAR OF SOLUTION ORIGINALLY CONTAINING $10^{-3.0}$ AND $10^{-3.5}$ MOLES DISSOLVED CO_2.

ephemeral springs of the Sierra Nevada, where the chief mineral being attacked is a high Na-feldspar. The correlation of calculation and prediction is interesting if not convincing.

Conclusions

The preceding brief discussion is probably all that is worth presenting at this stage concerning the role of silicate minerals in controlling natural water compositions. But it is perhaps enough to permit some tentative conclusions:

a. Silica in natural waters is commonly a major dissolved constituent and is derived from silicate minerals.

b. The silica in low temperature natural waters is seldom controlled by the pure silica phases, such as quartz and amorphous silica.

c. The reactions of waters with silicates, in terms of release of constituents to solution and the production of new solid phases, is so rapid that the waters remain in near-equilibrium with one or more phases at all times.

d. Because of silicate control, silica concentrations and ratios of cations to H^+ have certain genetic patterns. The simplest of these can be traced under well controlled conditions. Extension of investigations of this kind should eventually permit better prediction of water compositions.

e. The chief buffer system in many natural waters is CO_2 system, but silicate equilibria as well as carbonate equilibria must be considered. It is not inconceivable that the deviation of partial pressures of CO_2 from that of the atmosphere in many stream waters are controlled by silicate equilibria.

SUMMARY

The important but often neglected role of the solid phase in natural water systems has been stressed. In the first part of the paper, some aspects of the chemical and structural variation of $CaCO_3$ and $MgCO_3$ minerals and the influence of these factors on water composition have been discussed. In the second part of the paper, an attempt was made to show that silicate equilibria may exert the dominant control over a number of major constituents in natural waters, and must make an important contribution to hydrogen ion buffering.

ACKNOWLEDGEMENT

This work was supported by research grant from the National Science Foundation (GP4140), the Petroleum Research Fund of the American Chemical Society and the Public Health Service, USDHEW Washington, D.C., Research Grant No. TIES20.

LITERATURE CITED

1. Weber, W.J.Jr. and W. Stumm, J Am Water Works Assoc, $\underline{55}$, 1553 (1963).

2. Latimer, W.M., Oxidation Potentials, 2nd Ed., Prentice-Hall, New York, N.Y., (1952).

3. Jamieson, J.C., J Chem Phys, 21, 1385 (1953).

4. Brooks, R., et al, Proc Royal Soc (London) Philos Trans, 243A, 145 (1950).

5. Faivre, R., Comp Rend, 222, 227 (1946).

6. Cloud, P.E., U S Geol Surv Prof Paper No. 350 (1962).

7. Barnes, I., Geochim Cosmochim Acta, 29, 85 (1965).

8. Back, W., Intern Assoc Sci Hydrol, VIII Annee, No. 3, 43 (1963).

9. Weyl, P.K., J Geol, 69, 32 (1961).

10. Siever, R., et al, J Geol, 73, 39 (1965).

11. Schmalz, R.F. and K.E. Chave, Science, 139, 1206 (1963).

12. Lowenstam, H.A., J Geol., 62, 284 (1954).

13. Daniels, F., Geochim Cosmochim Acta, 22, 65 (1961).

14. Siegel, F.R., Bull Geol Soc Am, 69, 1643 (1958).

15. MacDonald, G.J.F., Am Mineralogist, 41, 744 (1956).

16. Holland, H.D., et al, The Coprecipitation of Metallic Ions With Calcium Carbonate, Annual Report, Contract A.P. (30-1) 2266, Princeton, N.J., (1960).

17. Graf, D.L. and J.R. Goldsmith, Geochim Cosmochim Acta, 1, 109 (1955).

18. Chave, K.E., J Geol, 62, 266 (1954).

19. Chave, K.E., J Geol, 62, 587 (1954).

20. Alderman, A.R. and H.C.W. Skinner, Am Jour Sci, 255, 561 (1957).

21. Graf, D.L. and J.R. Goldsmith, J Geol, 64, 173 (1956).

22. Langmuir, D., Stability of Carbonates in the System: $CaO-MgO-CO_2-H_2O$, PhD Thesis, Harvard University, Cambridge, Mass., (1964).

23. Goldsmith, J.R., et al, Geochim Cosmochim Acta, 7, 212 (1955).

24. Chave, K.E., et al, Science, 137, 33 (1962).

25. Faust, G.T., Am Mineralogist, 38, 4 (1953).

26. Skinner, B.J., Amer Mineralogist, 43, 159 (1958).

27. Graf, D.L., et al, Geol Soc Amer Bull, 70, 1610 (1959).

28. Deffeyes, K.S., et al, Science, 143, 687 (1964).

29. Barnes, I. and W. Back, U S Geol Survey Prof Paper No. 475-D, 179 (1964).

30. Hsu, K.J., J Hydrol, 1, 288 (1963).

31. Feth, J.H., et al, U S Geol Survey Water Supply Paper No. 1535 (1964).

32. Davis, S.N., Am J Sci, 262, 870 (1964).

33. Garrels, R.M. and C.L. Christ, Solutions, Minerals, and Equilibria, Harper and Row, New York, N.Y., (1965).

34. White, D.E., et al, U S Geol Survey Prof Paper No. 440-F (1963).

35. Hill, V.G. and A.C. Ellington, Econ Geol, 56, 533 (1961).

36. Bridgman, S., Am J Sci, 237, 7 (1938).

37. Hess, A., Personal Communication.

22

Reprinted from *U.S. Geol. Survey Jour. Research* 2:233-248 (1974)

WATEQ, A COMPUTER PROGRAM FOR CALCULATING CHEMICAL EQUILIBRIA OF NATURAL WATERS

By ALFRED H. TRUESDELL and BLAIR F. JONES,

Menlo Park, Calif., Washington, D.C.

Abstract. — The computer program, WATEQ, calculates the equilibrium distribution of inorganic aqueous species of major and important minor elements in natural waters using the chemical analysis and *in situ* measurements of temperature, pH, and redox potential. From this model, the states of reaction of the water with solid and gaseous phases are calculated. Thermodynamic stabilities of aqueous species, minerals, and gases have been selected from a careful consideration of all available experimental data. The program is written in PL-1 for IBM 360 computers.

The chemistry of water-rock interactions is determined in part by possible reactions with regard to the states of the water (undersaturated or supersaturated with respect to a solid phase or to a gas at a certain pressure). The reaction states may be calculated from an equilibrium chemical model of the water and from the stabilities of phases with which it may react. The examination of reaction states may suggest the origin of dissolved constituents and assist in the prediction of the chemical effects of ground-water production, recharge, and irrigation. Although the use of inorganic equilibrium models for the processes of mineral solution and precipitation cannot produce a complete description of these processes, an equilibrium model is a useful reference. It can indicate which processes are impossible for a given water-rock system and suggest which processes may control water compositions and which processes are so hindered by kinetic factors that the water compositions are indifferent to them.

Calculations of the states of saturation of natural waters with minerals are complicated by the consideration of all the factors which affect the activity of the ions involved in the solution equilibria. One simple approach for multicomponent water solutions is to assume the existence of complexes whose formation is described by mass-action expressions and to assume that the activity coefficients of simple ions and complexes can be described by equations depending only on the temperature and a function of the water composition, the ionic strength. The number of possible ions, complexes, and minerals and the use of iteration for the solution of simultaneous equations and for the calculation of activity coefficients practically necessitate the use of computer methods.

This report is an attempt to provide a general computer program, for the calculation of chemical equilibria in natural waters at low temperatures, that may be expanded and updated by the user as additional stability data on complexes and minerals become available. The complete computer program is available from the National Technical Information Service, Springfield, VA 22151, as document No. PB-220 464 at a cost of $1.45 per microfiche and $4.85 per paper copy. The study was financed in part by the Defense Advanced Research Projects Agency of the Department of Defense under Order 1813, Amendment 1.

Acknowledgments. — Our thanks are extended to Ivan Barnes, whose earlier program suggested the format, and to C. L. Christ, J. Haas, G. M. Lafon, F. J. Pearson, Jr., Y. Karaka and E. A. Jenne for data and for corrections to the program. We are especially grateful to Manuel Nathenson for checking the thermodynamic data. The thermodynamic approach has been influenced by Garrels and Christ (1965), Sillen and Martell (1964), and Denbigh (1955). Many readers find the approach familiar, and they may wish to omit the next sections in which the minimum thermodynamic theory necessary to explain the calculations is presented.

MASS-ACTION EQUILIBRIUM EQUATIONS

In a mixture at equilibrium, the activities of the chemical species present are related by a set of mass-action equilibrium equations (Garrels and Christ, 1965, p. 6, 342; Denbigh, 1955, p. 138, 307). For each possible reaction of the form,

$$aA + bB = cC + dD, \qquad (1)$$

in which lowercased letters are the stoichiometric coefficients of the chemical species represented by the uppercased letters, there is a mass action equation of the form,

$$K = \frac{[C]^c [D]^d}{[A]^a [B]^b} \qquad (2)$$

In this equation, K is the mass action or equilibrium constant, and the brackets represent activities. For equilibria involving low-pressure gases, the partial pressure of the gas may be used instead of activity, and for gas-aqueous solution equilibria, activities and partial pressures may be used in the same equation.

The equilibrium constants may be derived from experimental measurement of concentrations in a series of equilibrium mixtures of different total concentration with extrapolation to infinite dilution. Alternatively, the experimental concentrations may be corrected to activities by means of calculated activity coefficients (see later discussion). Useful compilations of experimentally derived equilibrium constants have been made by Sillen and Martel (1964), Barnes, Helgeson, and Ellis (1966), Ellis (1967) and Helgeson (1969).

The equilibrium constant for a reaction may also be derived from the standard free energy change of that reaction. For the reaction given by equation 1, the sum of the standard free energies of formation, ΔG°_f, of the products times their stoichiometric coefficients less that of the reactants times their stoichiometric coefficients is the standard free energy change of reaction:

$$\Delta G^\circ_r = c\Delta G^\circ_f, C + d\Delta G^\circ_f, D - (a\Delta G^\circ_f, A + b\Delta G^\circ_f, B). \quad (3)$$

This standard free energy change of reaction is related to the equilibrium constant of the reaction by the equation,

$$\Delta G^\circ_r = -2.303\ RT \log K \quad (4)$$

In which R is the gas constant and T is the absolute temperature. By the use of these equations, experimental equilibrium data may be related to thermochemical data derived from calorimetric measurements. Useful compilations of standard free energies of formation (and other thermochemical data) have been made by the National Bureau of Standards (Rossini and others, 1952; Wagman and others, 1968 and 1969) and by Latimer (1952), Garrels and Christ (1965), Robie and Waldbaum (1968) and Helgeson (1969).

No single source of equilibrium constants or thermochemical data is of sufficient scope or of recent enough publication to include all the data relevant to near-surface rock-water reactions. The data contained in WATEQ (table 1) are from a compilation in preparation by the authors of this program and Manuel Nathenson.

The effect of temperature and pressure on mass action equations will be considered in a later section.

Table 1.—*Reactions and thermodynamic data*

[Log K_{298} (logarithm of equilibrium constant at 298 K) and $\Delta H_{r,298}$ (heat of reaction at 298 K), unless otherwise noted, are calculated from free energies and enthalpies. Data source values are given for the reactions as considered by the original reference, not necessarily as printed here. R and W refer to Robie and Waldbaum (1968); 270-3 and 270-4 refer to Wagman and others (1968) and (1969), respectively. S°, standard state of entropy. Sources for ΔG_f and ΔH_f of individual ion species in solution are all from 270-3 or 270-4, except that Mg^{+2}, Ca^{+2}, Sr^{+2}, Ba^{+2}, and Li^+ are from Latimer (1952), $H_4SiO_4^\circ$ is from Helgeson (1969), and H^+ plus e^- are 0 by definition. ΔG and ΔH values are given in calories]

Reaction No.	Mineral or species name	Reaction	log K	ΔH_r	Data source
0	Fe^{+2}	$Fe^{+2} = Fe^{+3} + e^-$	−13.013	9,700	ΔG_f and ΔH_f from 270-4.
1	$FeOH^{+2}$	$Fe^{+2} + H_2O = FeOH^{+2} + e^- + H^+$	−15.473	20,115	$Fe^{+3} + H_2O = FeOH^{+} + H^+$, log $K = -2.46$; Lamb and Jacques as quoted in Langmuir (1969), ΔH_f from 270-4.
2	$FeOH^+$	$Fe^{+2} + H_2O = FeOH^+ + H^+$	−9.319	13,218	From ΔH_r and ΔS_r of magnetite hydrolysis (Sweeton and Baes, 1970).
3	$Fe(OH)_3^-$	$Fe^{+2} + 3H_2O = Fe(OH)_3^- + 3H^+$	−29.458	32,995	Do.
4	$FeSO_4^+$	$Fe^{+2} + SO_4^{-2} = FeSO_4^+ + e^-$	−8.886	15,920	ΔG_f and ΔH_f from 270-4.
5	$FeCl^{+2}$	$Fe^{+2} + Cl^- = FeCl^{+2} + e^-$	−11.600	18,152	Do.
6	$FeCl_2^+$	$Fe^{+2} + 2Cl^- = FeCl_2^+ + e^-$	−10.919	----------	ΔG_f from 270-4.
7	$FeCl_3^\circ$	$Fe^{+2} + 3Cl^- = FeCl_3^\circ + e^-$	−11.925	----------	Do.
8	$FeSO_4^\circ$	$Fe^{+2} + SO_4^{-2} = FeSO_4^\circ$	2.200	560	Log $K = 2.20$, $\Delta H_r = 560$ (Izatt and others, 1969).
9	Siderite	$FeCO_3 = Fe^{+2} + CO_3^{-2}$	−10.55	−5,328	Langmuir (1969), ΔH_f from R and W.
10	Magnesite	$MgCO_3 = Mg^{+2} + CO_3^{-2}$	−8.029	−6,169	Do.
11	Dolomite	$CaMg(CO_3)_2 = Ca^{+2} + Mg^{+2} + 2CO_3^{-2}$	−17.000	−8,290	Log $K_{298} = -17.0$ (Berner, 1967), $\Delta H_r = -8,290$ (Helgeson, 1969).
12	Calcite	$CaCO_3 = Ca^{+2} + CO_3^{-2}$	−8.370	−3,190	Log $K_{298} = -8.37$ (Berner, 1967), $\Delta H_r = -3,190$ (Helgeson, 1969).

Table 1.—Reactions and thermodynamic data—Continued

Reaction No.	Mineral or species name	Reaction	log K	ΔH_r	Data Source
13	$H_3SiO_4^-$	$H_4SiO_4^\circ = H_3SiO_4^- + H^+$	-9.930	8,935	Log $K=-9.929$, $\Delta H_r=8,935$ from log $K(T)$ expression (Ryzhenko, 1967).
14	$H_2SiO_4^{-2}$	$H_4SiO_4^\circ = 2H^+ + H_2SiO_4^{-2}$	-21.619	29,714	Log $K=-21.617$, $\Delta H_r=29,714$ from log $K(T)$ expression (Ryzhenko, 1967).
15	HPO_4^{-2}	$H^+ PO_4^{-3} = HPO_4^{-2}$	12.346	-3,530	ΔG_f and ΔH_f from 270-3.
16	$H_2PO_4^-$	$2H^+ + PO_4^{-3} = H_2PO_4^-$	19.553	-4,520	Do.
17	Anhydrite	$CaSO_4 = Ca^{+2} + SO_4^{-2}$	-4.637	-3,769	ΔG_f and ΔH_f from R and W.
18	Gypsum	$CaSO_4 \cdot 2H_2O = Ca^{+2} + SO_4^{-2} + 2H_2O$	-4.848	261	Do.
19	Brucite	$Mg(OH)_2 = Mg^{+2} + 2OH^-$	-11.204	850	ΔG_f and ΔH_f from R and W.
20	Chrysotile	$Mg_3Si_2O_5(OH)_4 + 5H_2O = 3Mg^{+2} + 2H_4SiO_4^\circ + 6OH^-$	-51.800	27,585	Log $K=-51.8$ (Hostetler and Christ, 1968), ΔH_r from R and W.
21	Aragonite	$CaCO_3 = Ca^{+2} + CO_3^{-2}$	-8.305	-2,959	ΔG_f and ΔH_f from R and W.
22	MgF^+	$Mg^{+2} + F^- = MgF^+$	1.820	4,674	Log $K=1.82$, $\Delta H_r=24$ (Sillen and Martell, 1964).
23	$CaSO_4^\circ$	$Ca^{+2} + SO_4^{-2} = CaSO_4^\circ$	2.309	1,650	Log $K=2.309$, $\Delta H_r=1,650$ (Bell and George, 1953).
24	$MgOH^+$	$Mg^{+2} + OH^- = MgOH^+$	2.600	2,140	Log $K=2.6$ (Hostetler, 1963); $\Delta H_r=2,140$ (Helgeson, 1969).
25	H_3BO_3	$H_3BO_3 = H^+ + H_2BO_3^-$	-9.240	3,224	Log $K=4.757$-log KW, $\Delta H_r=-10,121 -(\Delta H_r)_{KW}$ from log $K(T)$ expression (Mesmer and others, 1972).
26	NH_3°	$NH_4^+ = NH_3^\circ + H^+$	-9.252	12,480	ΔG_f and ΔH_f from 270-3.
27	Forsterite	$Mg_2SiO_4 + 4H_2O = 2Mg^{+2} + 2H_4SiO_4 + 4OH^-$	-27.694	4,870	ΔG_f and ΔH_f from R and W.
28	Diopside	$CaMgSi_2O_6 + 6H_2O = Ca^{+2} + Mg^{+2} + 2H_4SiO_4^\circ + 4OH^-$	-36.106	21,100	Do.
29	Clinoenstatite	$MgSiO_3 + 3H_2O = Mg^{+2} + H_4SiO_4^\circ + 2OH^-$	-16.658	6,675	Do.
30	$NaHPO_4^-$	$Na^+ + HPO_4^{-2} = NaHPO_4^-$	1.200	----------	Log $K=1.20$ obtained by calculation from data of Smith and Alberty (1956) by using $K_{equiv} = \gamma NaHPO_4^- / (\gamma Na^+ \gamma HPO_4^{-2})$ and K_{approx} and by assuming $\gamma HPO_4^{-2} = \gamma SO_4^{-2} = 0.25$, $\gamma Na^+ = 0.75$, and $\gamma NaHPO_4^- = \gamma Na^+$.
31	Tremolite	$Ca_2Mg_5Si_8O_{22}(OH)_2 + 22H_2O = 2Ca^{+2} + 5Mg^{+2} + 8H_4SiO_4^\circ + 14OH^-$	-139.426	90,215	ΔG_f and ΔH_f from R and W.
32	$KHPO_4^-$	$K^+ + HPO_4^{-2} = MgHPO_4^\circ$	1.090	----------	Log $K=1.09$ obtained by calculation from data of Smith and Alberty (1956) in a similar manner to $NaHPO_4^-$.
33	$MgHPO_4^\circ$	$Mg^{+2} + HPO_4^{-2} = MgHPO_4^\circ$	2.870	3,300	Log $K=2.87$ (Sillen and Martell, 1964), $\Delta H_r=3,300$ by analogy to $CaHPO_4^\circ$ data of Chughtai, Marshall, and Nancollas (1968).
34	$CaHPO_4^\circ$	$Ca^{+2} + HPO_4^{-2} = CaHPO_4^\circ$	2.739	3,300	Log $K=2.739$, $\Delta H_r=3,300$ (Chughtai and others, 1968).
35	HCO_3^-	$H_2CO_3^\circ = HCO_3^- + H^+$	-6.379	1,976	Log $K=-6.379$, $\Delta H_r=1,976$ from log $K(T)$ expression (Ryzhenko, 1963).
36	Sepiolite	$Mg_2Si_3O_{7.5}OH \cdot 3H_2O + 4.5H_2O = 2Mg^{+2} + 3H_4SiO_4^\circ + 4OH^-$	-40.079	26,532	$\Delta G_f = -1,105,600; S^\circ = 90.1$ (Christ and others, 1973).
37	Talc	$Mg_3Si_4O_{10}(OH)_2 + 10H_2O = 3Mg^{+2} + 4H_4SiO_4^\circ + 6OH^-$	-60.933	45,065	ΔG_f from Hostetler and others (1971); ΔH_f from R and W.
38	Hydromagnesite	$Mg_5(CO_3)_4(OH)_2 \cdot 4H_2O = 5Mg^{+2} + 4CO_3^{-2} + 2OH^- + 4H_2O$	-36.762	-25,520	ΔG_f and ΔH_f from Robie and Hemingway (1972).
39	Adularia	$KAlSi_3O_8 + 8H_2O = K^+ + Al(OH)_4^- + 3H_4SiO_4^\circ$	-20.573	30,820	ΔG_f and ΔH_f from R and W.

Table 1.—*Reactions and thermodynamic data*—Continued

Reaction No.	Mineral or species name	Reaction	log K	ΔH_r	Data source
40	Albite	$NaAlSi_3O_8 + 8H_2O = Na^+ + Al(OH)_4^- + 3H_4SiO_4^\circ$	-18.002	25,896	Do.
41	Anorthite	$CaAl_2Si_2O_8 + 8H_2O = Ca^{+2} + 2Al(OH)_4^- + 2H_4SiO_4^\circ$	-19.424	17,530	ΔG_f and ΔH_f from R and W.
42	Analcime	$NaAlSi_2O_6 \cdot H_2O + 5H_2O = Na^+ + Al(OH)_4^- + 2H_4SiO_4^\circ$	-12.701	18,206	Do.
43	K-mica	$KAl_3Si_3O_{10}(OH)_2 + 12H_2O = K^+ + 3Al(OH)_4^- + 3H_4SiO_4^\circ + 2H^+$	-49.102	67,860	Do.
44	Phlogopite	$KMg_3AlSi_3O_{10}(OH)_2 + 10H_2O = K^+ + 3Mg^{+2} + Al(OH)_4^- + 3H_4SiO_4^\circ + 6OH^-$	----------	----------	No data.
45	Illite	$K_{0.6}Mg_{0.25}Al_{2.3}Si_{3.5}O_{10}(OH)_2 + 11.2H_2O = 0.6K^+ + 0.25Mg^{+2} + 2.3Al(OH)_4^- + 3.5H_4SiO_4^\circ + 1.2H^+$	-40.267	54,684	ΔG_f and ΔH_f from Helgeson (1969).
46	Kaolinite	$Al_2Si_2O_5(OH)_4 + 7H_2O = 2Al(OH)_4^- + 2H_4SiO_4^\circ + 2H^+$	-36.921	49,150	Kaolinite $+6H^+ = 2Al^{+3} + 2H_4SiO_4 + H_2O$; log $K=7.185$ (Kittrick, 1966); ΔH_f from R and W.
47	Halloysite	$Al_2Si_2O_5(OH)_4 + 7H_2O = 2Al(OH)_4^- + 2H_4SiO_4^\circ + 2H^+$	-32.830	44,680	ΔG_f and ΔH_f from R and W.
48	Beidellite	$(Na,K,\tfrac{1}{2}Mg)_{0.33}Al_{2.33}Si_{3.67}O_{10}(OH)_2 + 12H_2O = 0.33(Na,K,\tfrac{1}{2}Mg)^+ + 2.33Al(OH)_4^- + 3.67H_4SiO_4^\circ + 2H^+$	-45.272	60,355	ΔG_f and ΔH_f from Helgeson (1969) for Na end member.
49	Chlorite	$Mg_5Al_2Si_3O_{10}(OH)_8 + 10H_2O = 5Mg^{+2} + 2Al(OH)_4^- + 3H_4SiO_4^\circ$	-89.563	54,760	ΔG_f and ΔH_f taken as average of Helgeson (1969) and Zen (1972).
50	Alunite	$KAl_3(SO_4)_2(OH)_6 = K^+ + 3Al^{+3} + 2SO_4^{-2} + 6OH^-$	-85.334	29,820	ΔG_f and ΔH_f from Hemley and others (1969).
51	Gibbsite (crystalline)	$Al(OH)_3 = Al^{+3} + 3OH^-$	-32.774	14,470	ΔG_f and ΔH_f from R and W.
52	Boehmite	$AlO(OH) + H_2O = Al^{+3} + 3OH^-$	-33.416	11,905	Do.
53	Pyrophyllite	$Al_2Si_4O_{10}(OH)_2 + 12H_2O = 2Al(OH)_4^- + 4H_4SiO_4^\circ + 2H^+$	-48.314	----------	$\Delta G_r = 65,900$ from data in tables 4 and 5 in Reesman and Keller (1968).
54	Phillipsite	$Na_{0.5}K_{0.5}AlSi_3O_8 \cdot H_2O + 7H_2O = 0.5Na^+ + 0.5K^+ + Al(OH)_4^- + 3H_4SiO_4^\circ$	-19.874	----------	Log $K=0.7$ for reaction phillipsite + $0.5K^+ = K$-feldspar $+ 0.5Na^+ + H_2O$; ΔG_f of K-feldspar from R and W; (Hess, 1966).
55	Erionite	$NaAlSi_{3.5}O_9 \cdot 3H_2O + 6H_2O = Na^+ + Al(OH)_4^- + 3.5H_4SiO_4^\circ$	----------	----------	No data.
56	Clinoptilolite	$(K,Na)AlSi_5O_{12} \cdot 3.5H_2O + 8.5H_2O = (K,Na)^+ + Al(OH)_4^- + 5H_4SiO_4^\circ$	----------	----------	Do.
57	Mordenite	$(Na,K)AlSi_{4.5}O_{11} \cdot 3H_2O + 8H_2O = (Na,K)^+ + Al(OH)_4^- + 4.5H_4SiO_4^\circ$	----------	----------	Do.
58	Nahcolite	$NaHCO_3 = Na^+ + HCO_3^-$	-0.548	3,720	ΔG_f and ΔH_f from Latimer (1952).
59	Trona	$NaHCO_3 \cdot Na_2CO_3 \cdot 2H_2O = 2H_2O + 3Na^+ + CO_3^{-2} + HCO_3^-$	-0.795	-18,000	From data on natron (this study), nahcolite (Latimer, 1952), and trona-nahcolite-soda in equilibrium at 21.1°C (Linke and Seidell, 1965, p. 925).
60	Natron	$Na_2CO_3 \cdot 10H_2O = 2Na^+ + CO_3^{-2} + 10H_2O$	-1.311	15,745	$Na_2CO_3 \cdot 10H_2O = Na_2CO_3 \cdot H_2O + 9H_2O$ (g); $\Delta G_r = 20,435$; $\Delta H_r = 113,218$ (Waterfield and others, 1968); ΔG_f and ΔH_f of thermonatrite computed in this study.
61	Thermonatrite	$Na_2CO_3 \cdot H_2O = 2Na^+ + CO_3^{-2} + H_2O$	0.125	-2,802	$Na_2CO_3 \cdot H_2O = Na_2CO_3 + H_2O$ (g); $\Delta G_r = 2,944$; $\Delta H_r = 14,037$; Waterfield and others (1968); ΔG_f of Na_2CO_3 from ΔH_f of Latimer (1952) and S° of Waterfield and others (1968).
62	Fluorite	$CaF_2 = Ca^{+2} + 2F^-$	-9.046	1,530	ΔG_f and ΔH_f from R and W.
63	Ca montmorillonite	$Ca_{0.17}Al_{2.33}Si_{3.67}O_{10}(OH)_2 + 12H_2O = 0.17Ca^{+2} + 2.33Al(OH)_4^- + 3.67H_4SiO_4^\circ + 2H^+$	-45.027	58,373	ΔG_f and ΔH_f from Helgeson (1969).
64	Halite	$NaCl = Na^+ + Cl^-$	1.582	918	ΔG_f and ΔH_f from R and W.
65	Thenardite	$Na_2SO_4 = 2Na^+ + SO_4^{-2}$	-0.179	-572	Do.

Table 1.—Reactions and thermodynamic data—Continued

Reaction No.	Mineral or species name	Reaction	log K	ΔH_r	Data source
66	Mirabilite	$Na_2SO_4 \cdot 10H_2O = 2Na^+ + SO_4^{-2} + 10H_2O$	-1.114	18,987	Do.
67	Mackinawite	$FeS + H^+ = Fe^{+2} + HS^-$	-4.648	----------	Log K=-17.566 (Berner, 1967).
68	CO_3^{-2}	$HCO_3^- = H^+ + CO_3^{-2}$	-10.330	3,550	ΔG_f and ΔH_f from 270-3.
69	$NaCO_3^-$	$Na^+ + CO_3^{-2} = NaCO_3^-$	1.268	8,911	Log K=-1.268 (Garrels and others, 1961), ΔH_r=-8,911 (Lafon, 1969).
70	$NaHCO_3^\circ$	$Na^+ + HCO_3^- = NaHCO_3^\circ$	-0.250	----------	Log K=0.25 (Garrels and Thompson, 1962).
71	$NaSO_4^-$	$Na^+ + SO_4^{-2} = NaSO_4^-$	0.226	2,229	Log K=0.226, ΔH_r=308 from log $K(T)$ expression (Lafon and Truesdell, 1971).
72	KSO_4^-	$K^+ + SO_4^{-2} = KSO_4^-$	0.847	3,082	Log K=0.847, ΔH_r=3,082 from log $K(T)$ expression (Truesdell and Hostetler, 1968).
73	$MgCO_3^\circ$	$Mg^{+2} + CO_3^{-2} = MgCO_3^\circ$	3.398	58	Log K=3.398 (Garrels and others, 1961); ΔH_r=58 (Lafon, 1969).
74	$MgHCO_3^+$	$Mg^{+2} + HCO_3^- = MgHCO_3^+$	0.928	10,370	$MgHCO_3^+ = MgCO_3^\circ + H^+$, log K=-7.86 (Hostetler, 1963); ΔH_r=+10,370 (Lafon, 1969).
75	$MgSO_4^\circ$	$Mg^{+2} + SO_4^{-2} = MgSO_4^\circ$	2.238	4,920	Log K=-2.238 (Hanna and others, 1971); ΔH_r=-4,920 (Helgeson, 1969).
76	$CaOH^+$	$Ca^{+2} + OH^- = CaOH^+$	1.400	1,190	Log K=1.40; ΔH_r=1,190 (Sillen and Martell, 1964).
77	$CaHCO_3^+$	$Ca^{+2} + HCO_3^- = CaHCO_3^+$	1.260	6,331	Log K=-1.26 (Garrels and Thompson, 1962); ΔH_r=-6,331 (Lafon, 1969).
78	$CaCO_3^\circ$	$Ca^{+2} + CO_3^{-2} = CaCO_3^\circ$	3.200	3,130	Log K=-3.2 (Garrels and Thompson, 1962); ΔH_r=-3,130 (Helgeson, 1969).
79	$Na_2CO_3^\circ$	$2Na^+ + CO_3^{-2} = Na_2CO_3^\circ$	0.672	----------	Log K=-0.672 (Garrels and Christ, 1965, p. 109)
80	$AlOH^{+2}$	$Al^{+3} + OH^- = AlOH^{+2}$	8.998	1,990	$Al^{+3} + H_2O = AlOH^{+2} + H^+$, log K=-5.00 (Hem and others 1973). ΔH_r=1,990 (Helgeson, 1969).
81	$Al(OH)_2^+$	$Al^{+3} + 2OH^- = Al(OH)_2^+$	18.235	----------	$Al^{+3} + 2H_2O = Al(OH)_2^+ + 2H^+$; log K=-9.76 (Hem and others, 1973).
82	$Al(OH)_4^-$	$Al^{+3} + 4OH^- = Al(OH)_4^-$	33.938	-9,320	$Al(OH)_3$ (microcryst)=Al^{+3}+3OH, log K=32.65; $Al(OH)_3$ (microcryst) +$H_2O = Al(OH)_4^- + H^+$; log K=-12.71. (Hem and Roberson, 1967); ΔH_f from 270-3.
83	AlF^{+2}	$Al^{+3} + F^- = AlF^{+2}$	7.010	----------	Log K=7.01 (Hem, 1968).
84	AlF_2^+	$Al^{+3} + 2F^- = AlF_2^+$	12.750	20,000	Log K=12.75 (Hem, 1968), ΔH_f from 270-3.
85	AlF_3°	$Al^{+3} + 3F^- = AlF_3^\circ$	17.020	2,500	Log K=17.02 (Hem, 1968), ΔH_f from 270-3.
86	AlF_4^-	$Al^{+3} + 4F^- = AlF_4^-$	19.720	----------	Log K=19.72 (Hem, 1968).
87	$AlSO_4^+$	$Al^{+3} + SO_4^{-2} = AlSO_4^+$	3.200	2,290	Log K=3.2 (Hem, 1968); ΔH_r=2,290 (Izatt and others, 1969).
88	$Al(SO_4)_2^-$	$Al^{+3} + 2SO_4^{-2} = Al(SO_4)_2^-$	5.100	3,070	Log K=5.1 (Hem, 1968); ΔH_r=3,070 (Izatt and others, 1969).
89	HSO_4^-	$H^+ + SO_4^{-2} = HSO_4^-$	1.987	4,910	Log K=-1.987, ΔH_r=-4,910 from log $K(T)$ expression (Lietzke and others, 1961).
90	SO_4^{-2}/H_2S	$SO_4^{-2} + 10H^+ + 8e^- = H_2S^+ + 4H_2O$	40.644	65,440	ΔG_f and ΔH_f from 270-3.
91	HS^-	$H_2S = H^+ + HS^-$	-6.994	5,300	Do.
92	S^{-2}	$HS^- = H^+ + S^{-2}$	-12.918	12,100	Do.
93	H_2O/O_2(gas)	$0.5H_2O = 0.25O_2(g) + H^+ + e^-$	-20.780	34,157	Definition.
94	HCO_3^-/CH_4(gas)	$HCO_3^- + 8e^- + 9H^+ = CH_4 + 3H_2O$	30.741	57,435	ΔG_f and ΔH_f from 270-3.

Table 1.—Reactions and thermodynamic data—Continued

Reaction No.	Mineral or species name	Reaction	Log K	ΔH_r	Data source
95	OH apatite	$Ca_5(PO_4)_3(OH) + 3H_2O = 5Ca^{+2} + 3HPO_4^{-2} + 4OH^-$	-59.421	17,225	OH apatite $= 5Ca^{+2} + 3PO_4^{-3} + OH^-$; log $K = -54.408$ (Brown, 1960); ΔH_f from R and W.
96	F apatite	$Ca_5(PO_4)_3F + 3H_2O = 5Ca^{+2} + 3HPO_4^{-2} + 3OH^- + F^-$	-67.243	19,695	ΔG_f and ΔH_f from Roberson (1966).
97	Chalcedony	$SiO_2 + 2H_2O = H_4SiO_4^°$	-3.523	4,615	Log K and ΔH_r obtained from data of R. O. Fournier and J. J. Rowe (in Fournier, 1973).
98	Magadiite	$NaSi_7O_{13}(OH)_3 \cdot 3H_2O + H^+ + 9H_2O = Na^+ + 7H_4SiO_4^°$	-14.300	-----------	Log $K = -14.3$ (Bricker, 1969).
99	Cristobalite	$SiO_2 + 2H_2O = H_4SiO_4^°$	-3.587	5,500	ΔG_f and ΔH_f from R and W.
100	Silica gel	$SiO_2 + 2H_2O = H_4SiO_4^°$	-3.018	4,440	Do.
101	Quartz	$SiO_2 + 2H_2O = H_4SiO_4^°$	-4.006	6,220	Do.
102	$Fe(OH)_2^+$	$Fe^{+2} + 2H_2O = Fe(OH)_2^+ + 2H^+ + e^-$	-20.173	-----------	$FeOH^{+2} + H_2O = Fe(OH)_2^+ + H^+$; log $K = -4.7$, Lamb and Jacques as quoted in Langmuir (1969).
103	$Fe(OH)_3^°$	$Fe^{+2} + 3H_2O = Fe(OH)_3^° + 3H^+ + e^-$	-26.571	-----------	$Fe(OH)_3^° = Fe(OH)_2^+ + OH^-$; log $K = 7.6$, Hem and Cropper as quoted in Langmuir (1969).
104	$Fe(OH)_4^-$	$Fe^{+2} + 4H_2O = Fe(OH)_4^- + 4H^+ + e^-$	-34.894	-----------	Rough estimate from $Fe^{+3} + 4OH^- = Fe(OH)_4^-$; log $K = 34.11$ in 3 M $NaClO_4$ solution (Langmuir, 1969).
105	$Fe(OH)_2^°$	$Fe^{+2} + 2H_2O = Fe(OH)_2^° + 2H^+$	-20.570	28,565	From ΔH_r and ΔS_r of magnetite hydrolysis (Sweeton and Baes, 1970).
106	Vivianite	$Fe_3(PO_4)_2 \cdot 8H_2O = 3Fe^{+2} + 2PO_4^{-3} + 8H_2O$	-36.000	-----------	Vivianite = $3Fe^{+2} + 2PO_4^{-3} + 8H_2O$; log $K = -36$ (Nriagu, 1972b).
107	Magnetite	$Fe_3O_4 + 8H^+ = 3Fe^{+3} + 4H_2O + e^-$	-9.565	-40,660	ΔG_f and ΔH_f from R and W.
108	Hematite	$Fe_2O_3 + 6H^+ = 2Fe^{+3} + 3H_2O$	-4.008	-30,845	Do.
109	Maghemite	$Fe_2O_3 + 6H^+ = 2Fe^{+3} + 3H_2O$	6.386	-----------	Maghemite + $3H_2O = 2Fe^{+3} + 6OH^-$; log $K = -77.6$ (Doyle as quoted in Langmuir, 1969).
110	Goethite	$FeO(OH) + H_2O = Fe^{+3} + 3OH^-$	-41.200	25,555	2 goethite = hematite + H_2O; $\Delta G_r = 545$ (Langmuir, 1971); ΔH_f from R and W.
111	Greenalite	$Fe_3Si_2O_5(OH)_4 + 5H_2O = 3Fe^{+2} + 2H_4SiO_4^° + 6OH^-$	-----------	-----------	No data.
112	$Fe(OH)_3$ (amorphous)	$Fe(OH)_3 + 3H^+ = Fe^{+3} + 3H_2O$	4.891	-----------	$Fe(OH)_3$ (amorphous) $= Fe^{+3} + 3OH^-$; log $K = -37.1$ (Langmuir, 1969).
113	Annite	$KFe_3AlSi_3O_{10}(OH)_2 + 10H_2O = K^+ + 3Fe^{+2} + Al(OH)_4^- + 3H_4SiO_4^°$	-85.645	62,480	ΔG_f and ΔH_f from Helgeson (1969).
114	Pyrite	$FeS_2 + 2H^+ + 2e^- = Fe^{+2} + 2HS^-$	-18.479	11,300	ΔG_f and ΔH_f from R and W.
115	Montmorillonite (Belle Fourche)	$(H,Na,K)_{0.28}Mg_{0.29}Fe^{+3}_{0.23}Al_{1.58}Si_{3.93}O_{10}(OH)_2 + 10.04H_2O = 0.28(H,Na,K)^+ + 0.29Mg^{+2} + 0.23Fe^{+3} + 1.58Al(OH)_4^- + 3.93H_4SiO_4^° + 0.23Fe^{+3}$	-34.913	-----------	Recalculated from data in table 2 of Kittrick (1971a) assuming hydrogen montmorillonite was dissolved in equilibrium with $Fe(OH)_3$ (amorph) rather than hematite.
116	Montmorillonite (Aberdeen)	$(H,Na,K)_{0.42}Mg_{0.45}Fe^{+3}_{0.34}Al_{1.47}Si_{3.82}O_{10}(OH)_2 + 9.16H_2O + 0.84H^+ = 0.42(H,Na,K)^+ + 0.45Mg^{+2} + 0.34Fe^{+3} + 1.47Al(OH)_4^- + 3.82H_4SiO_4^°$	-29.688	-----------	Recalculated from data in table 2 of Kittrick (1971b) assuming hydrogen montmorillonite was dissolved in equilibrium with $Fe(OH)_3$ (amorph) rather than hematite.
117	Huntite	$CaMg(CO_3)_4 = 3Mg^{+2} + Ca^{+2} + 4CO_3^{-2}$	-29.968	-25,760	ΔG_f and ΔH_f from Hemingway and Robie (1972).
118	Greigite	$Fe_3S_4 + 4H^+ + 2e^- = 3Fe^{+2} + 4HS^-$	-18.959	-----------	Log $K = -70.63$ (Berner, 1967).

Table 1.—Reactions and thermodynamic data—Continued

Reaction No.	Mineral or species name	Reaction	Log K	ΔH_r	Data source
119	FeS (precipitate)	$FeS + H^+ = Fe^{+2} + HS^-$	-3.915		Log $K = -16.833$ (Berner, 1967).
120	$FeH_2PO_4^+$	$Fe^{+2} + H_2PO_4^- = FeH_2PO_4^+$	2.700		Log $K = 2.7$ (Nriagu, 1972b).
121	$CaPO_4^-$	$Ca^{+2} + PO_4^{-3} = CaPO_4^-$	6.459	3,100	Log $K = 6.459$, $\Delta H_r = 3,100$ (Chughtai and others, 1968).
122	$CaH_2PO_4^+$	$Ca^{+2} + H_2PO_4^- = CaH_2PO_4^+$	1.408	3,400	Log $K = 1.408$, $\Delta H_r = 3,400$ (Chughtai and others, 1968).
123	$MgPO_4^-$	$Mg^{+2} + PO_4^{-3} = MgPO_4^-$	6.589	3,100	Log K adjusted from $CaPO_4^-$ by using analogy between $CaHPO_4^\circ$ and $MgHPO_4^\circ$; that is, log $K = 6.459 + (2.87 - 2.74) = 6.589$, $\Delta H_r = 3,100$ by analogy with $CaPO_4^-$.
124	$MgH_2PO_4^+$	$Mg^{+2} + H_2PO_4^- = MgH_2PO_4^+$	1.513	3,400	Log K adjusted from $CaH_2PO_4^-$ by using analogy between $CaHPO_4^-$ and $MgHPO_4^\circ$; that is, log $K = 1.408 + (2.87 - 2.74) = 1.513$, $\Delta H_r = 3,400$ by analogy with $CaH_2PO_4^-$.
125	$LiOH^\circ$	$Li^+ + OH^- = LiOH^\circ$	0.200	4,832	$\Delta G_r = -273$, $\Delta H_r = 4,832$ obtained by fitting best straight line in log K vs. $1/T$ plot of data in Sillen and Martell (1964).
126	$LiSO_4^-$	$Li^+ + SO_4^{-2} = LiSO_4^-$	0.640		Log $K = 0.64$ (Sillen and Martell, 1964).
127	NO_3^-/NH_4^+	$NO_3^- + 10H^+ + 8e^- = NH_4^+ + 3H_2O$	119.077	-187,055	ΔG_f and ΔH_r from 270-3.
128	Laumontite	$CaAl_2Si_4O_{12} \cdot 4H_2O + 8H_2O = Ca^{+2} + 2Al(OH)_4^- + 4H_4SiO_4$	-31.053	39,610	ΔG_f and ΔH_f from Zen (1972).
129	$SrOH^+$	$Sr^{+2} + OH^- = SrOH^+$	0.820	1,150	Log $K = 0.82$, $\Delta H_r = 1,150$ (Sillen and Martell, 1964).
130	$BaOH^+$	$Ba^{+2} + OH^- = BaOH^+$	0.640	1,750	Log $K = 0.64$, $\Delta H_r = 1,750$ (Sillen and Martell, 1964).
131	$NH_4SO_4^-$	$NH_4^+ + SO_4^{-2} = NH_4SO_4^-$	1.110		Log $K = 1.110$ (Sillen and Martell, 1964).
132	HCl°	$H^+ + Cl^- = HCl^\circ$	-6.100	18,630	Log $K = -6.1$, $\Delta H_r = 18,630$ (Helgeson, 1969).
133	$NaCl^\circ$	$Na^+ + Cl^- = NaCl^\circ$	-1.602		Log $K = -1.602$ (Hanna and others, 1971).
134	KCl°	$K^+ + Cl^- = KCl^\circ$	-1.585		Log $K = -1.585$ (Hanna and others, 1971).
135	$H_2SO_4^\circ$	$2H^+ + SO_4^{-2} = H_2SO_4^\circ$	-1.000		$H^+ + HSO_4^- = H_2SO_4^\circ$; log $K = -3$, (Sillen and Martell, 1964).
136	H_2O/O_2 (aqueous)	$0.5 H_2O = 0.25 O_2(aq) + H^+ + e^-$	-11.385		Eh = 0.70 from equation (1) of Sato (1960).
137	H_2CO_3	$CO_2(g) + H_2O = H_2CO_3$	-1.452	-5,000	ΔG_f and ΔH_f from 270-3.
138	$FeHPO_4^\circ$	$Fe^{+2} + HPO_4^{-2} = FeHPO_4^\circ$	3.600		Log $K = -3.6$ (Nriagu, 1972b).
139	$FeHPO_4^+$	$Fe^{+2} + HPO_4^{-2} = FeHPO_4^\circ + e^-$	-7.613		$Fe^{+3} + HPO_4^{-2} = FeHPO_4^+$; log $K = 5.4$ (Nriagu, 1971).
140	$Al(OH)_3$ (amorphous)	$Al(OH)_3 = Al^{+3} + 3OH^-$	-31.611	12,990	ΔG_f and ΔH_f from Latimer (1952).
141	Prehnite	$Ca_2Al_2Si_3O_{10}(OH)_2 + 8H_2O + 2H^+ = 2Ca^{+2} + 2Al(OH)_4^- + 3H_4SiO_4^\circ$	-11.695	10,390	ΔG_f and ΔH_f from Zen (1972).
142	Strontianite	$SrCO_3 = Sr^{+2} + CO_3^{-2}$	-11.789	2,361	ΔG_f and ΔH_f from R and W.
143	Celestite	$SrSO_4 = Sr^{+2} + SO_4^{-2}$	-6.349	-1,054	Do.
144	Barite	$BaSO_4 = Ba^{+2} + SO_4^{-2}$	-9.773	6,141	ΔG_f and ΔH_f from R and W.
145	Witherite	$BaCO_3 = Ba^{+2} + CO_3^{-2}$	-13.335	6,950	Do.
146	Strengite	$FePO_4 \cdot 2H_2O = Fe^{+3} + PO_4^{-3} + 2H_2O$	-26.400	-2,030	Log $K = -26.4$ (Nriagu, 1972b); ΔH_f from R and W.
147	Leonhardite	$Ca_2Al_4Si_8O_{24} \cdot 7H_2O + 17H_2O = 2Ca^{+2} + 4Al(OH)_4^- + 8H_4SiO_4^\circ$	-69.756	90,070	ΔG_f and ΔH_f from R and W.
148	$Na_2SO_4^\circ$	$2Na^+ + SO_4^{-2} = Na_2SO_4^\circ$	1.512	-2,642	Log $K = 1.512$, $\Delta H_r = 2,642$ from log $K(T)$ expression in Lafon and Truesdell (1971).

Table 1.—*Reactions and thermodynamic data*—Continued

Reaction No.	Mineral or species name	Reaction	log K	ΔH_r	Data source
149	Nesquehonite	$MgCO_3 \cdot 3H_2O = Mg^{+2} + CO_3^{-2} + 3H_2O$	4.999	−4,619	ΔG_f and ΔH_f from Robie and Hemingway (1972).
150	Artinite	$MgCO_3 \cdot Mg(OH)_2 \cdot 3H_2O = 2Mg^{+2} + CO_3^{-2} + 2OH^- + 3H_2O$	−17.980	498	ΔG_f and ΔH_f from Hemingway and Robie (1972).
151	H_2O/O_2 (aqueous)	$0.5H_2O = 0.25O_2(aq) + H^+ + e^-$	−21.495	33,457	ΔG_f and ΔH_f from 270-3.
152	H_2O	$H_2O = H^+ + OH^-$	−13.998	13,345	Do.
153	Sepiolite (precipitate)	$Mg_2Si_3O_{7.5}(OH) \cdot 3H_2O + 4.5H_2O = 2Mg^{+2} + 3H_4SiO_4^\circ + 4OH^-$	−37.212	----------	Log $K = -37.212$ (Wollast and others, 1968).
154	Diaspore	$AlOOH + H_2O = Al^{+3} + 3OH^-$	−35.121	15,405	ΔG_f and ΔH_f from 270-3.
155	Wairakite	$CaAl_2Si_4O_{12} \cdot 2H_2O + 10H_2O = Ca^{+2} + 2Al(OH)_4^- + 4H_4SiO_4^\circ$	−26.708	26,140	ΔG_f and ΔH_f from Zen (1972).
156	$FeH_2PO_4^{+2}$	$Fe^{+2} + H_2PO_4^- = FeH_2PO_4^{+2} + e^-$	−7.583	----------	$Fe^{+3} + H_2PO_4^- = FeH_2PO_4^{+2}$; log $K = -5.43$ (Nriagu, 1972b).

ACTIVITY COEFFICIENTS

In the limit of infinite dilution, a consequence of the definition of the standard state for ions in solution is that all ionic activities approach ionic concentrations and activity coefficients (defined as the ratios of activities to concentrations) approach unity. This property is useful in experimental studies where mass action expressions are written in which concentrations may be extrapolated to infinite dilution to yield equilibrium constants, but the property gives no clue to activity coefficients in real solutions of finite concentration. In real solutions of more than a few components, it is necessary to use single-ion activities and single-ion activity coefficients. These are formally defined by the equation,

$$a_i = \gamma_i m_i, \quad (5)$$

in which a_i, γ_i, and m_i are respectively the activity, the activity coefficient, and the molality of the ith ion. The convention that activities are dimensionless requires that single-ion activity coefficients have dimensions of molality $^{-1}$.

Single-ion activities and single-ion activity coefficients cannot be defined thermodynamically or exactly measured or calculated because measurement of the activity (and therefore the chemical potential) of a single charged ion would require the measurement of the finite free energy change of the solution resulting from a finite change in concentration of the single charged ion while the concentrations of all other ions and the electrical potential of the phase are held constant. This measurement obviously cannot be made. We must, therefore, use nonthermodynamic models to evaluate single-ion activity coefficients. The reader should be aware of the additional uncertainties introduced by this approach.

Two models have been used in WATEQ for the calculation of single-ion activity coefficients, the Debye-Hückel equation and the MacInnes assumption. These are not the only models available but are perhaps the most widely used and are generally consistent with the functions used to correct experimental determinations to infinite dilution. The Debye-Hückel theory provides an equation which describes single-ion activity coefficient behavior of ions in dilute solutions and which can be extended with adjustable parameters to more concentrated solutions. The MacInnes assumption provides information on the behavior of single-ion activities at higher concentrations with which to fit the parameters of the extended Debye-Hückel equation.

The Debye-Hückel theory

The Debye-Hückel theory considers the effect, on the free energy of a single ion, of electrical interactions with other ions by assuming that oppositely charged ions can be considered as forming a spherical shell around the ion. This assumption is valid only for very dilute solutions, and activity coefficients derived from the theory deviate increasingly from experimental results as the concentration increases. The original equation (Robinson and Stokes, 1955, p. 229) states that,

$$\log \gamma = -\frac{A z^2 \sqrt{I}}{1 + Ba\sqrt{I}} \quad (6)$$

where A and B are constants depending only on the dielectric constant, density, and temperature; z is the ionic charge; and I is the ionic strength (defined as half the sum of the products of the molality and the square of the charge of all ions in the solution) and contains one parameter, a, the "hydrated ion size" that must be estimated from experimental data. The extended form of the equation (Robinson and Stokes, 1955, p. 231),

$$\log \gamma = -\frac{A z^2 \sqrt{I}}{1 + Ba \sqrt{I}} + bI, \quad (7)$$

adds a second adjustable parameter, b, which allows for the effect of the decrease in concentration of solvent in concentrated solutions. This equation is used in WATEQ for major ions with a and b values calculated from experimental mean salt single-ion activity coefficients (see "The MacInnes Assumption") and for minor ions with values of a from Kielland (1937) and b set to zero. The constants A and B are calculated from the dielectric constant, density, and temperature by the equations (Hamer, 1968)

$$A = \frac{1.82483 \times 10^6 d^{1/2}}{(\epsilon T)^{3/2}} \text{ moles}^{-1/2} (10^3 \text{ g H}_2\text{O})^{1/2} \quad (8)$$

and

$$B = \frac{50.2916 \times 10^8 d^{1/2}}{(\epsilon T)^{1/2}} \text{ cm}^{-1} \text{ mole}^{-1/2} (10^3 \text{ g H}_2\text{O})^{1/2} \quad (9)$$

where d is the density of water (Keenan and Keyes, 1936), T is the absolute temperature, and ϵ is the dielectric constant of water (Malmberg and Maryott, 1956; Akerlof and Oshery, 1950).

The MacInnes assumption

In order to assign the adjustable parameters in equation 7, it is necessary to know the variation of single-ion activity coefficients with ionic strength in a single solution. Experimental values are available for the mean molal activity coefficients, $\gamma\pm$, of many salts, and if the activity coefficient of one ion can be calculated, then others may be derived from it. The MacInnes assumption (MacInnes, 1939) that the single-ion activity coefficients of K^+ and Cl^- are equal to each other and to the mean activity coefficient of KCl allows this to be done. By definition,

$$\gamma_+ \gamma_- \equiv \gamma_\pm^2 . \quad (10)$$

If

$$\gamma_\pm \text{KCl} = \gamma_{K^+} = \gamma_{Cl^-}, \quad (11)$$

then

$$\gamma_{Na^+} = \frac{\gamma_\pm^2 \text{NaCl}}{\gamma_\pm \text{KCl}}, \quad (12)$$

$$\gamma_{Ca^{+2}} = \frac{\gamma_\pm^3 \text{CaCl}_2}{\gamma_\pm^2 \text{KCl}}, \quad (13)$$

and

$$\gamma_{Br^-} = \frac{\gamma_\pm^2 \text{KBr}}{\gamma_\pm \text{KCl}}, \quad (14)$$

and so forth.

In deriving these mean salt activity coefficients, one must be careful to avoid solutions in which the ions are highly associated. In calculating $\gamma_{SO_4^{-2}}$, for example, $\gamma_{\pm K_2SO_4}$ cannot be used because of the formation of the KSO_4^- ion pair. In this calculation, the most reasonable values of $\gamma_{SO_4^{-2}}$ can be obtained from $\gamma_{\pm Cs_2SO_4}$, $\gamma_{\pm CsCl}$, and $\gamma_{\pm KCl}$ by the relation,

$$\gamma_{SO_4^{-2}} = \frac{\gamma_\pm^3 Cs_2SO_4 \; \gamma_\pm^2 KCl}{\gamma_\pm^4 CsCl} . \quad (15)$$

Even here, the results must be used with caution because Cs^+ and Cl^- may be weakly associated and $\gamma_{SO_4^{-2}}$ values derived in this way may be somewhat too high at high ionic strengths.

Values of a and b for major ions obtained from computer fitting of calculated mean salt activity coefficients as well as values of a for minor ions derived from Kielland (1937) are shown in table 2.

Table 2.—*Parameters of the Debye-Hückel equation*

Major ions[1]	a	b
Ca^{+2}	5.0	0.165
Mg^{+2}	5.5	.20
Na^+	4.0	.075
K^+	3.5	.015
Cl^-	3.5	.015
SO_4^{-2}	5.0	−.04
HCO_3^-	5.4	.0
CO_3^{-2}	5.4	.0

Minor ions[2]	a
$H_2BO_3^-$, NH_4^+	2.5
NO_3^-	3.0
OH^-, F^-, HS^-	3.5
$MgHCO_3^+$, $H_3SiO_4^-$	4.0
MgF^+, $Al(OH)_4^-$, AlF_4^-, $AlSO_4^+$, $Al(SO_4)_2^-$, HSO_4^-	4.5
$FeOH^{+2}$, $FeOH^+$, $FeSO_4^+$, $FeCl^{+2}$, $FeCl_2^+$, PO_4^{-3}, HPO_4^{-2}, S^{-2}, $LiSO_4^-$, Sr^{+2}, $SrOH^+$, Ba^{+2}, $BaOH^+$, $NH_4SO_4^-$	5.0
$H_2SiO_4^{-2}$, $CaPO_4^-$, $CaH_2PO_4^+$, $MgPO_4^-$, $MgH_2PO_4^+$, $NaCO_3^-$, $NaSO_4^-$, KSO_4^-, $H_2PO_4^-$, $NaHPO_4^-$, $KHPO_4^-$, $AlOH^{+2}$, $Al(OH)_2^+$, AlF^{+2}, AlF_2^+, $Fe(OH)_4^-$, $FeHPO_4^+$, $FeH_2PO_4^+$	5.4
Fe^{+2}, $CaOH^+$, $CaHCO_3^+$, Li^+	6.0
Fe^{+3}, Al^{+3}, H^+	9.0

[1] a and b values calculated from experimental mean salt single-ion activity coefficients.
[2] a values from Kielland (1937); b values set to zero.

Single-ion activity coefficients have been calculated for concentrated single-salt solutions by use of the Stokes-Robinson equation (Bates and others, 1970). Where comparisons are possible, these values agree reasonably with activity coefficients based on mean salt calculations. In table 3, values of single-ion activity coefficients used in WATEQ are compared with mean salt coefficients and with those calculated by Bates and others (1970).

The use of any model of single-ion activity coefficients based on experimental measurements made on single salt solutions requires the assumption that, at a given temperature, activity coefficients in simple solutions are equal to those in complex solutions of the same ionic strength. This assumption is reasonable in dilute solutions, but limited experimental work in concentrated (>1 molal) mixed electrolyte solutions indicates that it is not always true. The extent of deviation from ionic strength dependence is small except for ions that differ greatly in size and hydration such as H^+ and Cs^+. However, for models in which all ion associations are considered (as in WATEQ), these deviations have proved to be insignificant (Pytkowicz and Kester, 1969; Yeatts and Marshall, 1972). For further discussion and comparison of activity coefficient equations, see Truesdell and Jones (1969).

SOLUTION OF MASS ACTION AND MASS BALANCE EQUATIONS

Computation of solution species distribution is accomplished by means of a chemical model (Garrels and Thompson, 1962) which uses analytical concentrations, experimental solution equilibrium constants, mass balance equations, and the measured pH. The distribution of anionic weak acid species is calculated first from total analyzed concentrations, the pH, and activity coefficients of individual species, as illustrated by silicate equilibria,

$$H_4SiO_4 = H^+ + H_3SiO_4^- \qquad (16)$$

and

$$H_3SiO_4^- = H^+ + H_2SiO_4^{-2}. \qquad (17)$$

The concentration of each species is calculated from the total or analytical concentration, the pH, and the activity coefficients of the species. From the preceding equations,

$$K_1 = \frac{m_{H_3SiO_4^-} \, \gamma_{H_3SiO_4^-} \, 10^{-pH}}{m_{H_4SiO_4} \, \gamma_{H_4SiO_4}}, \qquad (18)$$

and

$$K_2 = \frac{m_{H_2SiO_4^{-2}} \, \gamma_{H_2SiO_4^{-2}} \, 10^{-pH}}{m_{H_3SiO_4^-} \, \gamma_{H_3SiO_4^-}}. \qquad (19)$$

The mass balance equation for total silica (silicic acid and silicate ions) is

$$m_{Si\ total} = m_{H_4SiO_4} + m_{H_3SiO_4^-} + m_{H_2SiO_4^{-2}}. \qquad (20)$$

The mass action equations can be combined with the mass balance equation to solve for $m_{H_4SiO_4}$, as follows:

$$m_{H_4SiO_4} = \frac{m_{Si\ total}}{1 + \gamma_{H_4SiO_4} \left(\frac{K_1 \, 10^{pH}}{\gamma_{H_3SiO_4^-}} + \frac{K_1 K_2 \, 10^{2pH}}{\gamma_{H_2SiO_4^{-2}}} \right)}; \qquad (21)$$

$m_{H_4SiO_4}$ is then substituted into the mass action equations to solve for $m_{H_3SiO_4^-}$ and $m_{H_2SiO_4^{-2}}$. The activity coefficients are calculated from the ionic strength by an iterative procedure. The same method is employed for phosphate, borate, and sulfide species and for the carbonate-bicarbonate distribution from pH and the alkalinity determination, after correction for other weak acid radicals (if the alkalinity has been corrected during the chemical analysis, this step may be bypassed in the program). The concentration of H_2CO_3 is calculated from the re-computed bicarbonate molality and the first dissociation constant of carbonic acid.

Table 3.—*Single-ion activity coefficients at 25°C from a two-parameter equation used in WATEQ compared with mean salt, Stokes-Robinson, and other single-ion activity coefficients*

Ionic strength		0.01	0.1	0.5	1.0	2.0	3.0	4.0
γ_{Na^+}	DH[1]	0.903	0.782	0.708	0.715	0.789	0.901	1.043
	MS[2]	.904	.786	.713	.716	.779	.896	1.062
	SR[3]		.783	.701	.697	.756	.870	1.038
γ_{K^+}	DH[1]	.900	.763	.642	.600	.570	.562	.563
	MS[2]	.901	.770	.649	.604	.573	.569	.577
	SR[3]		.773	.659	.623	.610	.626	.659
$\gamma_{Ca^{+2}}$	DH[1]		.670	.389	.266	.247	.289	.376	.509
	MS[2]		.680	.382	.266	.251	.291	.385	.553
	SR[3]			.380	.234	.210	.220	.265	.340
	Davies[4]	.661	.372	.288				
$\gamma_{Mg^{+2}}$	DH[1]		.674	.406	.292	.297	.389	.554	.822
	MS[2]		.685	.400	.289	.293	.380	.567	.945
	SR[3]			.390	.247	.230	.265	.350	.470
γ_{Cl^-}	DH[1]	.900	.763	.642	.600	.570	.562	.563
	MS[2]	.901	.770	.649	.604	.573	.569	.577
	SR[3]		.773	.661	.620	.590	.586	.591
$\gamma_{SO_4^{-2}}$	DH[1]	.667	.371	.205	.155	.112	.091	.077
	MS[2]	.653	.368	.214	.155	.108	.085	.070
$\gamma_{HCO_3^-}$	DH[1]	.905	.788	.692	.654	.623	.606	.596
	WBJ[5]	.904	.790	.692	.654	.627	.600	.580
$\gamma_{CO_3^{-2}}$	DH[1]	.671	.386	.229	.184	.150	.135	.126
	WBJ[5]	.668	.388	.230	.183	.154	....	....

[1] From parameters of Debye-Hückel equation (table 2).
[2] Mean salt.
[3] In chloride solutions (Bates and others, 1970); γ_{Cl^-} from NaCl solution.
[4] No adjustable parameters (Davies, 1962).
[5] From Walker, Bray, and Johnson (1927).

Calculation of the concentrations of ion pairs is accomplished by a procedure similar to that for weak acid species, but the analyzed or computed values for the anion concentrations are utilized in place of the pH and equilibrium association constants are employed. The calculations may be illustrated for the calcium ion species. The major ion-pairing reactions are

$$Ca^{+2} + OH^{-} = CaOH^{+}, \quad (22)$$

$$Ca^{+2} + HCO_3^{-} = CaHCO_3^{+}, \quad (23)$$

$$Ca^{+2} + CO_3^{-2} = CaCO_3^{\circ}, \quad (24)$$

and $\quad Ca^{+2} + SO_4^{-2} = CaSO_4^{\circ}. \quad (25)$

From equations 22–25, equilibrium constants for the association reactions are

$$K_1 = \frac{a_{CaOH^+}}{a_{Ca^{+2}} \, a_{OH^-}}, \quad (26)$$

$$K_2 = \frac{a_{CaHCO_3^+}}{a_{Ca^{+2}} \, a_{HCO_3^-}}, \quad (27)$$

$$K_3 = \frac{a_{CaCO_3^\circ}}{a_{Ca^{+2}} \, a_{CO_3^{-2}}}, \quad (28)$$

and $\quad K_4 = \frac{a_{CaSO_4^\circ}}{a_{Ca^{+2}} \, a_{SO_4^{-2}}}. \quad (29)$

From these equations the expressions,

$$m_{CaOH^+} = \frac{K_1 \, a_{OH^-} \, m_{Ca^{+2}} \, \gamma_{Ca^{+2}}}{\gamma_{CaOH^+}}, \quad (30)$$

$$m_{CaHCO_3^+} = \frac{K_2 \, a_{HCO_3^-} \, m_{Ca^{+2}} \, \gamma_{Ca^{+2}}}{\gamma_{CaHCO_3^{+2}}}, \quad (31)$$

$$m_{CaCO_3^\circ} = \frac{K_3 \, a_{CO_3^{-2}} \, m_{Ca^{+2}} \, \gamma_{Ca^{+2}}}{\gamma_{CaCO_3^\circ}}, \quad (32)$$

and $\quad m_{CaSO_4^\circ} = \frac{K_4 \, a_{SO_4^{-2}} \, m_{Ca^{+2}} \, \gamma_{Ca^{+2}}}{\gamma_{CaSO_4^\circ}} \quad (33)$

may be substituted into the mass balance equation for calcium as follows:

$$m_{Ca\,total} = m_{Ca^{+2}} + m_{CaOH^+} + m_{CaHCO_3^+} + m_{CaCO_3^\circ} + m_{CaSO_4^\circ}$$

to obtain an expression for free (uncomplexed) Ca^{+2} ion,

$$m_{Ca^{+2}} = \frac{m_{Ca\,total}}{1 + \gamma_{Ca^{+2}} \left(\dfrac{K_1 a_{OH^-}}{\gamma_{CaOH^+}} + \dfrac{K_2 a_{HCO_3^-}}{\gamma_{CaHCO_3^+}} + \dfrac{K_3 a_{CO_3^{-2}}}{\gamma_{CaCO_3^\circ}} + \dfrac{K_4 a_{SO_4^{-2}}}{\gamma_{CaSO_4}} \right)} \quad (34)$$

In actuality, these computations in WATEQ also include phosphate species. The computed concentration of free calcium ion, $m_{Ca^{+2}}$, is substituted back into the mass action expressions to solve for the concentrations of ion pairs. The concentrations assigned to ion pairs and weak acids reduce the concentrations of the free ions and change the ionic strength and therefore the activity coefficients. The corrected values are calculated by iteration. In each iteration, the program reduces, if necessary, the molalities of the free anions HCO_3^-, CO_3^{-2}, SO_4^{-2}, Cl^-, F^-, and PO_4^{-3} and recalculates the ionic strength and the activity coefficients. Then the calculations of free Ca^{+2} and Ca complexes along with similar calculations for Na, K, Mg, Fe, and H complexes are repeated. When the sums of all weak acids, complex ions, and free ions for all anions agree with the analytical values within 0.5 percent, the iteration is stopped.

ION RATIOS

When the chemical model is complete, it is useful to calculate molal concentration ratios and ion activity ratios for plotting on water composition and mineral stability diagrams, respectively. Comparison of these ratios with those of related waters can suggest possible origins of dissolved constituents and possible controls by mineral reactions. A number of these ratios are calculated in WATEQ.

ACTIVITY PRODUCTS AND SOLUBILITY PRODUCTS

The equilibrium of a solid phase with an aqueous solution can be characterized by a mass action equation. For a solid of formula AX which dissolves to form ions A^+ and X^-, this expression is

$$K = \frac{a_{A^+} a_{X^-}}{a_{AX}}, \quad (35)$$

where K is the equilibrium constant of solubility. If the solid is a pure substance, not a solid solution, its activity is equal to one because it is in its standard state (Garrels and Christ, 1965, p. 5), and the expression for the equilibrium constant reduces to the "solubility product,"

$$K_{SP} = a_{A^+} a_{X^-}. \quad (36)$$

In hydrolysis reactions, water is considered explicitly as part of the reaction. In the solution of quartz to form silicic acid, for example,

$$SiO_{2\,(quartz)} + 2H_2O = H_4SiO_4, \quad (37)$$

the water is written as part of the reaction, and its activity appears in the equilibrium expression.

A water sample when collected is usually no longer in contact with mineral phases, and these phases may not be accessible to observation. It is of interest then to determine with what mineral phases the water is saturated or nearly so. The calculated activities of the dissolved ions in a water may be combined to produce the appropriate activity product which may be compared with the solubility equilibrium constant to show the degree of saturation of the water with each mineral considered.

This comparison may be made by means of the ratio of the activity product to the equilibrium solubility product which is given in the program as "AP/K" and "LOG AP/K" and by means of the free energy change of the reaction ΔG_r (which is zero at equilibrium) and is given as "DELGR" in the program. These quantities are related by the expression

$$\Delta G_r = 2.303 RT \log(AP/K). \quad (38)$$

Some mineral formulas contain a relatively large number of atoms, and the ΔG_r values for these minerals will deviate from zero more rapidly with dilution of concentration than will those values for minerals with simple formulas. This deviation can be illustrated by comparing the activity product of dolomite, $a_{Ca^{+2}}\, a_{Mg^{+2}}\, a^2_{CO_3^{-2}}$ with that of calcite, $a_{Ca^{+2}}\, a_{CO_3^{-2}}$. If a water initially saturated with both minerals is diluted with pure water, ΔG_r dolomite will be twice ΔG_r calcite. To correct this, ΔG_r values are divided by the number of negative charges in the formula of the mineral and presented as (for want of a better label) "PER EQUIV" ΔG_r.

The compilation of a consistent set of stability constants for minerals suffers from several uncertainties. The standard enthalpy of formation and standard entropy of most minerals have been measured by calorimetric methods, and the standard free energy of formation calculated from these quantities is often referenced to the free energies of formation of the elements rather than to the free energies of formation of the ions formed on solution of the mineral. The combination of such values with those for solution species involving aqueous ions may lead to erroneous stability constants. The use of experimental solubility products or resulting free energy values is free from this inconsistency. The main uncertainty in the use of these data lies in the precise definition of reactants and products involved in the experiment and in the difficulty of reversing the equilibrium.

Because of these uncertainties, the logarithms of the maximum and minimum solubility products are calculated in WATEQ and presented in addition to the logarithm of the most probable value for visual comparison with the logarithm of the activity product. Because of space limitations only the most probable solubility product is used in calculating values of AP/K, log (AP/K), ΔG_r, and ΔG_r per equivalent. Enthalpy values and solubility products used in the program, together with the sources of all data, are given in table 1.

EFFECTS OF TEMPERATURE AND PRESSURE

In the relationships developed in the previous sections, temperature and pressure have been assumed to be constant and their effect on the equilibria has not been discussed. The great majority of experimental determinations of equilibrium constants and of free energy values has been made at 25°C and, particularly for solution equilibria, data at other temperatures may be entirely lacking. If experiments have been made over a wide range of temperatures or if complete thermochemical data are available for all species of a reaction, then the equilibrium constant may be expressed as a power function of the absolute temperature

$$\log K = A + BT + C/T + D \log T, \quad (39)$$

in which one or more coefficients may be zero. Where this type of expression was available in the literature, it has been used in WATEQ (table 4). If experimental determinations at only two or three temperatures are available, a linear dependence of log K with the reciprocal of the absolute temperature may be indicated (that is, B and D are zero in eq. 39) which is equivalent to a constant value of the enthalpy (heat content) change of the reaction, ΔH. This is expressed by the Van't Hoff relation,

$$\log K = \log K_{Tr} - \frac{\Delta H_{Tr}}{2.3R}\left(\frac{1}{T} - \frac{1}{Tr}\right), \quad (40)$$

in which Tr is the reference temperature (298.15 K (= 25°C) in WATEQ) and the constants A and C in equation 39 are equal to

$$\log K_{Tr} + \frac{\Delta H_{Tr}}{2.3RT} \text{ and } \frac{\Delta H_{Tr}}{2.3R},$$

respectively.

The enthalpy change of reaction can be obtained by determining the slope of a plot of experimental values of log K versus (1/T), from tabulated values of the standard enthalpy of formation of the species in the reaction by using a relation analogous to equation 3 or from direct measurements. The enthalpy of reaction at 25°C has been calculated for most of the equilibria used in WATEQ (table 1), and equation 40 is used to calculate the value of the equilibrium constant for the

Table 4. *Analytical expressions for log* $K(T)$ *used in WATEQ*

[T, in kelvins]

Identifier	Reaction	Expression for log $K(T)$	Reference
KT(13)	$H_4SiO_4° = H_3SiO_4^- + H^+$	$6.368 - 0.016346\ T - 3405.9/T$	Ryzhenko (1967).
KT(14)	$H_4SiO_4° = H_2SiO_4^{-2} + 2H^+$	$39.478 - 0.065927\ T - 12355.1/T$	Do.
KT(25)	$H_3BO_3° = H_2BO_3^- + H^+$	$1573.21/T + 28.6059 + 0.012078\ T$ $-13.2258 \log T + \log K_W$.	Mesmer, Baes, and Sweeton (1972).
KT(26)	$NH_4^+ = NH_3° + H^+$	$0.6322 - 0.001225\ T - 2835.76/T$	Wright, Lindsay, and Druga (1961).
KT(35)	$H_2CO_3° = HCO_3^- + H^+$	$8.153 - 0.02194\ T - 2382.3/T$	Ryzhenko (1963).[1]
KT(68)	$HCO_3^- = H^+ + CO_3^{-2}$	$5.388 - 0.02199\ T - 2730.7/T$	Do.
KT(72)	$K^+ + SO_4^{-2} = KSO_4^-$	$3.106 - 673.6/T$	Truesdell and Hostetler (1968).
KT(89)	$H^+ + SO_4^{-2} = HSO_4^-$	$-5.3505 + 0.0183412\ T + 557.2461/T$	Lietzke, Stoughton, and Young (1961).
KT(91)	$H_2S° = H^+ + HS^-$	$11.17 - 0.02386\ T - 3279/T$	D'yachkova and Khodakovskiy (1968).

[1] In more recent practice, the expressions based on the original work of Harned and co-workers (Harned and Owen, 1958) have been utilized, despite being limited to 50°C maximum temperature.

temperature of the water. For a few reactions in which data at temperatures other than 25°C were not available, the 25°C value of the equilibrium constant is used at all temperatures.

The effect of pressure has not been calculated in WATEQ because the necessity of inputing a measured pH value virtually limits WATEQ to surface and near-surface waters and because much necessary data is not available for ion pairs. Correlations suggested by Ellis and McFadden (1972) allow the calculation of the pressure effect on equilibria involving only minerals and simple ions (not ion pairs) to be made for temperatures to 250°C. These calculations suggest that pressure effects are not large for pressures less than a few hundred atmospheres.

REDOX REACTIONS

Oxidation-reduction equilibria have been treated in the same manner as other reactions in WATEQ. To achieve this, the measured Eh value, or the Eh value calculated from the measured concentration of dissolved oxygen, is converted to the negative logarithm of the conventional activity of the electron (or pE) by the relation,

$$pE \equiv Eh/(2.303RT/F), \quad (41)$$

in which F is the faraday and $2.303RT/F$ is the Nernst slope. pE is related to the conventional activity of the electron by

$$a_{e^-} = 10^{-pE} \quad (42)$$

This equation is similar to that assumed for pH, and because both measurements have an unknown liquid junction potential, the relations of pE to electron activity and of pH to hydrogen ion activity are equally uncertain. It is necessary, however, to use these relations despite the uncertainty. The standard free energy and enthalpy of the hydrated electron in aqueous solution are zero by convention. The conventional electron activity thus ranges from 10^{-20} to 10^{+20} while the actual electron activity is about 10^{-60} to 10^{-100}. These conventions are discussed by Sillen and Martell (1964) and by Truesdell (1968).

An advantage of the use of electron activity is that it is not necessary to set up separate redox equilibrium expressions. For example, the equilibrium between Fe^{+2} and Fe^{+3} is expressed by a conventional equilibrium constant,

$$K = \frac{a_{Fe^{+3}}\ a_{e^-}}{a_{Fe^{+2}}} \quad (43)$$

and the value of the equilibrium constant may be calculated from $G_f°, Fe^{+3}$, and $G_f°, Fe^{+2}$ ($G_f°$, electron = 0 by convention). Other redox equilibria are treated similarly, and the method of calculation of the concentration of ion pairs involving iron is the same as that involving metals that are not redox active.

In natural waters that contact the atmosphere the dissolved oxygen (DOX) content may have been measured in addition to or in place of the Eh. If the dissolved oxygen has been measured, it is read into the program after the normal data as a statement, "DOX = (parts per million of dissolved oxygen),". Two values of pE are calculated in WATEQ from the relation,

$$pE = -\log K - pH - 0.5 \log a_{H_2O} + 0.25 \log a_{DOX}, \quad (44)$$

in which log K values are from thermodynamic data ("PE CALC O") and from the empirical Eh-pH relation for waters in contact with the atmosphere (Sato, 1960). ("EMPIR PE O") and DOX activities are on a molal scale. If a DOX measurement is given without an Eh value, the value of PE CALC O is used throughout the program. If instead, EMPIR PE O is to be adopted, the statement "EMPOX = 1" is added to the optional data.

Separate analyses of reduced and oxidized species allow the calculation of pE values which may be compared with each other or the measured pE to estimate the degree of internal redox equilibrium. Two such pairs are sulfide-sulfate and ammonia-nitrate. The equilibrium between sulfide and sulfate can be written,

$$H_2S + 4H_2O = SO_4^{-2} + 10H^+ + 8e^-, \quad (45)$$

and the mass action expression can be rearranged to give

$$pE = (\log K + \log a_{SO_4^{-2}} - \log a_{H_2S} - 10pH - 4\log a_{H_2O})/8. \quad (46)$$

Similarly, the equilibrium between ammonium and nitrate yields the expression

$$pE = (-\log K + \log a_{NO_3^-} - \log a_{NH_4^+} - 10pH - 3\log a_{H_2O})/8. \quad (47)$$

These quantities, PE CALC S and PE CALC N, are calculated in WATEQ.

GAS PARTIAL PRESSURES

Although gas partial pressures are seldom measured in natural waters, in some solutions they may be calculated from the gas solubility constants and the water analysis. The partial pressures of CO_2, O_2, and CH_4 are calculated from the following equations,

$$\log P_{CO_2} = \log K + \log a_{HCO_3^-} + \log a_{H^+} - \log a_{H_2O}, \quad (48)$$

$$\log P_{O_2} = \log K' + 2\log a_{H_2O} + 4pH + 4pE, \quad (49)$$

$$\log P_{CH_4} = \log K'' + \log a_{HCO_3^-} - 9pH - 9pE - 3\log a_{H_2O}. \quad (50)$$

ACTIVITY OF WATER

The activity of water is calculated in WATEQ by the approximate relation (Garrels and Christ, 1965, p. 66)

$$a_{H_2O} = 1 - 0.017 \Sigma m_i. \quad (51)$$

where Σm_i is the sum of the molalities of dissolved anions, cations, and neutral species. The equation yields reasonable values if Σm_i is less than molal.

INPUT

Input to WATEQ consists of a complete chemical analysis of the water sample and field measurements of its temperature and pH. If available, measurements of Eh and dissolved oxygen as well as some trace element analyses may be included. In order to allow the inclusion of optional data, the last space on the first card is coded with ISTDATA which is the number of cards containing the necessary data including the normal chemical analysis and the sample description. Cards after the chemical analysis are used for optional data. A blank card must be included after each data set to separate data sets. The required data is coded in free field (that is, one space between each number) in the order indicated below. See list of identifiers in the complete computer program for detailed descriptions. Sample sets of data are given with the resulting printout after the program.

Card 1 .... Sample Description (79 spaces) and ISTDATA (space 80).
Card 2 .... TEMP, PH, EHM (in volts, code 9.9 if data is not available), FLAG (='PPM', 'MG/L', 'MEQ/L' or 'MOL').
Card 3 .... Chemical analysis in PPM, MG/L, and MEQ/L or MOL/L (set FLAG) in the order Ca, Mg, Na, K, Cl, SO_4, HCO_3, Fe, H_2S, CO_3, SiO_2, NH_4, B, PO_4, Al, F, NO_3.
Succeeding cards . . . . Other data (identifier, equality sign, numerical value, and comma) including: "DENS=☐," (if not specified, density is set equal to 1); if alkalinity is corrected for noncarbonate alkalinity "CORALK=1," (omitted if not corrected); electrical potential (volts) of the Eh cell including the calomel reference electrode "EHMC=☐,"; electrical potential (volts) of the Eh cell with Zobell's solution for calibration, "EMFZSCE=☐,"; parts per million of dissolved oxygen, "DOX=☐,"; and certain trace elements including Li (I=80), Sr (I=87), Ba (I=89) in the form, "CUNITS(I)=☐,". A semicolon in place of a comma follows the last data statement.
Last card .... Blank.

REFERENCES CITED

Akerlof, G. C., and Oshery, H. I., 1950, The dielectric constant of water at high temperatures and in equilibrium with its vapor: Am. Chem. Soc. Jour., v. 72, p. 2844–2847.

Barnes, H. L., Helgeson, H. C., and Ellis, A. J., 1966, Ionization constants in aqueous solutions: Geol. Soc. America Mem. 97, 401–413.

Bates, R. G., Staples, B. R., and Robinson, R. A., 1970, Ionic hydration and single ion activities in unassociated chlorides at high ionic strength: Anal. Chemistry, v. 42, No. 8, p. 867–871.

Bell, R. P., and George, J. H. B., 1953, The incomplete dissociation of some thallous and calcium salts at different temperatures: Faraday Soc. Trans., v. 49, p. 619–627.

Berner, R. A., 1967, Comparative dissolution characteristics of carbonate minerals in the presence and absence of aqueous magnesium ion: Am. Jour. Sci., v. 265, p. 45–70.

——1967, Thermodynamic stability of iron sulfides: Am. Jour. Sci., v. 265, p. 773–785.

Bricker, O. P., 1969, Stability constants and Gibbs free energies of formation of magadiite and kenyaite: Am. Mineralogist, v. 54, p. 1026–1033.

Brown, W. E., 1960, Behavior of slightly soluble calcium phosphates as revealed by phase-equilibrium calculations: Soil Sci., v. 90, p. 51–57.

Christ, C. L., Hostetler, P. B., and Siebert, R. M., 1973, Studies in the system MgO-SiO_2-CO_2-H_2O(III): the activity-product constant of sepiolite: Am. Jour. Sci., No. 293, p. 65–83.

Chughtai, A., Marshall, R., and Nancollas, G. H., 1968, Complexes in calcium phosphate solutions: Jour. Phys. Chemistry, v. 72, p. 208–211.

Davies, C. W., 1962, Ion Association: Washington, Butterworths, 190 p.

Denbigh, Kenneth, 1955, The Principles of Chemical Equilibrium: Cambridge, England, University Press, 491 p.

D'yachkova, I. B., and Khodakovskiy, I. L., 1968, Thermodynamic equilibria in the systems S-H_2O, Se-H_2O and Te-H_2O in the 25–300°C temperature range and their geochemical interpretatons: Geochemistry Internat., p. 1108–1125.

Ellis, A. J., 1967, The chemistry of some explored geothermal systems, in Barnes, H. L., ed., Geochemistry of hydrothermal ore deposits: New York, Holt, Rinehart, Winston, p. 465–514.

Ellis, A. J., and McFadden, I. M., 1972, Partial molal volumes of ions in hydrothermal solutions: Geochim. et Cosmochim. Acta, v. 36, p. 413–426.

Fournier, R. O., 1973, Silica in thermal waters: Laboratory and field investigations, in Proceedings of International Symposium on Hydrogeochemistry and Biochemistry, Japan 1970: Washington, D.C., J. W. Clarke, p. 122–139.

Garrels, R. M., and Christ, C. H., 1965, Solutions, minerals, and equilibria: New York, Harper & Row, 450 p.

Garrels, R. M., and Thompson, M. E., 1962, A chemical model for sea water at 250°C and one atmosphere total pressure: Am. Jour. Sci., v. 260, p. 57–66.

Garrels, R. M., Thompson, M. E., and Siever, Raymond, 1961, Control of carbonate solubility by carbonate complexes: Am. Jour. Sci., v. 259, p. 24–25.

Hamer, W. J., 1968, Theoretical mean activity coefficients of strong electrolytes in aqueous solutions from 0 to 100°C: U.S. Natl. Bur. Standards, Natl. Standard Reference Data Ser. 24, 271 p.

Hanna, E. M., Pethybridge, A. D., and Prue, J. E., 1971, Ion association and the analysis of precise conductimetric data: Electrochim. Acta, v. 16, p. 677–686.

Harned, H. S., and Owen, B. B., 1958, The physical chemistry of electrolytic solutions [3d ed.]: New York, Reinhold, 803 p.

Helgeson, H. C., 1969, Thermodynamics of hydrothermal systems at elevated temperatures and pressures: Am. Jour. Sci., Vo. 267, p. 729–804.

Hem, J. D., 1968, Graphical methods for studies of aqueous aluminum hydroxide, fluoride and sulfate complexes: U.S. Geol. Survey Water-Supply Paper 1827-B, 33 p.

Hem, J. D., and Roberson, C. E., 1967, Form and stability of aluminum hydroxide complexes in dilute solution: U.S. Geol. Survey Water-Supply Paper 1827-A, 55 p.

Hem, J. D., Roberson, C. E., Lind, C. J., and Polzer, W. L., 1973, Chemical interactions of aluminum with aqueous silica at 25°C: U.S. Geol. Survey Water-Supply Paper 1827-E, 57 p.

Hemingway, B. S., and Robie, R. A., 1973, A calorimetric determination of the standard enthalpies of formation of huntite, $CaMg_3(CO_3)_4$, and artinite, $Mg_2(OH)_2CO_3 \cdot 3H_2O$, and their standard Gibbs free energies of formation: U.S. Geol. Survey Jour. Research, v. 1, No. 5, p. 535–541.

Hemley, J. J., Hostetler, P. B., Gude, A. J., and Mountjoy, W. T., 1969, Some stability relations of alunite: Econ. Geology, v. 64, p. 599–612.

Hess, P. C., 1966, Phase equilibria of some minerals in the K_2O-NA_2O-Al_2O_3-SiO_2-H_2O system at 25°C and 1 atmosphere: Am. Jour. Sci., v. 264, p. 289–309.

Hostetler, P. B., 1963a, Complexing of magnesium with bicarbonate: Jour. Phys. Chemistry, v. 67, p. 720–721.

———1963b, The stability and surface energy of brucite in water at 25°C: Am. Jour. Sci., v. 261, p. 238–258.

Hostetler, P. B., and Christ, C. L., 1968, Studies in the system MgO-SiO_2-CO_2-H_2O (I): The activity-product constant of chrysotile: Geochim. et Cosmochim. Acta, v. 32, p. 485–497.

Hostetler, P. B., Hemley, J. J., Christ, C. O., and Montoya, J. J., 1971, Talc-chrysotile equilibrium in aqueous solutions: Geol. Soc. America Abs, with Programs v. 3, p. 605.

Izatt, R. M., Eatough, Delbert, Christensen, J. J., and Bartholomew, C. H., 1969, Calorimetrically determined log K, $\Delta H°$, and $\Delta S°$ values for the interaction of sulphate ion with several bi- and ter-valent metal ions: Chem. Soc. [London] Jour., sec. A, pt. 1, p. 47–53.

Keenan, J. H., and Keys, Frederick, 1936, Thermodynamic properties of steam: New York, John Wiley and Sons, Inc., 89 p.

Kielland, J., 1937, Individual activity coefficients of ions in aqueous solutions: Am. Chem. Soc. Jour., v. 59, p. 1675–1678.

Kittrick, J. A., 1966, Free energy of formation of kaolinite from solubility measurements: Am. Mineralogist, v. 51, p. 1457–1466.

———1971a, Stability of montmorillonites: I. Belle Fourche and Clay Spur montmorillonites: Soil Sci. Soc. America Proc., v. 35, p. 140–145.

———1971b, Stability of montmorillonites: II. Aberdeen montmorillonite: Soil Sci. Soc. America Proc., v. 35, p. 820–823.

Lafon, G. M., 1969, Some quantitative aspects of the chemical evolution of the oceans: Northwestern Univ., Ph. D. dissert., 136 p.

Lafon, G. M., and Truesdell, A. H., 1971, Temperature dependence of sodium sulfate complexing in aqueous solutions [abs.]: Am. Geophys. Union Trans., v. 52, p. 362.

Langmuire, Donald, 1969, The Gibbs free energies of substances in the system Fe-O_2-H_2O-CO_2 at 25°C, in Geological Survey research 1969: U.S. Geol. Survey Prof. Paper 650-B, p. B180–B183.

———1971, Particle size effect on the reaction geothite-hematite + water: Am. Jour. Sci, v. 271, p. 147–156.

Latimer, W. M., 1952, The oxidation states of the elements and their potential in aqueous solutions [2d ed.]: New York, Prentice-Hall, Inc., 392 p.

Lietzke, M. H., Stoughton, R. W., and Young, T. F., 1961, The bisulfate acid constant from 25 to 225° as computed from solubility data: Jour. Phys. Chemistry, v. 65, p. 2247–2249.

Linke, M. F., and Seidell, A., 1965, Solubilities of inorganic and metal-organic compounds: Washington, Am. Chem. Soc., v. 2, 1914 p.

MacInnes, D. A., 1939, The principles of electrochemistry: New York, Reinhold, 478 p.

Malmberg, C. G., and Maryott, A. A., 1956, Dielectric constant of water from 0° to 100°C: U.S. Natl. Bur. Standards Jour. Research, v. 56, p. 1–8.

Mesmer, R. E., Baes, C. F., Jr., and Sweeton, F. H., 1972, Acidity measurements at elevated temperatures: VI. Boric acid equilibria: Inorganic Chemistry, v. 11, No. 3, p. 537–543.

Nriagu, J. O., 1971, Solubility studies on vivianite and strengite: Geol. Soc. America Abs. with Programs, v. 3, p. 662.

———1972a, Solubility equilibrium constant of strengite: Am. Jour. Sci., v. 272, p. 476–484.

———1972b, Stability of vivianite and ion pair formation in the system $Fe_3(PO_4)_2$-H_3PO_4-H_2O: Geochim. et Cosmochim. Acta, v. 36, p. 459–470.

Pytkowicz, R. M., and Kester, D. R., 1969, Harned's rule behavior of NaCl-Na_2SO_3 solutions explained by an ion-association model: Am. Jour. Sci., v. 267, p. 217–229.

Reesman, A. L., and Keller, W. D., 1968, Aqueous solubility studies of high-alumina and clay minerals: Am. Mineralogist, v. 53, p. 929–942.

Roberson, C. E., 1966, Solubility implications of apatite in sea water, in Geological Survey research 1966: U.S. Geol. Survey Prof. Paper 550-D, p. D178–D185.

Robie, R. A., and Hemingway, B. S., 1973, The enthalpies of formation of nesquehonite, $MgCO_3 \cdot 3H_2O$, and hydromagnesite, $5MgO \cdot 4CO_2 \cdot 5H_2O$: U.S. Geol. Survey Jour. Research, v. 1, No. 5, p. 543–547.

Robie, R. A., and Waldbaum, D. R., 1968, Thermodynamic properties of minerals and related substances at 298.15°K (25.0°C) and one atmosphere (1.013 bars) pressure and at higher temperatures: U.S. Geol. Survey Bull. 1259, 256 p.

Robinson, R. A., and Stokes, R. H., 1955, Electrolyte solutions: London, Butterworths Scientific Publications, 559 p.

Rossini, F. D., Wagman, D. D., Evans, W. H., Levine, Samuel, and Jaffe, Irving, 1952, Selected values of chemical thermodynamic properties: U.S. Natl. Bur. Standards Circ. 500, 1,268 p. Part I. Tables, p. 1–822, reprinted in 1961.

Ryzhenko, B. N., 1963, Determination of dissociation constants of carbonic acid and the degree of hydrolysis of the CO_3^{-2} and HCO_3^- ions in solutions of alkali carbonates and bicarbonates at elevated temperatures: Geochemistry, No. 2, p. 151-163; translated from Geokhimya.

———1967, Determination of the hydrolysis of sodium silicate and calculation of dissociation constants of orthosilicic acid at elevated temperatures: Geochemistry Internat., v. 4, p. 99–107.

Sato, Motaki, 1960, Oxidation of sulfide ore bodies: Econ. Geology, v. 55, p. 928–1231.

Sillen, L. G., and Martell, A. E., 1964, Stability complexes of metal-ion complexes: Chem. Soc. [London] Spec. Pub. 17, 754 p.

Smith, R. M., and Alberty, R. A., 1956, The apparent stability constants of ionic complexes of various adenosine phosphates with monovalent cations: Jour. Phys. Chemistry, v. 60, p. 180–184.

Sweeton, F. H., and Baes, C. F., Jr., 1970, The solubility of magnetite and hydrolysis of ferrous iron in aqueous solutions at elevated temperatures: Jour. Chem. Thermodynamics, v. 2, No. 4, p. 479-500.

Truesdell, A. H., 1968, The advantage of using pE rather than Eh in redox equilibrium calculations: Jour. Geol. Education, v. 16, p. 17–20.

Truesdell, A. H., and Hostetler, P. B., 1968, Dissociation constants of KSO_4^- from 10° to 50°C: Geochim. et Cosmochim. Acta, v. 32, p: 1019–1022.

Truesdell, A. H., and Jones, B. F., 1969, Ion association in natural brines: Chem. Geology, v. 4, p. 51–62.

Wagman, D. D., Evans, W. H., Parker, V. B., Halow, I., Baily, S. M., and Schumm, R. H., 1968, Selected values of chemical thermodynamic properties: U.S. Natl. Bur. Standards Tech. Note 270-3, 264 p.

———1969, Selected values of chemical thermodynamic properties: U.S. Natl. Bur. Standards Tech. Note 270-4, 141 p.

Walker, A. C., Bray, U. B., and Johnson, John, 1927, Equilibrium in solutions of alkali carbonates: Am. Chem. Soc. Jour., v. 49, p. 1235–1256.

Waterfield, C. G., Linford, R. G., Goalby, B. B., Bates, T. R., Elyard, C. A., and Staveley, L. A. K., 1968, Thermodynamic investigation of disorder in the hydrates of sodium carbonate: Faraday Soc. Trans., v. 64, p. 868–874.

Wollast, Roland, Mackenzie, F. T., and Bricker, O. P., 1968, Experimental precipitation and genesis of sepiolite at Earth-surface conditions: Am. Mineralogist, v. 53, p. 1645–1662.

Wright, J. M., Lindsay, W. T., Jr., and Druga, T. R., 1961, The behavior of electrolytic solutions at elevated temperatures as derived from conductance measurements: Washington, D.C., U. S. Atomic Energy Comm. R & D rept. WAPD-TM-204, 32 p.

Yeatts, L. B., and Marshall, W. L., 1972, Solubility of calcium sulfate dihydrate and association equilibria in several aqueous mixed electrolyte salt systems at 25°C: Jour. Chem. and Eng. Data, v. 17, p. 163–168.

Zen, E-an, 1972, Gibbs energy, enthalpy, and entropy of ten rock-forming minerals, calculations, discrepancies, implications: Am. Mineralogist, v. 57, p. 524–553.

ERRATA

In Table 1, Reaction No. 6, the Mineral or species name, should read: "$FeCl_2^+$"

In Table 1, Reaction No. 32, the Reaction should read: "$K^+ + HPO_4^{-2} = KHPO_4^-$"

In Table 1, Reaction No. 49, the second line of the Reaction should read: "$+ 3H_4SiO_4 + 80H^-$"

In Table 1, Reaction No. 90, the Reaction should read: "$SO_4^{-2} + 10H^+ + 8e^- = H_2S + 4H_2O$"

In Table 1, Reaction No. 74, the ΔH_r should read: "1.037"

In Table 1, Reaction No. 113, the second line of the Reaction should read: "$3Fe^{+2} + Al(OH)_4^- + 3H_4SiO_4^0 + 60H^-$"

In Table 1, Reaction No. 115, the last line of the Reaction should read: "$+0.23Fe^{+3} + 1.58Al(OH)_4^- + 3.93H_4SiO_4^0 + 0.04H^+$"

In Table 1, Reaction No. 120, the Mineral or species name should read: "$FeH_2PO_4^+$"

Part IV

ISOTOPES IN GROUNDWATER

Editors' Comments
on Papers 23, 24, and 25

23 PEARSON and WHITE
Carbon-14 Ages and Flow Rates of Water in Carrizo Sand, Atascosa County, Texas

24 HITCHON and FRIEDMAN
Excerpts from Geochemistry and Origin of Formation Waters in the Western Canada Sedimentary Basin—I. Stable Isotopes of Hydrogen and Oxygen

25 DAVIS et al.
Geohydrologic Interpretation of a Volcanic Island from Environmental Isotopes

As applied to groundwater studies isotopic technology is a revolutionary development that has had minimal impact in North America. The development of this technology was nonevolutionary in the sense that it became available to the hydrologist only after the design of the modern mass spectrometer by Nier in 1940. However, despite the great hope of the 1960s, despite the obvious potential for significant contribution, and despite the several excellent studies demonstrating the feasibility of using isotopes in groundwater studies, isotopes have yet to become an integral part of North American hydrogeologic investigations. Over twenty years ago the feasibility of using ^{14}C to determine the age of groundwater was demonstrated by Munnich (1957); Vogel (1959); Munnich and Vogel (1959); Thatcher et al. (1961); and Hanshaw et al. (1965). Early applications of tritium in groundwater were made by von Buttlar and Wendt (1958) and Dincer and Davis (1967).

The papers by Friedman (1953), Craig (1961), Dansgaard (1964), and Gat (1971) provide the necessary background for understanding the technology and concepts available for the application of deuterium and oxygen isotopes in groundwater studies. The International Atomic Energy Agency, Vienna, has provided much of the leadership in using isotopes for humanitarian benefits, including hydrologic studies in many areas throughout the world. For example, B. R. Payne, head of the Section of Hydrology, prepared an informative review of isotope hydrology (1972) in which he outlined the basic principles and gave examples of the application to both groundwater and surface water hydrology.

The papers noted above, together with Paper 23, trace the development of the use of ^{14}C in studies of groundwater. Its first use in groundwater in the United States was in the Floridan aquifer to determine the velocity of groundwater (Hanshaw, Back, and Rubin, 1965), and to determine whether in the Brunswick, Georgia, area saltwater is a result of modern seawater encroachment or an intrusion of seawater into the aquifer during Pleistocene or earlier time (Hanshaw et al., 1965). This work was extended to study saltwater encroachment in other aquifers (Back et al., 1970) and to estimate the average hydraulic conductivity of the Floridan aquifer, kinetics of calcite dissolution, and rate of entropy change (Back and Hanshaw, 1971 and Hanshaw and Back, 1974). Winograd and Farlekas (1974) point out problems of ^{14}C-dating in aquifers containing carbonaceous material that can oxidize to carbon-dioxide gas and, therefore, give an erroneous age of the water.

In Paper 24, Hitchon and Friedman used deuterium and oxygen isotopes to identify sources of water and amount of mixing of water from the various sources of the Western Canada sedimentary basin. Deuterium has also been used successfully to identify water sources discharging from springs (Winograd and Friedman, 1972). Isotopic fractionation of rainfall responds to temperature differences resulting from altitude variation. The deviation of data from the worldwide meteoric line can also be used to identify processes of evaporation or bacterial action. Paper 25 demonstrates the use of deuterium, $^{18}O/^{16}O$, and tritium to identify sources of water and to evaluate water resources.

The $^{13}C/^{12}C$ fractionation processes have been used most effectively to identify sources of carbonate ions in groundwater systems to verify mass transfer calculations and to adjust ^{14}C ages. Presently the sulfur isotope ratio, $^{34}S/^{32}S$, has been used minimally in groundwater but the results are most encouraging. Because the sulfur-isotopic variations of the ocean through geologic time are reflected in the precipitated sulfate minerals, this fractionation value can be used to identify geologic formations from which the sulfate ion has dissolved in groundwater. Significant progress has been made in using the nitrogen isotope ratio, $^{15}N/^{14}N$, to identify sources of pollution in groundwater largely under the leadership of Kreitler et al. (1978). In a series of studies, Osmond et al. (1968, 1974) have been developing the use of uranium isotopes, ^{234}U and ^{238}U, to investigate sources, mixing, and age of groundwater.

Editors' Comments on Papers 23, 24, and 25

As more experience is gained, the use of isotopes will become a standard part of hydrogeologic investigations.

REFERENCES

Back, W., and B. B. Hanshaw, 1971, Rates of Physical and Chemical Processes in a Carbonate Aquifer, in *Nonequilibrium Systems in Natural Water Chemistry,* American Chemical Society, Washington, D.C., p. 77-93.

Back, W., B. B. Hanshaw, and M. Rubin, 1970, Carbon-14 Ages Related to Occurrence of Salt Water, *Am. Soc. Civil Engineers Proc., Jour. Hydraulics Div.* **96:**2325-2336.

Craig, H., 1961, Isotopic Variations in Meteoric Waters, *Science* **133:**1702-1703. (Reprinted as Paper 8 in *Geochemistry of Water,* Y. Kitano, Ed., Benchmark Papers in Geology, vol. 16, Dowden, Hutchinson & Ross, Stroudsburg, Pa., p. 159.)

Dansgaard, W., 1964, Stable Isotopes in Precipitation, *Tellus* **16:**436-468. (Reprinted as Paper 9 in *Geochemistry of Water,* Y. Kitano, ed., Benchmark Papers in Geology, vol. 16, Dowden, Hutchinson & Ross, Stroudsburg, Pa., pp. 160-192.)

Dincer, T., and G. H. Davis, 1968, Some Considerations on the Tritium Dating and Estimates of Tritium Input Function, *Internat. Assoc. of Hydrogeologists Memoirs,* **8:**276-286.

Friedman, I., 1953, Deuterium Content of Natural Water and other Substances, *Geochim. et Cosmochim. Acta* **4:**89-103. (Reprinted as Paper 7 in *Geochemistry of Water,* Y. Kitano, ed., Benchmark Papers in Geology, vol. 16, Dowden, Hutchinson & Ross, Stroudsburg, Pa., pp. 144-158.)

Gat, J. R., 1971, Comments on the Stable Isotope Method in Regional Groundwater Investigations, *Water Resources Research* **7:**980-993.

Hanshaw, B. B., and W. Back, 1975, Determination of Regional Hydraulic Conductivity through use of ^{14}C Dating Groundwater, *Internat. Assoc. of Hydrogeologists Memoirs,* **10-1:**195-196.

Hanshaw, B. B., W. Back, and M. Rubin, 1965, Radiocarbon Determinations for Estimating Groundwater Flow Velocities in Central Florida, *Science* **148:**494-495.

Hanshaw, B. B., W. Back, M. Rubin, and L. Wait, 1965, Relation of Carbon-14 Concentrations to Saline Water Contamination of Coastal Aquifers, *Water Resources Research* **1:**109-114.

Kreitler, C. W., S. E. Ragone, and B. G. Katz, 1978, Nitrogen-Isotope Ratios of Groundwater Nitrate, Long Island, New York, *Ground Water* **16:**404-409.

Munnich, K. O., 1957, Messungen des C^{14} gehaltes von hartem grundwasser, *Naturwissenschaften* **44:**32-33.

Munnich, K. O., and J. C. Vogel, 1959, Altersbestimmung von süsswasserkalkablagerungen, *Naturwissenschaften* **46:**168-169.

Osmond, J. K., H. S. Rydell, and M. I. Kaufman, 1968, Uranium Disequilibrium in Groundwater—An Isotope Dilution Approach in Hydrologic Investigations, *Science* **162:**997-999.

Osmond, J. K., M. I. Kaufman, and J. B. Cowart, 1974, Mixing Volume Calculations, Sources, and Aging Trends of Floridan Aquifer Water by Isotopic Methods, *Geochim. et Cosmochim. Acta* **38:**1083-1100.

Payne, B., 1972, Isotope Hydrology, *Advances in Hydroscience,* **8:**95–138.

Thatcher, L. M. Rubin, and G. Brown, 1961, Dating Desert Groundwater, *Science* **134:**105–106.

Vogel, J. C., 1959, Uber den isotopengehalt des kohlenstoffs in süsswasserkalkablagerungen, *Geochim. et Cosmochim. Acta* **16:**236.

Von Buttlar, H., and I. Wendt, 1958, Ground-Water Studies in New Mexico Using Tritium as a Tracer, *Am. Geophys. Union Trans.* **31:**660–668.

Winograd, I. J., and G. M. Farlekas, 1974, Problems in ^{14}C Dating of Water from Aquifers of Deltaic Origin—An Example from the New Jersey Coastal Plain, *Isotopes Techniques in Groundwater Hydrology 1974,* **2:**69–93.

Winograd, I. J., and I. Friedman, 1972, Deuterium as a Tracer of Regional Ground-Water Flow, Southern Great Basin, Nevada, and California, *Geol. Soc. America Bull.* **82:**3691–3708.

23

Reprinted from Water Resources Research 3:251-261 (1967)

Carbon 14 Ages and Flow Rates of Water in Carrizo Sand, Atascosa County, Texas[1]

F. J. PEARSON, JR.[2]

D. E. WHITE

Abstract. Ages of groundwater can be determined from the carbon 14 content of the carbonate dissolved in the water. The carbon 14 is derived from plant-produced CO_2 in the soil of the recharge area and is usually diluted by carbon 14-free carbonate dissolved from carbonate minerals in the soil and in the aquifer. Techniques based on the ratios of the stable carbon isotopes and on the over-all carbonate chemistry of the water can be used to correct for this dilution and to allow the calculation of true water ages. Water samples from wells in the Eocene Carrizo Sand in Atascosa and adjacent counties, Texas, were dated by this method. The ages of the water samples ranged from 0 years at the outcrop to 27,000 years 35 miles downdip. Based on the carbon 14 ages, the water velocities were about 8 feet per year 10 miles from the outcrop and 5.3 feet per year at 31 miles. Flow rates calculated from available hydrologic data are in agreement with carbon 14 results. (Key words: Geochemistry; geochronology; groundwater; carbon 14)

INTRODUCTION

One of the most important problems that must be solved by groundwater hydrologists is to estimate the quantity of water flowing through an aquifer. Accurate answers to this problem are becoming increasingly important as the need for water becomes more acute, and safe, sustained aquifer yields must be determined with greater precision.

To calculate the flow through an aquifer, the area through which flow takes place, the porosity, and the flow velocity must be known. Although in normal hydrologic practice flow velocity is not calculated explicitly, it is implicit in the various equations usually used in flow calculations. These calculations require knowledge of transmissibility or permeability, porosity, and hydraulic gradient, which must be measured at several points in the aquifer. Enough measurements of these parameters should be available so that their range of variation can be determined and limits of confidence placed on the calculated flow. To evaluate further the reliability of flow estimates, detailed pumpage and water-level data must be available. Thus, to characterize a large portion of an aquifer requires an extensive series of measurements and often considerable extrapolation. Furthermore, the mathematical derivations of the equations used in converting the well pumping and recovery data to transmissibility values assume certain idealized conditions. Among others, it is assumed that the aquifer is homogeneous and isotropic, and that the thickness of the aquifer supplying the test well is known. The natural deviations from these conditions will introduce errors of unknown size into the calculated transmissibility values and into flow rates derived from them.

Another method for determining flow rates has recently been developed, based on measurements of the carbon 14 content of the carbonate dissolved in the water. The method allows calculation of the time elapsed since the water left the recharge area, that is, calculation

[1] Publication authorized by the Director, U. S. Geological Survey.
[2] Present address: U. S. Geological Survey, Albany, N. Y.

of the age of the water. A number of such water age determinations can be used to delineate patterns and rates of flow with errors that are considerably smaller than those obtained from hydrologic data and are, moreover, of known magnitude.

The age and rate of flow of water in the Carrizo Sand in Atascosa and surrounding counties, south-central Texas (Figure 1), have been measured using the carbon 14 method. This paper presents the results and compares them with flow rates computed from conventional hydrologic data.

CARBON 14 DATING OF GROUNDWATER

Carbon 14 Dating. Carbon 14 has a half-life of about 5700 years [*Godwin*, 1962]. It is continuously formed in the upper atmosphere by the action of cosmic-ray-produced thermal neutrons on N^{14} [*Libby*, 1955, ch. 1], is oxidized to CO_2, and mixes rapidly with the atmospheric CO_2 reservoir. Any material that uses or reacts with atmospheric CO_2, plants or water, for example, will contain a quantity of carbon 14 that will be constant so long as the material is in equilibrium with the atmospheric reservoir. When the material is cut off from the reservoir, as when a plant dies or water leaves the aerated zone of an aquifer, the amount of carbon 14 in it will decrease by radioactive decay, and the amount remaining should be a function of the time since cutoff.

Groundwater passing through the aerated zone of the soil absorbs CO_2 from the soil at-

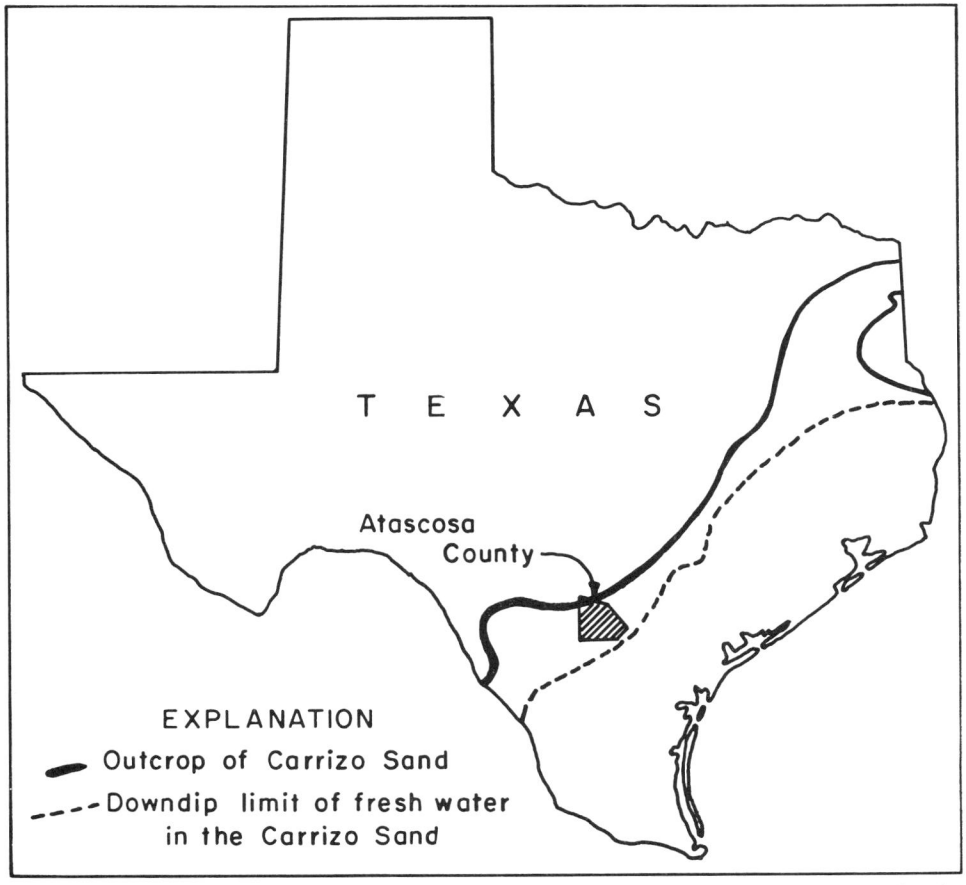

Fig. 1. Location of Atascosa County, outcrop and downdip limit of fresh water in the Carrizo Sand.

mosphere, which hydrolyzes to carbonate and bicarbonate ions. This CO_2 is from plant respiration and decay, and its carbon isotope composition is equivalent to that of plants [*Ingerson and Pearson*, 1964]. The carbon 14 content of the carbonate ions species dissolved in a sample of groundwater can thus be compared with the carbon 14 content of living plants to determine the time elapsed since the sample left the aerated zone of the soil, that is, the age of the water.

Correction factors. Groundwater also dissolves solid carbonate from limestone in the soil and in the aquifer and can dissolve additional CO_2 derived from organic material in the aquifer, if any is present. These sources of dissolved carbonate contain no carbon 14 because of their age, and they dilute the soil-air-derived, carbon-14-containing dissolved carbonate, making the water sample appear falsely old.

To obtain a valid water age, it is necessary to know what portion of the total carbonate content of a sample was initially derived from the soil air. The measured carbon 14 content of the total sample carbonate can then be corrected to that initial carbonate content and the corrected value used to calculate the true age.

If the diluting carbonate is derived from marine limestone, the ratio of the stable carbon isotopes C^{12} and C^{13} can be used to find the correction factor. Isotopic ratios are commonly reported as δ values, which are defined as

$$\delta = (R_{\text{sample}}/R_{\text{standard}} - 1) \times 1000 \quad (1)$$

where R is the ratio of C^{13} to C^{12} [*Craig*, 1957]. The standard referred to here is the Chicago PDB standard [*Craig*, 1957].

It is well known that the C^{13}/C^{12} ratio of marine limestone is fairly constant, with a δC^{13} value of about 0 per mille. Plant material also has a fairly constant C^{13}/C^{12} ratio, but with less C^{13} than marine limestone, having a δC^{13} value of about -25 per mille [*Ingerson and Pearson*, 1964, tab. 1, p. 272]. The δC^{13} value of a sample containing both plant and limestone derived material will be between 0 and -25 per mille, its specific value being determined by the relative amounts of carbon from each source. The δC^{13} value can thus be used to find the proportion of plant-derived carbon in the sample, the factor needed to correct the measured carbon 14 activity to determine the sample age.

If the amounts of plant-derived and limestone-derived carbon are designated as pl and ls, respectively, the correction factor can be written

$$P = pl/(pl + ls) \quad (2)$$

which is simply the ratio of plant-derived to total carbon in the sample. The δC^{13} value of the sample will be the weighted average of the δC^{13} values of the contributing components

$$\delta C_{sm}^{13} = [pl/(pl + ls)] \delta C_{pl}^{13}$$
$$+ [ls/(pl + ls)] \delta C_{ls}^{13} \quad (3)$$

Combining these two equations and solving for P gives the correction factor in terms of the various δC^{13} values

$$P = (\delta C_{sm}^{13} - \delta C_{ls}^{13})/(\delta C_{pl}^{13} - \delta C_{ls}^{13}) \quad (4)$$

This may be simplified by inserting the average values of δC^{13}_{pl}, $-25‰$, and δC^{13}_{ls} 0‰, i.e.

$$P = \delta C_{sm}^{13}/-25 \quad (5)$$

A difficulty with this scheme is the possible change both in the C^{13}/C^{12} ratio and the C^{14} content of the sample by isotope exchange. Unfortunately, there seems to be no theoretical method by which the δC^{13} value of the sample can be used both to correct for exchange and to determine the proportion of plant carbon in diluted materials. In practice, however, it is possible to use other information to judge what effect exchange may have on the correction factor and on the corrected sample ages. Such information can be obtained from the over-all carbonate chemistry of the water.

There are several reactions that affect the solubility of limestone in groundwater.

$$CO_{2aq} + H_2O + CaCO_{3sol}$$
$$= 2HCO_3^- + Ca^{++} \quad (6)$$

$$2\text{Na Clay}_{sol} + Ca_{aq}^{++}$$
$$= \text{Ca Clay}_{sol} + 2Na_{aq}^+ \quad (7)$$

These include the familiar reaction of dissolved CO_2 with solid $CaCO_3$, reaction 6, and the cation exchange reaction 7, which is thought to produce sodium bicarbonate waters of the type common to the Atlantic and Gulf Coastal Plains of North

America [*Foster*, 1950]. If the carbonate chemistry of the water is controlled exclusively by reaction 6, the total carbonate content at saturation will be twice the initial soil-air-derived carbonate content [*Ingerson and Pearson*, 1964, p. 270]. If clay minerals are present and reaction 7 takes place, the dissolved calcium content will decrease, allowing more $CaCO_3$ to dissolve, and the total dissolved carbonate content may then be greater than twice the initial carbon content. Also, gaseous CO_2, from the decomposition of plant material or of magmatic or petroleum origin, may be added to the water in the aquifer. Whatever its source, this additional CO_2 will react with $CaCO_3$ according to reaction 6 and further increase the total carbon content of the water.

Equation 8 is an expression for the correction factor based on the dissolved carbonate relations of the water

$$P' = C_{initial}/C_{total} \qquad (8)$$

It is simply the ratio of the concentration of the carbonate initially derived from the soil air to the total carbonate concentration, and, except for those waters in which gaseous CO_2 is introduced in the aquifer, should equal the correction factor from equation 4 or 5. The two expressions may be combined as equation 9.

$$(\delta C_{sm}^{13})/-25 = C_{initial}/C_{total} \qquad (9)$$

$$\log C_{total} = -\log(\delta C_{sm}^{13})/-25 + \log C_{initial} \qquad (10)$$

Equation 10 is the logarithmic form of the same equation arranged to be the equation of a straight line of slope -1, and y-intercept equal to the logarithm of the initial carbonate content.

Plotting data from a regional study of an aquifer in the form of equation 10 allows straight-line extrapolation to find a general average initial carbonate content. Moreover, the equation is valid only if no isotope exchange takes place, so that if the slope of the line of data is -1 as predicted by this equation, exchange cannot be significant.

RESULTS

Hydrogeology. The Carrizo Sand, of Eocene age, crops out nearly parallel to the Gulf Coast in southeastern Texas (Figure 1). The formation dips towards the coast and contains fresh water at depths as great as 5000 feet. The annual discharge of fresh water from the Carrizo, a major aquifer in Texas, is estimated to exceed 300,000 acre-feet.

In northern Atascosa County, the Carrizo Sand crops out as a band of rolling hills from 6 to 8 miles wide. The formation dips to the southeast at a rate of 100 to 130 feet per mile. The thickness of the formation increases from 600 to 700 feet near the outcrop to 1000 to 1300 feet in the southeastern corner of the county, where the top of the formation is 3600 to 3800 feet below sea level [*Alexander and White*, 1966, figs. 12 and 14].

The lithology of the Carrizo is remarkably uniform, consisting mostly of fine- to medium-grained quartz sand with minor amounts of clay, lignite, calcite, and pyrite. In the subsurface the Carrizo is readily identified on electric logs by the positive shift in the resistivity curves. Sands of higher transmissibility and higher resistivity are more abundant in the middle of the formation than in the upper and lower portions. The larger wells in Atascosa County, which yield from 800 to 1200 gallons per minute, are usually completed in these sands.

The U. S. Geological Survey has conducted pumping tests of 9 wells that tap Carrizo Sand in Atascosa County. Results of these tests, which are reported in detail by *Alexander and White* [1966, table 2], indicate a general decrease in both transmissibility and permeability with increased depth and distance from the outcrop. Permeability coefficients ranged from 375 gal/ft²/day near Poteet, less than 5 miles from the outcrop, to 80 gal/ft²/day at Campbellton near well Tx-214, about 30 miles from the outcrop. Laboratory analyses of 89 Carrizo Sand samples from the Winter Garden area to the west of Atascosa County gave permeability values that ranged from 3 to 1440 and averaged 286 gal/ft²/day [*Turner et al.*, 1960, p. 67]. Field coefficients determined from aquifer tests in that area averaged 200 gal/ft²/day [*Turner et al.*, 1960]. In the calculations below, a straight-line function decreasing from 400 gal/ft²/day in the middle of the outcrop to 80 gal/ft²/day at Campbellton was used [*Pearson*, 1966, Figure 6].

The porosity of the Carrizo Sand is estimated

to range from 30 to 40%. Laboratory analyses of core samples from the formation in southeastern Atascosa County give an average porosity of 38% (Caran Equipment Company, San Antonio, oral communication). Porosity from analyses of 89 samples from the Winter Garden area averaged about 40% [*Turner et al.*, 1960, p. 66]. A porosity of 35% was used in the calculations made below.

Development of the Carrizo Sand in Atascosa County began near the turn of the century. This development was confined largely to the north-central portion of the county, where many wells originally flowed. By 1930, annual discharge amounted to about 9500 acre-feet, and the piezometric head had declined 25 to 30 feet [*Lonsdale*, 1935, p. 49]. Predevelopment water levels are not available, and it was necessary to extrapolate to the 1900 piezometric profile using 1929–1930 water levels and reported head declines from 1900 to 1930 (Figure 2) [*Lonsdale*, 1935]. The extrapolated gradient, which decreases from about 10 feet per mile at the outcrop to 1 foot per mile near Campbellton, was used in the calculations made below.

Carbonate chemistry. Water samples from 15 wells in Atascosa and surrounding counties were collected for standard chemical analysis and for determination of the C^{13}/C^{12} ratios and C^{14} content of the dissolved carbonate. The pH and alkalinity of the wells on and near the Carrizo outcrop, Tx-90–94, Tx-210, were measured in the field. In these wells, a significant portion of the total carbonate was present as dissolved CO_2, and the field measurements minimized possible errors in the total carbonate determination due to loss of this CO_2 between the times of sampling and analysis [*Barnes*, 1964]. The pH and alkalinity data were used to calculate the total carbonate content and degree of saturation with respect to calcite of each sample [*Back*, 1961]. The carbonate contents of the samples are given in Table 1, and the analytical data by *Pearson* [1966].

The samples for radiocarbon measurement were collected by acidifying a large volume of water (100 to 200 liters) and collecting the evolved CO_2 in an NH_4OH solution, by a procedure similar to that described by *Feltz and Hanshaw* [1963]. The ammonium carbonate in solution was taken to the University of Texas Radiocarbon Dating Laboratory, where it was converted to benzene, the material on which radiocarbon measurements are made in that laboratory [*Pearson et al.*, 1965]. A small por-

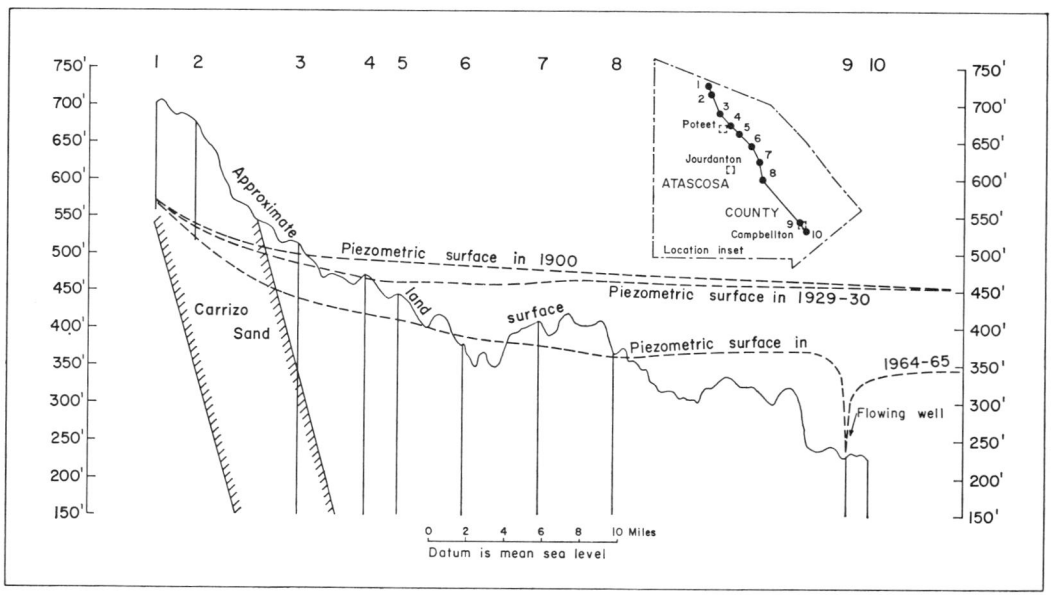

Fig. 2. Piezometric gradients, Carrizo Sand.

TABLE 1. Results of Analyses of Water from Carrizo Sand

Well Number*	Lab. Number†	C^{14} − (% Mdn ± 2σ)‡	δC^{13}, PDB	Total Carbonate, epm§	Fraction Dateable Carbonate		C^{14} age**
					P (eq. 5)	P' (eq. 8) ‖	
AL-6851-803	Tx-90	76.9 ± 2.0	−18.9	2.82 ± 0.25	0.756	0.784 ± 0.082	(98.1 ± 10.7%)
AL-6859-504	Tx-92	50.1 ± 1.4	−17.9	2.77 ± 0.20	0.716	0.798 ± 0.072	3,750 ± 1150
AL-6860-303	Tx-91	74.9 ± 1.4	−18.5	2.66 ± 0.20	0.740	0.831 ± 0.076	(90.1 ± 8.5%)
AL-7802-301	Tx-210	33.6 ± 0.8	−11.5	5.30 ± 0.32	0.460	0.417 ± 0.034	1,730 ± 670
AL-7805-104	Tx-93	17.1 ± 0.8	−11.8	5.89 ± 0.22	0.472	0.375 ± 0.025	6,300 ± 700
AL-7812-201	Tx-94	6.92 ± 0.82	−9.9	5.62 ± 0.14	0.396	0.393 ± 0.023	14,000 ± 1050
AL-7815-504	Tx-96	2.54 ± 0.40	−15.0	8.31 ± 0.16	0.600	0.266 ± 0.015	18,900 ± 1300
AL-7818-301	Tx-226	6.38 ± 0.50	−8.8, −8.9	4.60 ± 0.11	0.354	0.480 ± 0.028	16,200 ± 800
AL-7820-101	Tx-216	3.69 ± 0.28	−11.7	4.87 ± 0.11	0.468	0.454 ± 0.027	20,200 ± 770
AL-7822-202	Tx-214	1.36 ± 0.20	−9.8	9.46 ± 0.19	0.392	0.234 ± 0.014	22,900 ± 1300
Live Oak A-4	Tx-97	1.19 ± 0.34¶	−3.7	15.67 ± 0.30	0.148	0.141 ± 0.008	>18,000¶
RX-7822-801	Tx-215	2.06 ± 0.26¶	−9.1	6.24 ± 0.12	0.364	0.354 ± 0.020	>22,000¶
SU-7828-101	Tx-217	1.17 ± 0.18	−9.5	5.73 ± 0.11	0.380	0.386 ± 0.022	28,200 ± 1400
SU-7836-201	Tx-218	<0.62	−10.7	9.99 ± 0.20	0.428	0.221 ± 0.013	>28,200
SU-7838-101	Tx-219	<0.66	−10.6	15.48 ± 0.34	0.424	0.143 ± 0.008	>24,300

* For records of wells in Atascosa County, prefix AL-, see *Alexander and White*, 1966; in Live Oak County, see *Anders and Baker*, 1961; in La Salle County, prefix SU-, and McMullen County, prefix RX-, see *Harris*, 1965.
† University of Texas Radiocarbon Dating Laboratory sample numbers, see *Pearson et al.*, 1965.
‡ Error represents 95% confidence limit.
§ Includes dissolved CO_2, HCO_3^-, and $CO_3^=$; calculated from pH and alkalinity measurements [*Back*, 1961; *Barnes*, 1964]; errors based on estimated reliability of measurements.
‖ Using C initial = 2.21 ± 0.12 epm; see Figure 4 and text.
¶ Carbon 14 sample contaminated during collection; only limit age reported.
** Calculated using the expression Age = 8030 $e^{-(A^c)}$, where A^c is the corrected carbon 14 content and equals % Mdn/100 × P'.

tion was retained for the mass spectrometric C^{13}/C^{12} determinations. The amount of carbonate collected was in all samples greater than 90% of the amount expected from the chemical analyses, so that isotope fractionation effects during collection should not be significant.

The locations of the wells sampled and the results of the chemical analyses are shown in Figure 3, and references to the analyses themselves are given in Table 1. Figure 4 is a plot of the δC^{13} and total dissolved carbonate content data (Table 1) in the form of equation 10. This figure was made to aid in determining the correction factors to be applied to the C^{14} data, but it is also helpful in discussing the over-all carbonate chemistry of the water.

From Figure 3 it is evident that isochemical lines in this region trend northeast-southwest. The water in and near the outcrop is highly unsaturated with respect to calcite. The degree of saturation increases to the southeast to a line passing between wells Tx-93 and Tx-94, where it approaches 100%. The δC^{13} values of the dissolved carbonate range from about −18‰ in water in the outcrop to about −12‰ in well Tx-93. The dissolved sodium content of samples downdip to Tx-93 and Tx-94 is essentially constant and is less than, or at least not appreciably greater than, the calcium content.

The Carrizo water in the northwestern part of Atascosa County is thus of the calcium bicarbonate type, in which the chemistry of the dissolved carbonate is controlled by reaction 6, and the carbonate content of the calcite saturated water (ca. 5 epm) should equal twice that of the initial carbon derived from the soil-air.

To the southeast of the line where calcite saturation first becomes complete, there is a zone in which the total carbonate content increases slowly, but in which the sodium content increases rapidly and the calcium correspondingly decreases. In this zone, which extends to a line passing near well Tx-217, the cation exchange reaction (7) is taking place. The exchange of sodium for calcium produces waters of the sodium bicarbonate type and al-

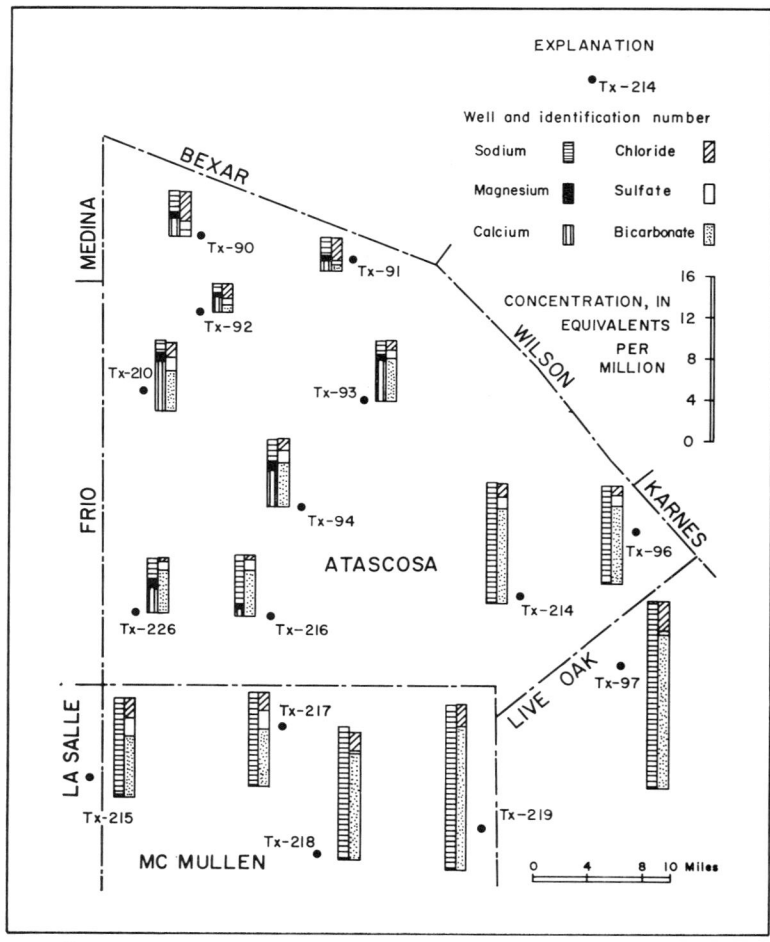

Fig. 3. Locations of wells sampled and bar graphs representing water analyses.

lows more $CaCO_3$ to be dissolved. As expected, the further solution of $CaCO_3$ is accompanied by decreasingly negative values of δC^{13}.

In the southeastern part of the area the water is of the high-sodium bicarbonate type. The water from wells Tx-214, 218, and 219 is equivalent to up-dip sodium bicarbonate water with carbonate of $\delta C^{13} \cong -12‰$ added. Such carbonate would result from the reaction of CO_2 from lignite in the aquifer with limestone (reaction 6).

On the basis of their chemical analyses alone, the samples from wells Tx-96 and Tx-97 would also be called high-sodium bicarbonate waters, although they were not produced by the same mechanism as those mentioned above. The δC^{13} value of well Tx-96 ($-15‰$) is too negative to have been produced by addition of lignite-derived CO_2 to a sodium-bicarbonate water. The well is located near a subsurface fault over a natural gas field [*Alexander and White*, 1966], suggesting that petroleum-derived CO_2, which is known to have δC^{13} values of -30 and below [*Pearson*, 1966, Table 1], may be reacting with limestone to produce the additional carbonate.

The too positive value of δC^{13} of well Tx-97 ($-3.7‰$) indicates that no additional plant-derived CO_2 has been introduced, but that the high bicarbonate content of the water results from continued sodium-for-calcium exchange (reaction 7), allowing solution of more calcite.

In a genetic classification, it would be grouped with waters of the sodium-bicarbonate type such as Tx-217, rather than with the high-sodium bicarbonate types which its general, but not its isotopic, chemistry resembles.

This mechanism of formation of waters of this type was suggested by *Foster* [1950], on the basis of laboratory work. The stable isotope data available here supported her ideas completely. For such a complete explanation of the carbonate chemistry of the water, the corrections to be applied to the C^{14} data to calculate water ages can be derived very simply.

Carbon 14 data. The experimental data, correction factors, and corrected C^{14} ages of the water samples are given in Table 1. The errors accompanying the water ages include the errors of the determinations of the initial and total carbonate contents and in the carbon 14 determinations. The limits given are two standard deviations (95% confidence limits). Figure 4 shows the δC^{13} and total carbonate data plotted in the form of equation 10. The lengths of the plotted lines represent the uncertainty in the experimental determinations of the total carbonate content [*Pearson*, 1966]. The dashed lines were drawn with the slope, -1, predicted by equation 10.

The four samples that plot in the upper right of Figure 4, Tx-96, -214, -218, and -219, are all high-sodium-bicarbonate waters. As such, part of their diluting carbonate has been derived from sources other than marine limestone, and equation 5 and hence equation 10 do not apply to them. Sample Tx-93 may also contain some plant- or petroleum-derived diluting carbonate, which would tend to make its plot fall slightly to the right of the line of equation 10. The position of sample Tx-226 is probably due to error in the total carbonate analysis of the water.

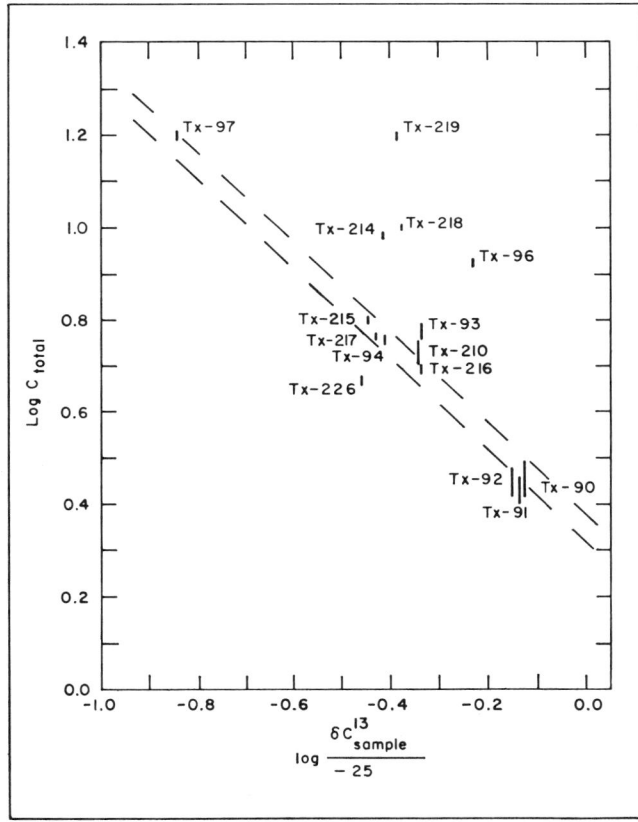

Fig. 4. Comparison of total carbonate content and δC^{13} values of water samples.

The amount of carbonate extracted for the carbon 14 measurement was greater than that calculated from the analysis, and the analysis is therefore probably low. The proper position for this sample should be above its present position in Figure 4, closer to if not in agreement with, the line of equation 10.

The agreement between the experimental data and equation 10 means that the assumptions on which the derivation of that equation are based are probably correct. The most important of these is the lack of demonstrable isotope exchange between dissolved and solid carbonate. The expected existence of such exchange has long been an argument against the use of C^{14} in groundwater studies. This work shows that the extent of such exchange, if it occurs at all, is too small to be measured and provides a method for monitoring exchange in future work.

In most groundwater studies the corrections to the measured carbon 14 values can best be made using δC^{13} information and equation 5, because δC^{13} values can be measured with greater precision than can dissolved carbonate contents. Equation 5 is not applicable to the high-sodium bicarbonate waters sampled in this work, so that for consistency the corrections to all samples are based on total carbonate contents and equation 8. According to equation 10, the value of $C_{Initial}$ needed to use equation 8 can be found from the y-intercept of a plot like Figure 5. The value from this figure is 2.21 ± 0.12 epm, which agrees with the value of ca. 2.5 epm expected from the total carbonate contents of the calcite saturated calcium bicarbonate water of the region.

Contoured water ages based on the data in Table 2 are shown in Figure 5. Points where the water ages could be expressed in round numbers were found by mechanical, linear interpolation between various pairs of wells. The possible error in the position of each point due to the errors in the sample ages was also determined. Contour bands were then drawn to include each interpolated point and its range of error, so that the width of the band represented the experimental uncertainty.

Comparison of results. To compare the C^{14} and conventional hydrologic data, rates of flow of water along a general northwest- trending line through eastern Atascosa County have been calculated (Figure 6). The rates from the C^{14} data were derived from the contoured water ages, with their ranges of error based on the widths of the age contours. The hydrologic rates were calculated at several points, using the estimated permeability gradient, a porosity of 35%, the extrapolated 1900 piezometric gradient (Figure 2), and Darcy's equation in the form

$$V = Pl/p \times 48.83 \qquad (19)$$

where V is flow velocity in feet per year at a given point, P the permeability in gal/ft²/day at the point of calculation, l the piezometric gradient in feet per foot, also at the point, and p the porosity fraction. The factor 48.83 converts the units of flow rate to feet per year.

The agreement between the hydrologic and C^{14} rates is gratifying. In the deeper portion of the aquifer, where the C^{14} contents of the samples were small (Table 1), contamination during sampling could have affected the calculated ages, giving an apparently high flow rate. The effect of contamination would not have been larger than the range shown, but to allow for it the C^{14} velocity line (Figure 6) was drawn through the lower portion of the range. Although slight contamination was possible, the disagreement could also have resulted from errors in the hydrologic data. In any case, the difference is too small to be attributable to fundamental errors in either method.

CONCLUSIONS

1. Rates of groundwater flow based on measured properties of the aquifer may be in error by unknown amounts because of (a) a number of measurements insufficient to reflect the range of variation of the properties of the aquifer; and (b) failure of the assumptions underlying the methods used to convert the field measurements to the parameters used for flow calculations.

2. The carbonate ion species dissolved in groundwater contain C^{14} derived from the soil air of the recharge area. Measurement of the C^{14} content of groundwater can be used to determine the travel time for recharge area to sampling location, or 'age' of the water, and hence the flow rate of the water. Techniques based on the C^{13}/C^{12} isotope ratios and the overall carbonate chemistry of the water can be used successfully to correct the measured C^{14} contents for the diluting effects of dissolved lime-

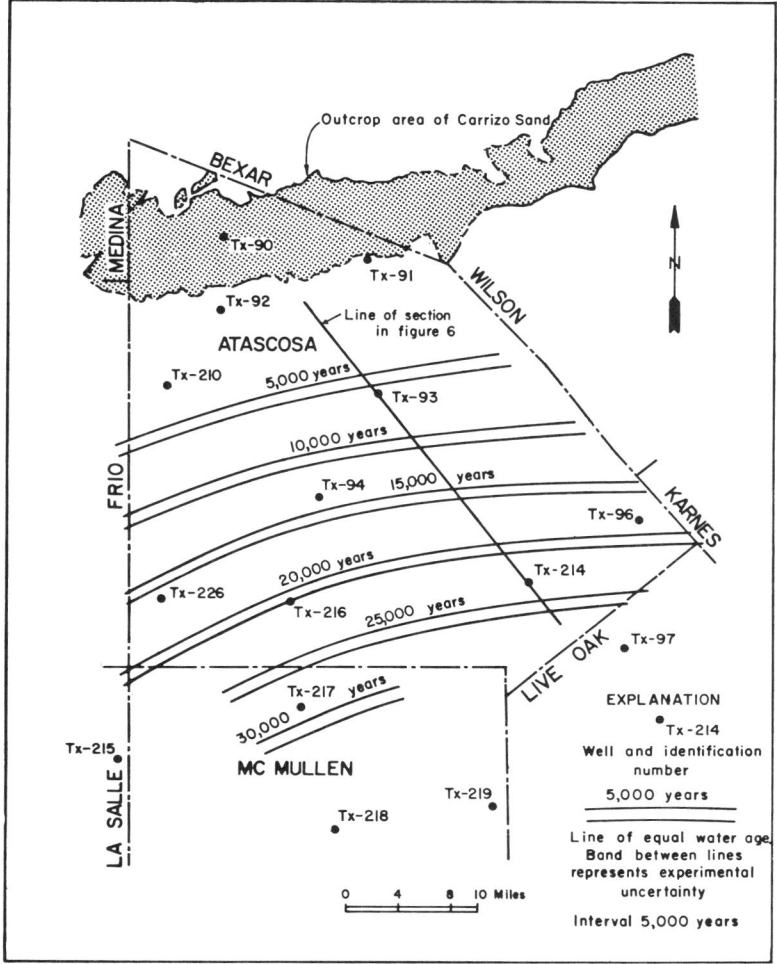

Fig. 5. Age of water in Carrizo Sand.

stone, so that true water ages can be calculated.

C^{14} data are advantageous because they require no subjective interpretation by the investigator and have known limits of error.

3. The prepumping rate of flow of water in the Carrizo Sand, Atascosa County, Texas, based on C^{14} data, decreases away from the outcrop until, at a distance of 30 miles and greater, it is stable at about 5.3 ft/yr. The flow rate calculated from available hydrologic data is in agreement with the C^{14} results.

Acknowledgments. We are grateful to Irving Friedman of the Branch of Isotope Geology of the U. S. Geological Survey, who provided us with the C^{12}/C^{13} analyses reported here. We have benefited from discussions of the work in progress and of this report with B. B. Hanshaw, Earl Ingerson, and A. G. Winslow.

REFERENCES

Alexander, W. H., Jr., and D. E. White, Ground-water resources of Atascosa and Frio Counties, Texas, *Texas Water Development Board Report*, 1966 (in press).

Anders, R. B., and E. T. Baker, Jr., Ground-water geology of Live Oak County, Texas, *Texas Board of Water Engineers Bull. 6105*, 1961.

Back, William, Calcium carbonate saturation in ground water, from routine analyses, *U. S. Geol. Surv. Water-Supply Paper 1535-D*, 1961.

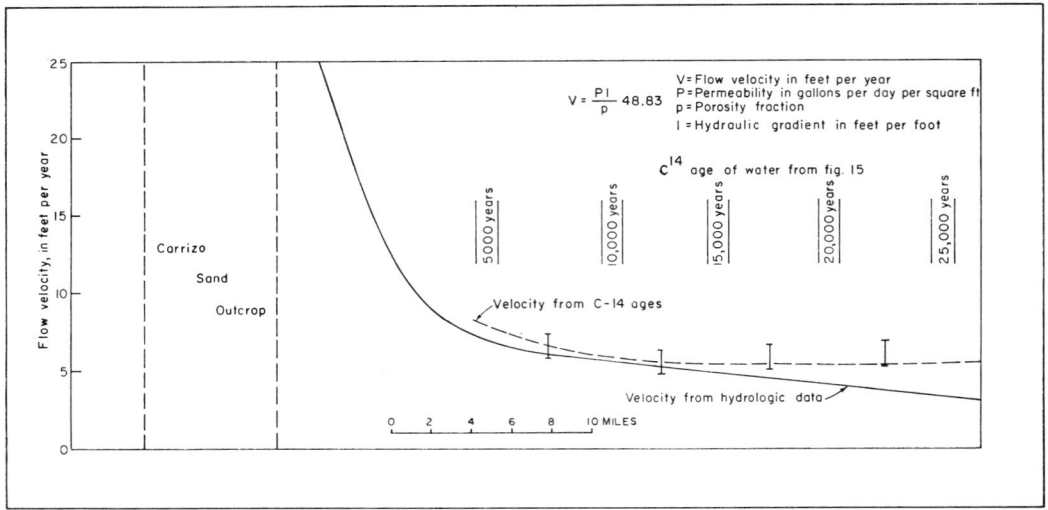

Fig. 6. Rates of flow of water in Carrizo Sand from C^{14} determinations and hydrologic data.

Barnes, Ivan, Field measurement of alkalinity and pH, *U. S. Geol. Surv. Water-Supply Paper 1535-H*, 1964.

Craig, Harmon, Isotopic standards for carbon and oxygen correction factors for mass-spectrographic analysis of carbon dioxide, *Geochim. Cosmochim. Acta, 12*, 133–149, 1957.

Feltz, H. R., and B. B. Hanshaw, Preparation of water samples for carbon 14 dating, *U. S. Geol. Surv. Circ. 480*, 1963.

Foster, Margaret D., The origin of high sodium bicarbonate waters in the Atlantic and Gulf Coastal Plains, *Geochim. Cosmochim. Acta, 1*, 33–48, 1950.

Godwin, H., Half-life of radiocarbon, *Nature, 195*, 984, 1962.

Harris, H. B., Ground-water resources of La Salle and McMullen Counties, Texas, *Texas Water Comm. Bull. 6520*, 1965.

Ingerson, Earl, and F. J. Pearson, Jr., Estimation of age and rate of motion of ground water by the C-14 method, in *Recent Researches in the Fields of Atmosphere, Hydrosphere, and Nuclear Geochemistry*, Sugawara Festival Volume, Maruzen Co., Tokyo, pp. 263–283, 1964.

Libby, Willard E., *Radiocarbon Dating*, 2nd ed., The University of Chicago Press, Chicago, 1955.

Lonsdale, John T., Geology and ground-water resources of Atascosa and Frio Counties, Texas, *U. S. Geol. Surv. Water-Supply Paper 676*, 1935.

Pearson, F. J., Jr., E. Mott Davis, M. A. Tamers, and Robert W. Johnstone, University of Texas radiocarbon dates III, *Radiocarbon, 7*, 296–314, 1965.

Pearson, F. J., Jr., Ground-water ages and flow rates by the carbon 14 method, unpublished Ph.D. thesis, University of Texas, 1966.

Turner, Samuel F., Thomas W. Robinson, and Walter N. White, Geology and ground-water resources of the Winter Garden District, Texas, 1948, *U. S. Geol. Surv. Water-Supply Paper 1481*, 1960.

(Manuscript received September 26, 1966; revised October 12, 1966.)

ERRATA

Page 256, line 9 in the table notes should read: "**Calculated using the expression Age = 8030 *ln* A^c, where . . . ."

Page 256, line 10 in the table notes should read: "100 x P'/% Mdn. . . ."

Page 259, line 15 from the bottom in the left hand column should read: "Table 1 are shown . . . ."

Copyright ©1969 by Microforms International Marketing Corporation
Reprinted from pages 1321-1330, 1339-1343, and 1346-1349 of Geochim. et Cosmochim. Acta **33**:1321-1349 (1969), by permission of Microforms International Marketing Corporation as exclusive licensee of Pergamon Press journal back files

Geochemistry and origin of formation waters in the western Canada sedimentary basin—I. Stable isotopes of hydrogen and oxygen*

BRIAN HITCHON
Research Council of Alberta, Edmonton, Canada

and

IRVING FRIEDMAN
U.S. Geological Survey, Denver, Colorado

(*Received* 16 *December* 1968; *accepted in revised form* 28 *May* 1969)

Abstract—Stable isotopes of hydrogen and oxygen, together with chemical analyses, were determined for 20 surface waters, 8 shallow potable formation waters, and 79 formation waters from oil fields and gas fields. The observed isotope ratios can be explained by mixing of surface water and diagenetically modified sea water, accompanied by a process which enriches the heavy oxygen isotope. Mass balances for deuterium and total dissolved solids in the western Canada sedimentary basin demonstrate that the present distribution of deuterium in formation waters of the basin can be derived through mixing of the diagenetically modified sea water with not more than 2·9 times as much fresh water at the same latitude, and that the movement of fresh water through the basin has redistributed the dissolved solids of the modified sea water into the observed salinity variations. Statistical analysis of the isotope data indicates that although exchange of deuterium between water and hydrogen sulphide takes place within the basin, the effect is minimized because of an insignificant mass of hydrogen sulphide compared to the mass of formation water. Conversely, exchange of oxygen isotopes between water and carbonate minerals causes a major oxygen-18 enrichment of formation waters, depending on the relative masses of water and carbonate. Qualitative evidence confirms the isotopic fractionation of deuterium on passage of water through micropores in shales.

INTRODUCTION

FORMATION waters are an important part of the hydrologic cycle because they are the medium within which crude oil and natural gas are found, and because diagenesis, cementation, and ore deposits within the sedimentary rocks are all features for which clues to their history may be sought by study of formation waters. Indeed, formation waters may be economic minerals in their own right, for example, as sources of bromine, iodine and calcium chloride. It is the intention of this series of papers to present the significant geochemical features of formation waters from the western Canada sedimentary basin based on a wide variety of parameters obtained from a suite of nearly ninety samples. Particular attention will be directed to those features which may be of value in elucidating the history of formation waters.

The western Canada sedimentary basin is structurally simple and stratigraphically well known. For a definitive detailed geological history the interested

---

* Joint publication of the Research Council of Alberta (Contribution No. 451), and the U.S. Geological Survey.

reader is referred to McCrossan and Glaister (1964). Rock volume and pore volume data for the plains region of the basin have been compiled by Hitchon (1968b). Abundant modern reliable data on the formation fluids are available, and their regional variations and geochemistry have been described (Hitchon et al., 1961; Hitchon, 1963a, b, c, 1964, 1968a). The fluid flow within this basin has been elucidated (Hitchon, 1963d, 1969a, b). Thus considerable information is available concerning the contained fluids and the rock matrix within which they move. It is against this background that the stable isotopes of hydrogen and oxygen in formation waters from Alberta, Canada, will be considered.

Previous Studies

Most early determinations and some current investigations of stable isotopes of hydrogen in formation waters have been made by very precise measurements of the density of the salt-free water. However, hydrogen isotopes may be fractionated during purification procedures, and furthermore, variations in the oxygen isotopes ratio also affect the density of the water. Users of the density method have all observed a more or less regular increase in deuterium content of formation waters with depth and have variously associated this increase with concomitant increases of the total dissolved solids content of the formation waters, the age of the rocks, and other factors. Anomalies have been ascribed to deep circulation of surface waters or to deuterium exchange with hydrocarbons. None of the users of the density method was able to measure differences in the oxygen isotopes, and since this omission may have seriously affected their results only data determined by mass-spectrometric methods will be discussed in detail.

Graf et al. (1965) and Clayton et al. (1966) have provided the only comprehensive studies of stable isotopes of both hydrogen and oxygen in formation waters. Baertschi (1953), Degens (1962) and Degens et al. (1964) have considered only oxygen isotopes, and Roth (1956) only hydrogen isotopes. A small fractionation of hydrogen isotopes resulting from passage of water through micropores in shales was postulated by Graf et al. (1965) and a major fractionation of the oxygen isotopes through exchange with carbonates of the reservoir rocks observed on the same samples by Clayton et al. (1966). Clayton and his co-workers also concluded that the water of formation waters is predominantly of local meteoric origin. This view differs from that of Degens et al. (1964), who attributed the changes of the oxygen isotope ratios to mixing of meteoric waters with original marine interstitial solutions. Other processes that might alter the ratio of the stable hydrogen isotopes, such as exchange with hydrogen sulphide which takes place easily and which is used in most commercial heavy water plants, were not considered in detail. The suite of samples studied in this paper offer an opportunity to examine statistically the effects of many variables which may cause fractionation of the stable isotopes of hydrogen and oxygen.

Sample Collection

The collection and analysis of formation waters (and other fluids) at reservoir temperatures and pressures would minimize physical and chemical changes that may occur when sampling and analysis are undertaken at temperatures and pressures close to those at the wellhead. Since collection is difficult, and analysis often impossible at reservoir temperatures and pressures, extreme care must be exercised to ensure that the sample is at least representative of formation water collected at normal wellhead temperature and pressure. The formation waters used in this study were collected at the wellhead, or at a non-operating treater or separator, thus prior heat treatment or chemical reactions were avoided. In two samples (79 and 80) about 30 ppm of corrosion inhibitor was added to the fluid deep in the well bore due to high hydrogen sulphide contents of the associated natural gas. Where free water was available, the samples for isotope analysis were collected in 4-ounce glass bottles with polyethylene inserts, direct from the wellhead, filled "heaping full" and the tops sealed with wax in the laboratory as soon as possible to minimize exchange with the air. In some instances only oil-water emulsions were recovered and these were removed to the laboratory and broken down by centrifugation or, in a few instances, by freezing at liquid nitrogen temperatures (samples 8, 9, 33, 40, 42, 43, 47 and 55). Separate portions of all samples were chemically analysed for major and minor components and this information compared with known representative chemical analyses. Only representative samples are included in this study. The formation waters from oil and gas fields come from strata ranging in age from Late Cretaceous to pre-Late Devonian (Granite Wash), and in depth to nearly 11,600 ft (Table 4). In addition to the formation waters, 20 surface waters and 8 shallow potable waters were sampled, details of which may be found in Tables 2 and 3, respectively.

EXPERIMENTAL

The deuterium concentrations were determined by converting samples of 0·01 ml to hydrogen gas by reaction with hot uranium metal (FRIEDMAN and WOODCOCK, 1957). The deuterium-hydrogen ratio in this gas was compared to the ratio in a standard gas, using a specially constructed mass spectrometer (FRIEDMAN, 1953). All samples were processed and analysed in replicate, the replicates agreeing to within ±0·1 per cent in 95 per cent of the samples analysed. The analyses are expressed as deuterium enrichments (plus δ values) or depletions (negative δ) relative to SMOW (Standard Mean Ocean Water, having a D/H ratio of about 158×10^{-6}, CRAIG, 1961b). Thus a sample having a δ value of -4 has 4 per cent less deuterium than does Standard Mean Ocean Water. The comparative study of isotope ratios of deuterium and oxygen-18 in natural waters by the International Atomic Energy Agency (HALEVY and PAYNE, 1967) indicates that the deuterium values reported in this paper will be about 1 per cent lighter than the mean lines for surface waters reported by CRAIG (1961b) and DANSGAARD (1954, 1960).

The oxygen-18 analysis on water samples was carried out by the method described by EPSTEIN and MAYEDA (1953). A sample of 10 ml of water was equilibrated with carbon dioxide at 25°C and an aliquot of the CO_2 was analysed for O^{18}. Some of the very saline waters were distilled to dryness before carrying out the equilibration. The carbonates were reacted with 100 per cent phosphoric acid to liberate CO_2 by the method described by MCCREA (1950).

Corrections to the mass spectrometric results were carried out as described by CRAIG (1957). In addition, dolomites were corrected for the difference in isotopic fractionation factor associated with the phosphoric acid reaction, as suggested by SHARMA and CLAYTON (1965). All O^{18} analyses are given relative to SMOW, as suggested by CRAIG (1961a, b). For the water and carbonate samples the O^{18} data are precise to ±0·1 per mille.

THEORETICAL CONSIDERATIONS

Within the surface regime of the hydrologic cycle, fractionation of the stable isotopes of hydrogen and oxygen in water takes place mainly through changes of state. The processes which fractionate the isotopes of hydrogen affect the isotopes of oxygen in a similar way. For surface waters there is a close relation between δD and δO^{18} when both are referred to Standard Mean Ocean Water (FRIEDMAN, 1953; CRAIG, 1961a). FRIEDMAN et al. (1964) have shown that for surface water, the temperature at which precipitation takes place, and which may be correlated with altitude and latitude, is the chief variable controlling fractionation of the isotopes of hydrogen and corresponding fractionation has been observed for the isotopes of oxygen (DANSGAARD, 1954, 1960, 1961).

Once surface water has penetrated into the subsurface, changes in the ratios of the stable isotopes of hydrogen and oxygen may take place through a variety of processes. The surface water may mix with waters of differing isotope ratios, as suggested by DEGENS et al. (1964) and the isotopic composition may also change because of exchange with other fluids (ROTH, 1956), or with the rock matrix (CLAYTON et al., 1966). In addition, the act of moving through shale micropores may affect the isotope ratio (GRAF et al., 1965). FRIEDMAN et al. (1964) have observed that water from shallow wells is probably one of the best sources of samples representative of the regional surface isotopic characteristics. Waters in the underground environment may be separated, rather artificially, into shallow, mainly potable, formation waters, representative of the regional surface isotopic characteristics, and deeper, variably saline formation waters. It is in the latter group that we must seek clues to the history of formation waters in sedimentary basins.

CLAYTON et al. (1966) observed that the variation in the deuterium content of formation waters among four sedimentary basins they studied was much greater than that within each basin. They also noted that the oxygen isotopic composition showed a large range within each basin. They considered several processes in order to account for the variation they observed, and concluded that the original water from the depositional basin had been lost during compaction and subsequent flushing and that the present formation water originated as precipitation over land, under climatic conditions not greatly different from those prevailing today. They thus rejected the thesis of DEGENS et al. (1964) that formation waters result from the mixing of marine and fresh water, noting that oxygen isotope ratios alone are insufficient to merit the conclusions reached by DEGENS et al. (1964). CLAYTON et al. (1966) presented convincing evidence that extensive oxygen isotopic exchange has taken place between the water and the rock matrix, and that the deuterium content of the formation waters has not been greatly altered by exchange or fractionation processes. The present authors believe that the data presented by both DEGENS et al. (1964) and CLAYTON et al. (1966) are equivocal and that, in fact, formation waters may also originate by mixing of surface waters, at the present latitude of the basin, with the modified marine (or non-marine) water present in the rocks, and that this mixing, together with extensive exchange of oxygen isotopes between the water and the rock matrix, results in the distinctive trends observed from basin to basin. The term "mixing" is used because it is the most pertinent to describe a process which is envisioned as a slow percolation of water through the basin, carrying with it the dissolved salts, and simultaneously changing both the composition of the inflowing water and the water in the basin. Neither diffusion nor flushing is an appropriate term because the former, strictly, implies only movement on a molecular scale, and the latter may be interpreted as a mass movement of water, sweeping all fluids before it.

CLAYTON et al. (1966) placed considerable emphasis on the trends of δD against δO^{18} for the basins they studied in their case for rejection of the mixing hypothesis of DEGENS et al. (1964). Figure 1 is based upon their graph of δD against δO^{18} and shows the interpretation that must be placed upon trends which fall within specific regions of the diagram, based on our assumption that mixing of surface water and sea water (SMOW) takes place. If mixing is unaccompanied by other processes causing differential isotope fractionation all formation waters would lie on the line Surface water—SMOW. If mixing is accompanied by a fractionation process differentially affecting only the hydrogen isotopes all formation waters will plot in the region above the proportional mixing line and with δO^{18} values less than SMOW. Similarly, if mixing is accompanied by a fractionation process which differentially affects the oxygen isotopes, all formation waters will plot in the region below the proportional mixing line and with δD values less than SMOW. It follows that if formation waters plot in the blank region of the diagram then even on the extension of the proportional mixing line, at least one process, other than mixing, must have occurred that altered isotopic compositions. At least two processes, other than mixing, are required if the two chemical species are differentially affected.

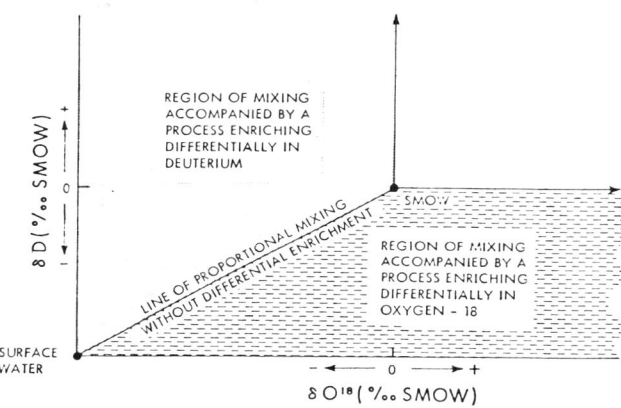

Fig. 1. Schematic plot of δD and δO^{18} indicating the effects of various mixing and fractionating processes for any specific sedimentary basin.

None of the Alberta samples analysed in this study fall in the blank region and only a few of those analysed by CLAYTON et al. (1966) are found there. They were predominantly from the Gulf Coast, where surface waters have hydrogen isotopic ratios fairly close to Standard Mean Ocean Water (FRIEDMAN et al., 1964). Thus, within the limits of sampling and experimental errors, and allowing for the possibility of a small fractionation of stable hydrogen isotopes by one of several processes noted by CLAYTON et al. (1966) which could place some of the Gulf Coast samples in the blank region, all formation waters examined so far lie within the region in which mixing of surface water with SMOW is accompanied by a process which differentially fractionates the oxygen isotopes. It follows that determination of the mass balance of deuterium in a sedimentary basin may indicate the degree of mixing to which the formation waters have been subjected.

During the past ten years much evidence has been accumulated to indicate that shales act as membrane ultrafilters (BERRY, 1958, 1967; BERRY and HANSHAW, 1960; BREDEHOEFT et al., 1963; GRAF et al., 1965; HANSHAW, 1962, 1964; HANSHAW and ZEN, 1965; WHITE, 1965). Water passes freely through the micropores but the dissolved salts are differentially removed. The deuterium concentration of a formation water is not necessarily related to the concentration of total dissolved solids. Nevertheless, in any sedimentary basin in which mixing has taken place, the mass balance of deuterium and the mass balance of dissolved solids obviously will be related.

The Alberta basin will be used as a model to consider the processes that have been described so far in terms of an actual sedimentary basin. In simplified terms the Alberta basin comprises a wedge of unfolded sedimentary rocks resting on buried Precambrian (Fig. 2). The thickness varies from zero at the exposed Precambrian Shield to over 16,000 feet adjacent to the folded strata of the foothills belt, which forms the eastern subdivision of the Canadian Cordillera.

Most of the sediments were deposited in marine environments, with the various marine incursions separated by periods of uplift, during some of which the hydrodynamic situation allowed influx of fresh water into the rocks as evidenced by structural studies of post-Middle Devonian strata which show several periods of

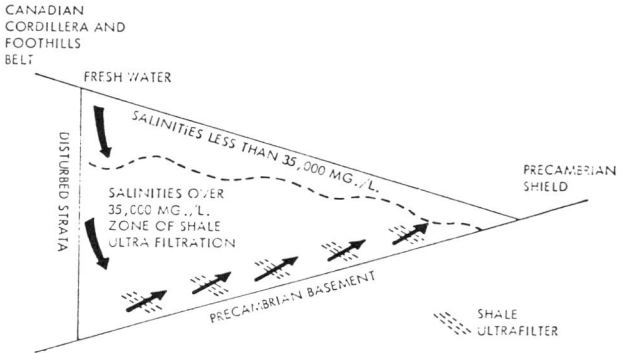

Fig. 2. Model of mixing and shale ultrafiltration mechanisms for the Alberta basin.

solution of the Middle Devonian halite (GORRELL and ALDERMAN, 1968). However, at no time prior to the culmination of the Laramide orogeny is it likely that there existed a hydraulic head greater than that generated by the present Canadian Cordillera. Paleogeographic studies indicate regional land elevations, and hence hydraulic head, decreasing westward (the opposite situation to the present) in pre-Laramide time, although regional land elevations were considerably lower. In addition, the sediments deposited in pre-Laramide time would have higher porosities and permeabilities then, than they have now, thus allowing more ease of fluid movement, but at the same time shale membrane efficiency would be lower. These various aspects of the complex history of the western Canada sedimentary basin suggest that the present salinity distribution of the formation waters is largely a post-Laramide feature. Paleolatitude studies indicate essentially the same latitude situation for western Canada during the post-Laramide period, and this suggests that the deuterium content of precipitation in western Canada has probably not been very much different from the present content since Laramide time (except possibly during Pleistocene glaciation). Whatever the paleolatitude position of western Canada there is no conclusive evidence which suggests that the composition of ocean water has changed significantly in the past 150 million years, although in pre-Jurassic time the composition may have been different. A few samples show sea waters along Canadian coasts have about 156 ppm deuterium (relative to SMOW = 157·6 ± 0·3 ppm absolute determination); the paucity of samples makes it difficult to determine the effect of dilution by continental run-off, but suggests that deuterium contents slightly lower than SMOW may also be reasonable (A. R. BANCROFT, personal communication, 1968). However, the water in the western Canada sedimentary basin immediately prior to the Laramide uplift was certainly isotopically no heavier than SMOW and chemically probably only slightly modified from sea water. During the period (post-Laramide) of effective mixing and shale ultrafiltration the fresh water recharge to the basin probably had a deuterium content similar to that at present.

The main recharge region is the foothills belt with discharge in the low lying areas adjacent to the Precambrian Shield (HITCHON, 1969a). Fluid flow is dominantly downward from the surface through much of the Cenozoic and Mesozoic strata and over much of this region salinities are less than 35,000 mg/l. Fluid flow in the deep strata is upward, out of the basin, and the result of this movement is an extensive cascade of increasingly saline waters differentially removed from the flowing formation water behind shale ultrafilter barriers, shown diagrammatically in Fig. 2.

We shall present evidence later to suggest that the cascade effect results in very efficient removal of the dissolved materials. Thus, if we are able to calculate the volumes of formation waters of different salinities we are in a position to estimate the amount of water passing through the shale ultrafilters. For example, if we have a portion of a sedimentary basin containing 1000 cubic miles of formation water with a salinity of 175,000 mg/l, or five times the concentration of sea water, we may estimate that with effective complete removal of the dissolved solids, 4000 cubic miles of sea water have passed through the space now occupied by the 1000 cubic miles of formation water. If we complete a calculation of this nature for formation waters of different salinities for an entire basin and we can be confident that most of the dissolved solids in the diagenetically modified sea water are still in the basin, by doing a mass balance on the dissolved solids in the formation waters and comparing this to the salinity of the sea water, then any excess volume of water over that of the total pore space of the basin should be equal to the volume of fresh water required to rearrange the dissolved solids. Formation waters with salinities less than sea water are included in the calculations by computing the volumes of fresh water and sea water required to produce the observed salinity.

The deuterium balance calculated for the basin based on the measured relation of deuterium to total dissolved solids and weighted for volumes of fluids of differing salinities should be equal to that obtained by mixing fresh water and sea water in the proportions just calculated. We can use the deuterium and total dissolved solids mass balances to estimate the membrane efficiency of the shales and the amount of water required to be passed through the basin in order to rearrange the total dissolved solids to give the variable concentration found at the time of sampling.

Variation of Deuterium Due to Mixing

The availability of reliable estimates of the pore volume in the western Canada sedimentary basin (Hitchon, 1968b), together with maps showing regional variations in salinity of the formation waters (Hitchon, 1964), allows determination of the volume of formation waters of differing salinities. A mass balance calculation using salinity variations weighted for their respective volumes indicates an average salinity of 46,400 mg/l for the entire western Canada sedimentary basin. The distribution of the individual major components indicates sea water modified by exchange of calcium and magnesium during diagenesis and by bacterial reduction of sulphate to yield H_2S and bicarbonate. However, the salinity suggests that essentially neither the remaining dissolved salts of the original sea water nor the products of diagenesis and bacterial action have been lost from the basin, and, in fact, some components (principally sodium and chloride) have been picked up from the rocks as a result of water movement through the basin.

The total pore volume of the western Canada sedimentary basin is 63,600 cubic miles, which represents, approximately, the volume of original modified sea water with a deuterium content of no less than 158 ppm (SMOW), and a total dissolved solids content of 35,000 mg/l. This has been redistributed within the basin by fresh water, which has a local mean deuterium content of 136 ppm, to yield an average weighted deuterium content of 142 ppm for the formation waters in the basin.

Forty-five thousand nine hundred cubic miles of water would be required to account for the 15,300-cubic-mile volume of formation waters with salinites greater than sea water, assuming perfect efficiency of the shale ultrafilters. The 15,300 cubic miles of formation waters contain 75 per cent of the total dissolved solids in the basin. This means essentially that the major portion of the dissolved solids in the sea water have been concentrated in a quarter of the pore space by the action of 45,900 cubic miles of fresh water. The remaining 25 per cent of the total dissolved solids have been extensively dispersed in the 48,300 cubic miles of pore space with formation waters whose total dissolved solids are less than 35,000 mg/l. Based on these figures the weighted average content of deuterium in the basin due to mixing should be 149 ppm. This is 7 ppm higher than the measured average for the basin. The measured average content of deuterium for the basin requires the mixing of about 184,000 cubic miles of fresh water. This could be interpreted to mean that the shales have at least 25 per cent membrane efficiency. The mass balance of deuterium and total dissolved solids demonstrates that the formation waters in the western Canada sedimentary basin can be derived by mixing of the diagenetically modified sea water present in the basin prior to uplift with about 2·9 times as much fresh water which passes through shale ultrafilters with at least 25 per cent membrane efficiency. If the deuterium content of the diagentically modified sea water present prior to the Laramide revolution was less than 158 ppm (SMOW) due to mixing with fresh water, the effect will be to reduce the volume of fresh water mixed in post-Laramide times that is required to produce the present salinity distribution and deuterium mass balance. Thus not more than 2·9 times as much fresh water will produce the observed situation. The possibility exists that some sedimentary basins may have been completely flushed by fresh water, as suggested by Clayton et al. (1966); indeed all variations in degree of flushing and mixing probably exist.

FRACTIONATION OF ISOTOPES

In addition to changes in concentration of deuterium and oxygen-18 brought about by mixing of fresh water and diagenetically modified sea water, oxygen isotopes are known to be extensively fractionated and fractionation of the hydrogen isotopes is possible.

Isotopic exchange of oxygen between water and carbonates is temperature dependent and a well known reaction. CLAYTON et al. (1966) have demonstrated exchange between formation waters and carbonates of the rocks. Because temperature and pressure increase rather uniformly with depth, formation waters at greater depths should contain more O^{18} than those at shallower depths, provided sufficient carbonate rocks are available to allow isotopic exchange. In the Alberta basin extensive volumes of carbonate rocks are found only at depth (Table 1) and so any enrichment of O^{18} in the shallower formation waters is most probably due to exchange with carbonate cements or, less likely because of smaller relative masses, to

Table 1. Some physical and geochemical parameters of the Alberta basin in relation to the 79 sets of data on formation waters from oil fields and gas fields of Alberta, Canada

Stratigraphic unit	Number of samples	Average depth of samples (ft)	Reservoir lithology	Per cent of samples with H_2S in associated natural gas	Volume of shales in Alberta basin (cu. miles)	Volume of carbonates in Alberta basin (cu. miles)
Tertiary and Cretaceous	28	3400	sandstone	15	110,000	—
Jurassic and Triassic	5	4200	sandstone	20	4600	240
Permian and Carboniferous	10	5200	carbonate	30	8900	15,000
Upper Devonian—Wabamun and Winterburn Groups	13	5700	carbonate	100	230	21,000
Upper Devonian—Woodbend Group and Beaverhill Lake Formation	21	7100	carbonate	76	26,000	22,000
Granite Wash	2	6300	sandstone	0	—	—

exchange with carbon dioxide, CO_3^{2-} or HCO_3^{2-} in the water. Most natural gases contain some carbon dioxide with a tendency to increased partial pressure of carbon dioxide in the deeper strata (HITCHON, 1963b) and so the possibility of exchange between the oxygens of water and carbon dioxide cannot be ruled out.

Fractionation of the stable hydrogen isotopes may take place between hydrogen sulphide and water, a reaction which is both pressure and temperature dependent and which takes place sufficiently readily to be of use in commercial heavy water plants. CLAYTON et al. (1966) have indicated other processes, such as exchange with the water of hydration in sedimentary minerals or between water and hydrocarbons, which may cause hydrogen isotope fractionation but reliable reaction rates are not available nor are the directions of fractionation known. Increased depth is accompanied by both increased temperature and partial pressure of hydrogen sulphide (HITCHON, 1963b). This component is absent from most of the shallower natural gases in Tertiary and Mesozoic strata and so any enrichment of deuterium in shallow formation waters must be due to factors other than exchange with hydrogen sulphide.

In addition to these processes, water from shallow groundwater wells is representative of regional surface isotope characteristics, which vary with altitude and latitude (FRIEDMAN et al., 1964; DANSGAARD, 1954, 1960, 1961) and therefore isotope fractionation will relate directly to latitude and altitude. Finally, since fluid potential is the main driving force isotopic fractionation through shale micropores will vary with fluid potential. No one parameter is likely to control the total fractionation of hydrogen and oxygen isotopes. In order to evaluate each process the nine parameters discussed above were examined by multiple regression analysis.

[Editors' Note: Material has been omitted at this point.]

IMPORTANCE OF RELATIVE MASSES OF MATERIALS IN ISOTOPIC EXCHANGE

The relative masses of the materials undergoing the exchange reaction is an important parameter in any study of isotopic exchange. The predominant fluid in the underground environment is formation water. The effect of the temperature dependent isotopic exchange between water and hydrogen sulphide will be insignificant simply because of the insignificant amount of hydrogen sulphide relative to that of the formation water. Thus variations in isotopic ratio of hydrogen (and oxygen) in formation waters will reflect exchange mechanisms which have affected the major mass of the formation water throughout the basin.

Table 6. Parameters used in the study of water samples rejected on the basis of their chemical composition

Sample No.	Production	δD (‰ SMOW)	δO^{18} (‰ SMOW)	Total dissolved solids (mg/l) (calculated) As sampled	Representative value	P_{H_2S} (psi)
R15	Oil	−6.2	+1.8	62,200	183,000	0
R18	Oil	−5.7	+4.3	128,000	175,000	0
R21	Oil	−10.7	−8.6	21,600	221,000	0
R52	Oil	−6.3	−17.0	1920	7000	0
R53	Gas	−15.2	−6.1	680	62,000	20
R57	Oil	−6.2	+7.4	89,600	214,000	0
R59	Gas	+11.9	−0.2	1610	75,000	1000
R61	Gas	+8.8	−2.3	40,200	69,000	940
R62	Gas	+5.5	−8.8	6420	24,000	100
R63	Gas	−10.4	−8.1	1010	7000	0
R65	Oil	−9.6	−11.0	6110	23,000	3
R77	Oil	−2.7	−4.3	1590	70,000	36
R84	Gas	−4.4	−2.8	2630	77,000	34

The importance of the relative masses of the materials undergoing isotopic exchange may be illustrated by examining the isotopic ratios in formation waters collected during the course of this study, but which, on the basis of chemical analyses, were rejected as non-representative. The pertinent data for these samples are given in Table 6. The samples contain very low total dissolved solids relative to representative formation waters from their respective oil fields or gas fields, and in some cases the proportions of the individual ions fail to agree with the best analysis available. In oil fields contamination with waterflood water is the most likely explanation for the low total dissolved solids content. The water collected from the gas fields is probably condensed water vapor which has been brought up with the natural gas and mixed with a minor amount of formation water. As might be anticipated, graphs of δD and δO^{18} against measured total dissolved solids (middle diagrams, Figs. 5 and 6, respectively) show that many of these rejected samples fall far outside the trends obtained by plotting these same variables using the 79 representative formation waters in Table 4 (top diagrams, Figs. 5 and 6, respectively). When δD and δO^{18} of these rejected samples are plotted against the respective representative total dissolved solids for each sample (bottom diagrams, Figs. 5 and

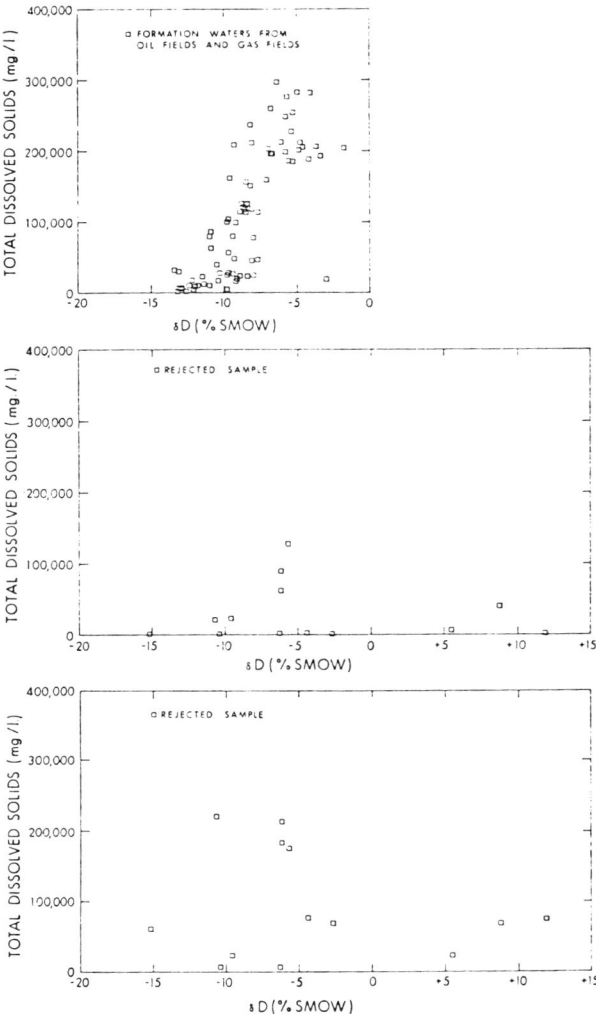

Fig. 5. Scatter diagrams showing relations of $\delta D(\%\ SMOW)$ to total dissolved solids for 79 representative formation waters used in this study (top diagram); to measured total dissolved solids of 13 non-representative formation waters (middle diagram); and to the representative value for total dissolved solids in the same 13 rejected samples (bottom diagram).

6, respectively), the adjustment to the correct salinity places all points on the δO^{18} graph within the trend obtained for the data in Table 4, but this is not true for the δD graph. This is interpreted to mean that the δO^{18} value on the rejected samples is correct, and that it is the δD values which are in error.

The most sensitive graph on which to check this assumption is that of δD against δO^{18}, for which the 79 representative samples had a correlation coefficient of 0·87. When plotted on this graph (Fig. 7) five of the samples fall within the trend for the representative data, one on the deuterium-deficient side and the remaining

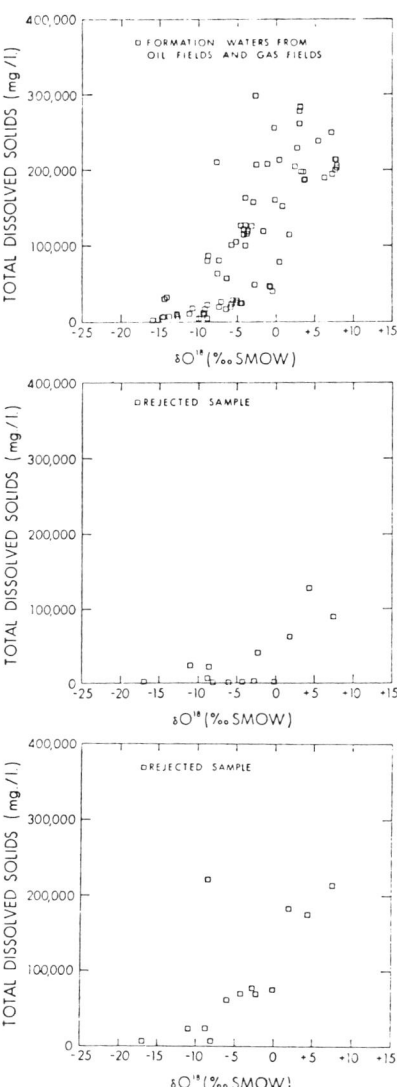

Fig. 6. Scatter diagrams showing relations of δO^{18}(‰ SMOW) to total dissolved solids for 79 representative formation waters used in this study (top diagram); to measured total dissolved solids of 13 non-representative formation waters (middle diagram); and to the representative value for total dissolved solids in the same 13 rejected samples (bottom diagram).

seven on the deuterium-enriched side. With the exception of one sample (R52), the greater the distance from the trend for the representative data, the greater the partial pressure of hydrogen sulphide. All samples falling within the trend for the representative data have zero partial pressures of hydrogen sulphide, and, with one exception, are from oil fields. With the exception of three samples, those falling outside the trend for the representative data are from gas fields. Only the oil fields are subjected to water flooding and because the surface waters lie at the deuterium-poor and oxygen-18-poor end of the trend for the representative data the effect of mixing formation water and fresh water used for water flooding in fields without hydrogen sulphide will be to move the data points down the trend for the representative data towards the region of the graph in which the surface and near-surface water data are found. The effect of water flooding would be exactly the same for samples associated with natural gases with high partial pressures of hydrogen sulphide. The reason that samples associated with high partial pressures of

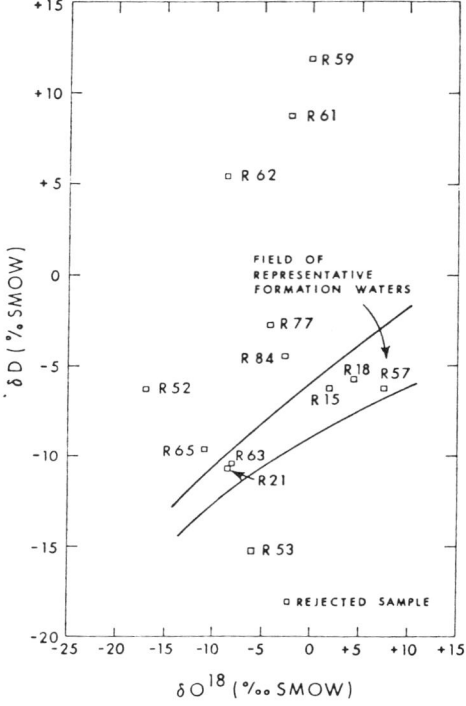

Fig. 7. Scatter diagram showing relation of δD (‰ SMOW) to δO^{18} (‰ SMOW) in 13 non-representative formation waters.

hydrogen sulphide and non-representative total dissolved solids fall outside the trend for the representative data is not because of incorrect δO^{18} values, but because the water sample has been subjected to exchange of deuterium with hydrogen sulphide consequent upon dilution of the formation water.

The amount of condensed water vapor found with natural gases when they are sampled at the surface is insignificant in relation to the amount of formation water with which the natural gas is associated in the reservoir. High partial pressures of hydrogen sulphide result in extreme enrichment of the small amount of condensed water in deuterium but not in O^{18}. The fact that one sample (R53) falls on the deuterium-deficient side is possibly an indication that the proportions of formation water, condensed water vapor and natural gas may be altered under differing temperature and pressure gradients during the passage of the fluids from the reservoir to the sampling point to yield either enriched or depleted samples, depending upon the exact physical conditions. The separation factor between H_2O/H_2S favors enrichment of the H_2O as the temperature decreases (BANCROFT, 1968), and this is undoubtedly an important feature in explaining some of the deuterium enrichment in the non-representative samples. This study of thirteen rejected samples demonstrates the importance of considering the relative amounts of materials involved in all isotopic exchange reactions. It also indicates the care which is required in sampling formation waters and suggests that two parameters should be measured to ensure confidence in samples to be used for isotope studies. These parameters are: (1) a chemical composition consistent with the water in the formation, and (2) δD and δO^{18} data which fall within a fairly narrow trend for that geographical region.

With the data available the isotope fractionation between water and carbonate rocks cannot be treated statistically as comprehensively as that between water and hydrogen sulphide. If the 79 sets of data in Table 4 are divided into two groups, samples from sandstone reservoirs and samples from carbonate reservoirs, a more precise indication of the relative importance of mixing and of exchange of the oxygen isotopes between water and the carbonate rocks can be obtained. Multiple regression analysis of the 35 waters from sandstone reservoirs indicates a multiple regression coefficient of 0·79, with temperature and fluid potential almost equally weighted—but with opposite signs in the regression equation. A similar analysis of the 44 waters from carbonate reservoirs showed temperature the dominant independent variable, but a multiple regression coefficient of 0·82. This suggests that oxygen isotopic exchange between water and carbonates is effective in all reservoirs because sufficient carbonate is present as either cement or host rock to allow equilibration.

To evaluate further the oxygen isotopic exchange between carbonate and water, oxygen isotopes were determined for a selected suite of rocks representing reservoirs from which the formation waters had been sampled. The measured δO^{18} of the carbonates, and subsequent calculations based on these values, are shown in Table 7. In making these calculations, we have assumed that the water and the reservoir carbonate were in equilibrium and the relative amount of sea water and fresh water in the strata can be found by consideration of the deuterium content of the formation water. The fourth and fifth columns show the measured δO^{18} of the formation water in contact with the carbonate at reservoir temperatures (taken from Table 4), and the calculated δO^{18} of the water that should be in equilibrium with the measured oxygen isotopic composition of the carbonate at the same reservoir temperature using the equilibrium relation $1000 \ln \alpha = 2 \cdot 7 \, (10^6 T^{-2}) - 2 \cdot 0$ (J. R. O'NEIL, personal communication, 1968). The discrepancies between the measured and calculated oxygen isotopic composition of the water are due to differences in the relative masses of carbonate and water in the various stratigraphic units. If, as a first approximation, the deuterium content of these formation waters is used as an indicator of the degree of mixing of the diagenetically modified sea water (SMOW) and the fresh water, and which has a local mean deuterium content of 136 ppm, then the proportion of fresh water in each sample can be calculated (Table 7, column 6). CRAIG (1961a) has calculated the regression relation between

[*Editors' Note:* Table 7 has been omitted.]

δD and δO^{18} for fresh waters throughout the world, and from this relation, and knowledge of the proportion of fresh water in the sample, we can calculate the δO^{18} value of the original formation water before oxygen isotopic exchange with the carbonates took place, and this value is shown in column 7 of Table 7. If we now assume that the original δO^{18} of the limestone was $+30‰$, a value representative of modern carbonates, then the change in oxygen isotopic composition of both the formation water and the limestone from their original values to their present values under the influence of both reservoir temperature and their relative masses can be computed (Table 7, columns 8 and 9). Although the δO^{18} values of both modern and ancient limestones vary within fairly narrow limits, the range of values is sufficiently small that even if extreme values were used in the calculations the conclusions would remain valid. By comparing the change in δO^{18} in the limestone with the change in δO^{18} in the formation water, the relative proportions of calcite and water which reacted to give the measured carbonate δO^{18} values can be calculated, and this proportional is shown in the final column of Table 7.

If we consider the values in two groups, those from the Cretaceous and those from the pre-Cretaceous, which have arenaceous and carbonate reservoirs, respectively, then among the Cretaceous samples those with the highest values and hence the largest mass of water, relative to carbonate, are from active fresh-water recharge areas (D, 60, 89, 55). Similarly, in the pre-Cretaceous samples, those from active freshwater recharge areas (87, 69, 73) have values greater than unity; all other samples range from 0·23 to 0·93 with an average of 0·53, indicating inverse proportions of carbonate and water. As with deuterium, the relative masses of the components undergoing isotopic exchange exerts a strong influence on the resulting isotopic composition.

DEUTERIUM FRACTIONATION BY SHALE MICROPORES

GRAF et al. (1965) have postulated a small isotopic fractionation of deuterium resulting from passage of water through micropores in shales. This postulation can be qualitatively checked using the more comprehensive set of data now available from Alberta. If, on a graph similar to the top diagram in Fig. 5, there are superimposed lines joining up points on the same flow paths, as in Fig. 8, then this diagram can be used to illustrate the effect of both mixing and shale micropore fractionation on the deuterium content of formation waters. In Fig. 8 the arrows point in the direction of flow, as determined from hydraulic head maps (HITCHON, 1969a, b). Those lines pointing, in general, from lower left to upper right are from regions where there is active freshwater recharge. In these regions the formation waters become more saline downflow and the deuterium content increases through mixing. All lines trending generally from upper right to lower left represent formation waters moving from the deeper parts of the Alberta basin, updip, out of the basin. In these formation waters the more saline waters are those on the upflow side of the shale ultrafilters and have consequently increased contents of both dissolved solids and deuterium. The formation waters that have passed through the ultrafilters are fresher and contain lower amounts of deuterium. In Part 2 of this

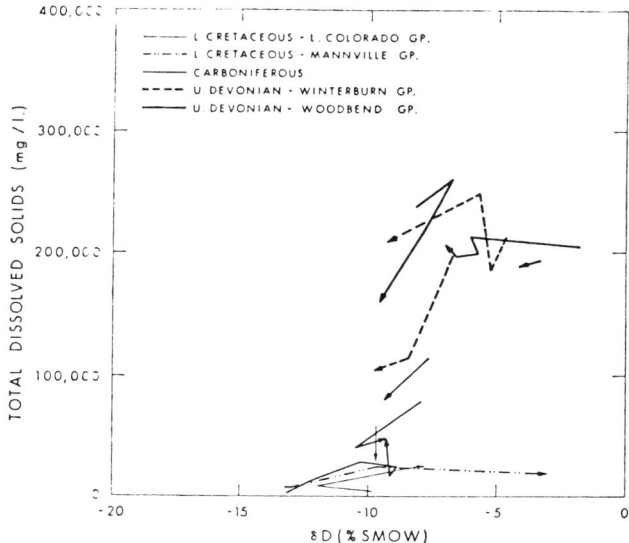

Fig. 8. Relation of δD and δO^{18} to flow directions in formation waters from Alberta, Canada.

series of papers (BILLINGS *et al.*, 1969) evidence will be presented to show that there is ionic fractionation, as well as changes in total dissolved solids.

Acknowledgments—Special acknowledgment is made to Dr. J. TÓTH, Research Council of Alberta, for many constructive ideas originating out of extensive discussions with the senior author on a multitude of topics related to hydrodynamics.

We appreciate the cooperation of the Oil and Gas Conservation Board of Alberta in collecting the formation waters and especially the able assistance of Mr. D. R. SHAW, Chief Chemist. Mr. MELVIN STROSHER of the Research Council of Alberta assisted in the field work and was responsible for sample preparation prior to isotope analysis and his thorough work is specially appreciated. The assistance of Dr. ROBERT RYE, Mr. JIM GLEASON and Mrs. JOY CHURCH in carrying out the isotopic analyses is most gratefully acknowledged. Critical and constructive reviews of the manuscript were provided by Dr. H. W. HABGOOD, Dr. R. GREEN, and Dr. A. VANDENBERG of the Research Council of Alberta; Dr. BRUCE B. HANSHAW and Dr. BLAIR F. JONES of the U.S. Geological Survey; Dr. DONALD L. GRAF, Department of Geology and Geophysics, University of Minnesota; Dr. H. ROY KROUSE, Physics Department, University of Alberta; Dr. G. K. BILLINGS, Department of Geology, Louisiana State University; Dr. PETER FRITZ, Department of Geology, University of Alberta; Dr. J. E. KLOVAN, Department of Geology, University of Calgary; and Mr. A. R. BANCROFT, Chalk River Nuclear Laboratories, Ontario. To all these reviewers the authors express their sincere thanks for their comments.

REFERENCES

BAERTSCHI P. (1953) Über die relativen unterschiede im $H_2^{18}O$-gehalt natürlicher wässer. *Helv. Chim. Acta* **36**, 1352–1369.

BANCROFT A. R. (1968) The Canadian approach to cheaper heavy water (1967). Atomic Energy of Canada Limited, Report AECL-3044.

BERRY F. A. F. (1958) Hydrodynamics and geochemistry of the Jurassic and Cretaceous systems in the San Juan basin, northwestern Colorado. Ph.D. Thesis, Stanford University.

BERRY F. A. F. (1967) Role of membrane hyperfiltration in origin of thermal brines, Imperial Valley, California. *Amer. Assoc. Petrol. Geol. Bull.* **51**, 454–455.

BERRY F. A. F. and HANSHAW B. B. (1960) Geologic field evidence suggesting membrane properties of shales. *Volume of Abstracts, Twenty-First Int. Geol. Congress, Copenhagen*, p. 209. Det Berlingske Bogtrykkeri, Copenhagen.

BILLINGS G. K., HITCHON B. and SHAW D. R. (1969) Geochemistry and origin of formation waters in the western Canada sedimentary basin. 2. Alkali metals. *Chem. Geol.* **4**, 211–223.

BREDEHOEFT J. D., BLYTH C. R., WHITE W. A. and MAXEY G. B. (1963) Possible mechanism for concentration of brines in subsurface formations. *Amer. Assoc. Petrol. Geol. Bull.* **47**, 257–269.

CLAYTON R. N., FRIEDMAN I., GRAF D. L., MAYEDA T. K., MEENTS W. F. and SHIMP N. F. (1966) The origin of saline formation waters, 1. Isotopic composition. *J. Geophys. Res.* **71**, 3869–3882.

CRAIG H. (1957) Isotopic standards for carbon and oxygen and correction factors for mass-spectrometric analysis of carbon dioxide. *Geochim. Cosmochim. Acta* **12**, 133–149.

CRAIG H. (1961a) Isotopic variations in meteoric waters. *Science* **133**, 1702–1703.

CRAIG H. (1961b) Standards for reporting concentrations of deuterium and oxygen-18 in natural waters. *Science* **133**, 1833–1834.

DANSGAARD W. (1954) The O^{18}-abundance in fresh water. *Geochim. Cosmochim. Acta* **6**, 241–260.

DANSGAARD W. (1960) O^{18} in atmospheric waters. *Summer Course on Nuclear Geology, Varenna*, 1960, pp. 150–166. Comitato Nazionale per L'Energie Nucleare, Lab. di Geol. Nucleare, Pisa, Italy.

DANSGAARD W. (1961) The isotopic composition of natural waters with special reference to the Greenland ice cap. *Medd. Grønland* **165**, 1–120.

DEGENS E. T. (1962) Geochemische untersuchungen von wässern aus der Ägyptischen Sahara. *Geol. Rundsch.* **52**, 625–639.

DEGENS E. T., HUNT J. M., REUTER J. H. and REED W. E. (1964) Data on the distribution of amino acids and oxygen isotopes in petroleum brine waters of various geologic ages. *Sedimentology* **3**, 199–225.

EPSTEIN S. and MAYEDA T. (1953) Variation of O^{18} content of waters from natural sources. *Geochim. Cosmochim. Acta* **4**, 213–224.

FRIEDMAN I. (1953) Deuterium content of natural waters and other substances. *Geochim. Cosmochim. Acta* **4**, 89–103.

FRIEDMAN I., REDFIELD A. C., SCHOEN B. and HARRIS J. (1964) The variation of the deuterium content of natural waters in the hydrologic cycle. *Rev. Geophys.* **2**, 177–224.

FRIEDMAN I. and WOODCOCK A. H. (1957) Determination of deuterium–hydrogen ratios in Hawaiian waters. *Tellus* **9**, 553–556.

GORRELL H. A. and ALDERMAN G. R. (1968) Elk Point Group saline basins of Alberta, Saskatchewan, and Manitoba, Canada. *Geol. Soc. Amer. Spec. Paper* **88**, 291–317.

GRAF D. L., FRIEDMAN I. and MEENTS W. F. (1965) The origin of saline formation waters, II. Isotopic fractionation by shale micropore systems. *Illinois State Geol. Surv.*, Circ. 393.

HALEVY E. and PAYNE B. R. (1967) Deuterium and oxygen-18 in natural waters: analyses compared. *Science* **156**, 669.

HANSHAW B. B. (1962) Membrane properties of compacted clays. Ph.D. Thesis, Harvard University.

HANSHAW B. B. (1964) Cation-exchange constants for clays from electrochemical measurements. *Clays, Clay Minerals* **12**, 397–421.

HANSHAW B. B. and ZEN E. (1965) Osmotic equilibrium and overthrust faulting. *Geol. Soc. Amer. Bull.* **76**, 1379–1386.

HITCHON B. (1963a) Geochemical studies of natural gas. Part I. Hydrocarbons in western Canadian natural gases. *J. Can. Petrol. Tech.* **2**, 60–76.

HITCHON B. (1963b) Geochemical studies of natural gas. Part II. Acid gases in western Canadian natural gases. *J. Can. Petrol. Tech.* **2**, 100–116.

HITCHON B. (1963c) Geochemical studies of natural gas. Part III. Inert gases in western Canadian natural gases. *J. Can. Petrol. Tech.* **2**, 165–174.

HITCHON B. (1963d) Composition and movement of formation fluids in strata above and below the pre-Cretaceous unconformity in relation to the Athabasca oil sands. *The K. A. Clark Volume*, pp. 63–74. Research Council of Alberta, Edmonton, Alberta.

HITCHON B. (1964) Formation fluids. In *Geological History of Western Canada*, Chap. 15. Alberta Soc. Petrol. Geol., Calgary, Alberta.
HITCHON B. (1968a) Geochemistry of natural gas in western Canada. *Natural Gases of North America*, Vol. 2, pp. 1995–2025. Amer. Assoc. Petrol. Geol., Tulsa, Oklahoma.
HITCHON B. (1968b) Rock volume and pore volume data for plains region of western Canada sedimentary basin between latitudes 49° and 60°N. *Amer. Assoc. Petrol. Geol. Bull.* **52**, 2318–2323.
HITCHON B. (1969a) Fluid flow in the western Canada sedimentary basin. 1. Effect of topography. *Water Resour. Res.* **5**, 186–195.
HITCHON B. (1969b) Fluid flow in the western Canada sedimentary basin. 2. Effect of geology. *Water Resour. Res.* **5**, 460–469.
HITCHON B., LEVINSON A. A. and REEDER S. W. (1969) Regional variations of river water composition resulting from halite solution, Mackenzie River drainage basin, Canada. *Water Resour. Res.* in press.
HITCHON B., ROUND G. F., CHARLES M. E. and HODGSON G. W. (1961) Effect of regional variations of crude oil and reservoir characteristics on *in situ* combustion and miscible-phase recovery of oil in western Canada. *Amer. Assoc. Petrol. Geol. Bull.* **45**, 281–314.
KROUSE H. R. and SASAKI A. (1968) Isotopic studies of springs in western North America. *Geol. Soc. Amer. Program with Abstracts*, 1968 Ann. Mtg., p. 167.
MCCREA J. M. (1950) On the isotopic chemistry of carbonates in a paleotemperature scale. *J. Chem. Phys.* **18**, 849–857.
MCCROSSAN R. G. and GLAISTER R. P. (1964) *Geological History of Western Canada*. Alberta Soc Petrol. Geol., Calgary, Alberta.
ROTH E. (1956) Problèmes relatifs à la production d'eau lourde en France. *Atti Congress Sci., Sez. Nucleare, III, Roma*, 1956, p. 337.
SASAKI A., KROUSE H. R. and COOK F. D. (1968) Sulphur, carbon and oxygen isotope variations in springs of western Canada. *Physics in Canada* **24**, 10.
SHARMA T. and CLAYTON R. N. (1965) Measurement of O^{18}/O^{16} ratios of total oxygen of carbonates. *Geochim. Cosmochim. Acta* **29**, 1347–1353.
SHAW D. R. (1968) Acid gas content of Alberta natural gases—to June 1, 1968. Oil and Gas Conservation Board, Calgary, Alberta.
WHITE D. E. (1965) Saline waters of sedimentary rocks. In *Fluids in Subsurface Environments—A Symposium*, pp. 342–366, Mem. 4. Amer. Assoc. Petrol. Geol.

Geohydrologic Interpretations of a Volcanic Island from Environmental Isotopes

GEORGE H. DAVIS[*]

CHANG KUN LEE

EDWARD BRADLEY AND B. R. PAYNE

Abstract. Isotopic analyses of ground waters collected from 1966 through 1968 on Cheju Island, Republic of Korea, have proved helpful in interpreting the general hydrologic relations and in developing a model of the ground-water flow system in the permeable volcanic rocks of the island. Comparison of tritium and stable isotope (D-^{18}O) contents provided a basis for a model of the ground-water flow regimes. Time variability of these isotopes determined by repeated samplings provided the basis for estimates of mean residence time of ground waters, ranging from about 2 to 9 years. Although the results are only approximate because simplified models were used, they can be of substantial assistance in planning further hydrologic investigations and appraising the feasibility of ground-water development.

INTRODUCTION

Cheju Island, Republic of Korea, is an elliptically shaped volcanic island about 80 km south of the Korean peninsula. It has traditionally been known to the Korean people as an isle of 'three many's and three naughts', namely many winds, many stones, and many women, as well as no thieves, no beggars, and no water. Actually there is water, but its availability and occurrence is not yet well understood. Between 1965 and 1969 a joint study of the application of environmental isotopes to the hydrology of Cheju Island was carried on by the Atomic Energy Research Institute, Seoul, Korea, and the International Atomic Energy Agency (IAEA), Vienna, Austria. Chang Kun Lee, principal investigator for the study, prepared two progress reports [*Lee et al.*, 1966, 1967] containing most of the data on which this report is based. Some of the most recent data are in IAEA files in Vienna.

[*] Formerly with the International Atomic Energy Agency, Vienna, Austria.

The environmental isotope study formed a part of a larger investigation under the direction of Chang Kun Lee of the Atomic Energy Research Institute of the Republic of Korea. This work was supported in part by Research Contract 367/RB with the IAEA, Application of Environmental Isotopes and Artificially Introduced Tracers in the Hydrology of a Volcanic Island (Cheju). In the study described in this paper sampling was carried out by Chang Kun Lee and colleagues, analyses for tritium (^{3}H) were performed by the IAEA Tritium Laboratory, Vienna, and analyses for deuterium (^{2}H) and ^{18}O were performed by Professor W. Dansgaard, University of Copenhagen.

The specific purposes of the study were to (1) characterize ground waters from different sources and to relate them to time and place of recharge, and (2) determine the nature of mixing of the various ground-water reservoirs and calculate their residence times.

An aspect of interest in the later topic was evaluation of the general order of magnitude of storage in the lens of fresh water, floating on denser salt water beneath the island. Attempts to develop this potential source of ground water by drilling wells had met with little success, and some doubt existed as to whether the lens contained fresh water in significant amounts [Lee et al., 1966, p. 27]. Such information should be useful in planning possible development of water supplies on inland parts of the island to relieve present (1968) population pressure in the coastal belt.

Geohydrology

Cheju Island has an area of 1792 square kilometers and is about 75 km long in an east-west direction and 32 km wide in a north-south direction. In the center of the island is an inactive volcano, Mt. Halla, which rises to 1950 meters above sea level. Mt. Halla is a symmetrical cone, but its slopes are interrupted by some 200 parasitic cones and are cut by numerous ravines and valleys.

The rocks of Cheju are almost entirely of volcanic origin with a predominance of basalt but including some andesitic rock types. Sedimentary rocks are limited generally to small deposits of reworked volcanic detritus along the coastal margins of the island. Because of the characteristic strong winds, soils are generally thin and poorly developed. Most of the volcanic rocks are vesicular and are cut by numerous joints and fissures. Lava tunnels and scoriaceous contact zones of individual lava flows are common and provide conduits for rapid ground-water flow.

Rainfall is abundant on Cheju Island; the 30-year normal (1931–1960) for Cheju City on the north coast, published by the World Meteorological Organization, indicates an annual average of 1440 mm. Undoubtedly this total is exceeded by a large amount at higher altitudes in the interior of the island. The heaviest precipitation occurs from June through September during the southwest monsoon season, when 60% of the annual total generally occurs. One or two typhoons usually strike the island during the rainy season and deliver as much as 150 mm of rain in a single day. The areal distribution of heavy rain is very erratic. Accordingly the annual totals show extreme variability among 16 precipitation stations where rainfall is measured regularly [Lee et al., 1966].

Despite the heavy rainfall there is little sustained streamflow on the island. Torrential runoff occurs during and immediately after rain, but the high permeability of the volcanic rocks encourages rapid infiltration. Locally spring fed streams flow for a few kilometers and then disappear. Along the southern coastal perimeter several springs form streams that flow for a few hundred meters and cascade over a cliff into the sea below. Most of the rainfall disappears rapidly into the permeable volcanic rocks. From the higher parts of the island it generally moves downward through joint planes and some pyroclastic beds and then laterally through sloping zones of scoria or similar material, emerging at the surface locally as high or medium altitude springs that give rise to small streams. These small streams disappear again, sometimes within a few hundred meters, where the rocks are highly permeable. Basalt layers dip toward the sea, whereas ash beds dip in various directions in conformance with local parasitic volcanic cones. Presumably a substantial part of the water supply infiltrates to a basal freshwater lens that discharges at springs around the perimeter of the island and perhaps offshore.

Isotope Sampling

Samples from 24 sources (sites 1–24, Figure 1) collected in the course of a reconnaissance in spring 1966 were analyzed for tritium, deuterium, and oxygen-18. The sources sampled included 14 springs, 8 wells, and 2 streams. Detailed information on the sources sampled is given by Lee et al. [1966]. This preliminary work indicated significant differences in both tritium and stable isotope contents of the water that appeared to relate to the time and location of recharge and the storage characteristics of the underground reservoirs. Most of the same sources were resampled in autumn 1966, following the summer monsoon, as a check on preliminary conclusions based on the initial sampling.

On the basis of the preliminary results some nine sources were selected for periodic resampling; some of the nine were sampled quarterly through 1967 and 1968. Subsequently four additional sites (sites 45, 46, 47, and 48), representing regimes not covered in the initial reconnaissance or replacing sources that had to be abandoned, were added to the periodic sampling. These data together with analyses of tritium, deuterium, and oxygen-18 content of precipitation at stations of the World Network of Isotopes in Precipitation at Pohang, Korea, and Tokyo, Japan [*International Atomic Energy Agency*, 1967] form the basis for the interpretations and conclusions presented here.

RESULTS OF ISOTOPE STUDIES

Isotope Relations

The environmental isotopes widely used in hydrologic work include the heavy stable isotopes of water, deuterium (D), and oxygen-18 (^{18}O) and the radioactive isotopes, tritium (^3H), and carbon-14 (^{14}C) found in the dissolved carbon species present in most natural waters. The principal heavy stable isotopes occur in natural waters in average concentrations of about 320 parts per million in the case of HDO, and about 2000 ppm in the case of $H_2^{18}O$. Small but significant variations occur in the stable isotope contents of natural waters that can be measured with high precision by mass spectrometry.

Tritium and ^{14}C are produced by cosmic ray bombardment of the upper atmosphere. These radioactive isotopes decay exponentially with half-lives of 12.26 and 5568 years, respectively. Since 1954 the natural input has been engulfed by injections of tritium and ^{14}C stemming from atmospheric testing of thermonuclear weapons that have occurred at variable rates. Since 1960 the tritium, deuterium, and ^{18}O contents of precipitation have been monitored on a worldwide basis [*International Atomic Energy Agency*, 1967]; the nearest stations in this World Network to Cheju are at Pohang, Korea, and Tokyo, Japan.

The spatial and time variability of input of the environmental isotopes in the hydrologic cycle and the decay of the radioactive isotopes are the key to their interpretation in hydrologic investigations. The trends of tritium and the stable isotopes in precipitation are now fairly well known, and with data from World Network stations input for given areas can be estimated with reasonable precision, thus providing a starting point for local hydrologic interpretation of isotopic data.

The special value of the stable isotopes in hydrologic studies stems from the fact that whenever water changes its form by condensation, evaporation, or freezing an isotope frac-

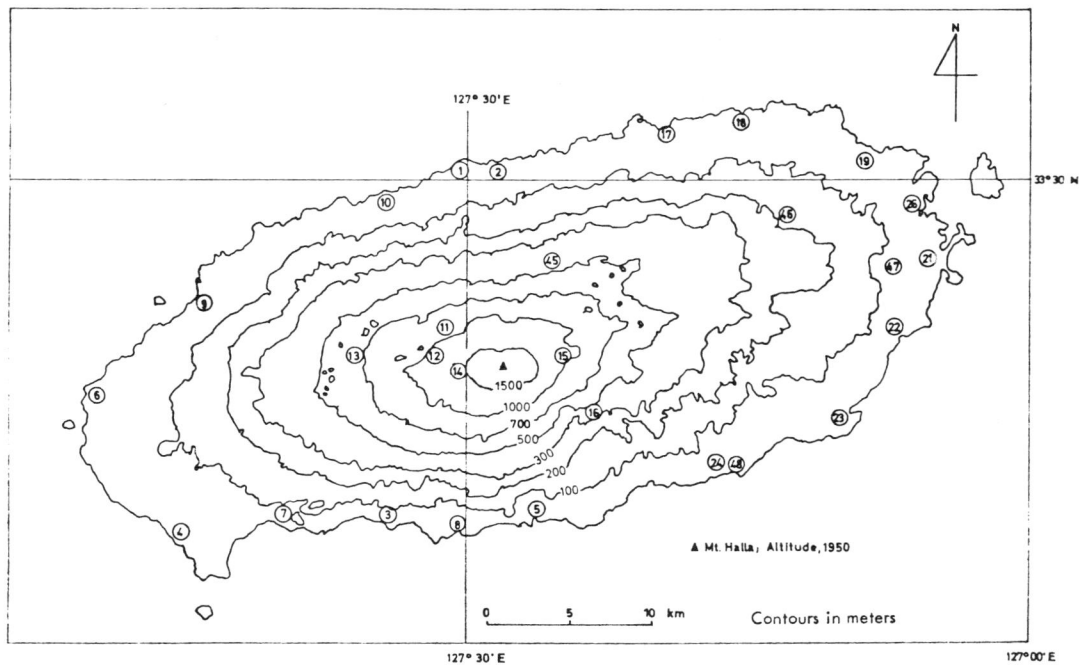

Fig. 1. Topographic map of Cheju Island showing location of selected sampling points.

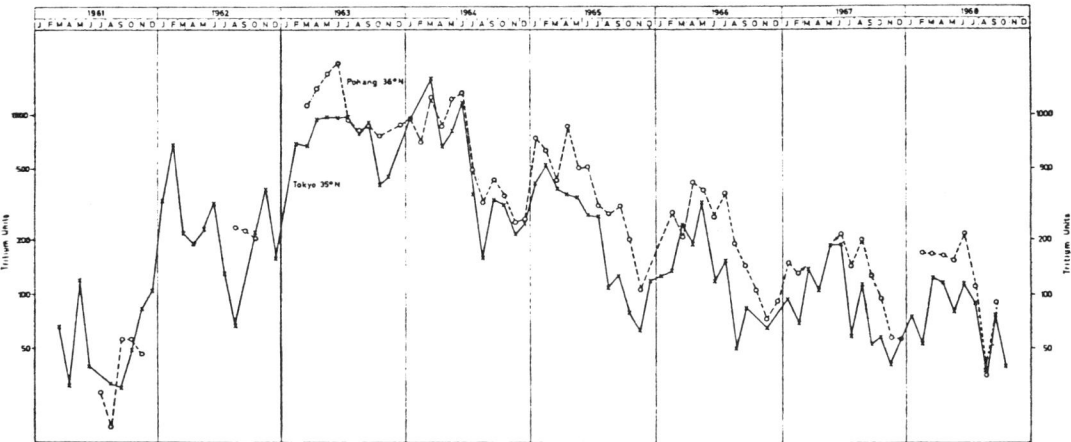

Fig. 2. Tritium content of precipitation, Pohang and Tokyo.

tionation occurs and the heavy isotopes tend to be enriched in the phase having lower vapor pressure. This effect is mainly governed by the temperature of evaporation and condensation; the lower the temperature, the greater the depletion in stable isotopes. In Korea the marked seasonal variation commonly observed elsewhere in stable isotope content of precipitation corresponding to the seasonal temperature variations is obscured by effects of seasonal variations in prevailing wind and source of moisture. Precipitation at higher altitudes on Mt. Halla are expected to be depleted in stable isotope content relative to that at sea level because of the orographic decrease in average temperature.

As hydrologic processes tend to smooth the variability of input, streams show less variability than precipitation; and ground waters, especially in granular materials subject to extensive mixing, show less variability than streams. However, in aquifers such as the volcanic rocks of Cheju, where the water flows through fractures and tubular openings, the input variability is preserved to a greater degree than is generally the case in granular aquifers.

The special value of tritium and ^{14}C in hydrologic studies is due to their property of radioactive decay, which can be used to date the time of recharge to a groundwater system. Once removed from contact with the atmosphere, these radioactive isotopes in ground water are not replenished, and they decay at a known consistent rate. Because of its short half-life of 12.26 years, the presence of measureable amounts of tritium is unequivocal evidence of a component of modern recharge in the water. If the tritium content is above that accountable to the natural prethermonuclear level of precipitation (about 5–10 TU in the northern hemisphere), then this is evidence of some post-1954 recharge.

As all the waters sampled on Cheju contained more than 50 TU, it was concluded that all contained large amounts of modern recharge. Accordingly ^{14}C with a half-life of 5568 years has not been used in the Cheju studies. It was also found that the D–^{18}O ratio did not vary significantly in the waters sampled; therefore in the following discussion the ^{18}O content is used as an index of both stable isotopes.

In reporting tritium, tritium units (TU) are used; 1 TU is a concentration of 1 atom of ^3H in 10^{18} atoms of H. A precision of ±5 to 10% of the reported value is obtained in routine analyses.

Deuterium and ^{18}O are expressed in deviation (δ) per mil ‰ from an arbitrary standard, Standard Mean Ocean Water, (SMOW) [Craig, 1961]. In routine analyses the precision for D is about ±2‰ and for ^{18}O, about ±0.2‰.

Isotopic Content of Precipitation

Records of monthly average deuterium and ^{18}O content of precipitation at Tokyo, Japan, and Pohang, Korea, are available from the initiation of the World Network of Isotopes in Precipitation in March 1961 through spring 1967. A continuous record for tritium in precipitation is available for Tokyo for the same period, but it is only available from March 1963 onward for Pohang. The period of record covers the large tritium maximum of 1963 and the subsequent decline characteristic of northern hemisphere stations [*International Atomic Energy Agency*, 1967]. Both records are consistent with other northern hemisphere stations that exhibit the same general trends and the characteristic annual decline of peak values by a constant ratio. Climatic records for Pohang, Tokyo, and Cheju Island indicate that all three are characterized by a marked deficiency of winter precipitation and very wet summers.

Tritium values in rainfall at Pohang are nearly always higher by significant amounts than values for comparable amounts at Tokyo by significant amounts (Figure 2). The average percent deviation of Tokyo with respect to Pohang is -24%. Presumably because of its marine location, the climatic parameters of Cheju correspond better with those of Tokyo than with those of Pohang, nearer to Cheju. For this reason the tritium record for Tokyo was used in preference to that of Pohang in extrapolating data to Cheju.

Dansgaard [1964] commenting on the Tokyo stable isotope record noted that D and ^{18}O plots of winter precipitation were above the trend line, whereas those plots for summer precipitation were below the trend line. He also observed a pronounced depletion in winter precipitation, due presumably to temperature, whereas during the rest of the year the D-^{18}O content was related to the amount of monthly precipitation. These effects may be ascribed to the fact that in winter the prevailing winds are from the northwest, from the Asiatic continent. The dry continental air collects moisture from the Sea of Japan (in Korea, East Sea of Korea) in a fast evaporation process accounting for the D-^{18}O plots above the trend line. In summer the prevailing wind is from the Pacific and rain, which commonly is very heavy, is closer to equilibrium with ocean water (i.e., $\delta D = 8\ \delta^{18}O$).

Figure 3 indicates a pronounced difference in ^{18}O values between Pohang and Tokyo. Unlike the situation for Tokyo, the most isotopically depleted precipitation at Pohang occurs in the summer during the season of maximum precipitation, and winter precipitation is notably enriched in stable isotope content. Figure 4 shows D-^{18}O content of precipitation at Pohang and Cheju Island ground waters. *Dansgaard* [1964, Fig. 19, p. 464] gives a similar diagram for Tokyo. The two records agree in one respect: on the D-^{18}O plot the points for winter precipitation are above the trend line and those for summer are below the trend line.

Both Tokyo and Pohang are affected by the regional climatic patterns but in different ways. In winter the Sea of Japan (East Sea of Korea) is evidently the source of moisture affecting both stations, but Pohang on the shore of the Sea of Japan (East Sea of Korea) receives precipitation that represents a first-stage condensa-

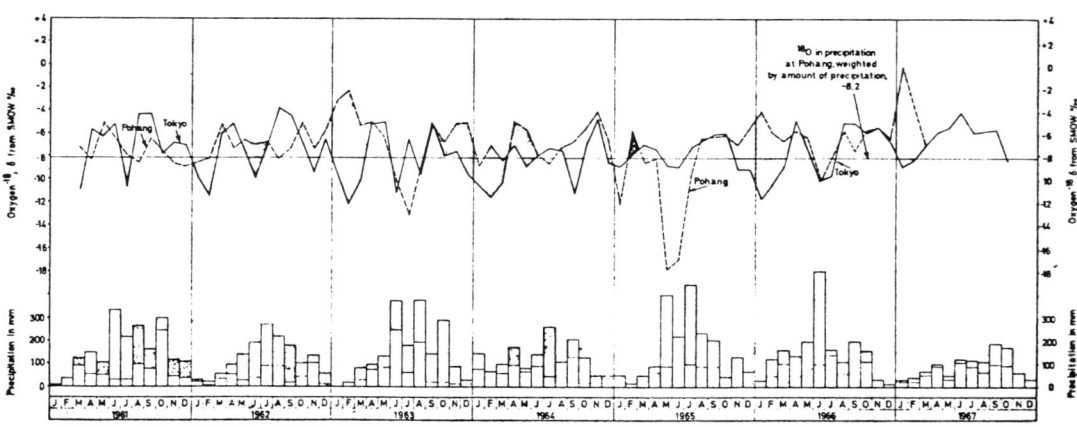

Fig. 3. Oxygen-18 content of precipitation, Pohang and Tokyo, shaded where Pohang exceeds Tokyo.

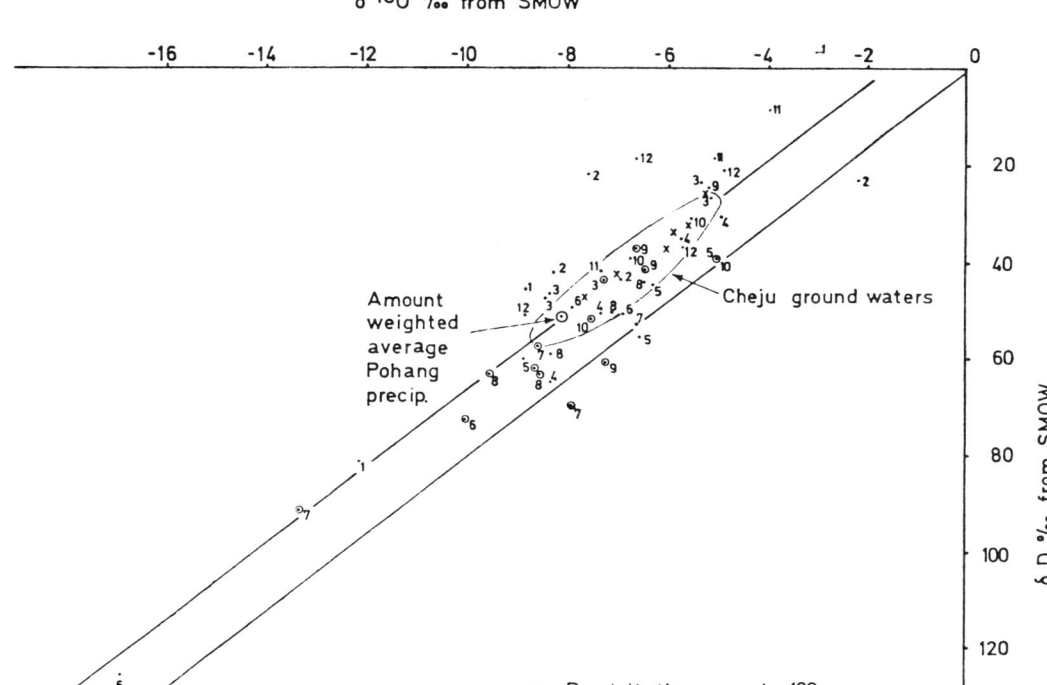

Fig. 4. Deuterium and oxygen-18 contents of precipitation at Pohang and in Cheju Island ground waters. Months indicated by numbers.

tion from the marine moisture; hence it is relatively close in stable isotope content to marine water. The same air masses before reaching Tokyo must cross high mountains and the width of Honshu. Considerable precipitation occurs west of Tokyo; thus the moisture is somewhat depleted in stable isotope content before condensation occurs at Tokyo and the rain is already a second-stage condensation. Interpretations of the isotopic relations of the waters of Cheju Island would be enhanced by establishing a reliable value for the long-term stable isotope input as well as any variations in stable isotope content due to altitude differences. Lacking long-term precipitation sampling at Cheju, the best estimate we can make is necessarily based on Cheju Island ground water. This estimate is somewhat complicated, however, by the transit of at least some waters from high altitude recharge to low altitude discharge points. Nor can we simply extrapolate from the records for Tokyo and Pohang, because many of the Cheju waters are less depleted in stable isotope content than rain at the two precipitation stations (evidently due to the large marine influence at Cheju).

Estimates of the stable isotope content of average recharge at sea level can be based on a widespread areal reconnaissance sampling of many sources in 1966 or on periodic sampling of fewer points. An average value for ^{18}O from the 1966 spring and autumn sampling of the wells and springs of the east coast, which are far removed from sources of high altitude recharge, suggests that the ^{18}O content of sea level recharge should be about $-6.5‰$. Repeated sampling at sites 6, 9, 23, and 48 show mean ^{18}O values of $-6.4‰$ to $-6.6‰$ (Table 1) and thus are in good agreement with the value obtained by the widespread reconnaissance. Sources consistently more depleted in ^{18}O, for example, sites 2, 45, 46, 11, 14, and 15 (high and medium altitude springs), suggest recharge from precipitation at higher altitudes.

Areal Variability of Isotope Contents of Ground Waters

The initial reconnaissance sampling of spring 1966 indicated fivefold classification of sources of waters based on their tritium and stable isotope contents. These sources of water are illustrated in Figure 5 and are related to mode of occurrence as follows:

A. Medium and high altitude springs (sites 11–16) characterized by generally high tritium content (140–265 TU) and wide spread of ^{18}O values. The high tritium content and wide spread of ^{18}O values were interpreted as indicating short transit time and poor mixing in general agreement with the hydrologic conditions previously presented in the Geohydrology section of this paper.

B. Streams (sites 3 and 5) characterized by tritium content close to 100 TU and water relatively enriched in ^{18}O (-5 to $-6‰$). This source is in good agreement, as might be expected, with antecedent rainfall at Pohang (Figures 2 and 3) taking into account the more marine environment of Cheju Island.

C. Large coastal springs (sites 1, 2, 4, 8, 9, and 10) characterized by tritium content generally less than 100 TU and ^{18}O content in the same general ranges as the medium and high altitude springs (-6.6 to $-7.8‰$). The wide spread of ^{18}O values corresponding to those of the higher altitude springs suggested a common source. The generally low tritium content was interpreted as indicating relatively longer transit than in the higher altitude sources.

D. Small springs and wells of the east coast (sites 18–20 and 22–24) characterized by tritium content less than 100 TU and ^{18}O values somewhat more enriched than in the large coastal springs. Stable isotope enrichment is believed due to an altitude effect reflecting generally lower terrain in the eastern part of the island. The generally low and uniform tritium content is interpreted as indicating relatively long transit time and discharge from a well mixed reservoir. In this group samples from one spring (site 17) and one well (site 21) had very high tritium contents for which there appears to be no logical explanation, since local precipitation records showed no heavy rains just prior to sampling and since it is unlikely that high altitude recharge could be conducted directly so far to only these two sites on the eastern part of Cheju Island.

E. Small capacity wells of the southwest coast (sites 6 and 7). These two shallow wells were markedly high in tritium content and indicated relative enrichment in ^{18}O, in the same range as the two streams sampled. The waters appear to represent response to local recharge and rapid turnover.

Although a rough relationship of stable isotope content and altitude of sampling site could be recognized, no reliable quantitative estimate of increasing depletion in stable isotope content with increasing altitude could be deduced. Presumably this is due in part to fairly free circulation of ground water from high altitude recharge areas to low altitude discharge points,

TABLE 1. ^{18}O Variability in Precipitation at Pohang and Tokyo and Ground Waters of Cheju Island

Source	Number of Samples	Mean, ‰	Standard Deviation
Pohang precipitation			
All months	69	−7.2	2.7
Months >100 mm	23	−7.8	1.8
Tokyo precipitation			
All months	77	−7.4	2.4
Months >100 mm	36	−7.3	2.4
Medium altitude springs (group A)			
Site 16 (400 meters)	7	−6.6	1.1
Site 45 (400 meters)	8	−7.1	0.3
Site 46 (roughly 200 meters)	3	−7.0	1.6
High altitude springs (group A)			
Site 11 (roughly 1000 meters)	6	−7.7	1.2
Site 14 (roughly 1400 meters)	4	−7.3	0.5
Site 15 (roughly 1000 meters)	7	−7.4	1.8
Stream (group B)			
Site 5	6	−6.8	1.0
Large coastal springs (group C)			
Site 2	9	−7.7	0.4
Site 9	5	−6.6	0.6
Small springs and wells of the east coast (group D)			
Site 23	6	−6.4	0.8
Site 48	3	−6.4	0.9
Site 47	3	−6.3	0.7
Shallow well, southwest coast (group E)			
Site 6	6	−6.5	1.6

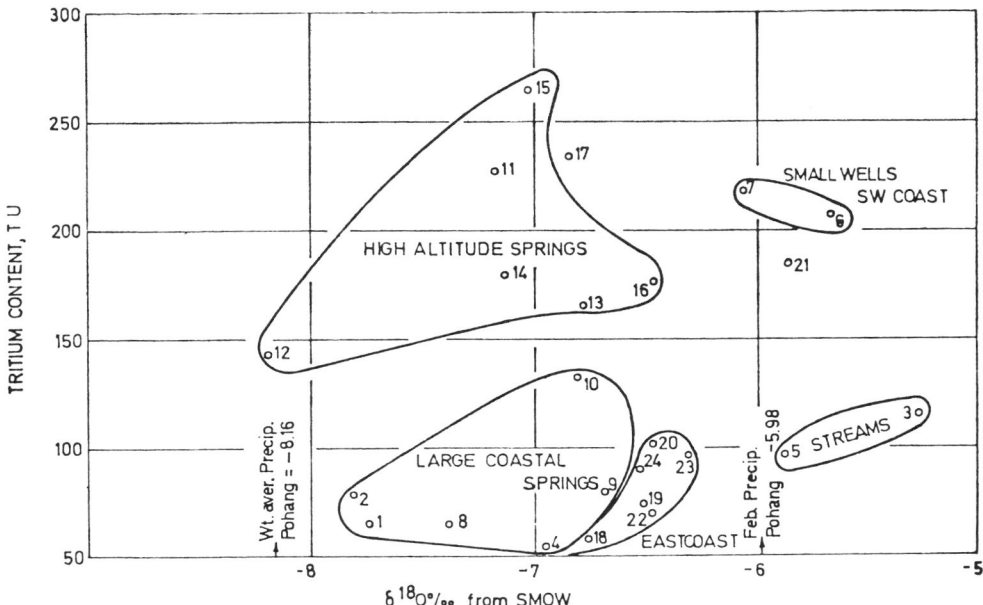

Fig. 5. Tritium versus oxygen-18 content of ground water sampled in spring 1966.

although of course the altitude of any discharge point represents a minimum altitude for precipitation recharging the ground waters. Subsequent time sampling of several sources indicates that the general altitude effect is largely obscured by seasonal variations in the stable isotope input.

Resampling of the original network of samples in autumn 1966 following the summer rainy season generally confirmed the areal relations noted above but indicated a pronounced time variability in some sources, which is discussed later. In general the high and medium altitude springs (group A) showed marked time variability in isotope content as compared with the large coastal springs (group C) and the small springs and wells of the east coast (group D). As might be expected the streams (group B) and the small wells of the southwest coast (group E) showed pronounced time variability.

It can readily be seen that a consolidated plot of all samples would result in a very complex graph owing to the nonsteady-state input of tritium and the seasonal variability in input of stable isotope content reflected in variability of output in some sources of ground water. Rather than presenting the time series data on a single graph, the tritium variability in the ground waters is shown in Figure 7, and the stable isotope variability is summarized in Table 1, as described in the following section.

Time Variability of Isotope Content

Over a period of several years the stable isotope content of precipitation represents a steady state input with random fluctuations about the mean (Figure 3). However, the short-term variability in the Korea–Japan region is pronounced. From the time precipitation reaches the ground it is subjected to mixing effects in runoff, in infiltrating the soil, and in moving through the saturated zone. All these processes tend to integrate the short-term variations in input, and in systems where infiltration through thick soil is the predominant form of recharge or the ground water occurs in granular materials, the stable isotope content of the ground waters generally approximates the long-term amount weighted average of the stable isotope content of the recharge.

On Cheju, however, recharge appears to be largely directed into fractures and tubular openings in the volcanic rocks, and ground water generally moves rapidly through these types of openings. As a result the original input variability is preserved to a considerable degree in

335

many of the ground waters. It stands to reason, however, that the greater the distance traveled by water underground and the larger the volume of the underground reservoir, the greater the damping of input variability as reflected in time series observations of ground-water systems. This feature can be used in a qualitative way to identify discharge from large well mixed systems and especially to distinguish discharge of the basal fresh water lens from that of superficial systems.

The stable isotope input as represented by the Tokyo and Pohang records is far more variable than that of Cheju ground waters. Even limiting the calculations to months of greater than 100 mm precipitation (the major recharge periods) although reducing the variability, still gives standard deviations far greater than the Cheju waters (Table 1). The same general relationship can be seen on the D–^{18}O diagram (Figure 4) where the ground waters all cluster near a mean value for the input. The variability of tritium content of the ground waters also can be used as an index of mixing, although the tritium input does not represent a steady state.

The time variability of the stable isotope content of ground water, as represented by the ^{18}O content (Table 1), provides support for preliminary conclusions with respect to mixing rates and storage characteristics of the subsurface reservoirs based on the 1966 reconnaissance sampling and hydrogeologic reasoning. If any of the sources sampled do tap a major fresh water lens, then this should be reflected in small time variability in ^{18}O content. Thus the low standard deviations of the samplings of the large coastal springs (group C), and the small springs and wells of the east coast (group D) ranging from 0.4 to 0.9, are in marked contrast to the standard deviations of the medium and high altitude springs (group A), the stream (group B), and the small well, southwest coast (group E) that generally exceed 1. The only marked exception to this is the record from site 45 in which the ^{18}O content is remarkably uniform, although the tritium content shows considerable variability. The explanation for this feature is not evident from the available data. Site 14 has relatively uniform ^{18}O also, but only four samples were collected, so it is not regarded as exceptional as site 45, despite its higher altitude.

Calculation of Residence Time of Ground Waters from Tritium Content

If the residence time of ground water and the discharge from a subsurface reservoir can be determined within reasonable limits, it is possible to calculate the reservoir's active storage capacity. Since the input of tritium can be calculated with fair precision and since its decay rate is known, tritium offers the most direct approach to the problem of estimating the residence time [*Dinger and Davis*, 1967]. Information required for the calculations includes: (1)

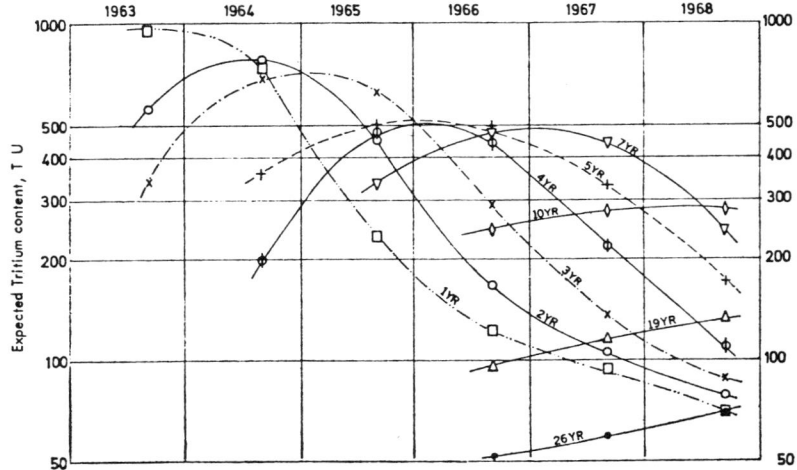

Fig. 6. Tritium transit time distribution functions calculated for Cheju Island.

EXPLANATION:

◊ Site No. 2
● Site No. 9 } Coastal springs
□ Site No. 16
+ Site No. 45 } Medium altitude springs
⊖ Site No. 11
⊙ Site No. 15 } High altitude springs

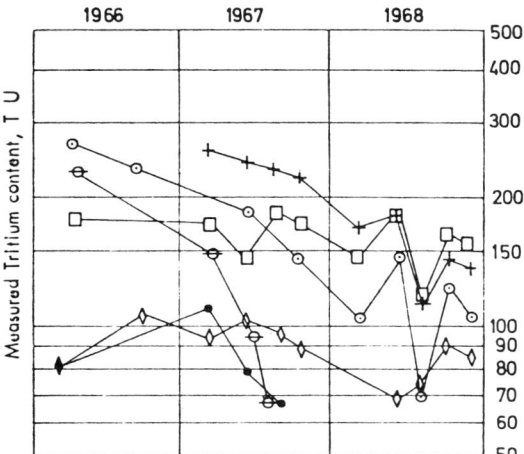

Fig. 7. Measured tritium content of selected samples of ground water.

tritium and water input based on records of tritium content and amount of precipitation; (2) periodic tritium analysis of discharge from the reservoir; and (3) a reasonably valid model of the reservoir system. For item 1, monthly rainfall totals at Cheju Island (1954–1967) and the monthly tritium content of precipitation monitored at Tokyo were used. The Tokyo tritium record (1961–1967) was extended back to 1954 by correlation with Ottawa, Canada. The extrapolated part of the record applies only to the pre-1961 period, and therefore has relatively little weight in most of the calculations. For item 2, 4 to 9 tritium analyses from 4 stations and 4 analyses from 2 stations were used (Figure 7). For item 3, judgment based on a rather limited knowledge of the geology and hydrology was used.

The model adopted for these trial calculations was that of a well mixed reservoir receiving additional recharge during months when precipitation exceeded 100 mm. Divergence of the tritium output from the calculated curves is a measure of the lack of agreement between the model assumed and the natural conditions.

In constructing the tritium output curves, the annual inputs were computed by multiplying the monthly precipitation (for months exceeding 100 mm) by the respective monthly tritium content of Tokyo precipitation and dividing the sum of these products by the sum of precipitation; thus

$$\langle C \rangle = \frac{\sum C_i P_i}{\sum P_i}$$

The resulting mean annual values for tritium content of recharge are shown in Table 2.

Transit time distribution functions were then calculated assuming triangular distributions with base periods of 3, 4, 5, 7, 10, 19, and 26 years (Figure 6). In a triangular distribution having a 3-year base it is assumed that the central year has a weight of 50% and the preceding and following years, 25% each. Similarly in a 5-year base period the weightings are 11, 22, 33, 22, and 11%, respectively. Longer periods are calculated in similar fashion. In the special cases of periods shorter than 3 years, a 1-year distribution represents piston flow, and a 2-year base period is a rectangular distribution, and the weighting assigned is 50% for each year. The mean residence time of water in the system corresponds to the central year of the triangular distribution. Thus a source of water agreeing with a 19-year base period distribution would have a mean residence time of about 10 years.

In Figure 6 the calculated curves having base periods indicated above can be compared with several actual records. In matching the curves both the absolute tritium content and the slope are important. It is readily seen that none match precisely, but this is not surprising in view of the inherent inaccuracies and assumptions. The curves are instructive, however, and they can be used for estimating residence times.

TABLE 2. Mean Annual Values for Tritium Content of Recharge

Year	$\langle C \rangle$, TU	Year	$\langle C \rangle$, TU	Year	$\langle C \rangle$, TU
1953	6	1958	185	1963	974
1954	84	1959	122	1964	737
1955	13	1960	57	1965	230
1956	67	1961	41	1966	120
1957	52	1962	166	1967	92
				1968	69

In general from the curves in Figure 7 we can confidently conclude that all the groundwater systems have substantial modern or post-bomb water in them. By comparing the probable residence times of waters from different sources we can estimate qualitatively comparative amounts of recent and relatively older waters according to indicated residence times. For example site 11, a high altitude spring which was believed from other evidence to have a short residence time, corresponds most closely to a 4-year transit time distribution curve, or a mean residence time of about 2 years. Site 15, another high altitude spring, corresponds roughly to a 5-year transit time ($2\frac{1}{2}$-year residence time).

Site 16, a medium altitude spring, on the other hand corresponds most closely to a distribution lying between 10 and 19 years, perhaps in the neighborhood of 15 years, and this figure would suggest a mean residence time of about $7\frac{1}{2}$ years. Thus an appreciable amount of older water may be mixed into this reservoir system. Site 45 fits best between the 5-year and 7-year distribution curves, giving about 3 years residence time. Thus it resembles the high altitude spring system more than site 16, although as noted below, its ^{18}O content is of low variability.

Site 2, and probably site 9 as well, both coastal springs, compare in shape with the 10-year distribution but in amount somewhere between the 19-year and 26-year distribution curves. A 17-year compromise position would give a $8\frac{1}{2}$-year residence time for the coastal spring group.

CONCLUSIONS

1. All water sampled contained substantial amounts of thermonuclear tritium (post-1954) indicating rapid circulation in the ground-water systems.

2. Areal sampling for tritium and stable isotopes indicated that ground waters could be classified into several flow regimes. This suggested a model including rapid recharge and discharge from springs high on the slopes of Mt. Halla and better mixing and longer residence in a principal fresh water lens that discharges at springs around the coastal perimeter of the island.

3. Time variability of tritium and ^{18}O contents as shown by periodic sampling supported the hypothesis of the model.

4. Calculations based on estimated tritium input and sampling of tritium content of discharge points indicated mean residence times ranging from 2 to $8\frac{1}{2}$ years underground. The longer residence times were associated with waters interpreted from other evidence as representing discharge from an extensive well mixed fresh water lens.

This information, which strengthens knowledge about the occurrence, availability, and movement of ground water, may be useful to supplement other hydrogeologic data and to serve as a basis for more detailed isotopic studies of the hydrologic regime of Cheju Island. It can also be used as a guide for similar studies of comparable geohydrologic settings.

Acknowledgments. The authors wish to acknowledge valuable assistance from L. L. Thatcher, U. S. Geological Survey, Denver, Colorado (formerly at the International Atomic Energy Agency) who proposed the project and helped guide it in the early stages, and J. T. Callahan, U. S. Geological Survey, Seoul, Korea, for comments on the island's geohydrology.

The authors also thank Yong Kyu Lim, Atomic Energy Research Institute, Seoul, Korea, and Gi Young Nahm, Office of Geological Survey of Korea, for assistance in project field activities.

REFERENCES

Craig, H., Standard for reporting concentration of deuterium and oxygen-18 in natural waters, *Science, 133,* 1833–1834, 1961.

Dansgaard, W., Stable isotopes in precipitation, *Tellus, 16*(4), 436–68, 1964.

Dinçer, T., and Davis, G. H., Some considerations on the tritium dating and the estimates of tritium input function, *Mémoires of the Congress of Istanbul,* pp. 276–286, International Association of Hydrogeologists, 1967.

International Atomic Energy Agency, Tritium and other environmental isotopes in the hydrological cycle, 83 pp., *Tech. Rep. Ser. 73,* International Atomic Energy Agency, Vienna, 1967.

Lee, C. K., Y. K. Lim, G. Y. Nahm, I. Lee, and M. H. Jeen, Report on the application of environmental isotopes and artificially introduced tracers in the hydrology of a volcanic island (Cheju Island), Atomic Energy Research Institute, Seoul, Korea, 1966.

Lee, C. K., D. H. Kim, Y. K. Lim, and G. Y. Nahm, Report on the application of environmental isotopes and artifically introduced tracers in the hydrology of a volcanic island (Cheju Island), Atomic Energy Research Institute, Seoul, Korea, 1967.

(Manuscript received September 4, 1969.)

Part V

HEAT AND MASS TRANSPORT

Editors' Comments
on Papers 26 Through 29

26 CARTWRIGHT
Tracing Shallow Groundwater Systems by Soil Temperatures

27 BREDEHOEFT and PINDER
Mass Transport in Flowing Groundwater

28 SCHWARTZ and DOMENICO
Simulation of Hydrochemical Patterns in Regional Groundwater Flow

29 ANDERSON
Excerpts from *Using Models to Simulate the Movement of Contaminants Through Groundwater Flow Systems*

Every groundwater system is essentially an energy system composed of three forms of energy—potential, thermal, and chemical. The science of hydrogeology is largely directed toward understanding the distribution and transformation of this energy. The papers in Parts I through IV demonstrate how chemical hydrogeology has continuously developed through a simple descriptive phase to a level in which the emphasis was on understanding the hydrochemical systems. We are now in the third phase emphasizing predictions over space and time. The law of mass action and free energy provide the reference point for understanding deviation from equilibrium. Mass-balance relationships are used both intuitively and overtly to identify the controlling chemical reactions and physical processes. These two approaches provide the basis for spatial predictions; but predictions over time require the addition of nonequilibrium thermodynamics and kinetic relations. All these concepts, together with principles from other disciplines, can be coupled with the physical aspects covered in *Physical Hydrogeology* (Freeze and Back, 1983) most effectively by use of mass-transfer and mass-transport equations.

Current goals of chemical hydrogeology are to predict flow paths, travel times, and concentration patterns of dissolved constituents. These predictions require an application of the theory of mass transport through porous media. There are three components to the process: an advective component whereby chemical species are carried along by flowing groundwater; a spreading component due to hydrodynamic dispersion (which includes diffusion); and a mobilization and retardation component due to the chemical interactions between water, rock, gases, and dissolved constituents. The basic advection-dispersion theory was developed in the field of fluid mechanics by scientists involved in the fundamental analysis of flow through porous media (de Josselin de Jong, 1958; Saffman, 1959; Bear, 1961; Scheidegger, 1961). The modern concepts of hydrodynamic dispersion were clearly articulated by Bachmat and Bear (1964). In the hydrogeological world, the results of early dispersion experiments were reported by Simpson (1962) and Skibitzke and Robertson (1963). Ogata and Banks (1961) provided one of the first analytical solutions to the one-dimensional advection-dispersion equation.

Rather than trace these early developments of the laboratory-scale concept of dispersion, we have chosen to emphasize the application of mass-transport principles on a regional scale. The movement of water is in itself a transport of mass and a transformation of potential energy to thermal energy. Understanding of the process was gained largely through application of potential theory as amply demonstrated by papers in *Physical Hydrogeology*. Significance of distribution of thermal energy in aquifers has been elucidated by investigating the transport of heat by flowing groundwater. The equations of flow were first presented in the hydrogeological literature by Stallman (1963). Schneider (1964) and Cartwright (1968; Paper 26) have applied the theory to field situations and developed methodology whereby temperature regimes can be utilized to interpret hydrogeological conditions.

However, the transport of chemical species, rather than heat, is attracting the current research activity of hydrogeological mass transport. The goal of much of this research is to develop numerical simulation models that can predict the concentrations of various chemical constituents in aquifers or throughout regional groundwater flow systems over some prescribed period of time. The development of equations for chemical mass transport at aquifer scale was pioneered by Shamir and Harleman (1967), Reddell and Sunada (1970), and Bredehoeft and Pinder (Paper 27); however, none of these studies included the effects of chemical reactions. Schwartz and Domenico (Paper 28) were the first to incorporate chemical reactions by developing a model capable of simulating a spatial pattern of the most important aqueous species in a regional groundwater system and applying the model to an actual field system. Pickens and Lennox (1976) provide an extensive suite of simulations in the same spirit.

In recent years, attention has turned to an environmental application of water chemistry: the assessment of groundwater contamination. Robertson (1974) provides an example of the technique used in the analysis of a complex waste-transport system. The model includes the effects of advective transport, hydrodynamic dispersion, radioactive decay, and reversible linear absorption. Nelson (1977) addresses the question of how best to interpret groundwater contamination predictions for environmental impact assessment. As the final paper for these volumes, we have selected an excerpt from a most perceptive review by Anderson (Paper 29). She clearly articulates the present state of development of chemical transport models and outlines both the scientific problems of the present and the goals for the future.

REFERENCES

Bachmat, Y., and J. Bear. 1964, The General Equations of Hydrodynamic Dispersion, *Jour. Geophys. Research* **69:**2561–2567.

Bear, J., 1961, Some Experiments on Dispersion, *Jour. Geophys. Research* **66:**2455–2467.

Cartwright, K., 1968, Thermal Prospecting for Groundwater, *Water Resources Research* **4:**395–401.

De Jong, G. de Josselin, 1958, Longitudinal and Transverse Diffusion in Granular Deposits, *Am. Geophys. Union Trans.* **39:**67–74.

Freeze, R. A., and W. Back, 1983, *Physical Hydrogeology,* Benchmark Papers in Geology, vol. 72, Hutchinson Ross Publishing Co., Stroudsburg, Pa., 448p.

Nelson, R. W., 1977, Evaluating the Environmental Consequences of Groundwater Contamination. 1. An Overview of Contaminent Arrival Distributions as General Evaluation Requirements, *Water Resources Research* **14:**409–415.

Ogata, A., and R. B. Banks, 1961, A Solution of the Differential Equation of Longitudinal Dispersion in Porous Media, *U.S. Geol. Survey Prof. Paper 411-A,* 7p.

Pickens, J. F., and W. C. Lennox, 1976, Numerical Simulation of Waste Movement in Steady Groundwater Flow Systems, *Water Resources Research* **12:**171–180.

Reddell, D. L., and D. K. Sunada, 1970, Numerical Simulation of Dispersion in Groundwater Aquifers, *Colorado State Univ. Hydrology Paper No. 41,* 79p.

Robertson, J. B., 1974, Digital Modeling of Radioactive and Chemical Waste Transport in the Snake River Plain Aquifer at the National Reactor Testing Station, Idaho, *U.S. Geol. Survey Open-File Report IDO-22054,* 41p.

Saffman, P. G., 1959, A Theory of Dispersion in a Porous Medium, *Jour. Fluid Mechanics* **6:**321–349.

Scheidegger, A. E., 1961, General Theory of Dispersion of Porous Media, *Jour. Geophys. Research* **66:**3273–3278.

Schneider, R., 1964, Relation of Temperature Distribution to Groundwater Movement in Carbonate Rocks of Central Israel, *Geol. Soc. America Bull.* **75:**209–216.

Shamir, U. Y., and D. R. F. Harleman, 1967, Numerical Solution for Dispersion in Porous Mediums, *Water Resources Research* **3:**557–581.
Simpson, E. S., 1962, Transverse Dispersion in Liquid Flow through Porous Media, *U.S. Geol. Survey Prof. Paper 411-C,* 30p.
Skibitzke, H. E., G. M. Robertson, 1963, Dispersion in Groundwater Flowing through Heterogeneous Materials, *U.S. Geol. Survey Prof. Paper 386-B,* 8p.
Stallman, R. W., 1963, Computation of Groundwater Velocity from Temperature Data, *U.S. Geol. Survey Water-Supply Paper 1544A,* pp. 36–46.

Copyright ©1974 by the American Geophysical Union
Reprinted from Water Resources Research 10:847–855 (1974)

Tracing Shallow Groundwater Systems by Soil Temperatures

KEROS CARTWRIGHT

Illinois State Geological Survey, Urbana, Illinois 61801

Circulating water is known to affect the temperature of the rock through which it flows. Analysis of existing theoretical equations suggests that shallow groundwater flow will also affect the surface soil temperature. Soil temperatures therefore might be used to delineate small, shallow groundwater flow systems; both recharge and discharge zones might be distinguished. Temperature data taken in the field during summer and winter support the theoretical conclusions and agree with the groundwater flow patterns inferred from hydrologic data. In general, winter soil temperature profiles have less data scatter than profiles of summer temperatures. The greater the rate of groundwater flow, the greater the temperature difference between recharge and discharge zones. Temperature variations ranged from 0.75°C in clayey glacial till to 5°C in sand. The effect on soil temperature also decreases with increasing horizontal distance between the recharge and the discharge zone. The effect of horizontal flow in a shallow confined aquifer is also predictable from theoretical models and can be shown by field studies. The amount of heat redistributed depends on the thermal properties of the aquifer and on overburden and the velocity of fluid flow.

The description of groundwater flow systems is of fundamental interest to the hydrologist seeking a solution to both theoretical and applied problems. *Hubbert* [1940] presented the theoretical framework of fluid flow from which was derived the basic theory for simulating flow systems on the digital computer [e.g., *Tóth*, 1963; *Freeze and Witherspoon*, 1966, 1967, 1968]. Descriptions of flow systems, especially shallow ones based on field data, for comparison with theoretical models, have been made by numerous investigators, among them *Williams* [1966, 1968], *Meyboom* [1966, 1967], and *Tóth* [1966]. This paper suggests that the use of soil temperatures may add an additional field technique to help the hydrologist describe small flow systems when piezometers and wells are either inadequate or absent. Field studies presented here show that the direction of groundwater movement has a significant effect on soil temperatures.

The distribution of temperature within the lithosphere can be significantly affected by the movement of groundwater. *Stallman* [1960] presented the basic equation for the simultaneous transfer of heat and water and suggested that the equation might be useful in determining the rates of groundwater movement and the permeability of formations. *Stallman* [1965] and *Bredehoeft and Papadopulos* [1965] presented a hypothetical model and compared it with field measurements, and *Cartwright* [1970] applied the Bredehoeft and Papadopulos solution to groundwater discharge from the Illinois basin.

The horizontal movement of groundwater in shallow aquifers has also been shown to affect soil temperatures [*Cartwright*, 1968, 1971; *Frolov*, 1968, p. 186; *Birman*, 1969; *Krčmář and Mašín*, 1970]. The aquifer acts as a heat sink or source, depending upon the season of the year, and the heat is exchanged between the aquifer and the land surface.

The greater the velocity of water movement through a porous medium and the resulting larger mass movement in the system, the greater the effect of the fluid movement on the aquifer temperature [*Cartwright*, 1971] and hence on the soil temperature. In shallow flow systems in which the rate of movement of water is low, only the effect of vertical movement will be reflected in the soil temperature.

Figure 1 is a generalized picture of the movement of water in a shallow flow system during the summer months. Warm water enters the recharge zone, moves downward, and thus warms the soil; as the water moves horizontally, it comes into equilibrium with the general thermal regime; as the water moves upward toward the point of discharge, the soil is cooled. In winter the water entering the system cools the soil, and water at the discharge point warms it.

The temperature of the soils within the flow system depends on the velocity of groundwater movement, the thermal properties of the materials, heat gained or lost to the atmosphere, and geothermal heat added to the system. The measured soil temperatures at the recharge and discharge areas are functions of the vertical velocity, and the soil temperatures in between are functions of the horizontal velocity and the distance of travel.

SIGNIFICANCE OF GROUNDWATER MOVEMENT

The effect of the movement of groundwater on the thermal regime has been discussed by *Stallman* [1960, 1965], *Bredehoeft and Papadopulos* [1965], *Parsons* [1971], and others. The effect of the variation in thermal regime on soil temperatures has been discussed by *Cartwright* [1968]. The following mathematical model is modified from several of the above papers. The effect of heat dispersions is assumed to be negligible; only heat movement by mass transport and conduction is considered.

The problem may be divided into two parts: the area of vertical movement of groundwater and the area of horizontal movement of groundwater (Figure 2). The basic equation for the simultaneous transfer of heat and fluid is [*Stallman*, 1960]:

$$\frac{\partial^2 T}{\partial x^2} + \frac{\partial^2 T}{\partial y^2} + \frac{\partial^2 T}{\partial z^2} - \frac{c_w \rho_w}{k}\left(\frac{\partial v_x T}{\partial x} + \frac{\partial v_y T}{\partial y} + \frac{\partial V_z T}{\partial z}\right) = \frac{c\rho}{k}\frac{\partial T}{\partial t} \quad (1)$$

For one-dimensional vertical flow of both heat and fluid the equation can be reduced to

$$\frac{\partial^2 T}{\partial z^2} - \left(\frac{c_w \rho_w v_z}{k}\right)\frac{\partial T}{\partial z} = \frac{1}{\alpha}\frac{\partial T}{\partial t} \quad (2)$$

Equation (2) applies only if (1) flow of fluid occurs in the ver-

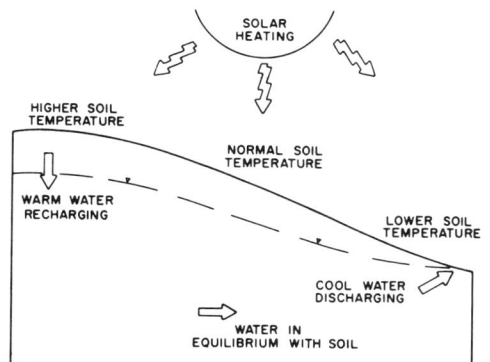

Fig. 1. Generalized small groundwater flow system during summer; temperature conditions are reversed in winter.

tical direction (positive taken downward) parallel to the z axis with steady and uniform flow, (2) the temperature of fluid and rock at all points in the system is equal at all times, and (3) the conductive flow of heat occurs in the vertical direction parallel to the z axis and the thermal properties are constant in time and space.

The annual heating and cooling of the land surface can be described as a sinusoidal temperature wave that is approximated by the following equation:

$$T_s = f(t) = T_m + T_r \sin(2\pi t/P) \quad \text{at } z = 0 \quad (3)$$

Time t is normally taken as zero at the spring crossover when $T_s = T_m$. The temperature at infinity can be assumed to be affected by thermal conditions other than the surface temperature fluctuation, and it is equal to the mean ambient temperature T_{mz} at that depth:

$$\lim_{z \to \infty} T_z = T_{mz} \quad (4)$$

If the lower temperature measurement is at the base of the confining layer, T_{mz} will equal T_A.

For the boundary conditions stated in (3) and (4), *Stallman* [1965] gives the following solution:

$$T_z = T_{mz} + \{T_r \exp\left[-z[((E^2 + F^4/4)^{1/2} + F^2/2)^{1/2} - F]\right]\}$$
$$\cdot \{\sin\left[(2\pi t/P) - z((E^2 + F^4/4)^{1/2} - F^2/2)^{1/2}\right]\} \quad (5)$$

where $E = \pi/\alpha_o P$ and $F = c_w \rho_w v_z / 2k_o$. Equation (5) applies if the three conditions specified following (2) are met, if the earth materials are fully saturated, continuous, and uniform from the surface downward, and if the origin of the z axis (depth variable) in the coordinate system is taken at the land surface and is positive downward. The mean temperature T_{mz} at the measuring depth z is a function of z, and it may be approximated by a linear function equal to the general temperature gradient. It has been shown [*Cartwright*, 1968] that shallow groundwater temperatures are almost equal to the mean ambient surface temperature; therefore $T_{mz} \simeq T_m$. Equation (5) describes the simultaneous vertical flow of heat and fluid in uniform porous media; it can also be used to describe the flow of heat between the land surface and a shallow aquifer provided the effect of the sinusoidal wave is negligible at the interface between the confining layer and the aquifer (z_2 in Figure 2) and the aquifer temperature is equal to a constant.

The equation for conduction of solar energy into the earth found in most heat conduction texts [e.g., *Ingersoll et al.*, 1954, p. 47] is derived from the Fourier conduction equation, which for one-dimensional flow reduces to

$$\alpha(\partial^2 T/\partial z^2) = \partial T/\partial t \quad (6)$$

When (6) is solved subject to the boundary condition given by (3) and (4), it yields

$$T_{zs} = T_{mz} + T_r \exp\left[-z(E)^{1/2} \sin\right]$$
$$\cdot \left[(2\pi t/P) - z(E)^{1/2}\right] \quad (7)$$

where T_{zs} represents the temperature resulting from surface effects. This describes the fluctuation of temperature around surface soil temperature. The observed soil temperature will be the result of the combined effect of the processes described by (7) and the heat added from underground sources.

The effect of horizontal fluid flow on soil temperature can be found by determining the temperature of the aquifer, assumed to be constant at any point, and adding the effects of conduction of periodic heating and cooling of the land surface to the steady flow of heat between the aquifer and the land

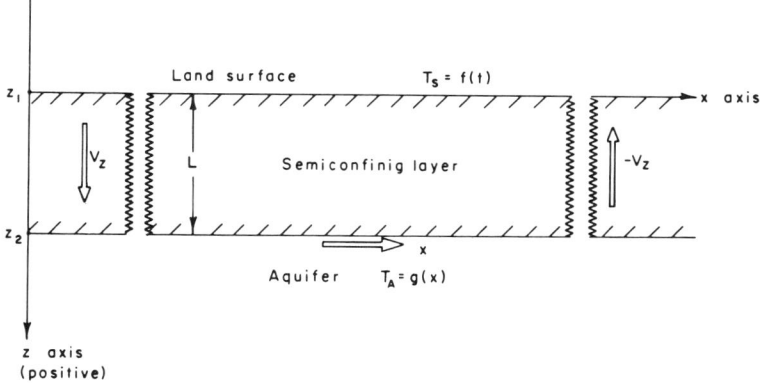

Fig. 2. Conceptual model of a groundwater flow system used in the analysis of shallow temperature relations; z is positive downward.

surface. The distribution of temperature resulting from a sinusoidal temperature fluctuation at the surface is given by (7). The equation for the steady flow of heat between the land surface and the aquifer was given by [Cartwright, 1968, equation 15]:

$$T_{za} = -\frac{z_1 T_A - z_2 T_m}{L} + \frac{(T_A - T_m)z}{L} \quad (8)$$

where T_{za} is the temperature resulting from aquifer effects; T_A is the temperature at z_2, the aquifer-confining layer interface; T_m is the mean surface soil temperature; and L is the thickness of the confining layer. Equation (8) describes the soil temperature distribution for steady flow of heat by conduction between the aquifer and the land surface. Since the temperature is not a function of time, we can assume that

$$T_{za} = T_{mz} \quad (9)$$

Substituting (8) into (7) for a specific depth of measurement yields

$$T_z = +T_r \exp[-z(E)^{1/2}] \sin[(2\pi t/P) - z(E)^{1/2}]$$
$$- \frac{z_1 T_A - z_2 T_m}{L} + \frac{(T_A - T_m)z}{L} \quad (10)$$

where $L = z_2 - z_1$.

The temperature T_A on the lower boundary, however, is a function of distance traveled and may be approximated from (1), assuming that (1) all fluid flow is in the horizontal plane parallel to the x axis and the flow is uniform and steady, (2) the temperatures of the rock and the fluid are equal at all times, (3) the flow of heat by conduction is in the vertical direction parallel to the z axis and the thermal properties are constant in time and space, and (4) all the geothermal and solar heat is redistributed by the moving water. The distribution of temperature T with respect to distance x along the lower boundary (top of the aquifer) is described by Cartwright [1971] as

$$T_A = g(x) = T_{A \text{ at } z=0} + \frac{k_A x}{c_w \rho_w v_x} \frac{\partial^2 T}{\partial z^2} \quad (11)$$

If the aquifer temperature T_A and the surface soil temperature T_m vary slowly with time and distance, the system can be considered almost at equilibrium at any point; therefore T_s and T_A may be considered constant, and (10) is valid at any point where T_s and T_A may be defined. Recharge from the surface to a shallow aquifer, which commonly occurs, could reduce the observed soil temperature differences.

FIELD STUDIES

In the field studies a thermistor at the tip of an insulated aluminum-tipped probe, similar to that used in temperature surveying for groundwater [Cartwright, 1968], was used. The probe was inserted 100 cm into the soil, and the temperature was read after coming to equilibrium with the soil, roughly after about 5 min. Some of the data presented here were originally taken for other purposes, and the probe in those tests was inserted to lesser depths.

The depth to the zone of saturation can have a significant effect on the soil temperature measurement in well-drained, sandy soils. A correction factor T_c (Figure 3) was calculated by using the steady state model of Cartwright [1968, equation 15] for the depth to the top of an aquifer:

$$T_c = (\Delta T)z/D \quad (12)$$

The 'corrected' temperature then is equal to the observed temperature T_z minus the correction factor T_c:

$$T_{cor} = T_z - T_c \quad (13)$$

In winter when the aquifer is warmer than the land surface, T_c will be negative.

Cartwright's model assumed that the material above the aquifer was saturated and not an air-water boundary, which may considerably affect the temperature distribution. The

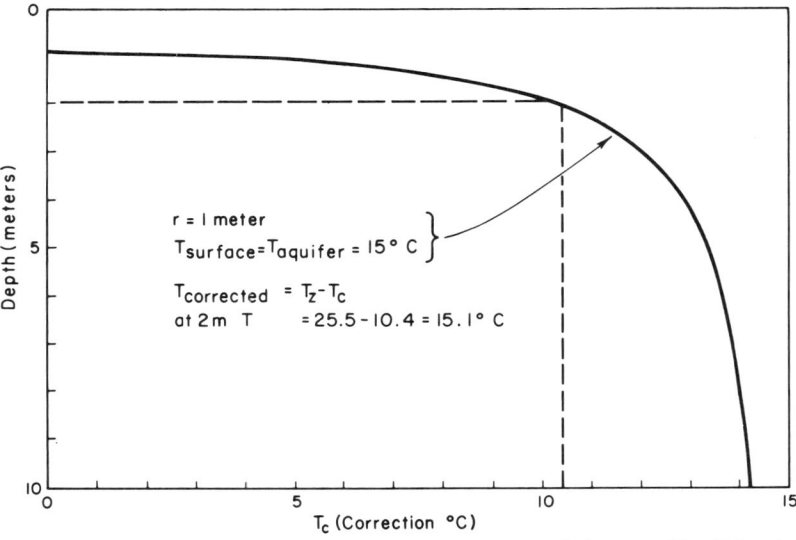

Fig. 3. Example of the correction factor to be used where permeable soils overlie the water table, which varies in depth (see (15) and (16)). The corrected temperature is determined by finding the correction factor for the depth and subtracting it from the observed temperature; T_c is negative in winter and positive in summer.

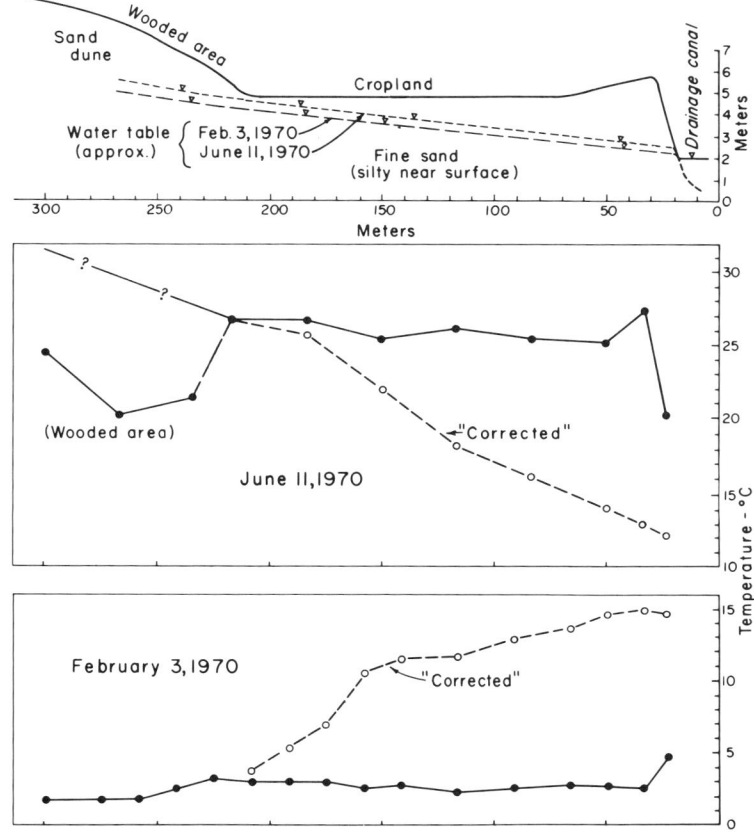

Fig. 4. Observed and corrected soil temperature profiles (measured at a depth of 1 m) and geologic cross section at Havana profile 1 (NE¼NW¼ sec. 23, T. 21 N., R. 5 W., Mason County, Illinois).

correction changes the value measured in the field to one equivalent to a value assuming a water table at a constant depth, the depth of measurement. This correction in effect exaggerates the anomaly.

In areas of sharp relief, topographic correction could also be made in a manner similar to that applied for heat flux measurement on the ocean floor [*Lachenbruch*, 1969]. However, no topographic corrections were applied to data in this study because the areas in question have low relief, making any correction insignificant.

Highly permeable soils. Two series of soil temperature profiles were made in the Havana area of central Illinois (Figures 4 and 5). The temperatures shown are for a depth of 1 m. This area, sometimes referred to as the Havana lowland, is a broad sand plain bordering the Illinois River. The sites studied lie along the east side of a sand dune on the Wisconsinan outwash terraces, which parallel a drainage canal (the hydrogeology of the region was described by *Walker et al.* [1965]). The dune areas generally are heavily wooded and from the surface downward are composed mostly of sand. The flatter cropped areas also are sandy, but the surface soil is silty. The total depth of sand is approximately 50 m. Most precipitation on the area either is apparently evapotranspired or else infiltrates to the groundwater reservoir, since there is very little natural surface drainage. Although recharge occurs over the entire area, the dunes serve as the principal recharge area; the silty soil retards recharge on the flat cropped areas and increases evaporation.

The effect of water movement on the soil temperature is shown in Figures 4 and 5, in which the two lower graphs show temperatures in summer and winter along a line from dune to drainage canal, shown in profile at the top of the figure. The temperatures observed at a depth of 1 m show a decrease of 5°C in summer from recharge to discharge points and an increase of 3°C in winter along the lines of profile. The correction for corrected temperature (13) difference for each season is about 15°C. A steady change in corrected temperatures, more pronounced in summer than in winter, occurs along each line from the main recharge area in the dunes toward the point of discharge into the drainage canal. The dune area is anomalously cool (uncorrected temperature) in the summer owing to the constant shading from solar heat by heavy underbrush and trees. The groundwater gradient at Havana profile 1 site (Figure 4) was established by observations of seeps along the canal and the depth to the water at the toe of the sand dune.

Havana profile 2 (Figure 5) is 1 mi south along the same dune-canal system. It displays a temperature distribution similar to that of Havana profile 1, but the observed temperatures between recharge and discharge zones exhibit a

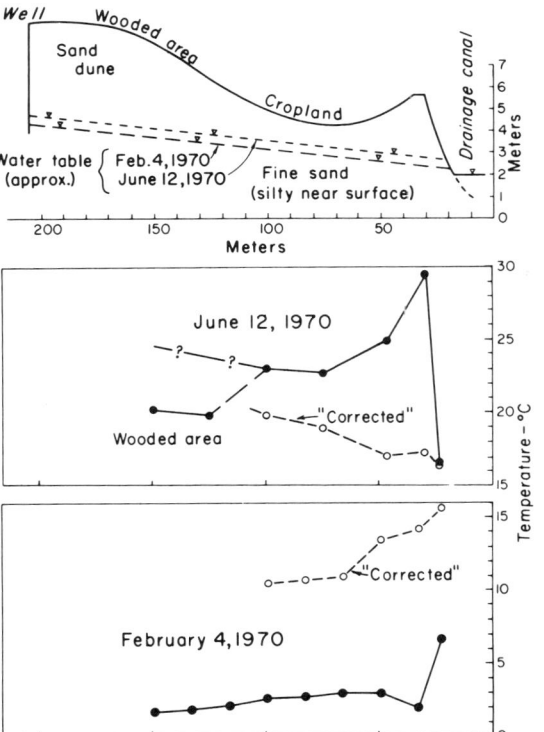

Fig. 5. Observed and corrected soil temperature profiles (measured at a depth of 1 m) and geologic cross section at Havana profile 2 (NW¼NW¼ sec. 26, T. 21 N., R. 5 W., Mason County, Illinois).

difference of about 1° more. The groundwater gradient was established by observing seeps along the canal and the water level in the domestic well on the left of the figure.

These results can be compared with T_z, estimated by using (5), which approximately describes the field problem if the flow is assumed to be vertical (upward) in the discharge zone. The following values for constants were used: $z = 100$ cm, $v_z = -10^{-4}$ cm/s, $\alpha_o = 6 \times 10^{-3}$ cm^2/s, $k_o = 4 \times 10^{-3}$ cal/s cm °C, $c_w = 1$ cal/g °C, $\rho_w = 1$ g/cm^3, $P = 3.15 \times 10^7$ s yr, $T_r = 12$°C, and $T_m = 12$°. Substituting these values into the equation gives a T_z of 13°C in June and 11°C in February. This compares with corrected field values in profiles 1 and 2 of 12° and 16°C (the actual observed temperatures were 20° and 16°C), respectively, in June and 15° and 16°C (the actual observed temperatures were 5° and 7°C), respectively, in February.

There is closer agreement between the calculated and the corrected field temperatures along Havana profile 1 than along profile 2, probably because the discharging water has traveled a greater distance through the aquifer along profile 1 and is thus closer to being in thermal equilibrium with the aquifer. In neither case is the flow vertical in the discharge zone, as may account for some of the differences between the observed and the calculated temperatures. Nevertheless, the observed and calculated temperatures are reasonably close.

Poorly permeable soils. Two series of soil temperature profiles for a depth of 1 m were made near Champaign, Illinois (Figures 6 and 7). The area is one of relatively uniform clayey glacial till overlying a major sand and gravel aquifer encountered at a depth of about 150 ft. Sand lenses, most of them at depths below 80 ft, are present in the till; some of them are moderately continuous and capable of supplying water for domestic consumption. The sites are located near the crest of the Champaign moraine.

Small road cuts (Figure 6), even though they are near topographic highs, are often points of intermittent groundwater discharge in Illinois [*Williams,* 1966]. Seeps are found in the roadside ditches near the crest of the hill where the road is cut to a maximum depth into the till. The road cut shown in Figure 6 is along a country road, cut approximately 3 m into a clayey till at the crest of a glacial moraine. The soil temperature profile shows a 1½°–2°C decrease in the summer and a 2°–3°C increase in the winter from the general land surface to the road ditch. Away from the road cut there is also a

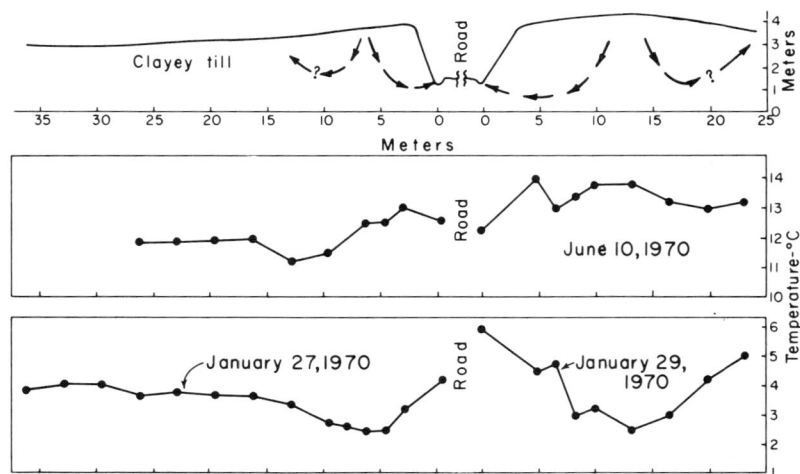

Fig. 6. Observed soil temperature profiles (measured at a depth of 1 m) and geologic cross section at a road cut along the crest of the Champaign moraine, east edge of SW¼SW¼ sec. 3 and NW¼NW¼ sec. 10, T. 19 N., R. 8 E., Champaign County, Illinois.

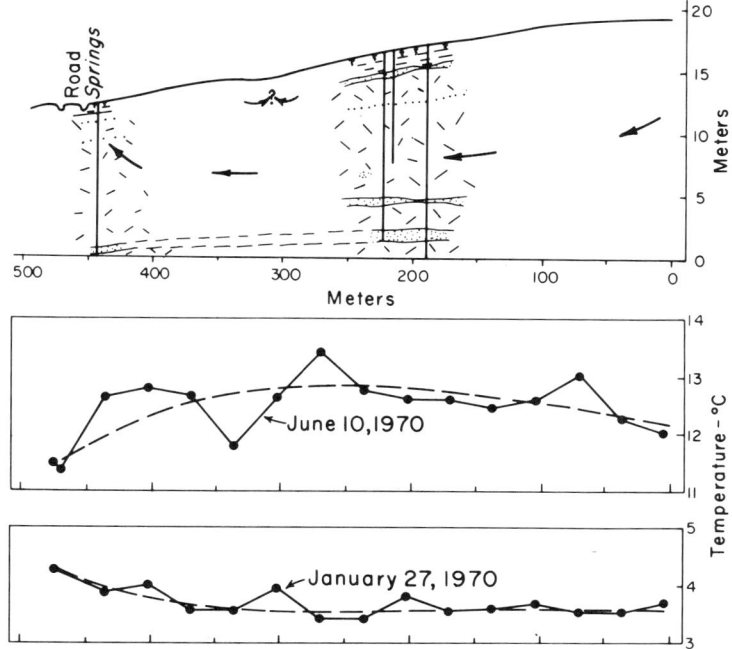

Fig. 7. Observed soil temperature profiles (measured at a depth of 1 m) and geologic cross section immediately west of the crest of the Champaign moraine, SW¼SW¼ sec. 3, T. 19 N., R. 8 E., Champaign County, Illinois. The dashed line is an empirical fit of the data.

marked temperature change from the highest points on the land surface toward lower ground, suggesting groundwater flow in these directions. This is most pronounced in the January 29 data (Figure 6). The effect of both downward and upward movement of groundwater is most clearly seen in the winter data.

The field situation of the area in Figure 6 is similar to that at Havana; however, the groundwater flow is in a clayey glacial till rather than in sand. Equation (5) again approximately describes the field problem. The same values of constants used at Havana can be used here except for the following: $v_z = 10^{-6}$ cm/s, $\alpha_o = 5 \times 10^{-3}$ cm²/s, and $k_o = 2.5 \times 10^{-3}$ cal/s cm °C. When these values are substituted into (5), the soil temperature difference calculated from normal soil temperature (second term on the right side) if no water is moving (8) is 3.2°, which is very close to the temperature observed.

Figure 7 shows a section perpendicular to that in Figure 6 and at some distance from the road cut to avoid that minor flow system. The profile extends from the moraine crest to a group of springs near the break in slope of the moraine front. The summer temperature profile shows about 1°C decrease in soil temperature, and the winter profile shows a 0.75°C increase. There are no data that suggest definite points of recharge. The dashed lines represent an empirically fit line through the points; the data suggest only a slight horizontal gradient in the soil temperature profile, much less pronounced than the gradient present in more permeable material at Havana.

If it is assumed that the principal zones of horizontal groundwater flow are the thin sand stringers in the till and that the discharge at the small springs on the left side of Figure 7 is the result of vertical leakage through the till, (5) approximates the field situation. If the previous values of velocity and thermal constants are used, there should be a 1°C increase in soil temperature in February and a 1°C decrease in the summer. These are very close to the temperature differences measured in the field.

Shallow buried aquifers. The effect of horizontal flow of groundwater on soil temperature in a buried aquifer is clearly shown by the examples used in temperature prospecting for groundwater [*Cartwright,* 1968]. Equation (10) shows that a shallow aquifer affects the shallow soil temperatures, and (11) shows that the temperatures increase slightly in the direction of groundwater flow. The latter effect is related to the hydraulic properties of the aquifer.

The clearest example of the effect of groundwater flow in an aquifer is shown in Figure 8, which depicts a moderately permeable aquifer that receives recharge from the glacial highland to the east and presumably discharges groundwater into the alluvial deposits of the Sangamon River to the south. The geometry and hydraulic parameters of the aquifer are moderately well known. The aquifer is in a small valley on the Illinoian till plain that carried outwash from the Shelbyville morainic system (Wisconsinan); a small creek occupies the valley today. At present, wells are withdrawing water from the aquifer at only two points, one north and the other west of Niantic. The aquifer underlies 4.5–6 m of Holocene stream alluvium, and the permeability increases from 6.6×10^{-2} cm/s north of Niantic to 1.6×10^{-1} cm/s to the southwest. Groundwater flows from the northeast to the northwest at an average gradient of 8.8×10^{-4} cm/cm; on the basis of a hydraulic conductivity of 10^{-1} cm/s, this results in a Darcy velocity of 8.8×10^{-5} cm/s (27.8 m/yr).

Soil temperatures, measured at a depth of 45 cm during summer, are shown on the map (Figure 8). Soil temperatures in a longitudinal profile over the center of the deposit vary from 13.9°C near the recharge area to the northeast to 16.3°C 7 km southwest near the area where water is discharged into the Sangamon River. The lateral variation of 2.4°C over a distance of 7 km gives a horizontal variation of 0.3×10^{-3} °C/m. Data taken during the winter months are less complete; however, the data available suggest a similar temperature increase to the southwest. If the source of heat added to the system is assumed to be geothermal, the heat absorbed by the aquifer is about 10% of the geothermal heat flux estimated for this area [*Cartwright*, 1971].

The aquifer at Mulberry Grove (Figure 9) is less well known. The maps show the winter (Figure 9a) and summer (Figure 9b) soil temperatures measured at a depth of 45 cm. Two wells located near Hurricane Creek in the southwestern part of the surveyed area were in use at the time of the surveys and are near the assumed natural discharge point of the aquifer. The aquifer is in the floodplain of the creek, has a maximum thickness of 3.3 m, and is generally less than 30 m wide. It is generally underlain by bedrock and overlain by silty clay alluvium. The character of the deposits varies rapidly from very fine-grained sand to coarse sand mixed with some gravel. The hydraulic conductivity at the well field is about 2.5×10^{-2} cm/s.

In the surveyed area the measured soil temperatures in both winter and summer increase by approximately 0.4°C along the flow path to the discharge point. This temperature increase occurs over a distance of about 300 m, giving a horizontal variation of 1.3×10^{-3} °C/m, 4 times greater than that measured at Niantic. This greater horizontal variation probably is due in part to increased groundwater velocities that result from pumping in the well field for 25 yr prior to the survey.

The hydraulic properties of the other aquifers studied by *Cartwright* [1968] in groundwater prospecting are less well known, although their geometry is well established. The surveys clearly indicate that soil temperatures increase in the direction of groundwater flow, although this phenomenon was not under consideration in the 1968 study. Where water level data were not available, the various longitudinal profiles of temperature presented in 1968 when they are compared with groundwater flow patterns interpreted from topographic maps show that soil temperatures increase in the direction of groundwater flow.

SUMMARY

The results obtained in the field studies agree well with the theoretical analysis. The location of the regions where discharge is occurring is shown in the field results by increased soil temperatures in winter and decreased soil temperatures in summer. Observed temperatures of the soil at a depth of 1 m showed temperature variations from 0.75°C in clayey till to 5°C in sand. The regions where recharge occurs are generally not reflected in the soil temperatures; this fact probably is due

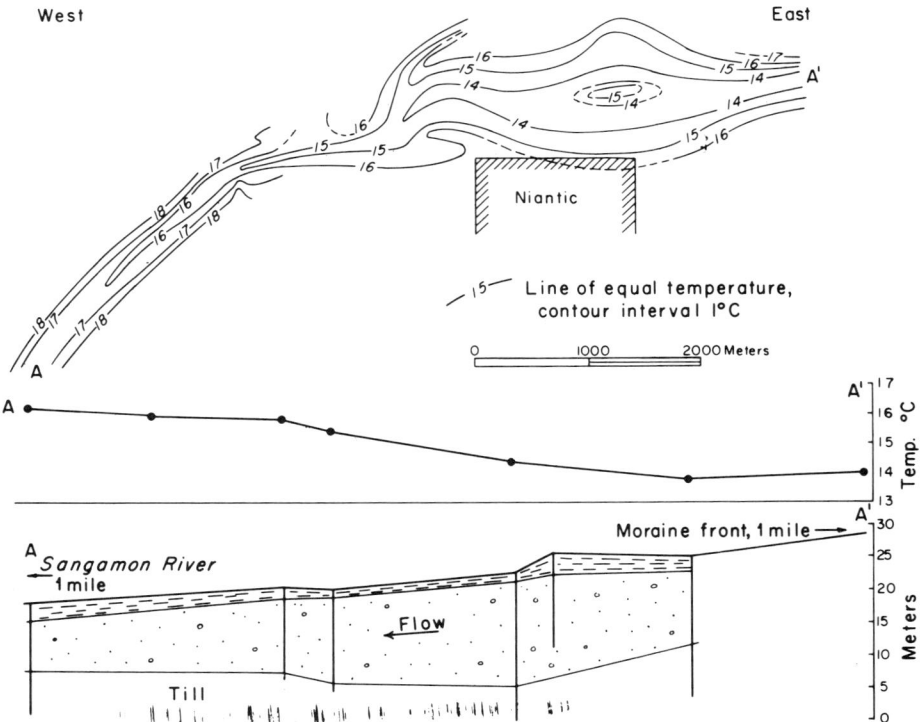

Fig. 8. Soil temperatures map and cross section at Niantic, Illinois (measured at a depth of 45 cm during the summer), illustrating the effect on soil temperature of flow (east to west) in a shallow aquifer (after *Cartwright* [1968]).

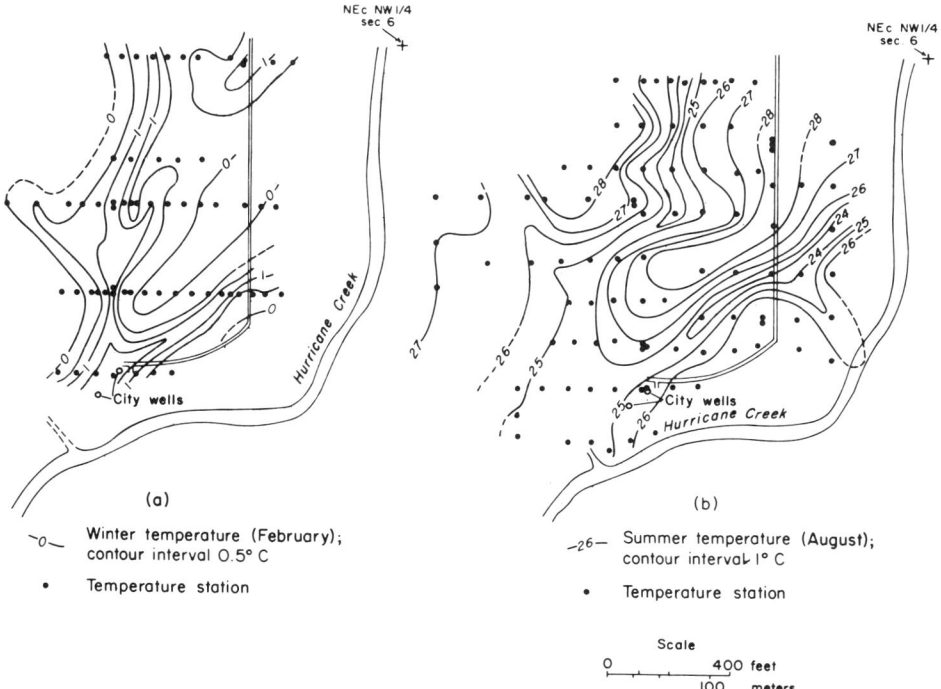

Fig. 9. Soil temperatures at Mulberry Grove, Illinois (measured at a depth of 45 cm), for (a) winter and (b) summer. Both sets of soil temperatures show an increase in temperature of 0.4°C near the discharge point (after *Cartwright* [1968]).

in large part to the intermittent nature of recharge events. Measurements probably would have to be made during the recharge event to obtain significant results.

Horizontal movement of groundwater absorbs and redistributes a portion of the geothermal heat, the result being higher soil temperatures in the direction of groundwater movement. The horizontal temperature variation is related to the permeability of the underlying aquifer; however, there are too many other variables involved to permit temperature variations to be used as a practical method of calculating permeability.

Notation

- t time since flow started.
- P period of temperature fluctuation at the surface.
- T temperature at any point at time t.
- T_s temperature of the land surface at any time.
- T_m mean surface soil temperature.
- T_r one-half annual temperature fluctuation (amplitude).
- T_z temperature at any point z below the land surface at time t.
- T_c temperature correction for different depths to the water table or aquifer.
- T_A temperature of the aquifer at some point x at the top of the aquifer.
- ΔT temperature difference between the land surface and the aquifer at some time t and at point x, equal to $T_s - T_A$.
- v velocity of fluid flow, direction specified by subscripts x, y, and z.
- x, y, z Cartesian coordinates.
- d depth at which soil temperature is measured.
- L thickness of overburden, equal to $z_2 - z_1$.
- D depth to water table.
- A subscript representing aquifer.
- o subscript representing overburden.
- c specific heat of rock-fluid complex.
- c_w specific heat of water.
- k thermal conductivity of the rock-fluid complex.
- ρ density of rock-fluid complex.
- ρ_w density of water.
- α thermal diffusivity, equal to $c\rho/k$.

References

Birman, J. H., Geothermal exploration for ground water, *Geol. Soc. Amer. Bull.*, 80(4), 617–630, 1969.

Bredehoeft, J. D., and I. S. Papadopulos, Rates of vertical groundwater movement estimated from the earth's thermal profile, *Water Resour. Res.*, 1(2), 325–328, 1965.

Cartwright, K., Temperature prospecting for shallow glacial and alluvial aquifers in Illinois, *Ill. Geol. Surv. Circ. 433*, 1–41, 1968.

Cartwright, K., Groundwater discharge in the Illinois basin as suggested by temperature anomalies, *Water Resour. Res.*, 6(3), 912–918, 1970.

Cartwright, K., Redistribution of geothermal heat by a shallow aquifer, *Geol. Soc. Amer. Bull.*, 82(11), 3197–3200, 1971.

Freeze, R. A., and P. A. Witherspoon, Theoretical analysis of regional groundwater flow, 1, Analytical and numerical solution to the mathematical model, *Water Resour. Res.*, 2(4), 641–656, 1966.

Freeze, R. A., and P. A. Witherspoon, Theoretical analysis of regional groundwater flow, 2, Effect of water table configuration and subsurface permeability variation, *Water Resour. Res.*, 3(2), 623–634, 1967.

Freeze, R. A., and P. A. Witherspoon, Theoretical analysis of regional groundwater flow, 3, Quantitative interpretations, *Water Resour. Res.*, *4*(3), 581-590, 1968.

Frolov, N. M., *Gidrogeotermiia (Hydrogeothermics)*, p. 186, Nedra, Moscow, 1968.

Hubbert, M. K., The theory of ground-water motion, *J. Geol.*, *48*(8), 785-944, 1940.

Ingersoll, L. R., O. J. Zobel, and A. C. Ingersoll, *Heat Conduction With Engineering, Geological and Other Applications*, revised ed., p. 47, University of Wisconsin Press, Madison, 1954.

Krčmář, B., and J. Mašín, Prospecting by the geothermic method, *Geophys. Prospect.*, *18*(2), 255-260, 1970.

Lachenbruch, A. H., The effect of two dimensional topography on superficial thermal gradients, *U.S. Geol. Surv. Bull. 1203-E*, 1-86, 1969.

Meyboom, P., Unsteady ground-water flow near a willow ring in hummocky moraine, *J. Hydrol.*, *4*, 38-62, 1966.

Meyboom, P., Mass-transfer studies to determine the ground-water regime of permanent lakes in hummocky moraine of western Canada, *J. Hydrol.*, *5*, 117-142, 1967.

Parsons, M. L., Ground-water thermal regime in a glacial complex, *Water Resour. Res.*, *6*(6), 1701-1720, 1971.

Stallman, R. W., Notes on the use of temperature data for computing groundwater velocity, 6th Assembly on Hydraulics, *Rep. 3*, pp. 1-7, Soc. Hydrotech. de France, Nancy, France, 1960. (Reproduced in Methods of Collecting and Interpreting Ground-Water Data, compiled by Ray Bentall, *U.S. Geol. Surv. Water Supply Pap. 1544-H*, 36-46, 1963.)

Stallman, R. W., Steady one-dimensional fluid flow in a semi-infinite porous medium with semisolid surface temperature, *J. Geophys. Res.*, *70*(12), 2821-2827, 1965.

Tóth, J., A theoretical analysis of groundwater flow in small drainage basins, *J. Geophys. Res.*, *68*(16), 4795-4812, 1963.

Tóth, J., Mapping and interpretation of field phenomena for groundwater reconnaissance in a prairie environment, Alberta, Canada, *Bull. Int. Ass. Sci. Hydrol.*, *11*(2), 1-49, 1966.

Walker, W. H., R. E. Bergstrom, and W. C. Walton, Ground-water resources of the Havana region in west-central Illinois, *Groundwater Rep. 3*, 61 pp., Ill. Geol. Surv. and State Water Surv. Coop., Urbana, 1965.

Williams, R. E., Shallow hydrogeology of glacial drifts in northeastern Illinois, Ph.D. thesis, 179 pp., Univ. of Ill., Urbana, 1966.

Williams, R. E., Ground-water flow systems and related highway pavement failure in cold mountain valleys, *J. Hydrol.*, *6*, 183-193, 1968.

(Received July 27, 1973;
revised April 1, 1974.)

Mass Transport in Flowing Groundwater

JOHN D. BREDEHOEFT

GEORGE F. PINDER*

The mass transport equation and the equation of motion have been coupled and solved numerically for a saturated isothermal groundwater system in which there are no chemical reactions. A case history of groundwater contamination at Brunswick, Georgia, illustrates the use of this physical–chemical model in predicting and controlling the future movement of contaminants.

Hydrologists are becoming increasingly interested in optimizing the use of groundwater reservoirs, not only through making the maximum use of the quantity of water available but also by managing the quality of water in the system. Efforts currently underway include predicting and controlling the movement of a saltwater–freshwater interface, recharging water of differing quality into an aquifer and predicting the resultant quality changes in the system in both time and space, and predicting quality changes in an aquifer due to changing irrigation patterns and irrigation efficiency. Certainly any consideration of waste disposal in the subsurface must involve a prediction of the resultant chemical changes in the fluid in both time and space.

The prediction of changes in groundwater quality in a complex hydrologic system generally requires simulation of the field problem, making use of deterministic models. In the most general case, the complete physical–chemical description of moving groundwater must include chemical reactions in a multicomponent fluid and requires the simultaneous solution of the differential equations that describe the transport of mass, momentum, and energy in porous media.

The difficulties encountered in solving this set

* Now at the Department of Civil and Geological Engineering, Princeton University, Princeton, New Jersey 08540.

of equations for real problems have forced hydrologists and reservoir engineers to consider simplified subsets of the general problem. The equation of motion for single component groundwater flow, which describes the rate of propagation of a pressure change in an aquifer, has been solved for many different initial and boundary conditions. To describe the transport of miscible fluids of different densities, such as salt water and freshwater, the mass transport equation and the equation of motion have been coupled and solved numerically. Numerical solutions have also been obtained for the heat transport equation and the equation of motion, particularly for convection problems.

The long-range objective of our research program is to treat the general problem of the transport of fluid, dissolved solids, and heat in aquifer systems. In this paper we will limit ourselves to the isothermal case with no chemical reactions. We will develop within the generally accepted framework of transport phenomena a set of differential equations that provides a physical–chemical description of a saturated groundwater system and we will demonstrate their field application through a case history.

The workings of a given system can be examined with this mathematical model provided that (1) the appropriate boundary and initial conditions are specified, (2) the parameters that describe the system are known, and (3) the set of equations can be solved. It has been possible by the use of numerical techniques to solve the equations describing changes in concentration in space and time for rather general field situations. It seems possible that, although difficulties exist, these methods can be extended to systems that include chemical reactions.

The system is described by ρ_i, the mass per unit volume of solution of constituent $i(ML^{-3})$, and p, the fluid pressure $(ML^{-1}t^{-2})$. The system will be analyzed in the classical way by considering the conservation of mass and the equation of motion, in a system with isothermal conditions. We generally treat the system from the macroscopic or continuum viewpoint, in which we do not attempt a description within the pore space.

Conservation of Mass

The conservation of mass for a system with several chemical constituents dissolved in a moving liquid is given in terms of the fluxes by the continuity equation

$$\frac{\partial}{\partial t}(\epsilon \rho_i) = -\nabla \cdot (\rho_i \mathbf{v}_i \epsilon) + \sum_{k=1}^{s} R_{ik} + W_i \quad (1)$$

where

- R_{ik}, rate of production of constituent i in reaction k, expressed as mass per unit volume of solution per unit time, $ML^{-3}t^{-1}$;
- s, number of reactions taking place in the system;
- ϵ, effective porosity of the porous media, L^0;
- $\mathbf{v}_i$, mass velocity of constituent i, Lt^{-1};
- $W_i(x, y, z) = \sum_{j=1}^{r} Q_i(x_j, y_j, z_j)\rho_i^* \delta \cdot (x - x_j)(y - y_j)(z - z_j)$;
- Q_i, source, $L^3 t^{-1}$;
- $\delta(x - x_j)(y - y_j)(z - z_j)$, Dirac delta function;
- r, number of sources and sinks;
- ρ_i^*, mass concentration of constituent i in the source or sink fluid, ML^{-3}.

The mean velocity of the solution is defined as

$$\rho \mathbf{v} = \sum_{i=1}^{n} \rho_i \mathbf{v}_i$$

or

$$\mathbf{v} = \frac{1}{\rho} \sum_{i=1}^{n} \rho_i \mathbf{v}_i$$

Following *Cooper* [1966] we assume that $\mathbf{v}$ is the sum of a Darcy velocity $\mathbf{v}_D$ plus the grain velocity $\mathbf{W}_g$ or $\mathbf{v} = \mathbf{v}_D + \mathbf{W}_g$. We further assume that $\mathbf{v}_D$ is given by Darcy's law [*Hubbert*, 1940, 1956]

$$\mathbf{v}_D \epsilon = \mathbf{q} = -k/\mu (\nabla p - \rho \mathbf{g})$$

where $\mathbf{v}$ is the mass average velocity of the fluid (Lt^{-1}), $\mathbf{g}$ is gravitational acceleration (Lt^{-2}), k is intrinsic permeability (L^2), and μ is dynamic viscosity $(ML^{-1}t^{-1})$.

The barycentric mass velocity of the ith species is $\mathbf{j}_i = \rho_i (\mathbf{v}_i - \mathbf{v})$. It follows, therefore, that

$$\rho_i \mathbf{v}_i = \mathbf{j}_i + \rho_i \mathbf{v} = \mathbf{j}_i + \rho_i (\mathbf{v}_D + \mathbf{W}_g) \quad (2)$$

Substituting 2 into 1 we obtain

$$\frac{\partial}{\partial t}(\epsilon \rho_i) = -\nabla \cdot [\epsilon \mathbf{j}_i + \epsilon \rho_i (\mathbf{v}_D + \mathbf{W}_g)] + \sum_{k=1}^{s} R_{ik} + W_i$$

If we further assume, as we do in the flow equations, that the deformation of the medium is essentially only vertical, then

$$\frac{\partial}{\partial t}(\epsilon \rho_i) = -\nabla \cdot \epsilon \mathbf{j}_i - \nabla \cdot \epsilon \rho_i \mathbf{v}_D - \frac{\partial}{\partial z}(\epsilon \rho_i w_g) + \sum_{k=1}^{s} R_{ik} + W_i \quad (3)$$

where $\mathbf{W}_g = \mathbf{k} w_g$ and $\mathbf{k}$ is the unit vector in the z direction. A further expansion of 3 gives

$$\epsilon \frac{\partial \rho_i}{\partial t} + \rho_i \frac{\partial \epsilon}{\partial t}$$

$$= -\nabla \cdot \epsilon \mathbf{j}_i - \nabla \cdot \epsilon \rho_i \mathbf{v}_D - \epsilon \rho_i \frac{\partial w_g}{\partial z}$$

$$- \rho_i w_g \frac{\partial \epsilon}{\partial z} - \epsilon w_g \frac{\partial \rho_i}{\partial z} + \sum_{k=1}^{s} R_{ik} + W_i \quad (4)$$

If we then define the substantial derivative moving with the grains as $D/Dt = \partial/\partial t + \mathbf{W}_g \cdot \nabla$, (4) becomes

$$\epsilon \frac{D\rho_i}{Dt} + \rho_i \frac{D\epsilon}{Dt} = -\nabla \cdot \epsilon \mathbf{j}_i - \nabla \cdot \epsilon \rho_i \mathbf{v}_D - \epsilon \rho_i \frac{\partial w_g}{\partial z} + \sum_{k=1}^{s} R_{ik} + W_i \quad (5)$$

Again, following the development of *Cooper* [1966, p. 4787], for the flow equation we can say $\partial w_g/\partial z = 1/J\, DJ/Dt \simeq 1/\Delta z\, D\Delta z/Dt$, where J is the Jacobian of the transformation. Given the effective stress law [*Hubbert and Rubey*, 1959] $\sigma_z = \sigma_T - p$, where σ_z is the vertical effective grain to grain stress $(ML^{-1}t^{-2})$ and σ_T is the vertical total stress $(ML^{-1}t^{-2})$. If we assume that the total load of the overburdened remains constant, we find that $Dp/Dt = -D\sigma_s/DT$, and if we note that $1/\Delta z \, D\Delta z/Dt = -\alpha D\sigma_s/Dt$, where α is the compressibility of the medium (LT^2M^{-1}), we obtain

$$\partial w_\sigma/\partial z = \alpha Dp/Dt \quad (6)$$

Substituting (6) into (5) yields

$$\epsilon\rho_i\alpha \frac{Dp}{Dt} + \rho_i \frac{D\epsilon}{Dt} + \epsilon \frac{D\rho_i}{Dt}$$
$$= -\nabla \epsilon j - \nabla \cdot \epsilon\rho_i \mathbf{v}_D + \sum_{k=1}^{s} R_{ik} + W_i \quad (7)$$

The mass velocity j_i for constituent i relative to the mass average velocity of a fluid in a porous media can be expressed in terms of hydrodynamic dispersion and is [*Bear et al.*, 1968, p. 310]

$$\mathbf{j}_i = -\rho \mathbf{D}_i/\epsilon \cdot \nabla(\rho_i/\rho)$$

where $\mathbf{D}_i$ is the hydrodynamic dispersion coefficient for component i (L^2t^{-1}). Substituting for both $\mathbf{j}_i$ and $\mathbf{v}_D$ in (7) yields

$$\epsilon\rho_i\alpha \frac{Dp}{Dt} + \frac{D}{Dt}(\epsilon\rho_i)$$
$$= \nabla \cdot \rho \mathbf{D}_i \cdot \nabla\left(\frac{\rho_i}{\rho}\right) - \nabla \cdot \rho_i \frac{\mathbf{k}}{\mu} \cdot (\nabla p - \rho \mathbf{g})$$
$$+ \sum_{k=1}^{s} R_{ik} + W_i \quad (8)$$

which is our general transport equation for any constituent i in the system.

If we further assume that

$$w_\sigma \frac{\partial p}{\partial z} \ll \frac{\partial p}{\partial t}, \quad w_\sigma \frac{\partial \epsilon}{\partial z} \ll \frac{\partial \epsilon}{\partial t}, \quad w_\sigma \frac{\partial \rho_i}{\partial z} \ll \frac{\partial \rho_i}{\partial t}$$

equation 8 can be simplified to

$$\epsilon\rho_i\alpha \frac{\partial p}{\partial t} + \frac{\partial}{\partial t}(\epsilon\rho_i) = \nabla \cdot \rho \mathbf{D} \cdot \nabla\left(\frac{\rho_i}{\rho}\right)$$
$$- \nabla \cdot \rho_i \mathbf{q} + \sum_{k=1}^{s} R_{ik} + W_i \quad (9)$$

where $\mathbf{q}$ is the flux defined by Darcy's law.

Flow Equation

The equation of motion for the flow of groundwater has received careful attention by a number of workers including *Theis* [1935], *Hubbert* [1940, 1956], *Jacob* [1940, 1950], and *Cooper* [1966].

Following *Cooper*'s [1966, p. 4787, equation 15] development of the flow equation we can write for the conservation of fluid in a volume element, which may deform only vertically,

$$-\nabla \cdot \rho \mathbf{q} + \sum_{i=1}^{n} W_i$$
$$= \epsilon \frac{D\rho}{Dt} + \rho \frac{D\epsilon}{Dt} + \rho\epsilon \frac{\partial w_\sigma}{\partial z} \quad (10)$$

The conservation of the solid skeleton in a fixed volume element in which we assume only vertical deformation of the aquifer skeleton is given by

$$-\frac{\partial}{\partial z}[\rho_s(1-\epsilon)w_\sigma] = \frac{\partial}{\partial t}[\rho_s(1-\epsilon)]$$

where ρ_s is the density of the solid grains (ML^{-3}), which reduces to

$$\frac{D\epsilon}{Dt} = \frac{1-\epsilon}{\rho_s} \frac{D\rho_s}{Dt} + (1-\epsilon) \frac{\partial w_\sigma}{\partial z}$$

If we accept *Jacob*'s (1950, p. 329) assumption that the volume of individual grains remains constant during deformation of the medium, which seems reasonable for isothermal conditions, i.e., $D\rho_s/Dt = 0$, then by substituting from above into (10) we obtain

$$-\nabla \cdot \rho \mathbf{q} + \sum_{i=1}^{n} W_i = \epsilon \frac{D\rho}{Dt} + \rho \frac{\partial w_\sigma}{\partial z} \quad (11)$$

If we assume isothermal conditions the dilatation due to pressure is given by (6)

$$\partial w_\sigma/\partial z = \alpha Dp/Dt$$

Substituting into (11) we obtain

$$-\nabla \cdot \rho \mathbf{q} + \sum_{i=1}^{n} W_i = \epsilon \frac{D\rho}{Dt} + \rho\alpha \frac{Dp}{Dt}$$

and introducing Darcy's law

$$\nabla \cdot \left[\rho \frac{\mathbf{k}}{\mu} \cdot (\nabla p - \rho \mathbf{g})\right]$$
$$+ \sum_{i=1}^{n} W_i = \epsilon \frac{D\rho}{Dt} + \rho\alpha \frac{Dp}{Dt} \quad (12)$$

Introducing a first-order equation of state relating fluid density to pressure and concentration we have

$$\rho = \rho_0 + \rho_0 \beta_p (p - p_0)$$
$$+ \frac{1}{V_0} \sum_{i=1}^{n} (m_i - m_{i0})$$

where

β_p, compressibility coefficient of the fluid, $LM^{-1}t^2$;
ρ_{i0}, mass per unit volume of solution of species i at a reference pressure, ML^{-3};
$\rho_0 = \sum_{i=1}^{n} \rho_{i0}$;
m_i, mass of species i in the reference volume V_0, M;
m_{i0}, mass of species i in the reference volume at the reference pressure, M;
V_0, reference volume of the fluid, L^3.

Taking the substantial time derivative yields

$$\frac{D\rho}{Dt} = \rho_0 \beta_p \frac{Dp}{Dt} + \frac{1}{V_0} \sum_{i=1}^{n} \frac{Dm_i}{Dt} \quad (13)$$

Substituting (13) into (12) we obtain

$$\nabla \cdot \left[\rho \frac{k}{\mu} \cdot (\nabla p - \rho g) \right] + \sum_{i=1}^{n} W_i$$
$$= \rho \alpha \frac{Dp}{Dt} + \rho_0 \epsilon \beta_p \frac{Dp}{Dt} + \frac{\epsilon}{V_0} \sum_{i=1}^{n} \frac{Dm_i}{Dt} \quad (14)$$

If we assume that

$$w_s \frac{\partial p}{\partial z} \ll \frac{\partial p}{\partial t} \qquad w_s \frac{\partial m_i}{\partial z} \ll \frac{\partial m_i}{\partial t}$$

equation 14 becomes

$$\nabla \cdot \left[\rho \frac{k}{\mu} \cdot (\nabla p - \rho g) \right] + \sum_{i=1}^{n} W_i$$
$$= \rho \alpha \frac{\partial p}{\partial t} + \rho_0 \epsilon \beta_p \frac{\partial p}{\partial t} + \frac{\epsilon}{V_0} \sum_{i=1}^{n} \frac{\partial m_i}{\partial t} \quad (15)$$

In the above development we assume that (1) permeability k is independent of pressure, temperature, and concentration, (2) there is no change in volume upon mixing fluids of different ionic concentrations, and (3) the proportionality constants β_p and α are independent of pressure and fluid composition.

The above system of partial differential equations, which describes the pressure distribution and the transport of chemical constitutents, forms a description for an isothermal groundwater system both physically and chemically, thus serving as a unified framework in which to view the system.

The fact that we restricted ourselves to an aquifer in which we assumed the only significant driving force to be hydraulic head simplifies the problem. This restriction eliminates the necessity to consider the Onsager relationships and the coupling between the various forces that can produce transport.

Rose [1966] argues strongly that the coupling effect should indeed be considered in the treatment of reservoir problems, and points out that these considerations are generally neglected by reservoir engineers. *Groenevelt and Bolt* [1969], in their discussion of the coupling of forces and fluxes in a soil–water system, indicate the magnitude of coefficients associated with each flux. *Olsen* [1969] also has shown that temperature gradients, chemical gradients, and electrical gradients can produce significant flow in clayey type deposits. We have sidestepped these issues by restricting ourselves to an aquifer and making the assumptions given above.

AQUIFER CONTAMINATION AT BRUNSWICK, GEORGIA: A CASE HISTORY OF TRANSPORT

It is difficult to visualize how a solution of the transport equations can, in fact, be applied to real field problems. We have, therefore, attempted to analyze the movement of contaminants in the Principal Artesian aquifer at Brunswick, Georgia, to illustrate our point. At Brunswick, the U.S. Geological Survey has made extensive hydrologic studies, which include documentation of contaminant movement in the aquifer [*Wait*, 1965; *Wait and Gregg*, 1967].

Brunswick hydrology. The Principal Artesian aquifer in the Brunswick area is composed of permeable zones in the Ocala and underlying limestones of Claiborne age. At least three more or less isolated zones are recognized. These zones are the upper water-bearing zone, which is approximately 100 feet thick and generally contains freshwater; the lower water-bearing zone, which ranges in thickness from approximately 20 to 100 feet and which generally contains freshwater; and the brackish water zone, which underlies the lower zone.

The permeable zones are separated from one another by dense limestones, which act hydrologically as confining layers for each of the units. The units that act as aquifers are highly permeable; caverns have been detected in each.

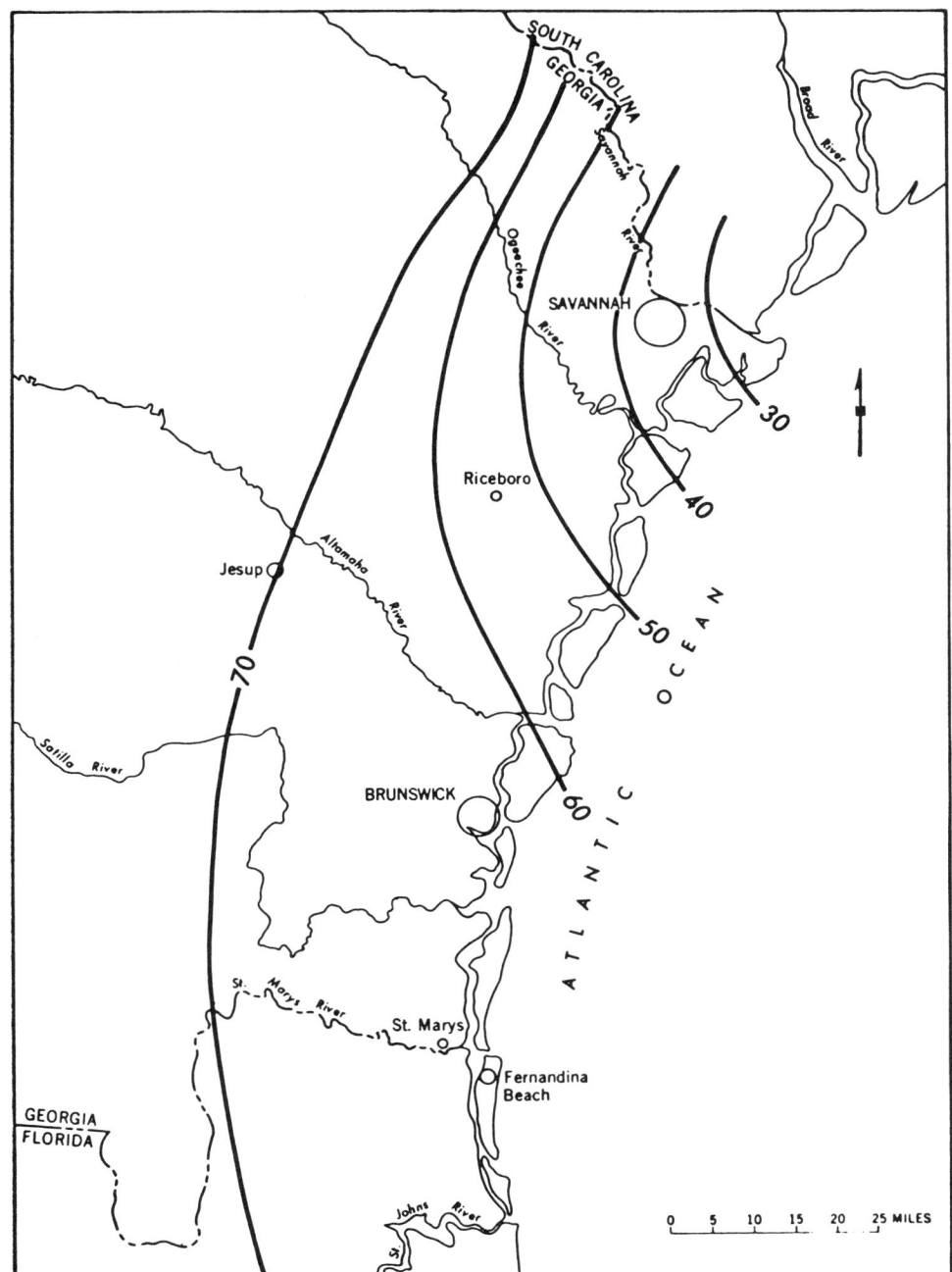

Fig. 1. Potentiometric surface (1880–1900) representing virgin conditions in the aquifer [from *Warren*, 1944].

Overlying the Ocala limestone is approximately 500 feet of Hawthorn formation composed of clayey silt, silty clay, sand, sandy limestone, and phosphatic sandy limestone. One bed of silty clay seems to be reasonably continuous over the coastal plain of Georgia. This bed, which can be as thick as 200 feet, forms the major confining layer for the limestone system.

The limestone crops out in the highlands of central Georgia, approximately 150 miles northwest of Brunswick. To the east, beneath the Atlantic, the extent of the aquifer is virtually unknown. Freshwater was, however, produced from equivalent limestones some 30 miles or so east of Jacksonville, Florida.

The aquifer is extensively developed in the Brunswick area. The first wells into the Principal Artesian aquifer in the coastal area were drilled in Savannah in 1885. Figure 1 shows the potentiometric surface reported by *Warren* [1944], which represents virgin conditions for the aquifer. Brunswick, along with other coastal communities, drilled wells into the aquifer shortly after 1885.

By 1959 the pumpage at Brunswick had reached 94 mgd (million gallons per day) (145 ft^3 sec^{-1}). This rate remained approximately constant through 1959, 1960, 1961, and 1962. Steady flow seems to have been established within a year at Brunswick. Approximately 70% of the total withdrawals comes from the upper water-bearing zone; 30% comes from the lower water-bearing zone. Figure 2 shows the potentiometric surface for the upper water-bearing zone, October 1962 [from *Wait and Gregg*, 1967], which represents the hydraulic head in the aquifer during 1959–1962.

In December 1962, pumpage in the Bruns-

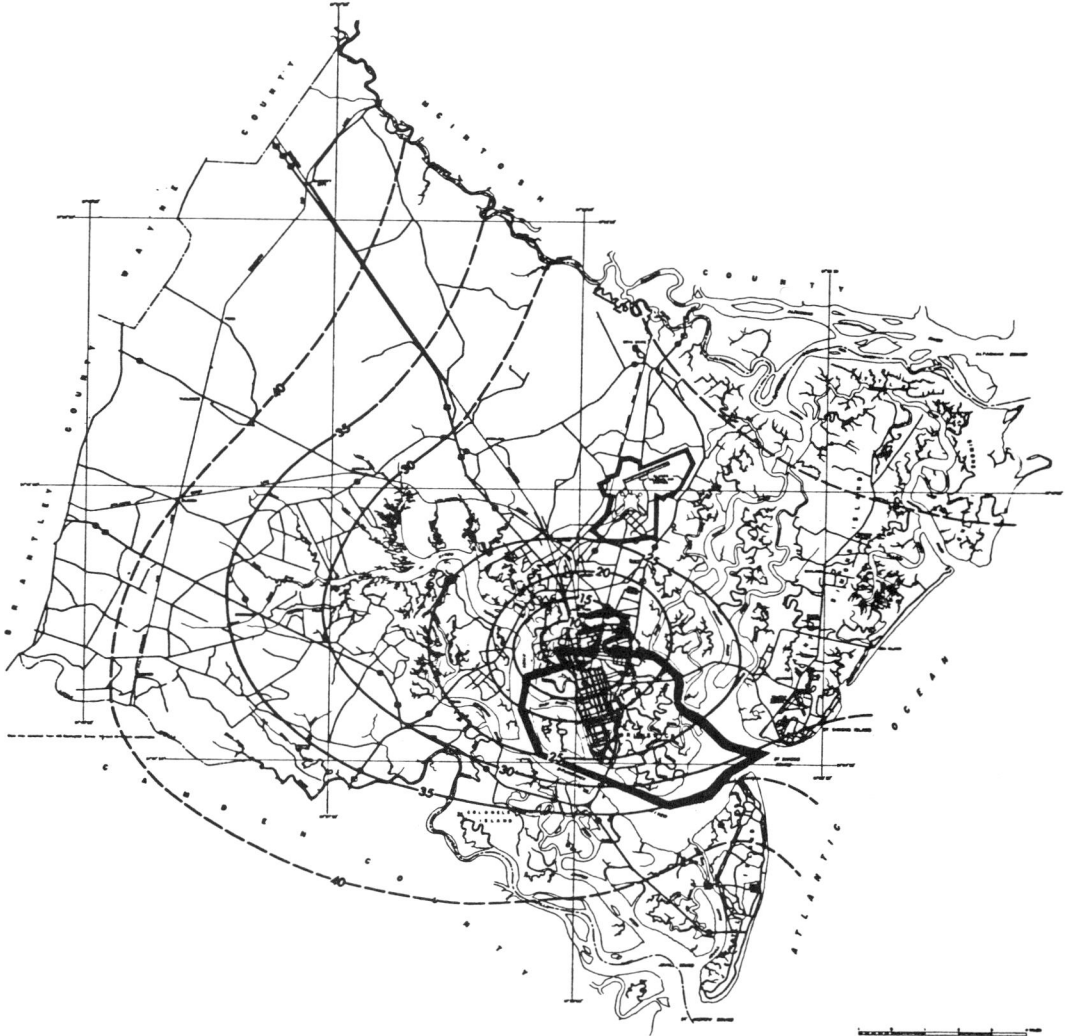

Fig. 2. Upper water-bearing zone potentiometric surface, October 1962 [from *Wait and Gregg*, 1967].

wick area was increased from 94 to 125 mgd (145–193 ft³ sec⁻¹), a rate which has remained approximately constant to the present. Figure 3 shows the potentiometric surface for the upper zone, December 1963 [*Wait and Gregg*, 1967], which represents a new steady flow condition.

Contamination. Both the upper and lower water-bearing zones show areas of saltwater contamination. This contamination was first detected in the early 1960's. Extensive investigations over the past 10 years have defined the areas of contamination as well as the movement. Figures 4–7 are a series of maps showing the distribution of chloride in the upper water-bearing zone for the years 1962, 1966, 1969, and 1970 [*Wait and Gregg*, 1967].

These maps include the hydrologist's interpretation of data. The 1962 mapping is highly interpretative and probably not too accurate, as later test drilling indicated.

The data suggest that contamination is occurring from two point sources in the upper water-bearing zone. Data for the lower zone

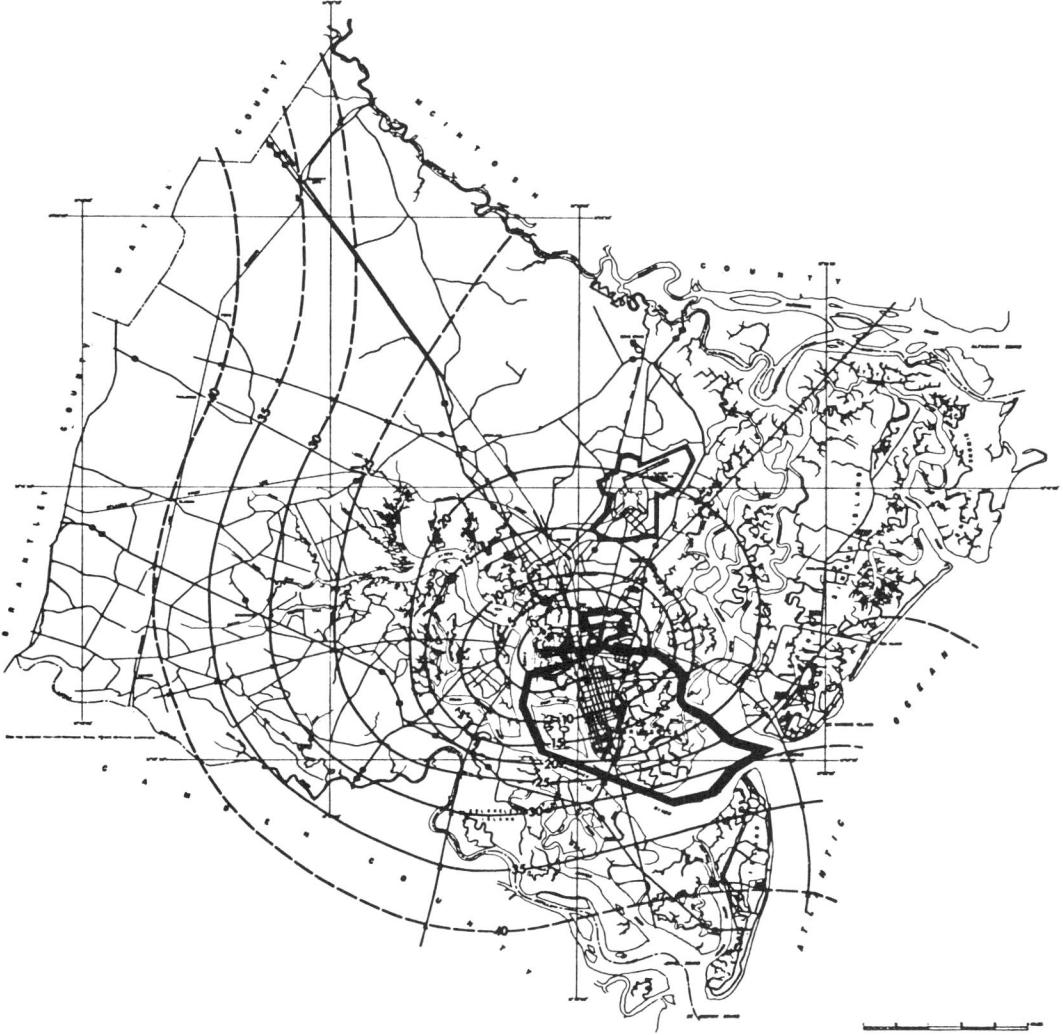

Fig. 3. Upper water-bearing zone potentiometric surface, December 1963 [*Wait and Gregg*, 1967].

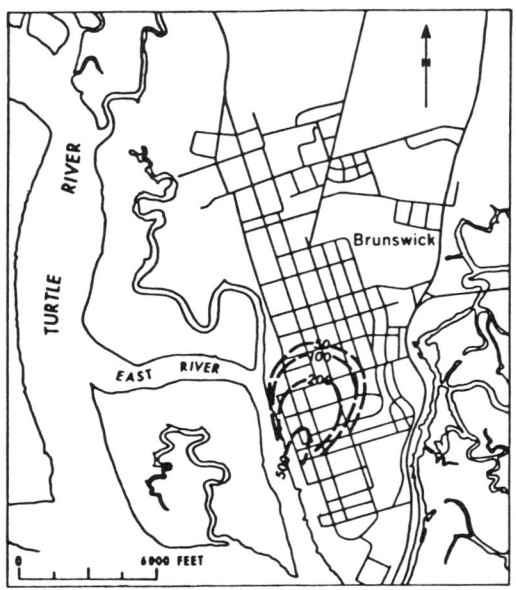

Fig. 4. Upper water-bearing zone observed chloride distribution (mgl^{-1}), 1962 [*Wait and Gregg*, 1967].

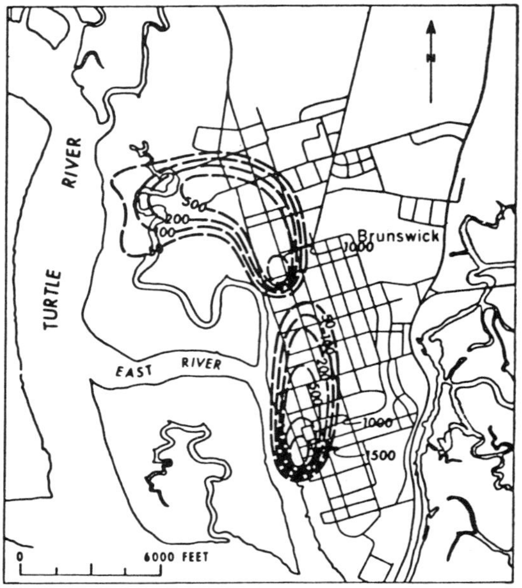

Fig. 6. Upper water-bearing zone observed chloride distribution (mgl^{-1}), 1969.

show a similar pattern. One would immediately think that seawater from the Brunswick River is infiltrating through old unplugged wells. This possibility has been checked exhaustively, and no evidence of old wells has been found.

Interestingly, the contaminated water has a higher temperature than the groundwater in the surrounding areas. The source of the water now seems to be the underlying brackish water zone, the water migrating upward through two

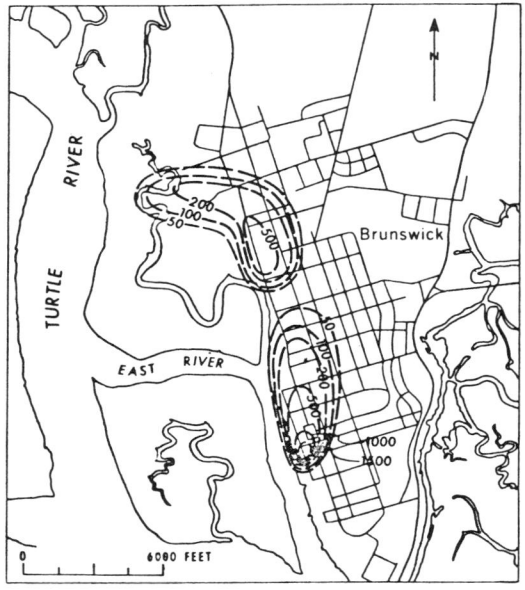

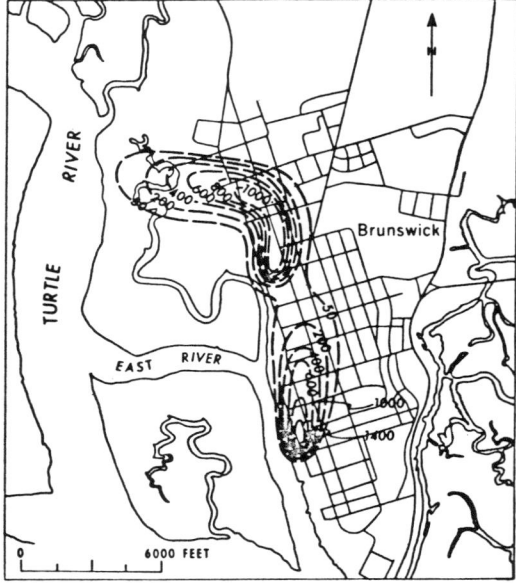

Fig. 5. Upper water-bearing zone observed chloride distribution (mgl^{-1}), 1966.

Fig. 7. Upper water-bearing zone observed chloride distribution (mgl^{-1}), 1970.

natural conduits. These conduits may be associated with a fault.

Under virgin conditions the upper, lower, and brackish zones had approximately the same heads. Once large-scale pumping developed significant drawdown, differences in head were produced between the various layers. It is this difference in head that causes the flow from the brackish zone to the lower and upper zones. In the southern part of the two contaminated areas the head in the brackish zone is 8 feet above the head in the lower zone. In both areas of interest the brackish zone contains water with approximately 2000 mgl^{-1} (milligrams per liter) of chloride.

Transport model. Because the water quality problem at Brunswick was a result of chloride contamination, energy transport and chemical reactions could be neglected and the system was modeled by the mass transport equation and the flow equation. The concentrations observed in the aquifer were sufficiently low so that a constant density fluid could be assumed.

The assumption of a constant density fluid allows us to solve the flow equation in terms of hydraulic head rather than pressure. If we consider a thickness b of saturated porous media, two-dimensional flow through the entire thickness of the aquifer can be described by modifying (15) to read

$$\nabla \cdot \mathbf{T} \cdot \nabla h = S \frac{\partial h}{\partial t}$$
$$+ Q(x, y) - \frac{K_z}{m}(H_{wt} - h) \quad (16)$$

where

$\mathbf{T} = b\mathbf{k}\rho g/\mu$;
$S = b(\rho g\alpha + \rho_0 g \epsilon \beta_p)$;
$h = (p/\rho g) + z$;
$-K_z/m(H_{wt} - h)$, steady state leakage into the aquifer;
$Q(x, y)$, volumetric flux component of W_i;
$\mathbf{T}$, transmissivity, L^2t^{-1};
K_z, vertical hydraulic conductivity of the confining layer, Lt^{-1};
z, elevation;
H_{wt}, water table head assumed constant in time, L;
m, thickness of the confining layer, L;
S, aquifer storage coefficient, L^0;

The mass average velocity of flow through the aquifer relative to the grains is given by

$$\mathbf{v} = \mathbf{q}/\epsilon = -\mathbf{K}/\epsilon \nabla h$$

where $\mathbf{K}$ is the hydraulic conductivity of the aquifer (Lt^{-1}). In two dimensions the mass average flux through the entire thickness of aquifer, in which we assume the vertical components of flow are negligible, is given by

$$\mathbf{q}^* = -\mathbf{T} \cdot \nabla h \quad (17)$$

where $\mathbf{q}^*$ is the mass average flux (L^2t^{-1}) (by flux we mean here the rate of flow per unit thickness of aquifer relative to the grains), and the ∇ operator is defined in the two-dimensional case as

$$\nabla \equiv \frac{\partial}{\partial x}\mathbf{i} + \frac{\partial}{\partial y}\mathbf{j}$$

A further simplification is made by assuming that the change in porosity is not significant in the artesian system. With this assumption the change in mass concentration of chloride is given from (9) by

$$-\nabla \cdot \rho_{Cl}\mathbf{q} + \nabla \cdot \mathbf{D} \cdot \nabla \rho_{Cl}$$
$$= \frac{\partial \epsilon \rho_{Cl}}{\partial t} + \epsilon \rho_{Cl}\alpha \frac{\partial p}{\partial t} + Q\rho_{Cl}*$$

where ρ_{Cl} is the mass of chloride per unit volume of solution (ML^{-3}), $\rho_{Cl}*$ is the mass of chloride per unit volume of the source or sink solution, and Q is the rate of production of the source or sink (t^{-1}). In a manner comparable to (16) we can write that a two-dimensional equation for the change in mass concentration of chloride is given from (9) by

$$-\nabla \cdot \rho_{Cl}\mathbf{q}^* + \nabla \cdot \mathbf{D}^* \cdot \nabla \rho_{Cl}$$
$$= \frac{\partial(\epsilon b \rho_{Cl})}{\partial t} + b\epsilon\rho_{Cl}\alpha \frac{\partial p}{\partial t} + Q\rho_{Cl}*$$

which can be written

$$-\left(q_x^* \frac{\partial \rho_{Cl}}{\partial x} + q_y^* \frac{\partial \rho_{Cl}}{\partial y}\right)$$
$$- \rho_{Cl}\left(\frac{\partial q_x^*}{\partial x} + \frac{\partial q_y^*}{\partial y}\right)$$
$$+ \frac{\partial}{\partial x}\left(D_{xx}^* \frac{\partial \rho_{Cl}}{\partial x} + D_{xy}^* \frac{\partial \rho_{Cl}}{\partial y}\right)$$
$$+ \frac{\partial}{\partial y}\left(D_{yy}^* \frac{\partial \rho_{Cl}}{\partial y} + D_{yx}^* \frac{\partial \rho_{Cl}}{\partial x}\right)$$
$$= \frac{\partial(\epsilon b \rho_{Cl})}{\partial t} + \epsilon\rho_{Cl}\alpha \frac{\partial p}{\partial t} + Q\rho_{Cl} \quad (18)$$

where b is the thickness of the aquifer (L) and $D^* = Db$ is an effective dispersion coefficient for the entire thickness aquifer (L^3t^{-1}).

This coupled set of partial differential equations (16–18) describe the movement of chlorides in the upper water-bearing zone at Brunswick.

The hydrodynamic dispersion coefficient is generally thought to be a function of the velocity; molecular diffusion is negligible compared to hydrodynamic dispersion for most actual field situations. Both longitudinal and transverse dispersion must be considered. *Scheidegger* [1960] gives the relationship

$$D_{ij} = \alpha_{ijmn} v_m v_n / v$$

where α_{ijmn} is the dispersivity of the medium (L), $v_m v_n$ are the components of the velocity in the m and n directions (LT^{-1}), and v is the magnitude of the velocity (LT^{-1}) for an isotropic media. Scheidegger pointed out that

$$\alpha_{1111} = \alpha^*$$
$$\alpha_{1122} = \alpha^{**}$$
$$\alpha_{1212} = \tfrac{1}{2}(\alpha^* - \alpha^{**})$$
$$\alpha_{1221} = \tfrac{1}{2}(\alpha^* - \alpha^{**})$$

where α^* is the longitudinal dispersivity (L) and α^{**} is the transverse dispersivity (L). The longitudinal and transverse dispersion coefficients are related to the dispersivities by $D_L = \alpha^* v$ and $D_T = \alpha^{**} v$. We have taken $\alpha^{**} = 0.3\alpha^*$. The fact that the principal components of the tensor D change orientation as the velocity field changes means that we must carry all the cross-product terms in our solution of (13); we have followed *Reddell and Sunada* [1970] in our development of this aspect of the problem.

Some care is necessary in solving the equations. For the flow equation we use finite difference techniques in conjunction with an iterative alternating direction method to solve the resulting set of simultaneous equations [*Pinder and Bredehoeft*, 1968; *Bredehoeft and Pinder*, 1970]. Once the head is computed, computation of the mass average flux $\mathbf{q}^*$ from (17) is straightforward.

The mass balance equation (18) poses other problems. As long as the velocity term is sig-

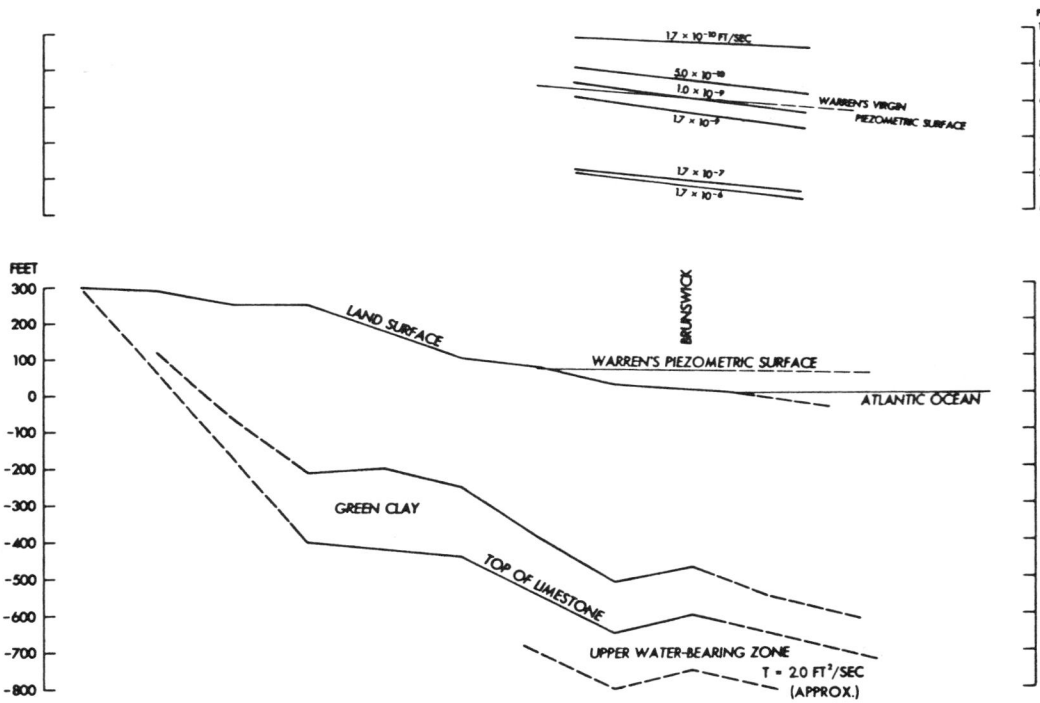

Fig. 8. Schematic section of the aquifer.

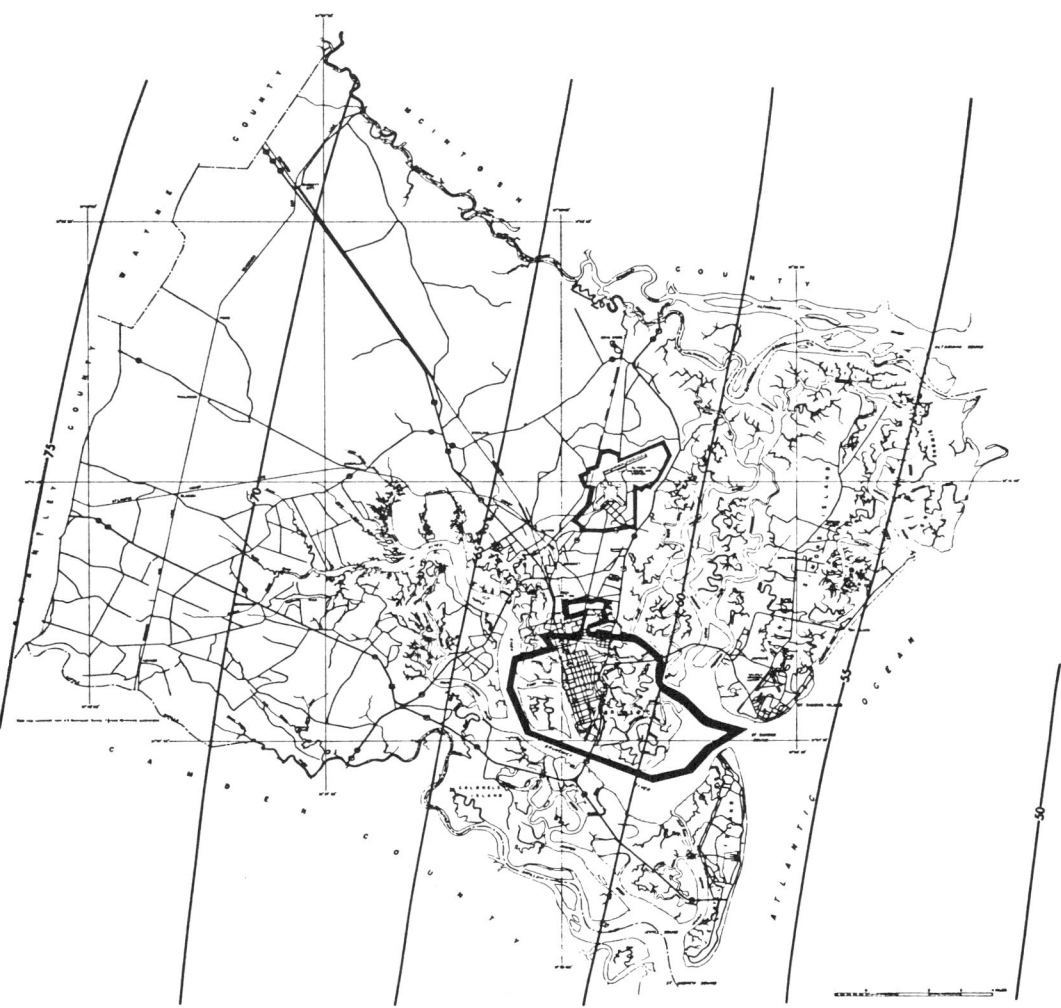

Fig. 9. Computed virgin potential map.

nificant, (18) behaves like a hyperbolic equation and finite difference methods lead to numerical dispersion. For this reason we solve this equation by using the method of characteristics. The method of solution is generally similar to that used by *Pinder and Cooper* [1970] and *Reddell and Sunada* [1970]. However, since we neglect changes in density in our flow equation, it is not necessary to iterate between equations 16–18.

Brunswick model results. The mass average flux must be obtained from the flow equation in order to solve for the mass transport. This, in turn, dictates that absolute values of head be computed rather than the drawdown. The aquifer system is depicted schematically in Figure 8.

A steady flow analysis for virgin conditions was first attempted. The following parameters were assumed to be known in the model: (1) aquifer transmissivity, which varied over the model area, (2) thickness of the silty 'green' clay, which again varied in thickness over the model area, and (3) water table altitude.

The segment of the aquifer that was modeled extended from south of Jacksonville, Florida, to north of Savannah, Georgia, and from the outcrop in central Georgia to approximately 70 miles into the ocean. Modeling such a larger area was facilitated by increasing the grid spacing away from Glynn county.

Using the above input and assuming the outcrop to be a constant head boundary one can compute virgin head distributions for the upper water-bearing zone. The head computations are sensitive to the vertical hydraulic conductivity of the confining layer K_z. Figure 8 shows the effects of several vertical hydraulic conductivities. A value of 1.7×10^{-6} ft sec^{-1} was used as an initial guess as this was the value obtained in laboratory analysis; the results indicate this to be an obviously high value. At a value of 1.0×10^{-9} ft sec^{-1} the results approach the virgin potentiometric surface reported by *Warren* [1944]. Figure 9 shows the computed virgin potential map and can be compared with Figure 1.

Since the system reaches steady flow quickly, the hydrology can be approximated by two steady flow periods: the period prior to December 1962 and the period after December 1962. This approximation reduces the computations in the hydrologic model. The model is, however, designed for the more general situation of continually changing head and concentration.

Figure 10 shows the results of simulating conditions prior to December 1962. A comparison between the model results and the observed val-

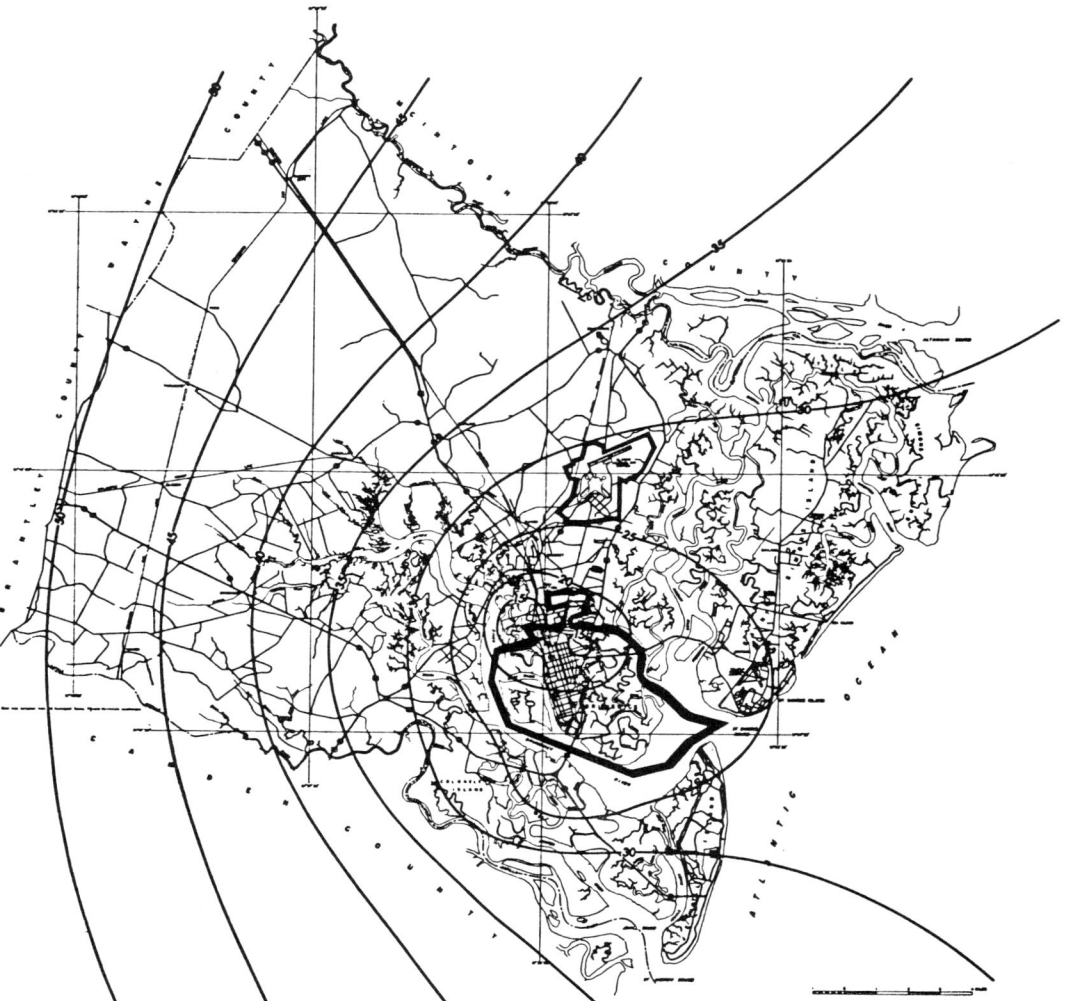

Fig. 10. Computed upper water-bearing zone potential for 1962.

ues shows some differences. It would appear that there is a low permeability zone between Brunswick and St. Simons Island to the east. However, in the vicinity of the areas of contamination (our main area of interest), the model results are good.

The period following December 1962 was simulated and the results are presented in Figure 11. The best match for the virgin conditions and the 1962 and 1963 potential surfaces was obtained using 0.9 of the values indicated in Figure 8. In the final analysis a variable transmissivity was used with a value of 1.8 ft² sec⁻¹ in the immediate Brunswick area.

The thickness of the confining layer was varied, and the vertical hydraulic conductivity was considered uniform and given the value 9×10^{-10} ft sec⁻¹.

Quality model results. It is apparent that since the mass average flux of fluid is an integral part of the mass transport equation, the hydrologic model must be accurate before one can hope to simulate the water quality. Once a satisfactory hydrologic model was obtained we attempted to incorporate the movement of the chloride. Figures 12–15 show the computed chloride distributions for the years 1962, 1965, 1967, and 1970. These maps can be compared

Fig. 11. Computed upper water-bearing zone potential for 1963.

Fig. 12. Computed chloride distribution (mgl⁻¹), 1962.

with the observed chloride distributions (Figures 4–7). We believe the fit of computed to observed data to be reasonably good.

Three parameters are unknown in the quality model: (1) porosity ϵ, (2) the characteristics

Fig. 14. Computed chloride distribution (mgl⁻¹), 1967.

mixing length α^* necessary to formulate the dispersion coefficient, and (3) the quantity of leakage at the two points of contamination. The best results were obtained with (1) a porosity of $\epsilon = 0.35$, (2) a characteristic mixing length

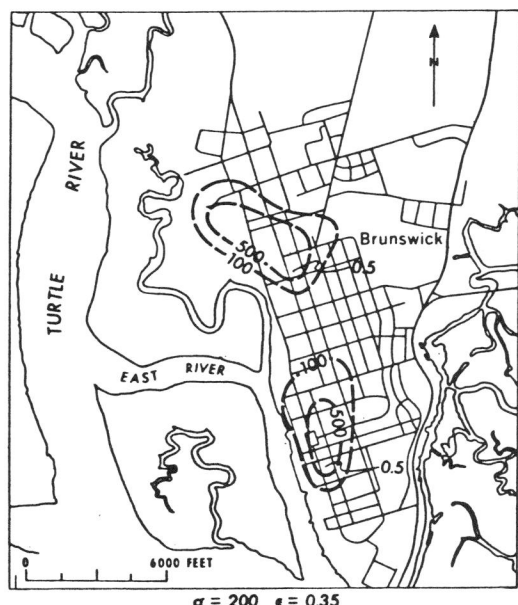

Fig. 13. Computed chloride distribution (mgl⁻¹), 1965.

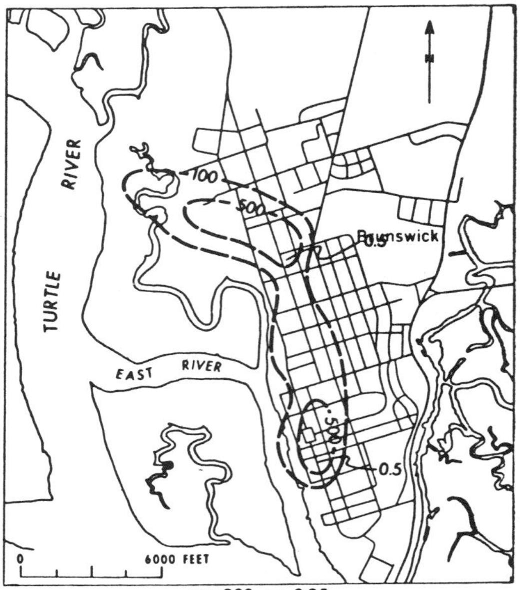

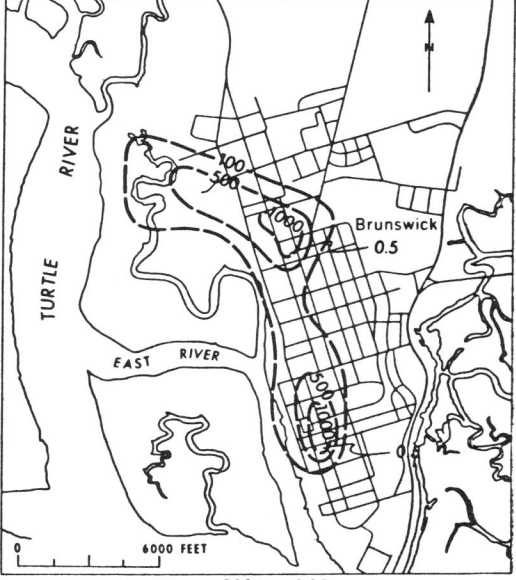

Fig. 15. Computed chloride distribution (mgl⁻¹), 1970.

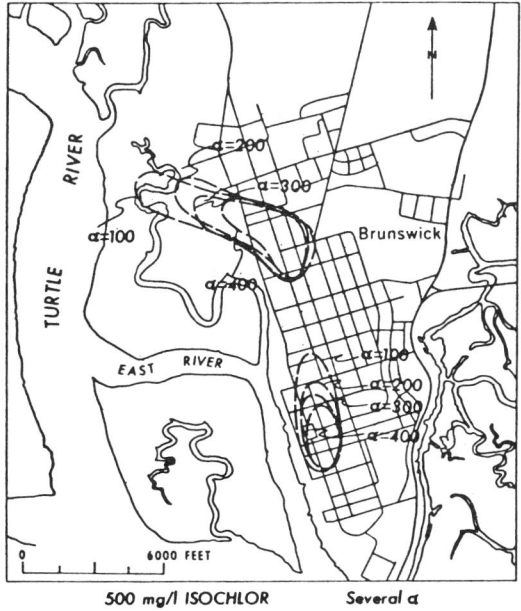

Fig. 16. Simulated 1970 results showing the 500 mgl^{-1} isochlor for several α^*.

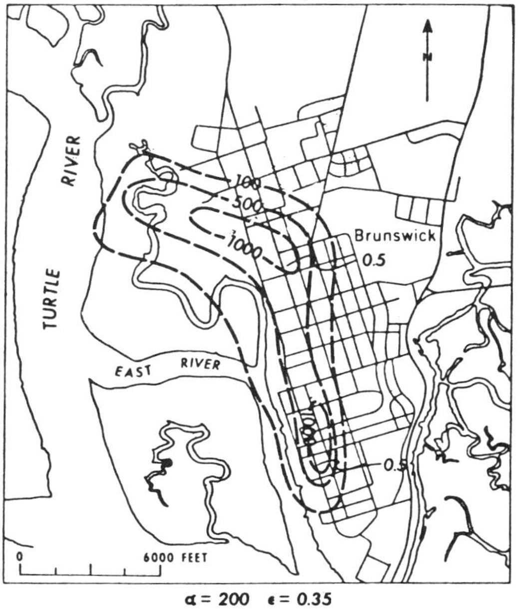

($\alpha^* = 200$ feet), and (3) a quantity of leakage of 0.5 ft^3 sec^{-1} of water with 2000 mgl^{-1} of chloride concentration starting in 1958 and main-

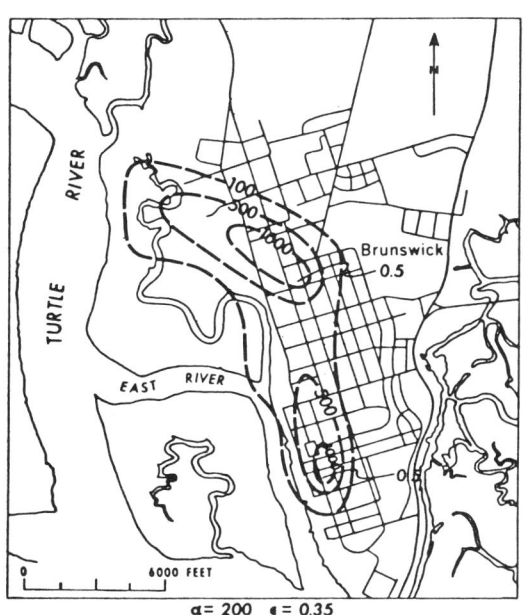

Fig. 17. Computed chloride distribution (mgl^{-1}), 1975.

Fig. 18. Computed chloride distribution (mgl^{-1}), 1980.

tained at this level through the period of analysis.

Several values of α^*, the characteristic longi-

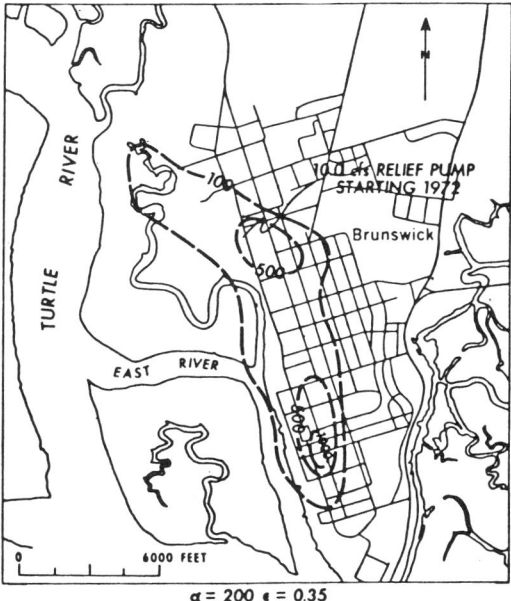

Fig. 19. Computed 1975 chloride distribution (mgl^{-1}) with 10 ft^3 sec^{-1} (cfs) of protective pumping initiated in 1972.

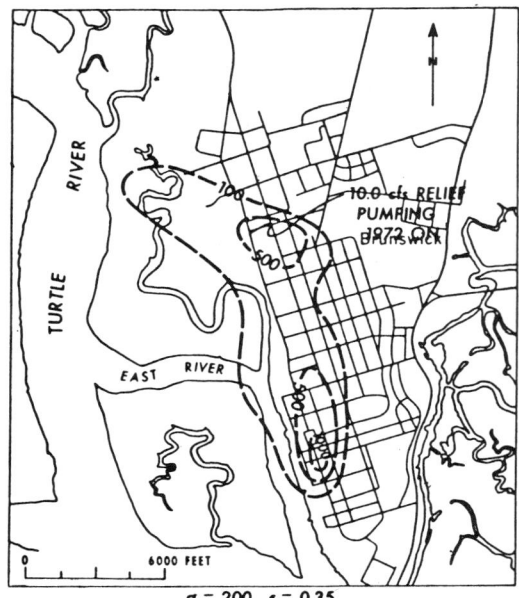

Fig. 20. Computed 1980 chloride distribution (mgl^{-1}) with 10 ft^3 sec^{-1} protective pumping initiated 1972.

tudinal mixing length, were simulated. Figure 16 gives the simulated results for 1970 showing the position of the 500 mgl^{-1} isochlor for differing α^*. The computed results are sensitive to α^*. Although the differences are subtle, the response of the model to changes in α^* can be separated from the influence of changes in porosity.

Once a fit to the historical data was obtained, the projection for 1975 and 1980 shown in Figures 17 and 18 was made. By 1980 a marked increase in the chloride concentration in the major well field will have occurred.

The most feasible measure to protect the well field is to put interceptive pumping in the aquifer between the source of contamination and the pumping center. We simulate the effect of 10 ft^3 sec^{-1} (cfs) of protective pumping. On the assumption that this pumping was initiated in 1972, Figure 19 shows the predicted chloride distribution for 1975. The pumping would be effective, as a comparison of Figures 17 and 19 indicates. Although 10 ft^3 sec^{-1} may appear to be a large amount of pumping, one average well will produce this much water. (One well at the Brunswick Pulp and Paper Company was tested at 29 ft^3 sec^{-1}.) Probably the most effective measure would be to distribute the protective pumping over a larger area by using two or perhaps three wells. The chloride distribution for 1980 with the protective pumping was also computed and is shown in Figure 20. The 1980 results are not materially changed from 1975, thus indicating that a steady state situation is essentially established.

NOTATION

b, aquifer thickness, L;
D_i, hydrodynamic dispersion coefficient for component i, L^2t^{-1};
D^*, effective dispersion coefficient for the entire thickness of the aquifer, L^3t^{-1};
g, gravitational acceleration, Lt^{-2};
h, hydraulic head, L;
H_{wt}, water table head assumed constant in time, L;
j_i, mass flux of i relative to the mass average velocity, ML^2t^{-1};
k, intrinsic permeability, L^2;
K, hydraulic conductivity, Lt^{-1};
K_z, vertical hydraulic conductivity of the confining layer, Lt^{-1};
m, thickness of the confining layer, L;
m_i, mass of species i in the reference volume V_0, M;
m_{i0}, mass of species i in the reference volume at the reference temperature and pressure, M;
n, total number of constituents in the system;
p, fluid pressure, $ML^{-1}t^{-2}$;
q, specific discharge of the fluid, LT^{-1};
q^*, mass average flux, L^2t^{-1};
Q, volumetric source, L^3t^{-1};
Q_i, source or sink, L^3t^{-1};
r, number of sources and sinks;
R_{ik}, rate of production of constituent i in reaction k; expressed as mass per unit volume of solution per unit time, $ML^{-3}t^{-1}$;
s, number of reactions taking place in the system;
S, aquifer storage coefficient, L^0;
T, transmissivity of the aquifer, L^2t^{-1};
v, mass average velocity of the fluid, Lt^{-1};
v_D, Darcy's velocity, Lt^{-1};
v_i, mass average velocity of constituent i, LT^{-1};
V_i^*, volume per unit mass of component i, $M^{-1}L^{-3}$;
V_0, reference volume of the fluid, L^3;
W_g, grain velocity, Lt^{-1};
z, elevation above an arbitrary datum, L;
α_{ijmn}, dispersivity of the medium, L;
α, compressibility of the medium, $M^{-1}Lt^2$;
α^*, longitudinal dispersivity, L;
α^{**}, transverse dispersivity, L;
β_p, compressibility coefficient of the fluid, $LM^{-1}t^2$;
$\delta(x - x_i)(y - y_i)(z - z_i)$, Dirac delta function;
ϵ, effective porosity of the porous media, L^0;

μ, dynamic viscosity, $ML^{-1}t^{-1}$;
ρ_i, mass per unit volume of solution of constituent i, ML^{-3};
ρ_i^*, partial mass density of source or sink fluid, ML^{-3};
ρ_{i0}, mass per unit volume of solution of species i at a reference temperature and pressure, ML^{-3};
ρ_{Cl}, mass of chloride per unit volume of solution, ML^{-3};
ρ_{Cl}^*, mass of chloride per unit volume of the source or sink solution, ML^{-3};
ρ_s, density of the solid grains of the porous media, ML^{-3}.

Acknowledgment. We want to express our thanks to Harlan Counts of the U.S. Geological Survey who helped greatly with the hydrologic model for the Brunswick area, and to Hilton Cooper and Arnold Klute whose suggestions were especially helpful.

REFERENCES

Bear, J., D. Zaslavsky, and S. Irmay, *Physical Principles of Water Percolation and Seepage*, Unesco, Paris, 1968.

Bredehoeft, J. D., and G. F. Pinder, Digital analysis of areal flow in multiaquifer groundwater systems: A quasi three-dimensional model, *Water Resour. Res.*, 6(3), 883–888, 1970.

Cooper, H. H., Jr., The equation of groundwater flow in fixed and deforming coordinates, *J. Geophys. Res.*, 71(20), 4785–4790, 1966.

Hubbert, M. K., The theory of ground-water motion, *J. Geol.*, 48(8), 785–944, 1940.

Hubbert, M. K., Darcy's law and the field equations of flow of underground fluids, *Trans. Amer. Inst. Mining, Metal. Petrol. Eng., 207*, 222–239, 1956.

Hubbert, M. K., and W. W. Rubey, Role of fluid pressure in mechanics of overthrust faulting, 1, Mechanics of fluid-filled porous solids and its application to overthrust faulting, *Bull. Geol. Soc. Amer.*, 70, 115–166, 1959.

Jacob, C. E., On the flow of water in an elastic artesian aquifer; *EOS Trans. AGU*, 20, 574–586, 1940.

Jacob, C. E., *Engineering Hydraulics*, edited by H. Rouse, chap. 5, pp. 321–386, John Wiley, New York, 1950.

Olsen, H. W., Simultaneous fluxes of liquid and charge through saturated kaolinite, *Soil Sci. Soc. Amer. Proc.*, 33, 338–344, 1969.

Pinder, G. F., and J. D. Bredehoeft, Application of the digital computer for aquifer evaluation, *Water Resour. Res.*, 4(5), 1069–1093, 1968.

Pinder, G. F., and H. H. Cooper, Jr., A numerical technique for calculating the transient position of the saltwater front, *Water Resour. Res.*, 6(3), 875–882, 1970.

Reddell, D. L., and D. K. Sunada, Numerical simulation of dispersion in ground water aquifer, *Hydrol. Pap. 41*, pp. 1–79, Colo. State Univ., Fort Collins, 1970.

Rose, W., Reservoir engineering: Reformulated, *Proc. 25th Tech. Conf. Petrol. Prod.*, 431, 431, 23–67, 1966.

Scheidegger, A. E., *The Physics of Flow through Porous Media*, rev. ed., University of Toronto Press, Toronto, Ont., 1960.

Theis, C. V., The relation between the lowering of the piezometric surface and the rate and duration of discharge of a well using ground-water storage, *EOS Trans. AGU*, 16, 519–524, 1935.

Wait, R. L., Geology and occurrence of fresh and brackish ground water in Glynn County, Georgia, *U.S. Geol. Surv. Water Supply Pap. 1613-E*, 1965.

Wait, R. L., and D. O. Gregg, Hydrology and chloride contamination of the principal artesian aquifer in Glynn County, Georgia, open-file report, U.S. Geological Survey, Washington, D.C., 1967.

Warren, M. A., Artesian water in southeastern Georgia with special reference to the coastal area, *Bull. 49*, Ga. Geol. Surv., Atlanta, 1944.

(Received June 5, 1972;
revised July 7, 1972.)

Simulation of Hydrochemical Patterns in Regional Groundwater Flow

FRANKLIN W. SCHWARTZ

PATRICK A. DOMENICO

The chemical state of a regional groundwater system in which several different processes are acting is described by a partial equilibrium simulation model that incorporates mass transfer rates and reaction kinetics. Quantitative analysis with the model indicates that mineral dissolution, saturation constraints on dissolution, the degree of saturation, partial pressures of CO_2 gas, and reaction kinetics in relation to the residence time of the groundwater flow play different roles in determining the spatial distribution of ionic constituents. Simultaneous evaluation of several geochemical processes permits the study of interdependent phenomena such as shifts in equilibrium concentrations resulting from the addition of common ions by cation exchange or sulfate reduction processes. The utility of the model is demonstrated by applying it to the groundwater reservoir in the Upper Kettle Creek, Ontario, Canada, where a favorable comparison is achieved between the real and the theoretical hydrochemical patterns.

The study of the chemical evolution of groundwater in a regional flow system has generally followed two paths. On the one hand is the descriptive approach, which relies on the analysis of field data to interpret changes in the chemical character of the groundwater by invoking well-known physical, chemical, and biological processes. For example, the six-variable graphical method developed by *Piper* [1944] and later extended by *Back* [1966] to the concept of hydrochemical facies is frequently used to describe the variation of major ions in groundwater.

On the other hand is the mechanistic approach, which starts with the processes themselves to generate chemical patterns and to verify observations made in the field. Several investigators including *Sillén* [1961], *Back* [1961], *Back et al.* [1966], *Garrels and MacKenzie* [1967], and *Kramer* [1967] have used this approach to abstract the details of natural water systems by chemical equilibrium models. Although these models tell us nothing about the rate at which mass is transported, they provide the reference base for transport models that entail chemical reactions. Equilibrium models identify the controlling reactions because aqueous constituents can be in equilibrium with respect to one or more minerals and out of equilibrium with respect to others. *Helgeson* [1968] has extended the equilibrium models to include mass transfer in such a way that a groundwater system can be described as a partial equilibrium system.

The principal objective of this study is to solve a material transfer problem by developing a nonequilibrium simulation model capable of generating spatial patterns of the most important dissolved aqueous species in a regional groundwater system. A second objective is to demonstrate the applicability of the model by using it to achieve a clearer understanding of the roles that physical, chemical, and biological processes play in the chemical evolution of groundwater from an actual field area, the Upper Kettle Creek watershed, Ontario.

MATHEMATICAL FORMULATION OF THE PROBLEM

Assumptions and working equations. A reasonably accurate conceptualization of material transport in regional groundwater flow is given in Figure 1. Material enters the groundwater

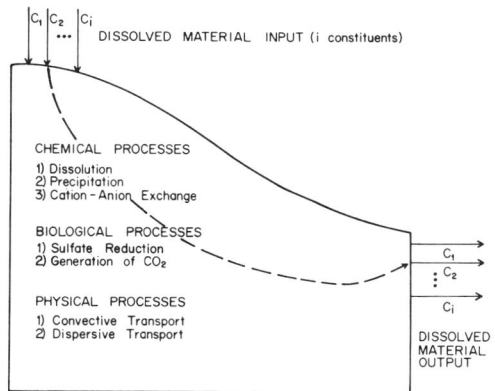

Fig. 1. Conceptualization of mass transfer in regional groundwater flow.

as the dissolved ionic constituents of recharge, is transported through the system, and exits as groundwater discharge. Moreover, important chemical and biological processes operate within the flow system to modify the concentration of dissolved solids. Mathematically, this problem may be stated by the molar form of the mass balance equation for the ith constituent in a multicomponent system:

$$\nabla \cdot (D_i \nabla c_i) - (\mathbf{v} \cdot \nabla c_i + c_i \nabla \cdot \mathbf{v}) + \sum_{j=1}^{n} R_{ij} = \frac{\partial c_i}{\partial t} \quad (1)$$

where D_i is the dispersion coefficient, t is time, c_i is the molal concentration of species i, $\mathbf{v}$ is the velocity of groundwater movement, and R_i is a generation term.

To arrive at a complete chemical description of this complex system, it is necessary to solve an array of mass balance equations of the form given by (1), one for each of the chemical constituents. This array expresses the fact that the concentration variation of the ith species results from dispersion, convection, and the spontaneous generation of that constituent within the system from n chemical or biological processes. In other words, the processes depicted in Figure 1 are incorporated into the array, where the velocity terms account for convective transport, the first term on the left accounts for dispersive transport, and the last term on the left is a chemical source term that must account for all the significant chemical and biological processes under consideration.

A complete analysis of material transport should include all the processes depicted in Figure 1. The model applied in this paper considers only complex reactions in a moving medium and requires certain assumptions. First, it is assumed that the ion concentrations are not materially affected by chemical perturbations such as seasonal variations in the partial pressure of CO_2 gas and that they can vary spatially but not temporally. This assumption appears valid from a practical point of view in that the sampling error associated with repeated chemical analyses at a single point may often mask the small variations in chemistry that might take place, at least for periods of time up to 5 or 10 years. A more restrictive assumption is required in the case of dispersive transport. In general, the dispersion coefficients are a complex function of ionic concentration, aqueous flow velocity, structure of the porous media, and temperature. Experimental measurements of the dispersion coefficients for many of the common ionic species are in general not available. Therefore as a first approximation it will be assumed that dispersive transport is negligible. This assumption of 'plug flow' has often been invoked in the design of chemical processing systems [*Himmelblau and Bischoff*, 1970].

Any further modification of (1) requires further information on the velocity field $v(x, y, z)$. Two equations are generally required to determine this field. The first of these, the continuity equation for a nondeforming porous medium, is expressed by

$$r \, \partial \rho_w / \partial t = -[\mathbf{v} \cdot \nabla \rho_w + \rho_w \nabla \cdot \mathbf{v}] \quad (2)$$

where r is the porosity of the medium and ρ_w is the density of water. Because the variation in space of the fluid mass density ρ_w is generally very small in a shallow groundwater system, it is reasonable to assume that the density of the fluid is independent of the aqueous constituents. In other words, the material transfer problem is one of forced rather than free, or natural, convection. For forced convection the second equation is merely Darcy's law

$$\mathbf{v} = -K(x, y, z) \nabla \phi \quad (3)$$

where $K(x, y, z)$ is the hydraulic conductivity and ϕ is the hydraulic head.

The groundwater model applied in this paper is one developed by *Tóth* [1963] and *Freeze and Witherspoon* [1966, 1967]. In its simplest form (Figure 2), it comprises a two-dimensional vertical section with a real impermeable boundary at the bottom, theoretical impermeable side boundaries, and a water table boundary along the top. Since the flow is assumed to be steady, the upper boundary, or water table, is fixed in space, and local fluctuations are either unimportant or have been averaged to give a mean water table position. The groundwater basin is assumed to be isotropic but nonhomogeneous with respect to hydraulic conductivity.

For the assumptions stated above, the continuity equation (2) reduces to

$$\nabla \cdot \mathbf{v} = 0 \quad (4)$$

or, by substitution of Darcy's law, to

$$\nabla \cdot [K(x, y, z) \nabla \phi] = 0 \quad (5)$$

which is Richards' equation. For isotropic media, (5) reduces to Laplace's equation.

The solution of either Laplace's or Richards' equation with appropriate boundary conditions describes the distribution of the hydraulic head throughout the region of interest. The velocity distribution is readily obtainable, since it is proportional to the gradient of the hydraulic head. Hence the material transfer problem is solved by first solving the flow problem independent of chemical considerations and incorporating the results into the mass balance equations. The mass balance equations are now of the form

$$\mathbf{v} \cdot \nabla c_1 = \sum_{j=1}^{n} R_{1j}$$

$$\mathbf{v} \cdot \nabla c_2 = \sum_{j=1}^{n} R_{2j}$$

$$\vdots \quad (6)$$

$$\mathbf{v} \cdot \nabla c_i = \sum_{j=1}^{n} R_{ij}$$

Although certain characteristics of the natural system have been lost through simplification, the complex reaction phenomena in a moving

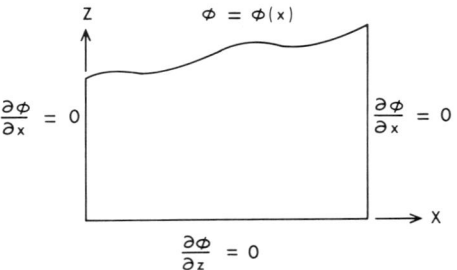

Fig. 2. Schematic representation of a groundwater basin.

medium have been retained and are of first rank importance in shallow flow systems of the kind described in this paper.

Mathematical methods. The Flopat 2D regional flow model developed by *Freeze* [1967] was modified for this investigation to calculate the velocity at a finite set of grid points over the region. An added program segment uses the calculated values of hydraulic head around each node point to determine the gradient of hydraulic head in the x and z coordinate directions. The vertical and horizontal velocity components at each node are then calculated by simple application of the Darcy equation. Details of the finite difference model are given by *Freeze* [1967], *Freeze and Witherspoon* [1966, 1967], and especially by *Freeze* [1969].

In our simulation model the mass balance equations are solved numerically by taking small steps along the flow paths. Because the effects of dispersion are considered to be negligible, each flow path is accounted for separately.

The region is discretized by a grid consisting of rectangular subareas. Groundwater flow across any one of the subareas is approximated by a linear extrapolation of the velocity vector across the rectangle. The mineralogic composition and the processes occurring within a particular subarea are assumed to be invariant. However, it is possible to construct a two-dimensional representation of a complex multi-layered groundwater system whose mineralogy and processes vary in space from one subarea to another.

The generation term R_i for any ionic constituent is in part a function of the ionic strength of the solution and the concentration of ions associated through chemical equilibria relationships. Thus it is necessary to take many small steps along each flow path because R_i is dependent on the chemical state of the system at every point in the system. For example, with the process of dissolution, ion activity products are compared to the equilibrium constants at the end of each small step to test whether the R_i term requires adjustment for the next step.

An explicit numerical method is a convenient way to solve the array of first-order differential equations. The concentration of each constituent at the end of the first step along the flow path is calculated by the simple Euler approximation

$$c_i^2 = c_i^1 + \frac{1}{v_s} \cdot \sum_{j=1}^{n} R_{ij}^1 \cdot \Delta s \quad (7)$$

where Δs is the step size in the direction of flow and c_i^2 is the concentration of the ith constituent at the beginning of the second step.

The large truncation error associated with the Euler method makes it only a crude approximation to the real solution [Gear, 1971]. The more accurate midpoint method is used for the remaining steps along the flow path:

$$c_i^{m+1} = c_i^{m-1} + \frac{1}{v_s} \sum_{j=1}^{n} R_{ij}^m \cdot 2\Delta s \quad (8)$$

Qualitatively, these approximations state that the concentration of a particular ionic constituent at the end of the step is equal to the concentration at the beginning of the step increased or diminished by the acting processes. The amount of variation in concentration across the small step is directly proportional to the rate of the geochemical processes embedded in the R term and inversely proportional to the flow velocity.

It is apparent from looking at the system of equations (8) that there is coupling within the rate term R. Consequently, a numerical value for the R terms requires the simultaneous evaluation of all the chemical and biological processes. In a series of three papers [Helgeson, 1968; Helgeson et al., 1969, 1970] an algorithm is presented that provides a practical way to calculate the rate term R_i for each of the processes. A new variable $\bar{R}$ is defined as the rate of change of concentration with respect to the reaction progress variable ξ [Helgeson, 1968],

$$\bar{R}_i = dc_i/d\xi_j \quad (9)$$

where the subscripts i and j designate the ith species and the jth process. The variable ξ is analogous to time and describes the quantity of mass transferred in one step of the process. The rate term R is related to $\bar{R}$ as follows:

$$R_i = \frac{dc_i}{dt} = \frac{dc_i}{d\xi_j} \cdot \frac{d\xi_j}{dt} = \bar{R}_i \cdot \frac{d\xi_j}{dt} \quad (10)$$

The net reaction rate term $\sum R_{ij}$ is calculated by summing R_{ij} over the n processes. Thus the functional relationship between the two variables ξ and t relates the amount of mass transferred by one reaction cycle of a given process to the residence time of the groundwater in the system.

The solution of the mass balance equations can be thought of most simply as a mass accounting procedure wherein any set of chemical and biological processes progresses by taking mass from reactants, transferring it to the products, and further redistributing it according to liquid-liquid equilibria relationships.

The last problem to be dealt with is that of reaction kinetics, a topic about which very little is known. Actual kinetic rate equations cannot be deduced theoretically but must be derived experimentally. However, the complexities evident in most groundwater systems, for example, a wide variation in mineral structure and abundance and concentrated flow along fractures, make the extrapolation of results from laboratory studies difficult. This problem is being resolved by determining the rates of physical and chemical processes through the detailed investigation of actual flow systems [Back and Hanshaw, 1971].

In this study the rates of reaction are specified empirically by choosing ξ_j and t functions. Thus the rate of mass transfer for each of the geochemical processes can be related to time.

It has often been noted that many of the dissolution reactions take place most rapidly within the soil zones. An exponential function of the following form probably best approximates the observed dissolution kinetics of the system:

$$\xi_{\mathrm{dis}} = B(1 - e^{-t/a}) \quad (11)$$

where B is a maximum number of reaction steps,

t is time in days, and a is characteristic time. For small values of the exponent t/a, ξ increases very rapidly, this increase indicating that many reaction steps take place over a short period of time. The number of reaction steps per unit of time decreases as the value of the exponent becomes >1 and ultimately reaches a maximum value B. The parameter a controls the shape of the exponential function. The relative reaction rates $\bar{R}$ are assumed in all cases to remain constant with respect to time.

All the remaining kinetic functions for the biological and chemical processes, for example, sulfate reduction and cation exchange, are also assumed to be of exponential form, either increasing or decreasing according to the nature of the physical system.

The geochemical and biological processes traditionally considered important in altering the chemistry of groundwater systems are obviously not specified on the basis of rigorous theoretical or experimental evidence but on intuitive grounds. In this way the solution of these mass balance equations differs from the solution of more explicitly posed boundary value problems in which the differential equations are specified exactly. A trial and error selection process ensues until there is satisfactory correspondence between the model output and measured characteristics of the real system. A sequence of computer operations involved in the simulation are shown in the flow chart of Figure 3. The input data include the following: the ionic concentration of recharge; the specified saturation levels for calcite, dolomite, gypsum, and halite; the initial partial pressure of CO_2 gas in the soil zone; kinetic and control parameters for mineral dissolution, sulfate reduction and cation exchange; matrix equations for dissolution and sulfate reduction; discretization parameters, the number of columns and rows and the distance between rows and columns; the x-z velocity at each node in the region; the mineralogic composition for each rectangular subarea; and an integer code for each subarea to indicate which of the processes are acting.

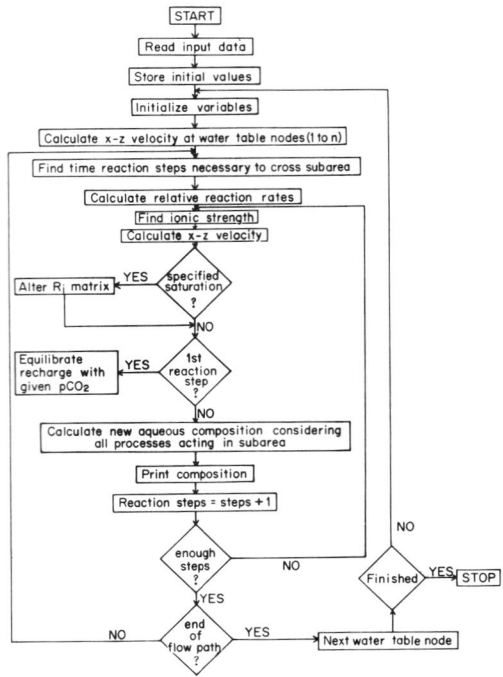

Fig. 3. Computer flow chart of the simulation model.

PROCESS ANALYSIS

Perhaps one of the most important uses of this model is as a tool to explore the controls that the processes and physical parameters exert on the overall composition of groundwaters. Because the aqueous chemistry at any point in a groundwater system represents the totality of processes and interactions, it is generally very difficult to assess the relative contribution of each process and interaction to that total. A particular process can be implicated but not verified as contributing to the observed concentration. The effects of the processes and interactions can easily be separated with the simulation model. The typical porous medium considered thoughout this section consists of the following minerals: calcite, dolomite, and plagioclase (labradorite composition) with small amounts of gypsum and halite. These simulations use pK values of 8.32, 16.71, and 10.49 for calcite, dolomite, and bicarbonate at 10°C, respectively. A simple one-dimensional flow model is assumed.

Mineral dissolution. Dissolution of a mineral phase proceeds in the absence of chemical equilibrium between the aqueous constituents and that mineral phase. When the dissolution of all minerals composing the porous medium

TABLE 1. Variation in Aqueous Composition with Unconstrained Dissolution along a Flow Path

Ion Concentration, mg/l						
Ca	Mg	Na	HCO_3	SO_4	Cl	pH
3.7	0.3	0.8	8.9	1.6	1.0	5.80
13.0	1.8	1.3	44.1	2.5	1.6	6.04
23.4	3.5	1.9	83.1	3.4	2.4	6.37
33.9	5.2	2.6	122.2	4.4	3.1	6.59
44.3	6.9	3.2	161.2	5.3	3.8	6.78
54.7	8.6	3.8	200.2	6.3	4.5	6.94
65.1	10.3	4.4	239.1	7.3	5.3	7.11

The reaction proceeds downward from the initial values, given in the first horizontal column.

proceeds without interruption as it does in the soil zones of recharge areas, the concentration of those ionic species not further redistributed by liquid-liquid equilibria must increase. Consider the following simulation run (Table 1). A dilute recharge solution enters the typical porous medium at an initial partial pressure of CO_2 gas of 5.0×10^{-2}.

All the cations and $SO_4^=$ and Cl^- ions are not further redistributed by liquid-liquid equilibria. Their rate of increase is in part directly proportional to the mineral abundance and dissolution rates. The concentration of ionic species such as HCO_3^- and H^+ is controlled in part by complex liquid-liquid equilibria relationships and does not necessarily increase, as in the case of H^+. Thus any decrease of any major ionic species indicates that one or more additional processes are occurring.

Solubility constraints. The degree of mineral saturation with respect to a mineral phase, defined as the ion activity product–solubility constant ratio, is specified as input to the model. Model output suggests that the degree of saturation with respect to calcite and dolomite is not a particularly sensitive input variable. In other words, relatively large variation in the specified saturation levels produces a small variation in the actual calculated concentration of ionic species. The simulation runs of Table 2 indicate the magnitude of the variation in Ca^{++} and HCO_3^- ion concentrations and pH for up to a twofold increase in the degree of saturation with respect to calcite.

It is clear that doubling the degree of calcite saturation results in relatively small increases in Ca^{++} and HCO_3^- ions and pH. Conversely, the calculation of the percent of saturation of calcite and dolomite from chemistry measurements is an extremely sensitive calculation, since only small variations in Ca^{++}, Mg^{++}, HCO_3^- ion concentrations and pH are required to produce large variations in the percent of saturation.

An input variable to the model that is more sensitive than calcite-dolomite saturation levels is the partial pressures of CO_2 gas encountered in the soil zones of recharge areas. The simulation runs of Table 3 compare the compositional differences for five solutions at the calcite saturation level of 1.0 for differing partial pressures of CO_2 gas. An increase in the partial pressure of CO_2 gas acts to lower the pH of the groundwater. This in effect allows more calcite to be dissolved before the point of equilibrium is reached. Any process that will lower the pH of recharge will tend to increase the concentration of Ca^{++}, Mg^{++}, and HCO_3^- ions in groundwaters. The effect of some varieties of air pollution, for example, excess sulfur dioxide and nitrogen oxide, is to lower the pH of rainfall to as low as 3.9 [*Likens et al.,* 1972].

TABLE 2. Comparison of Ion Concentrations to Varying Degrees of Calcite Saturation

Calcite Saturation	Ion Concentration, mg/l		
	Ca	HCO_3	pH
1.0	76.5	282.0	7.30
1.2	78.6	289.7	7.34
1.4	80.7	297.5	7.37
1.6	82.8	305.2	7.41
1.8	84.9	313.0	7.46
2.0	85.9	316.9	7.47

TABLE 3. Comparison of Ion Concentrations to Varying Partial Pressures of CO_2

$pCO_2 \times 10^{-2}$	Ion Concentration, mg/l		
	Ca	HCO_3	pH
3.0	55.7	203.7	7.57
5.0	76.5	282.0	7.30
7.0	91.1	336.7	7.13
9.0	103.6	383.6	7.02
11.0	115.0	426.5	6.94

TABLE 4. Comparison of Concentration Variations as a Result of Common Ion Effects

Saturation		Ion Concentration, mg/l			pH
Calcite	Dolomite	Ca	Mg	HCO$_3$	
1.0	0.23	76.5	12.2	282.0	7.30
1.0	1.0	49.5	37.7	309.1	7.43
1.6	0.53	82.8	13.2	305.2	7.41
1.6	1.0	72.2	22.6	313.3	7.46
2.0	0.82	85.9	13.7	316.9	7.46
2.0	1.0	83.9	15.4	318.9	7.48

The fact that groundwater is in equilibrium with respect to a mineral phase (for example, calcite) does not imply that the concentration of those aqueous species related to the equilibrium (for example, Ca^{++}, H^+, and HCO_3^-) will remain unchanged. The effect of adding these common ions from the dissolution of other minerals or other processes is to alter the equilibrium concentrations. The particular direction of the equilibrium shift depends on which of the common ions is most abundant. Simulation results summarized in Table 4 demonstrate the theoretical variation of Ca^{++}, Mg^{++}, HCO_3^-, and H^+ ions in solutions in equilibrium with calcite at three different saturation levels and then in equilibrium with respect to dolomite at a saturation level progressing to 1.0.

With a constant calcite saturation level of 1.0, continued dissolution of dolomite with a small amount of gypsum causes a very significant decrease in Ca^{++} ion concentration and increases in the pH, Mg^{++}, and HCO_3^- ion concentrations. If calcite saturation levels are >1.0, the common ion effects are in general reduced. The solution more quickly approaches the point of dolomite equilibrium when the solution becomes supersaturated with respect to calcite. Table 4 shows also how sensitive Mg^{++} ion concentrations in the three cases are to the degree of calcite saturation in a solution in equilibrium with respect to calcite and dolomite. A comparison of the three simulation runs at the point of dolomite equilibrium indicates that Mg^{++} ion concentration is inversely proportional to the calcite saturation level.

Sulfate reduction and cation exchange. The ability of sulfate reduction and cation exchange to modify almost completely the composition of groundwaters has been pointed out in many hydrochemical studies. It is well known that sulfate reduction decreases the concentration of $SO_4^=$ ions and increases the concentration of HCO_3^-. Increasing concentrations of HCO_3^- in groundwaters in equilibrium with calcite and dolomite will also force decreases in the concentration of Ca^{++}, Mg^{++}, and pH because of the common ion effect. The magnitude of this decrease is dependent on the total amount of $SO_4^=$ ion available for reduction and the amount of organic material.

Similarly, cation exchange, considered in this discussion to be Ca^{++} and Mg^{++} adsorbed and Na^+ released, not only alters the cation chemistry of the solution but should also increase the concentrations of HCO_3^- and increase the pH through the calcite and dolomite equilibrium relationships. Water with the initial composition designated 1 (Table 5) is allowed to

TABLE 5. Trends of Compositional Variation with Sulfate Reduction and Cation Exchange

Sample	Ion Concentration, mg/l						pH
	Ca	Mg	Na	HCO$_3$	SO$_4$	Cl	
1	82.8	13.2	5.4	305.2	9.0	6.4	7.41
2	75.5	22.6	15.0	310.6	24.0	17.8	7.44
3	73.3	22.1	15.0	328.2	3.1	17.8	7.43
4	32.6	9.6	99.8	313.9	24.0	17.8	7.78

proceed to composition 2 by mineral dissolution only. Superimposing the processes of sulfate reduction and Ca^{++}, Mg^{++} for Na^+ ion exchange on the dissolution yields water with the compositions described in 3 and 4, respectively.

More important than the actual magnitude of the compositional variations is their trend. The magnitude of the variation is controlled by kinetic factors in the case of ion exchange and by the amount of $SO_4^=$ ion available for reduction in the case of sulfate reduction.

Velocity field. The velocity and direction of groundwater flow are also very important factors controlling the pattern of chemical evolution. Because the chemical and biological processes are spatially distributed parameters, the direction of groundwater flow determines the order in which the processes occur.

The velocity of groundwater flow is important when it is compared with the duration of the chemical and biological processes. If the kinetics driving the system toward chemical equilibrium are much faster than the rate of flow through the system, variations in the flow velocity will have no influence on the hydrochemistry. On the other hand, residence time in the system becomes important when the kinetics of individual processes are slow in comparison to the flow velocities. In general, the wide variation in the apparent reaction rates suggests that natural groundwater systems represent both these types.

APPLICATION OF THE MODEL TO THE UPPER KETTLE CREEK WATERSHED, ONTARIO, CANADA

A properly designed and calibrated simulation model provides a most effective framework in which to formulate a hypothesis concerning the processes acting in any groundwater basin. A practical example of the application of the two-dimensional convective transfer model is provided by the Upper Kettle Creek watershed, a small glaciated area typical of several watersheds located in the hummocky moranic area on the north shore of Lake Erie (Figure 4).

Hydrogeology and details of groundwater flow. Three hydrostratigraphic units, an upper

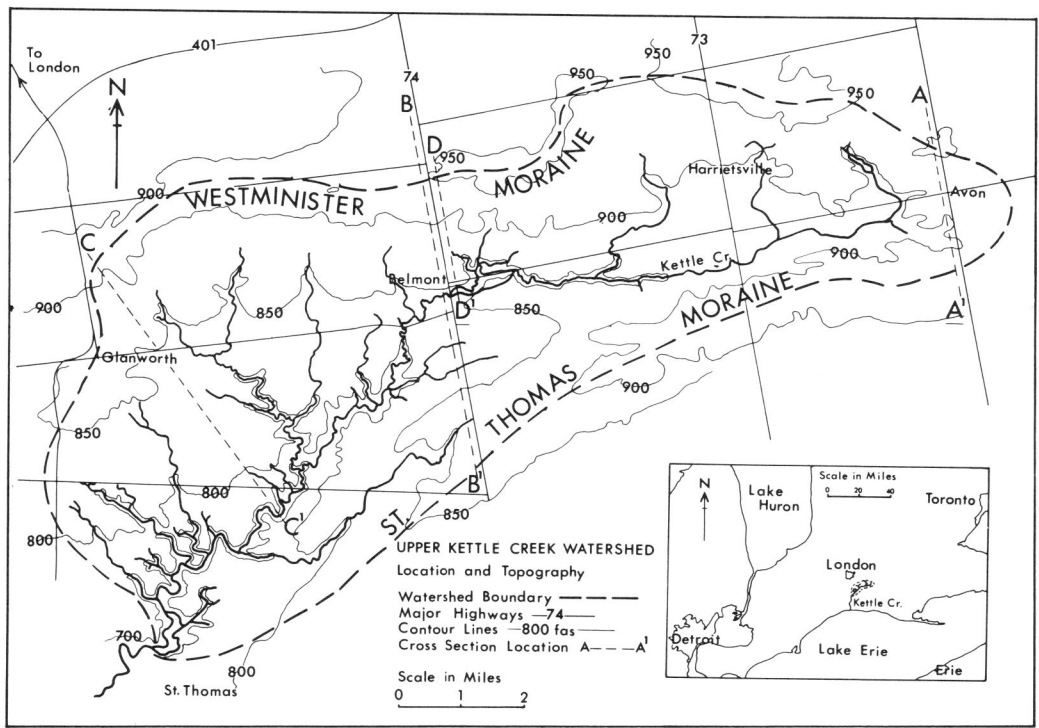

Fig. 4. Location and regional topography of the Kettle Creek watershed, Ontario, Canada.

till unit, a lower drift unit, and a bedrock unit, can be identified within the watershed. The upper till unit, composed predominately of the clay-rich Port Stanley till, Late Wisconsin in age with minor accumulation of recent alluvium, completely blankets the watershed to depths ranging from 100 to 200 feet. The underlying lower drift unit, which ranges in thickness from 50 to 90 feet, is composed of the sandy Catfish Creek till, also Late Wisconsin in age, and a thin layer of glaciolacustrine sediments stratigraphically separating the Port Stanley and Catfish Creek tills. This discontinuous layer of glaciolacustrine sediments with a maximum thickness of 15 feet is composed of well-bedded silts, sands, and clays and locally is an important aquifer for domestic and industrial water supplies. The third hydrostratigraphic unit is the Dundee limestone of Middle Devonian age.

A generalized description of flow within the field area was obtained theoretically by solving the two-dimensional form of Richards' equation with appropriate boundary conditions. It was not possible to verify the simulated flow patterns because of the prohibitive cost involved in installing the necessary instrumentation. Published piezometric data for wells within the watersheds were not of sufficient accuracy to be useful.

Figure 5 illustrates the three hydrostratigraphic units and the theoretical flow patterns along the section D-D¹. Immediately apparent in Figure 5 are the two zones of groundwater recharge and discharge as well as extensive areas of lateral flow.

Hydrochemistry. During the study approximately 300 samples of precipitation, groundwater, and surface water were collected for chemical analysis. Measurements of pH were made at the sampling sites. Samples were analyzed for HCO_3^- ion concentration within 1-2 days after sampling. The remaining ion concentrations were determined by the Analytical Services Section of the Inland Waters Branch, Ottawa.

The chemical composition of precipitation (Table 6) was measured to characterize the mass inflow to the groundwater system. The average pH of rainfall in the Kettle Creek basin is somewhat lower than 5.7, the pH of pure water in equilibrium with atmospheric concentrations of CO_2 gas [*Garrels and Christ*, 1965].

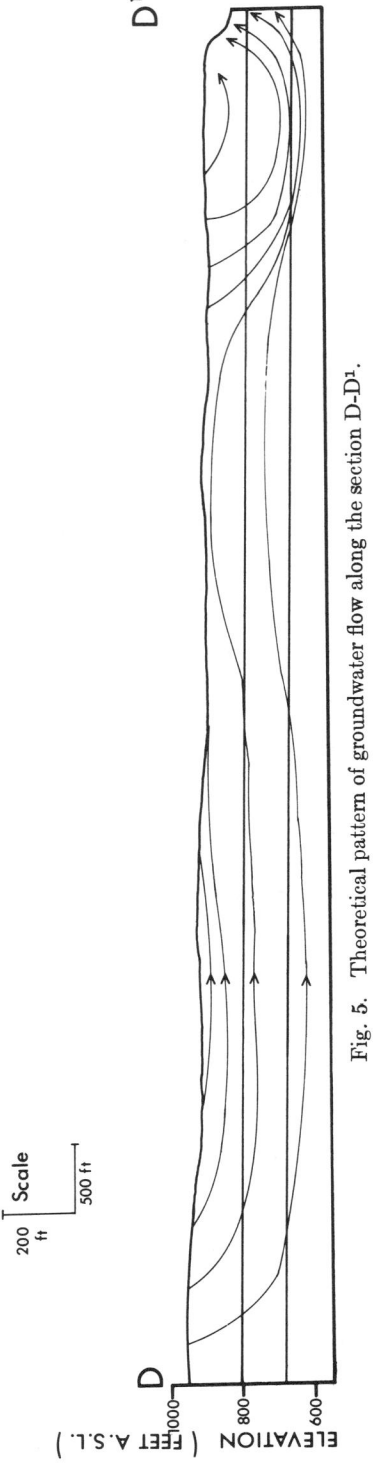

Fig. 5. Theoretical pattern of groundwater flow along the section D-D¹.

TABLE 6. Comparison of Average Rainfall and Snowmelt Chemistry

	No. of Samples	Ion Concentration, mg/l							
		Ca	Mg	Na	K	HCO$_3$	SO$_4$	Cl	pH
Snowmelt	12	4.1	0.4	1.0	0.2	9.0	2.2	1.5	6.46
Rainfall	7	3.5	0.5	0.4	0.1	3.5	6.7	2.3	5.03

Coupled with the relatively high SO$_4^=$ ion concentrations, this low value of pH probably indicates abnormally high SO$_2$ gas concentrations resulting from air pollution. The seasonal variation of pH and SO$_4^=$ concentration could be related to differences in summer-winter air circulation patterns or differences in the relative efficiency of rain and snow in removing SO$_2$ gas from the atmosphere.

A representative description of the two-dimensional variation of ion concentrations in the groundwater flow is presented in Figure 6. This representative figure was obtained by first dividing the x-z region along the section D-D^1 into three lateral and six vertical zones. All the chemical data obtained from wells along or closely adjacent to the section are inserted in an appropriate zone and then averaged. Representation of the total flow system is assured by selecting the three lateral zones in areas of recharge, lateral flow, and lateral flow-discharge.

A trend of decreasing anion concentrations with depth is apparent. The same general decrease with depth is evident with respect to Ca^{++} and Mg^{++} ion concentrations. Only Na$^+$ ion concentrations increase with depth.

In general, all groundwater samples were either saturated or slightly supersaturated with

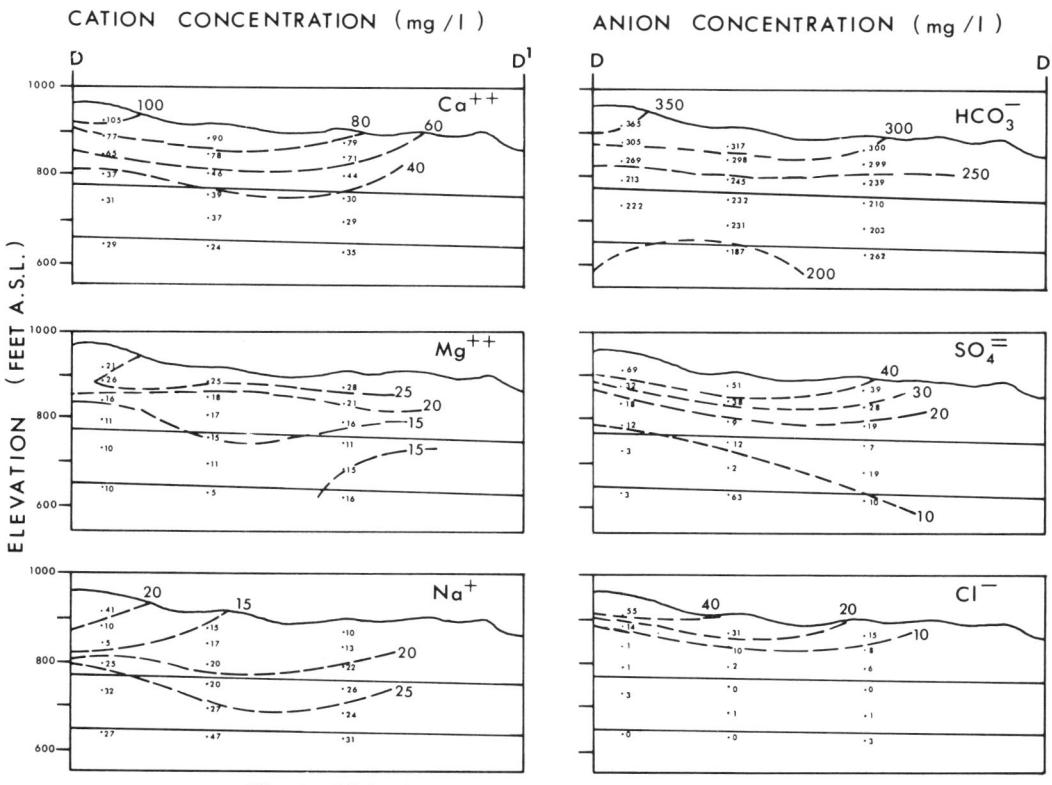

Fig. 6. Major ion concentration along the section D-D^1.

TABLE 7. Downstream Variation in Major Ion Concentrations, June 12, 1971

Distance from Stream Head, miles	Ion Concentration, mg/l							
	Ca	Mg	Na	K	HCO$_3$	SO$_4$	Cl	pH
1.0	88.1	18.7	5.7	0.7	305.	40.8	10.5	8.08
3.0	78.3	19.8	8.5	1.4	256.	50.9	18.1	7.90
5.5	76.1	18.7	9.9	2.4	248.	48.4	20.0	7.80
8.0	76.1	18.2	8.6	2.6	276.	31.9	16.9	7.98
10.0	53.5	16.4	11.3	2.2	200.	31.4	17.1	8.39
12.5	52.0	16.8	16.0	3.2	195.	31.2	30.2	8.08

respect to calcite and undersaturated with respect to gypsum at 10°C, the average groundwater temperature. The saturation levels with respect to dolomite are quite variable, ranging from supersaturation to undersaturation, the latter state being most prevalent.

During periods with little or no precipitation, streamflow is maintained by the flow of groundwater into the stream channels. Results of chemical analyses indicate a marked variation in ion concentrations in a downstream direction. The low-flow composition of the surface water (Table 7) very closely resembles shallow groundwater in the upper till unit. Similarly, the decreasing composition of Ca^{++}, Mg^{++}, HCO$_3^-$ and the increasing composition of Na$^+$ ion in the downstream direction is analogous to the chemical evolution of groundwater along the flow system. Deepening the stream valleys in the downstream direction enables progressively deeper groundwaters to enter the Kettle Creek.

Discussion of processes in the Kettle Creek watershed. Achieving a satisfactory correspondence of the real and simulated systems can help define the nature and extent of processes operating within the groundwater system. Comparison of the theoretical hydrochemical patterns obtained from the model (Figure 7) with the representative field patterns (Figure 6) indicates that the trends and magnitude of the spatial variation in ion concentrations are quite similar. Point to point comparisons between Figures 6 and 7 indicate that the best fit is achieved with Ca^{++}, Mg^{++}, and HCO$_3^-$ ion concentrations. However, there is no case for which the predicted concentrations lie outside the range of observed data.

Table 8 summarizes the input parameters for the mass balance equation as determined from the field investigations. The idealization of the dissolution process with calcite, dolomite, gypsum, halite, and plagioclase (labradorite composition) is valid as a first approximation in spite of the mineralogic diversity of the natural system. The relative dissolution rates for minerals in each of the hydrostratigraphic units were determined as the product of the relative mineral abundance and the relative reaction rates. Relative reaction rates, 1.0, 0.30, 1.0, 2.0, and 0.001 for calcite, dolomite, gypsum, halite and plagioclase, respectively, were found to be most suitable. The kinetic parameters for dissolution as defined by (13) were $B = 20,000$, $a = 12,000$ days, and $d\xi = 1 \times 10^{-4}$. The compositional decreases along the flow paths of the actual system indicate that one or more additional processes are taking place. These processes become important soon after the groundwater reaches equilibrium with respect to calcite and dolomite.

Simulation results indicate that only the calcite and dolomite saturation constraints are required on the dissolution process. Calcite saturation levels of 1.6 were necessary to achieve satisfactory concentration levels of Mg^{++} ion. As was discussed previously, lower calcite saturation levels result in relatively high concentrations of Mg^{++} ion. The significant decrease in Ca^{++} ion concentrations along the flow paths of the actual system cannot be attributed to the common ion effect. The decreasing concentrations of HCO$_3^-$ and Mg^{++} ions suggest that some other process is involved.

A process of cation and anion exchange simulated the decrease of the major anions and Ca^{++} and Mg^{++} ions with depth, as was observed in the real system. Because these exchange processes vary considerably, it is not within the scope of the model to present an explicit, qualitative description of the phenomena. How-

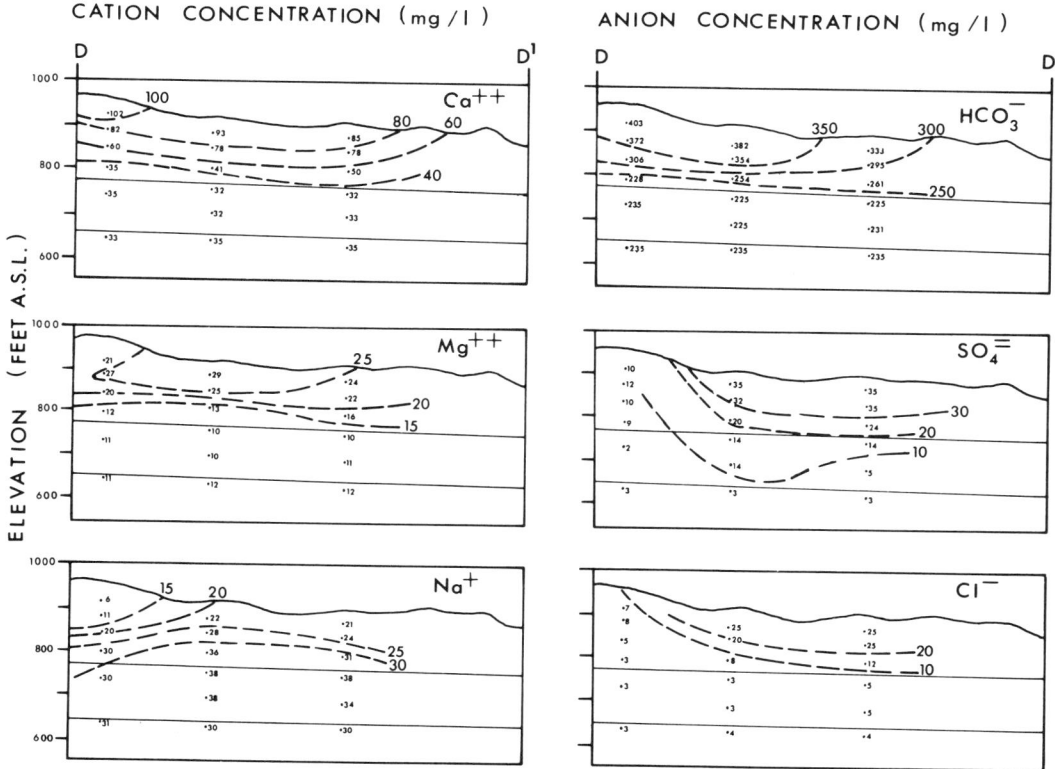

Fig. 7. Simulated ion concentration along the section D-D¹.

ever, satisfactory results were achieved by considering that Ca^{++} and Mg^{++} are exchanged with Na^+ and H^+, whereas HCO_3^-, Cl^-, and $SO_4^=$ ions are exchanged for OH^-. At the same time, conditions of solution electroneutrality and specified levels of calcite saturation are maintained.

The best simulation results were achieved with an exponentially increasing kinetic function to control the exchange process. Because the amount of secondary joint and fracture porosity of the upper till unit diminishes with depth, the exchange processes become increasingly important with depth. The compact clay-rich character of the upper till unit suggests that the exchange process should be restricted to this unit.

The model considered sulfate reduction to take place within the lower drift and bedrock units. This process accounts for the quantities of H_2S gas in water samples from many of the wells >250 feet deep. The same exponential function of (13) and kinetic parameters de-

TABLE 8. Summary of Input Parameters and Processes

Unit	Porous Media Composition, %					Hydraulic Conductivity, cm/sec
	Calcite	Dolomite	Gypsum	Halite	Plagioclase	
Upper till	23	12	1	1	10	0.0007
Lower drift	18	25	1	1	10	0.0021
Bedrock (estimated)	50	20	0	0	5	0.0056

The input parameters are the velocity field (Figure 5), snowmelt chemistry (Table 6), and porous media composition.

scribed earlier for mineral dissolution also defined the rate of sulfate reduction within the model. Thus the rate of sulfate reduction is assumed to decrease exponentially in the direction of flow.

Partial pressures of CO_2 gas varying linearly from 8×10^{-2} along the basin boundaries to 4×10^{-2} near the Kettle Creek resulted in the most satisfactory concentration levels of Ca^{++} and HCO_3^- ions in the shallow groundwater.

Conclusions

In a complex natural system it is often not possible to differentiate between the processes because of the multiplicity of the chemical and biological interactions. However, quantitative analysis of the processes within a model framework indicates the manner in which they interact as well as the relative importance of several of the parameters. Process analysis by the simulation model reveals the following facts:

1. Mineral dissolution is the most important mass generation process within the system. Any decrease in the concentration of major cations or anions indicates that the dissolution process is being modified by one or more other processes.

2. Saturation constraints are important in the chemical evolution of groundwaters. However, relatively large changes in the specified saturation levels for calcite and dolomite in the model result in relatively small variation in the simulated concentrations of Ca^{++}, H^+, and HCO_3^- ions at the calcite equilibrium point and Ca^{++}, Mg^{++}, H^+ and HCO_3^- ions at the dolomite equilibrium point.

3. The partial pressure of CO_2 gas in the soil zones of recharge areas is most effective in controlling the groundwater composition. Any process that lowers the initial pH of the groundwater will ultimately produce higher concentrations of Ca^{++}, Mg^{++}, and HCO_3^- ions in solution.

4. Equilibrium with respect to a mineral phase does not imply that the concentration of ionic species related to the equilibrium remains unchanged. The effect of adding common ions from other dissolution or processes such as cation exchange or sulfate reduction is to shift the equilibrium concentration.

5. The direction of groundwater flow determines the order in which the spatially distributed processes occur. The change in chemical composition along a flow path depends in part on the duration of the processes acting in relation to the residence time of the flow.

The results of a simulation from the Upper Kettle Creek watershed emphasize the practical usefulness of the model in providing a framework to interpret the nature of chemical variability in the groundwater. The discrepancies between the actual and the predicted chemical patterns arise in the overall simplifications necessary to model this complex system. Numerical values of physical parameters such as hydraulic conductivity or relative mineral abundance cannot be specified at each point within the region. They are most often extrapolated from a limited number of measurements made on the system. Still other parameters, such as the kinetic functions and the partial pressures of CO_2 gas, are unknown and of necessity must be estimated. In addition, the processes considered within the framework of the model are idealized somewhat and lose some of the flexibility and sensitivity of the natural processes. In general, however, a satisfactory correspondence of the real and idealized system can be achieved by a trial and error procedure. It follows that this simulation like all other forms of simulation is inherently nonunique.

Computer time required for a simulation run is most dependent on the number of flow lines and solid or dissolved reaction products. A moderately complex problem involving 15 flow lines and 18 solid or dissolved products requires approximately 8 min of execution time on the University of Illinois IBM 360 model 75 computer.

Acknowledgment. This study was supported by the Groundwater Subdivision, Inland Waters Branch, Environment Canada.

Notation

a, characteristic time, T;
B, maximum number of reaction steps;
c_i, molal concentration of species i, M/L^3;
D_i, dispersion coefficient of species i, L^2/T;
K, hydraulic conductivity, L/T;
m, superscript for arbitrary time position, T;
n, arbitrary number of geochemical processes;
R_{ij}, molal rate of generation of species i and process j, M/TL^3;
r, porosity of the porous medium;
s, distance along the streamlines, L;
t, time, T;

v, velocity of flow, L/T;
x, y, z, coordinate directions, L;
Δs, step size, L;
ρ_w, density of water, M/L^3;
ϕ, total hydraulic head, L;
ξ, progress variable.

References

Back, W., Calcium carbonate saturation in groundwater from routine analysis, *U.S. Geol. Surv. Water Supply Pap. 1535-D*, 1961.

Back, W., Hydrochemical facies and groundwater flow patterns in northern part of Atlantic Coastal Plain, *U.S. Geol. Surv. Prof. Pap. 498-A*, 1966.

Back, W., and B. Hanshaw, Rates of chemical and physical processes in a carbonate aquifer, in *Nonequilibrium Systems in Natural Water Chemistry*, edited by J. D. Hem, pp. 77–93, American Chemical Society, Washington, D. C., 1971.

Back, W., R. Cherry, and B. Hanshaw, Chemical equilibrium between the waters and minerals of a carbonate aquifer, *Nat. Speleol. Soc. Bull., 28*, 1966.

Freeze, R. A., Program FLOPAT 2D, Inland Waters Br., Environ. Can., Ottawa, Ont., Jan. 1967.

Freeze, R. A., Theoretical analysis of regional groundwater flow, *Sci. Ser. 3*, Inland Waters Br., Environ. Can., Ottawa, Ont., 1969.

Freeze, R. A., and P. Witherspoon, Theoretical analysis of regional groundwater flow, 1, Analytical and numerical solutions to the mathematical model, *Water Resour. Res., 2*, 641–656, 1966.

Freeze, R. A., and P. Witherspoon, Theoretical analysis of regional groundwater flow, 2, Effect of water table configuration and subsurface permeability variation, *Water Resour. Res., 3*, 623–634, 1967.

Garrels, R. M., and C. L. Christ, *Solutions, Minerals, and Equilibria*, Harper and Row, New York, 1965.

Garrels, R. M., and F. T. MacKenzie, Origin of the chemical composition of some springs and lakes, in *Equilibrium Concepts in Natural Water Systems*, edited by R. F. Gould, American Chemical Society, pp. 222–242, Washington, D. C., 1967.

Gear, C. W., *Numerical Initial Value Problems in Ordinary Differential Equations*, pp. 1–253, Prentice-Hall, Englewood Cliffs, N. J., 1971.

Helgeson, H. C., Evaluation of irreversible reactions in geochemical processes involving minerals and aqueous solutions, 1, Thermodynamic relations, *Geochim. Cosmochim. Acta, 32*, 853–877, 1968.

Helgeson, H. C., R. M. Garrels, and F. MacKenzie, Evaluation of irreversible reactions in geochemical processes involving minerals and aqueous solutions, 2, Applications, *Geochim. Cosmochim. Acta, 33*, 455–481, 1969.

Helgeson, H. C., T. H. Brown, A. Nigrini, and T. Jones, Calculation of mass transfer in geochemical processes involving aqueous solutions, *Geochim. Cosmochim. Acta, 34*, 569–592, 1970.

Himmelblau, D. M., and K. Bischoff, *Process Analysis and Simulation*, John Wiley, New York, 1970.

Kramer, J. A., Equilibrium models and the composition of the Great Lakes, in *Equilibrium Concepts in Natural Water Systems*, edited by R. F. Gould, pp. 243–254, American Chemical Society, Washington, D. C., 1967.

Likens, G. E., F. Bormann, and N. Johnson, Acid rain, *Environment, 14*(2), 33–40, 1972.

Piper, A. M., A graphical procedure in the geochemical interpretation of water analyses, *Eos Trans. AGU, 25*, 914–923, 1944.

Sillén, L. G., The physical chemistry of sea water, in *Oceanography*, edited by M. Sears, pp. 549–581, AAAS, Washington, D. C., 1961.

Tóth, J., A theoretical analysis of groundwater flow in small drainage basins, *J. Geophys. Res., 68*, 4795–4811, 1963.

(Received September 26, 1972;
revised January 12, 1973.)

29

Copyright ©1979 by The Chemical Rubber Co., CRC Press, Inc.

Reprinted by permission from pages 97-98, 139-146, and 148-156 of *CRC Critical Reviews in Environmental Control* **9**:97–156 (1979)

USING MODELS TO SIMULATE THE MOVEMENT OF CONTAMINANTS THROUGH GROUNDWATER FLOW SYSTEMS

Mary P. Anderson
Department of Geology and Geophysics
University of Wisconsin
Madison, Wisconsin

I. INTRODUCTION AND SCOPE

Concern over the potential for migration of wastes in the subsurface has generated a great deal of interest in the mechanisms responsible for contaminant transport through groundwater systems. Increased attention by both researchers and regulatory agencies to subsurface contamination can be attributed to the growing popularity of waste disposal by means of landfills, sludge lagoons, and deep injection wells[1,2] and to the concern over subsurface disposal of low to high level radioactive wastes.[3,4] To prevent the deterioration of groundwater quality, it has become necessary to develop a methodology for monitoring, analyzing, and predicting the movement of contaminants through the subsurface. The hope of formulating such a methodology has motivated the development of predictive tools in the form of mathematical models designed to simulate the transport of contaminants through groundwater systems. This review analyzes recent advances in this type of model development. A few of the models to be discussed treat transport through both the unsaturated and the saturated zones, but transport through the unsaturated zone is not considered in detail here. Moreover, the models to be discussed were developed to simulate migration through porous media and equivalent porous media, rather than through systems of rock fractures. This review covers material published through September, 1978.

The components of the transport process are advection, dispersion (including diffusion), and chemical reactions. In this paper, the term advection is used to refer to what most groundwater hydrologists call convection. Properly speaking, convection refers to movement as a consequence of a temperature differential within the medium. In the present context, advection refers to movement as a result of differences in head. Dispersion refers to mixing and spreading caused in part by molecular diffusion and microscopic variations in velocities within individual pores. For many field problems, these effects are negligible in comparison with dispersion caused by large scale heterogeneities within the aquifer. In the presence of large scale heterogeneities, dispersion occurs as contaminants move selectively around the less permeable units. However, when advection is weak, mechanical dispersion is negligible relative to molecular diffusion. Molecular diffusion can be important in low velocity systems, especially where high level radioactive waste is the contaminant.

Slichter,[5] in the course of experiments designed to measure groundwater velocity using salt as a tracer, was among the first to note that dispersion affects the transport of contaminants through a porous medium. Quantitative descriptions of dispersion were attempted in the 1950s by Taylor[6] and Aris,[7] among others. Their approach was based on a description of dispersion in capillaries and was reviewed by Bear[8] and by Fried and Combarnous,[9] who concluded that it has limited utility for the study of dispersion in porous media. Statistical models of randomly distributed capillaries were constructed by several researchers including Scheidegger,[10] de Josselin de Jong,[11] and Saffman.[12,13] This work laid the foundation for the development of the mathematical framework used in the derivation of the macroscopic dispersion equation presented by Bear,[8] among others.[14-20] Current efforts aimed at incorporating the dispersion process into the study of contaminant transport in porous media are centered around attempts to generate solutions to mathematical models based on the macroscopic dispersion equation. Many analytical solutions to simplified forms of this equation have been derived, but only a few of these may be useful for application to field problems. In most cases of practical interest, a numerical solution is required.

In certain problems, the dispersive component of the transport process is considered to be negligible in comparison with advection. In this case, either a mixing cell model may be formulated to compute concentrations, or it may be sufficient to determine the configuration of flow lines and estimate travel times from the velocity distribution. Thus, there are two approaches to the contaminant transport problem as applied to groundwater systems.[21] Either an advection-dispersion model is constructed in which the effects of dispersion are included in addition to advection, or an advection model is developed. Both types of models may incorporate chemical reactions. The models to be reviewed fall into these two major categories — advection-dispersion models and advection models. The section of the review on advection-dispersion models is further subdivided into sections dealing with one and two-dimensional areal and profile models and three-dimensional models. The section covering advection models is subdivided into distributed parameter models and lumped parameter models. For the most part, applications of contaminant transport models have been restricted to consideration of conservative (nonreacting) ions, but a few models incorporate simple chemical reactions.

The major problems associated with modeling contaminant movement in groundwater systems where the effects of dispersion are considered to be significant are measuring dispersivity in the field and overcoming numerical difficulties which can arise when solving the dispersion equation. Incorporating chemical reactions into both advection-dispersion models and advection models is another difficult problem.

[Editors' Note: Material has been omitted at this point.]

IV. INCORPORATING CHEMICAL REACTIONS INTO CONTAMINANT TRANSPORT MODELS

The reaction term in Equation 2 represents the net rate of transfer of a solute species from the liquid phase to other phases or to other ionic or molecular forms in solution. It includes the effects of all the chemical processes and reactions operating in the system, including precipitation and solution, co-precipitation, oxidation and reduction, adsorption and desorption, ion exchange, complexation, nuclear decay, ion filtration, and gas generation.[91] To achieve a complete chemical description of a system, it is necessary to solve a set of mass balance equations, such as a set of the form given by Equation 2, where one equation is needed for each of the chemical constituents. Rubin and James[105] discussed the formulation of the problem in detail and developed a solution technique for solving such a system of equations generated, using a Galerkin finite element approach. Primarily they dealt with equilibrium-controlled ion exchange reactions, but also considered a problem involving the precipitation-dissolution of gypsum and the formation or dissociation of the neutral calcium sulfate complex. These two reactions may occur along with sodium-calcium exchange in certain types of leaching problems.

To date, only relatively simple chemical processes such as adsorption and radioactive decay have been incorporated into most contaminant transport models (Section II.B.3, II.B.4, and III.B.1). This can be attributed to the difficulties encountered when solving transport models for conservative constituents and to the problems of incorporating geochemical data into contaminant transport models. Although the types of possible reactions may be known in a general way and numerous case histories of contamination problems have been documented,[206,207] the reactions are difficult to quantify for purposes of contaminant transport modeling. In reference to contamination of groundwater by nitrate, Kaufman[208] noted that

Although the literature is replete with case studies dealing with the occurrence of nitrate pollution, and the general theory of the nitrogen cycle in nature has been understood for many decades, the coupling of theory and field experience in a manner to permit the engineering of management systems is sadly lacking.

Similar statements could be made for other contaminants contributing to groundwater pollution except that for many contaminants even the theoretical framework has not yet been established. For example, relatively little is known concerning the behavior of bacteria and organic compounds in groundwater.[91,209] Jackson et al.[89] observed that although conceptual models seem to explain some features of the hydrogeochemistry of aquifers, "our knowledge of the mechanisms of the processes themselves is primitive."

To date, emphasis has been on incorporating ion exchange into contaminant transport models through the use of the distribution coefficient (Sections II.B.3 and III.B.1). Distribution coefficients are commonly determined in the laboratory by allowing a solution containing the contaminant to equilibrate with a sample of porous media. However, Back and Cherry[91] observed that

... values so obtained may or may not be useful in the prediction of contaminant movement under field conditions. There is a need for development of sampling and testing methodologies that will enable distribution functions to be determined on borehole samples under conditions where the physical and chemical integrity of the sample is maintained. Even if this is accomplished, however, the distribution function approach (sometimes known as the isotherm approach) will still be limited in its usefulness because it takes into account only the relatively fast mass transfer processes. *(page 156 of Reference 91)*

Reaction rates must be viewed relative to the groundwater velocity. If reaction rates are much faster than the flow rate, reactions are equilibrium-controlled, but if the residence time in the system is comparable to or larger than the reaction rate, kinetic reactions must be considered. The wide range in reaction rates reported for groundwater systems suggests that both equilibrium-controlled and rate-controlled reactions are possible.

Groundwater quality changes can be considered in terms of natural or stressed systems, that is, systems influenced by artificial contaminants. The most sophisticated water quality models have been applied to natural groundwater systems dominated by reactions involving calcium, magnesium, and sulfate ions, that is, for calcite-dolomite-gypsum aquifers.[210-213] The chemical evolution of groundwater along a flow path in these natural systems has been simulated using mass balance and mass action relationships.

Schwartz and Domenico[212] developed a nonequilibrium model for a regional groundwater system using mass balance equations of the form

$$q \cdot \nabla c_i = \sum_{j=1}^{n} R_{ij} \qquad i = 1,m \qquad (26)$$

where q is the Darcy velocity, c_i is the molal concentration of species i, R_i is the reaction term for species i, m is the number of species, and n is the number of chemical processes in which species i participates. The reaction rates were specified empirically, using exponential functions. Empirical constants and the relative contribution of each chemical reaction to the total water chemistry were determined by a trial and error selection process whereby simulated results were matched to observed concentrations. The model was applied to a groundwater basin in Ontario where the hydrostratigraphy consisted of till overlying limestone. The concentrations of calcium, magnesium, sodium, bicarbonate, sulfate, and chloride were calculated along several flow paths through a representative cross section of the watershed.

Plummer[213] used mass balance equations of the form,

$$\sum_{j=1}^{\phi} \alpha_j \beta_{cj} = \Delta m_c \qquad c = 1,n \qquad (27)$$

where ϕ is the number of phases (reactants and products), α_j is the number of moles of the jth phase in the reaction obtained by simultaneous solution of mass balance equations for n constituents, β_{cj} is the stoichiometric coefficient of the cth constituent in the jth phase, and Δm_c is the change in moles (per km of H_2O) of the cth constituent in the aqueous phase along the path. A mass transfer scheme[214,215] was used to refine the estimates of α_j. Isotopic balances were also used to remove ambiguity where there existed multiple choices of controlling reactions. The model was applied to simulate the evolution of groundwater quality along a flow path in the Floridan Aquifer. Several reactions that simulated the observed natural water chemistry were identified through the use of the model, and the derived mass balance coefficients (α_j) were demonstrated to be in part functions of the apparent rates of reaction along the flow path.

These types of geochemical models have not been applied to stressed systems in general because of uncertainties in the types of reactions that occur in such systems. One of the few examples of the application of a geochemical model to a stressed system was given by Kunkle and Shade.[216] They used a water quality model[217] which computes the activities of dissolved species in the calcite-dolomite-gypsum water system, the saturation indices of these minerals, and the dissolved carbon dioxide partial pressures. On the basis of the results of the model, they concluded that groundwater in a carbonate aquifer beneath a landfill in Ohio was at near saturation with respect to calcite and dolomite. Therefore, the alkalinity would be expected to vary only within narrow limits. In reality, alkalinity (expressed as mg/ℓ $CaCO_3$) increased from background concentrations of less than 100 mg/ℓ to over 260 mg/ℓ beneath the landfill. These changes suggest the occurrence of other chemical reactions superimposed on the carbonate equilibrium system. Kunkle and Shade[216] concluded that sulfate reduction was the primary controlling reaction inasmuch as sulfate concentrations were lower in groundwater beneath the landfill (200 mg/ℓ) than in the background water (1000 to 1200 mg/ℓ), and sulfate reduction would also account for the high alkalinities.

Yet sulfate reduction was estimated to occur at least 5 years before the leachate reached the aquifer. Therefore, it was postulated that the reducing environment of the landfill was transferred to the carbonate aquifer by the upward migration of oxygen from the aquifer to the landfill in response to a gas pressure gradient. The consequent reducing conditions in the aquifer activated the onset of sulfate reduction. However, the occurrence of relatively low concentrations of total dissolved solids, hardness, and chloride beneath the landfill could not be explained by sulfate reduction. Furthermore, hydrogeologic evidence suggested that the decrease could not be attributed to local dilution. The occurrence of co-precipitation between calcium and other dissolved species was suggested as a possible explanation, but no data were available to support this hypothesis.

It is evident that the kinds of reactions occurring in stressed systems require detailed laboratory and field investigations such as those by Griffin et al.[218,219] Back and Cherry[91] suggested that it will be necessary to shift attention from regional studies to more detailed field studies of small areas characterized by shallow water tables and simple hydrogeology so that it may be possible "to obtain adequate hydrologic, mineralogic, bacteriologic, and inorganic and organic chemical data needed to test computer and experimental models." Legrand[14] called for "a synthesis and a compilation of the attenuation habits of major contaminants in a variety of hydrogeologic settings."

Back and Cherry[91] identified the obstacles to development of geochemical models of groundwater systems to be

1. The lack of adequate and valid geologic data
2. The dearth of nonequivocal laboratory data on chemical mechanisms and kinetics of solution-precipitation reactions, and the role of trace elements and adsorption phenomena in these reactions
3. The lack of information on the occurrence of bacteria, gases, and organic compounds in groundwater and their behavior in the saturated zone
4. The lack of a theoretical framework that can relate results of laboratory experiments to field conditions and the difficulty in transferring field experience and data to theoretical models

Improvement in field measurements and development of new field techniques along with improvement in the theoretical framework is necessary before geochemical models can be routinely applied to contaminant transport problems. For example, the use of environmental isotopes in field studies may be helpful in identifying possible chemical reactions.[91,213,220] Measurement of oxidation-reduction potentials is important to the understanding of geochemistry of heavy metals and dissolved organic compounds in aquifers, but new measurement techniques are needed to resolve measurement discrepancies and to measure redox potentials in systems with slow reactions. Results from several studies suggest that redox reactions in groundwater systems obey the Nernst equation,[91] but other attempts to use the Nernst equation to correlate measured Eh values in groundwater with the concentration of the dissolved species have failed.[221] Theoretically, it will be necessary to evaluate the significance of the Nernst equation to the reaction term of the mass transport equation and to explore ways of dealing with situations where the Nernst equation is not applicable.

V. SUMMARY AND CONCLUSIONS

The application of advection-dispersion models is still in the developmental stage. Several models, particularly the U.S. Geological Survey's finite difference-MOC model,[26] have been applied to field situations and have been used to predict the effects of alternative management strategies, but it is difficult to put a great deal of confidence in the predictions in view of (1) the uncertainties involved in determining the input data — especially velocity distributions and dispersivities — and (2) the inherent numerical difficulties involved in solving the models. These difficulties have resulted in imperfect model calibration in several instances. However, models cannot be expected to reproduce perfectly the complexities of real world situations. After all, models are abstractions of reality, and in spite of all the uncertainties, a model can be useful as a tool to aid in the planning process.

Accepting the inherent limitations of models in general, it is still pertinent to ask whether advection-dispersion models as currently used are worthwhile predictors. Groundwater modeling in the past has relied on the use of deterministic models in which each parameter has a well-defined physical meaning. Recently, investigators have recognized the fact that hydraulic conductivity, although physically well defined, may take on the character of a fitted parameter in models where the hydraulic conductivity distribution is adjusted by trial and error during model calibration. Likewise in advection-dispersion models of contaminant transport, dispersivity is usually little more than a fitted parameter. However, unlike hydraulic conductivity, dispersivity assumes a heuristic character because the physical meaning of dispersivity is not clearly understood. For example, in the absence of a rationale for adjusting the spatial variations in dispersivity, it is routinely assumed to be a constant. However, in reality it is likely that spatial variations occur. Moreover, the assumption that the medium is isotropic with respect to dispersivity is commonly made when in fact anisotropy may be important. In order to specify physically meaningful values for dispersivity, it will be imperative to determine the distribution of hydraulic conductivity both areally and vertically, where in the past it has been sufficient to determine a spatially averaged value. This is necessary because the nature and extent of heterogeneities within the aquifer determine the magnitude of the dispersion, and dispersivity must be a function of variations in hydraulic conductivity. It will then become necessary to obtain water quality samples at the same scale for which dispersivity is defined.

It should be recognized that the present generation of advection-dispersion models are not entirely deterministic. An assessment of the reliability of these models must await advances in theory and improvements in field techniques including (1) completion of research aimed at obtaining a quantitative definition of dispersivity, (2) development of methods for determining the distribution of heterogeneities within the subsurface and ways of determining the areal and vertical variations of hydraulic conductivity, (3) improvements in field methods for collecting water quality data in three dimensions, (4) development of methods for measuring velocity in the field and ways of estimating spatial variations in porosity for use in calculating velocity indirectly using Darcy's Law.

It is evident that progress will depend a great deal on improvements in our ability to describe the system in the field. In the past, incorrect predictions of models in general have been the result of incorrect definition of the properties of the medium and poor choice of boundary conditions rather than the mathematical modeling scheme.[3,146] At a recent conference on groundwater quality measurement and prediction,[158] the key to future progress was recognized to be improvement in "laboratory and field methods of measurement, sampling and analysis."[158a] It is appropriate that we emphasize field and laboratory studies now because

It is increasingly clear that our ability to solve the appropriate equations has outstripped our ability to describe the physical characteristics of groundwater systems. In many situations we simply do not have the appropriate data to put into the models to answer the questions in a meaningful way. *(page 13 of Reference 222)*

Moreover, a deterministic water quality model "demands data more accurate than any used to date."[190] According to Bredehoeft et al.,[3]

We need, as a minimum, the permeability and porosity of the media and the hydraulic head gradients all in three dimensions. In addition, we need to know the sorptive characteristics of the media along all paths, and we need to estimate the variable rates at which the solidified wastes will enter the transporting fluids. Needed, in particular, is information on the distribution and extent of major heterogeneities. The need for such data severely taxes both the available data base and the technology for generating it. Most of the requisite available data have such large error limits that their usefulness in predictive models is limited. (Philip).[223]

This statement was made with respect to predicting the migration of high level radioactive wastes, but the same kinds of data are needed in other contaminant transport problems as well.

The need for accurate field data is paramount, but we cannot afford to neglect at least two important theoretical aspects of the contaminant transport problem — the need for a theoretical definition of dispersivity and for improvements in the numerical techniques used to solve the dispersion equation. An understanding of the physical meaning of dispersivity will require theoretical input, so that it will be possible to quantify dispersivity as a function of variations in hydraulic conductivity.

Operational advection-dispersion models include the U.S. Geological Survey MOC model,[26] the Intera finite difference model,[49] and a number of finite element models, most notably those developed at Princeton University,[74,120] the University of Waterloo,[46,88] and the U.S. Geological Survey.[48] Van Genuchten[224] listed several others. Appel and Bredehoeft[225] summarized the status of modeling within the U.S. Geological Survey. Recent revisions in the U.S. Geological Survey MOC model[26,73] include several features which improve the accuracy of the solution, but introduce restrictions on the size of the time step required for a stable solution. An analysis of several test problems suggested that the error in the mass balance will be less than 10% for most applications, although comparison with an analytical solution showed some numerical dispersion for a radial flow problem.[73] Galerkin finite element models which use higher order basis functions are relatively immune to numerical difficulties, but require large numbers of equations for an accurate solution to field size problems. Therefore, continued research is needed in identifying numerical procedures which are accurate and computationally efficient for solving large systems of equations.

While there certainly is a need for advection-dispersion models for use in detailed studies of contaminant migration from point sources, there also is a need for models which are simple both conceptually and operationally for use in regional size problems. The effects of dispersion in regional problems will be small compared to advection, and the use of a model which neglects dispersion, thereby avoiding the problem of dispersivity, is valid. Moreover, Gillham et al.[165] concluded that the use of advection-dispersion models to study regional contaminant migration "may not be a useful endeavor."

Flow models divorced from water quality models are useful in estimating the minimum time required for a contaminant to arrive at a certain location in a flow system such as a well or stream. Flow models can also be useful in determining the probable path of migration of a contaminant. However, in order to predict the concentration of a contaminant at a point within the flow system or to predict the average concentration of the contaminant in a system influenced by several contaminant sources, a flow model must be coupled to a water quality model. Coupled flow-water quality models can be classified as distributed parameter models or lumped parameter models.

Only limited success has been reported for basin wide distributed parameter models. Even with the best possible calibration, considerable divergence often exists between simulated and observed concentrations. Calibration could very likely be improved by the use of more and better field data. A smaller scale model[193] designed to predict the migration of leachate from landfill sites has been more successful at reproducing field data,[195] probably because of less variability and better control of the input data. Hassan et al.[189] found that a lumped parameter water quality model applied to determining concentrations of total dissolved solids in large groundwater basins gave results that were "too broad for meaningful interpretation." However, lumped parameter models have been used with apparent success to model nonpoint source pollution of groundwater systems. These models are calibrated by adjusting the amount and concentration of the recharge to the water table. Although the volume of the contaminant applied to the land can usually be estimated with some measure of confidence, the volume of the contaminant which reaches the water table is an unknown with reasonable estimates varying over a relatively wide range. Calibrations for lumped parameter models seem to be better than calibrations for distributed parameter models because in a lumped parameter model, data must be fitted for only one node, while it is difficult to adjust all the parameters in a distributed parameter model in such a way that all the nodes are calibrated. However, if a long time record of observed concentrations is available, it can be expected that the calibration of the lumped parameter model will be valid for certain types of prediction. The calibrated lumped parameter model at the very least can predict relative changes in concentrations under various proposed management strategies and can provide insight into the sensitivity of the system to various stresses.

In general, advection-dispersion models are applied to small scale problems where a detailed description of the contaminant plume is desired, while advection models are applied to regional problems where dispersive transport is relatively small. However, in a few cases, both types of models have been developed to study the same problem. A problem which has received attention by both types of modelers is the build-up of total dissolved solids in groundwater in irrigated agricultural basins. Konikow and Bredehoeft[30] and Helweg and Labadie[32] described applications of an advection-dispersion model to this problem, while others[189-191,197,198] have used advection models. The test of the calibration usually takes the form of graphs showing the temporal variation of observed and simulated concentrations of total dissolved solids at a number of points in the basin. Exact calibration has not been possible for either type of model, although the advection-dispersion models appear to perform somewhat better. This may be due to the fact that for this type of problem, dispersion, although small relative to convection, is not negligible. In fact, at least one advection model[189,190] incorporates the effects of dispersion empirically. However, the superior calibration of the advection-dispersion models for the cases tested may be a result of the availability of better field data.

The use of the Konikow-Bredehoeft model[30,32] is an example of the application of an advection-dispersion model to a regional problem. However, it is also true that advection models have been applied to problems of contaminant migration from point sources.[157,180-184,193] In most of these applications, the contaminant is assumed to move with the groundwater, and the goal is merely to estimate approximate travel paths and travel times rather than to define the shape of the contaminant plume in detail. However, the results of at least one advection model reportedly compare favorably with results from an advection-dispersion model,[195] although the results have not been published in detail in the open literature. Of course, migration of contaminants from point sources has been studied by means of advection-dispersion models as well.[16,29,34-36,48,50,74,113]

In the future, it is expected that both advection-dispersion and advection models will continue to be useful in studying groundwater contamination problems. Advection-dispersion models will be necessary to predict contaminant migration when toxic chemicals are to be disposed of in the subsurface. Successful use of an advection-dispersion model will require a great number and variety of detailed field data. When nontoxic chemicals are involved, it is likely that the expense of acquiring the necessary data for input to an advection-dispersion model will not be tolerable, and in the absence of these data, advection models will be adequate as management tools. It will be necessary to establish a methodology for incorporating chemical reactions into both types of models.

The clearest message contained in the existing body of work dealing with the use of models to simulate the movement of contaminants in groundwater systems is that both types of models, but especially advection-dispersion models, require more detailed and extensive field data than any type of groundwater model used to date. Back and Cherry[91] maintained that the acquisition of reliable data is the "critical limiting factor" in the application of advection-dispersion models to field situations, and according to Hassan et al.[189] in reference to the calibration of an advection model, "The most important task in ground water quality modeling is the development of input data required for the verification of the model." Thus, while a number of contaminant transport models are currently operational, realization of their full potential must await the resolution of problems in acquiring detailed field data, theoretically defining dispersivity, solving the dispersion equation numerically, and incorporating chemical reactions into the modeling scheme.

REFERENCES

[*Editors' Note:* Only the references cited in the preceding excerpts are reprinted here.]

1. **Ballentine, R. K., Reznek, S. R., Hall, C. W.**, Subsurface Pollution Problems in the United States, EPA Technical Studies Report TS-00-72-02, 1972.
2. **Rima, D. R., Chase, E. B., Myers, B. M.**, Subsurface Waste Disposal by Means of Wells — a Selective Annotated Bibliography, U.S. Geological Survey Water Supply Paper 2020, 1971, 305.
3. **Bredehoeft, J. D., England, A. W., Stewart, D. B., Trask, N. J., and Winograd, I. J.**, Geologic Disposal of High-Level Radioactive Wastes — Earth Science Perspectives, Geol. Survey Circular 779, 1978, 15.
4. **deLaguna, W.**, Importance of deep permeable disposal formations in location of a large nuclear-fuel reprocessing plant, in *Subsurface Disposal in Geologic Basins — a Study of Reservoir Strata*, Galley, J. E., Ed., American Association of Petroleum Geologists, Tulsa, Oklahoma, 1968, 21.
5. **Slichter, C. S.**, Field measurements of the Rate of Movement of Underground Water, U.S. Geological Survey, Water Supply Paper 140, 1905, 122.
6. **Taylor, G.**, Dispersion of soluble matter in solvent flowing slowly through a tube, *Proc. R. Soc. London Ser. A*, 219, 186, 1953.
7. **Aris, R.**, On the dispersion of a solute in a fluid flowing through a tube, *Proc. R. Soc. London Ser. A*, 235, 67, 1956.
8. **Bear, J.**, *Dynamics of Fluids in Porous Media*, Elsevier, New York, 1972, 764.
9. **Fried, J. J. and Combarnous, M. A.**, Dispersion in porous media, in *Advances in Hydroscience 7*, 1971, 169.
10. **Scheidegger, A. E.**, Statistical hydrodynamics in porous media, *J. Appl. Phys.*, 25, 994, 1954.
11. **De Josselin de Jong, G.**, Longitudinal and transverse diffusion in granular deposits, *Trans. Am. Geophys. Union*, 39, 67, 1958.
12. **Saffman, P. G.**, A theory of dispersion in a porous medium, *J. Fluid Mech.*, Part 3, 6, 321, 1959.
13. **Saffman, P. G.**, Dispersion due to molecular diffusion and macroscopic mixing in flow through a network of capillaries, *J. Fluid Mech.*, Part 2, 7, 194, 1960.
14. **Bachmat, Y. and Bear, J.**, The general equations of hydrodynamic dispersion in homogeneous, isotropic, porous mediums, *J. Geophys. Res.*, 69, 2561, 1964.
15. **Blake, T. R. and Garg, S. K.**, On the species transport equation for flow in porous media, *Water Resour. Res.*, 12, 748, 1976.
16. **Bredehoeft, J. D. and Pinder, G. F.**, Mass transport in flowing groundwater, *Water Resour. Res.*, 9, 194, 1973.
16a. **Gray, W. G.**, A derivation of the equations of multiphase transport, *Chem. Eng. Sci.*, 30, 229, 1975.
17. **Konokow, L. F. and Grove, D. B.**, Derivation of Equations Describing Solute Transport in Ground Water, U.S. Geological Survey, Water Resources Investigations, 77-19, 1977, 35.
18. **Ogata, A.**, Theory of Dispersion in a Granular Medium, U.S. Geological Survey Professional Paper, 411-I, 1970, 34.
19. **Scteidegger, A. E.**, General theory of dispersion in porous media, *J. Geophys. Res.*, 66, 3273, 1961.
20. **Whitaker, S.**, The transport equations for multiphase systems, *Chem. Eng. Sci.*, 28, 139, 1973.
21. **Fried, J. J.**, *Groundwater Pollution*, Elsevier Scientific, Amsterdam, 1975, 330.
26. **Konikow, L. F. and Bredehoeft, J. D.**, Computer model of two-dimensional solute transport and dispersion in ground water, U.S. Geological Survey Techniques of Water Resources Investigations, Book 7, C2, 1978.
29. **Konikow, L. F.**, Modeling Chloride Movement in the Alluvial Aquifer at the Rocky Mountain Arsenal, Colorado, U.S. Geological Survey Water Supply Paper 2044, 1977, 43.
30. **Konikow, L. F. and Bredehoeft, J. D.**, Modeling flow and chemical quality changes in an irrigated stream-aquifer system, *Water Resour. Res.*, 10, 546, 1974.
32. **Helweg, O. J. and Labadie, J. W.**, Linked models for managing river basin salt balance, *Water Resour. Res.*, 13, 329, 1977.
34. **Robson, S. G.**, Feasibility of Digital Water-Quality Modeling Illustrated by Application at Barstow, Ca., U.S. Geological Survey Water Research Investigation Report 46-73, 1974, 66.
35. **Robson, S. G.**, Application of Digital Profile Modeling Techniques to Ground-Water Solute Transport at Barstow, Calif., U.S. Geological Survey Water Supply Paper 2050, 1978, 28.
36. **Schwartz, F. W.**, On radioactive waste management: model analysis of a proposed site, *J. Hydrol.*, 32, 257, 1977.

46. **Segol, G.**, A three-dimensional Galerkin—finite element model for the analysis of contaminant transport in saturated-unsaturated porous media, in *Finite Elements in Water Resources*, Gray, W. G., Pinder, G. F., and Brebbia, C. A., Eds., Pentech Press, London, 1977, 2.123.
48. **Grove, D. B.**, The use of Galerkin Finite-Element Methods to Solve Mass-Transport Equations, U.S. Geological Survey, Water Resources Investigations, 77-49, 1977, 61.
49. Intercomp. Resource Development & Engineering, Inc., A model for calculating effects of liquid waste disposal in deep saline aquifers, Part 1 — Development, Part 2 — Documentation, U.S. Geological Survey Water Research Investigations, 76-61, 1976.
50. **Ahlstrom, S. W., Foote, H. P., Arnett, R. C., Cole, C. R., and Serne, R. J.**, Multicomponent mass transport model: theory and numerical implementation (discrete-parcel-random-walk version), *Battelle BNWL-2127 (UC-70, UC-32)* prepared for ERDA and the Atlantic Richfield Hanford Co., 1977.
73. **Konikow, L. F.**, Revised computer model of solute transport in flowing ground water, *Trans. Am. Geophys. Union EOS*, 59, (Abstr.), 281, 1978.
74. **Pinder, G. F.**, A Galerkin-finite element simulation of groundwater contamination of Long Island, *Water Resour. Res.*, 9, 1657, 1973.
88. **Pickens, J. F. and Lennox, W. C.**, Numerical simulation of waste movement in steady groundwater flow systems, *Water Resour. Res.*, 12, 171, 1976.
89. **Jackson, R. E., Merritt, W. F., Champ, D. R., Gulens, J., and Inch, K. J.**, The distribution coefficient as a geochemical measure of the mobility of contaminants in a ground water flow system, in *Proc. IAEA Advisory Group Meeting*, Cracow, Poland, 1977.
91. **Back, W. and Cherry, J. A.**, Chemical aspects of present and future hydrogeologic problems, in *Advances in Groundwater Hydrology*, Saleem, Z. A., Ed., American Water Research Association, Minneapolis, 1976, 153.
105. **Rubin, J. and James, R. V.**, Dispersion-affected transport of reacting solutes in saturated porous media: Galerkin method applied to equilibrium-controlled exchange in unidirectional steady water flow, *Water Resour. Res.*, 9, 1332, 1973.
113. **DeMarsily, G., Ledoux, E., Barbreau, A., Marot, J.**, Nuclear waste disposal: can the geologist guarantee isolation?, *Science*, 197, 519, 1977.
120. **Segol, G., Pinder, G. F., and Gray, W. G.**, A Galerkin finite element technique for calculating the transient position of the saltwater front, *Water Resour. Res.*, 11, 343, 1975.
145. **Legrand, H. E.**, Patterns of contaminated zones of water in the ground, *Water Resour. Res.*, 1, 83, 1965.
146. **Theis, C. V.**, Aquifer and models, in *Proc. Natl. Symp. Groundwater Hydrol.*, Marino, M. A., Ed., American Water Resources Association, Urbana, 1967, 138.
157. **Cherry, J. A., Grisak, G. E., and Clister, W. E.**, Hydrogeologic studies at a subsurface radioactive-waste management site in West-Central Canada, in *Underground Waste Management and Artificial Recharge*, American Association of Petroleum Geologists, 1973, 436.
158. Water Research Centre: University of Reading, *Groundwater Quality — Measurement, Prediction and Protection — Papers and Proceedings*, Berkshire, England, 1977, 856.
158a. **Jackson, R. E.**, Conference reports: groundwater quality. Report on the conference entitled, "Groundwater Quality-Management, Prediction and Protection", University of Reading, England, tb3f11976, *Geosci. Can.*, 4, 89, 1977.
165. **Gillham, R. W., Hendry, M. J., Cherry, J. A., Frind, E. O., and Pucovsky, G. M.**, Studies of the agricultural contribution to nitrate enrichment of groundwater and the subsequent nitrate loading to surface waters. I. Field Investigations of the Processes Controlling the Transport of Nitrate in Groundwater, University of Waterloo, 203, 1978.
180. **Freeze, R. A.**, Subsurface hydrology at waste disposal sites, *IBM J. Res. Dev.*, 16, 117, 1972.
181. **Kirkham, D. and Affeck, S. B.**, Solute travel times to wells, *Ground Water*, 15, 231, 1977.
182. **Kirkham, D. and Sotres, M. O.**, Casing depths and solute travel times to wells, *Water Resour. Res.*, 14, 237, 1978.
183. **Cushman, J. and Kirkham, D.**, Solute travel times to wells in single or multiple layered aquifers, *J. Hydrol.*, 37, 169, 1978.
184. **Nelson, W. R.**, Evaluating the environmental consequences of groundwater contamination. I—IV. *Water Resour. Res.*, 14, 409, 1978.
189. **Hassan, A. A., Kleinecke, D. C., Johanson, S. J., and Pierchaea, C. E.**, Mathematical Modeling of Water Quality for Water Resources Management, California Department of Water Resources, Southern District, District Report, 1974.
190. **Weber, E. M. and Hassan, A. A.**, Role of models in groundwater management, *Water Resour. Bull.*, 8, 198, 1972.
191. **Maddaus, W. O. and Aaronson, M. A.**, A regional groundwater resource management model, *Water Resour. Res.*, 8, 231, 1972.

193. **Elzy, E., Lindstrom, T., Boersma, L., Sweet, R., and Wicks, P.**, Analysis of the Movement of Hazardous Waste Chemicals in and from a Landfill Site via a Simple Vertical-Horizontal Routing Model, Special Report 414, Oregon State University, Corvallis, 1974, 109.
195. **Crutcher, A. J. and Sykes, J. F.**, Modelling sanitary landfill leachates using a chemical mass balance technique, *Trans. Am. Geophy. Union EOS*, 58(Abstr.), 386, 1977.
197. **U.S. Bureau of Reclamation**, Prediction of Mineral Quality of Irrigation Return Flow: Vol. 1, EPA-600/2-77-179e, 1977, 59.
198. **McLin, S. G. and Gelhar, L. W.**, Hydrosalinity Modeling of Irrigation Return Flow in the Mesilla Valley, New Mexico, paper presented at the Int. Conf. on Managing Saline Water for Irrigation, Texas Technical University, Lubbock, 1976.
206. **Todd, D. K. and McNulty, D. E. O.**, *Polluted Groundwater: A Review of the Significant Literature*, Water Information Center, Inc., Huntington, New York, 1976, 179.
207. **Summers, W. K. and Spiegel, Z.**, *Ground-Water Pollution: A Bibliography*, Ann Arbor Science Publication, Inc., Ann Arbor, 1974, 83.
208. **Kaufman, W.J.**, Chemical pollution of groundwaters, *J. Am. Water Works Assoc.*, 66, 152, 1974.
209. **Hall, E. S.**, Some chemical principles of groundwater pollution, in *Groundwater Pollution in Europe*, Cole, J. A., Ed., Water Information Center, New York, 96, 1972.
210. **Back, W. and Hanshaw, B. B.**, Comparison of chemical hydrogeology of the carbonate peninsulas of Florida and Yucatan, *J. Hydrol.*, 10, 330, 1970.
211. **Back, W. and Hanshaw, B. B.**, Rates of physical and chemical processes in a carbonate aquifer, in *Non Equilibrium Systems in Natural Water Chemistry*, Advances in Chemistry Series 106, American Chemical Society, Washington, D. C., 1971, 77.
212. **Schwartz, F. W. and Domenico, P. A.**, Simulation of hydrochemical patterns in regional groundwater flow, *Water Resour. Res.*, 9, 707, 1973.
213. **Plummer, L. N.**, Defining reactions and mass transfer in part of the Floridan aquifer, *Water Resour. Res.*, 13, 801, 1977.
214. **Truesdall, A.H. and Jones, B. F.**, WATEQ — a computer program for calculating chemical equilibria of natural water, *J. Res. U.S. Geol. Surv.*, 2, 233, 1974.
215. **Plummer, L. N., Jones, B. F., and Truesdell, A. H.**, WATEQF — A Fortran IV Version of WATEQ, U.S. Geological Survey Water Resources Investigations, 76-13, 1976, 61.
216. **Kunkle, G. R. and Shade, J. W.**, Monitoring ground-water quality near a sanitary landfill, *Ground Water*, 14, 11, 1976.
217. **Wigley, T. M. L.**, A Computer Program for Water Quality Analysis, Technical Note 15, University of Waterloo, Canada, 1972, 16.
218. **Griffin, R. A., Cartwright, K., Shimp, N. F., Steele, J. D., Ruch, R. R., White, W. A., Hughes, G. M., and Gilkeson, R. H.**, Attenuation of pollutants in municipal landfill leachate by clay minerals. I. Column leaching and field verification, Illinois State Geological Survey Environmental Geology Notes, 78, 1976, 34.
219. **Griffin, R. A., Frost, R. R., Au, A. K., Robinson, G. D., and Shimp, N. F.**, Attenuation of pollutants in municipal landfill leachate by clay minerals. II. Heavy metal adsorption, Illinois State Geological Survey Environmental Geology Notes 78, 1977, 47.
220. **Reardon, E.J. and Fritz, P.**, Computer modelling of groundwater ^{13}C and ^{14}C isotope compositions, *J. Hydrol.*, 36, 201, 1978.
221. **Gulens, J., Champ, D. R., and Jackson, R. E.**, Redox processes in ground water in, *Hydrological and Geochemical Studies in the Perch Lake Basin: A Report of Progress*, Barry, P. J., Ed., Atomic Energy of Canada Limited, Chalk River Nuclear Laboratories, AECL-5836, 1977, 67.
222. **Bredehoeft, J. D.**, Status of quantitative groundwater hydrology, in *Advances in Groundwater Hydrology*, Saleem, Z. A., Ed., American Water Resources Association, Minneapolis, 1976, 8.
223. **Philip, J. R.**, Some remarks on science and catchment prediction, in *Prediction in Catchment Hydrology; Natl. Symp. Hydrology*, Chapman, T. G. and Dunin, F. X., Eds., Australian Academy Science, Canberra, 1975, 23.
224. **VanGenuchten, M. T.**, Simulation models and their application to landfill disposal siting; a review of current technology, in *Proc. 4th Annu. Hazardous Waste Management Symp.*, San Antonio, Texas, 1978.
225. **Appel, C. A. and Bredehoeft, J. D.**, Status of Ground-Water Modeling in the U.S. Geological Survey, U.S. Geological Survey Circular 737, 1976, 9.

ERRATUM

Page 142, line 22 should read: "and experimental models." Legrand[145]. . . ."

AUTHOR CITATION INDEX

Aaronson, M. A., 394
Adams, C. S., 78, 231
Adams, J. A., 178
Addington, A. R., 231
Adey, W., 171
Adkins, W. S., 78
Affeck, S. B., 394
Ahlstrom, S. W., 394
Akerlof, G. C., 290
Alberty, R. A., 292
Alderman, A. R., 275
Alderman, G. R., 326
Alexander, G. B., 140
Alexander, W. H., Jr., 307
Alter, C. M., 126
Anders, R. B., 307
Anderson, G. M., 140
Anderson, E. T., 128
Andrussow, N., 21
Appel, C. A., 395
Aris, R., 393
Arnett, R. C., 394
Arnston, R. H., 231
Au, A. K., 395

Baas Becking, L. G. N., 5, 176
Bachmat, Y., 342, 393
Back, W., 5, 6, 11, 94, 95, 140, 176, 177, 192, 206, 231, 255, 275, 276, 296, 307, 342, 383, 394, 395
Badiozamani, D., 94, 158, 171
Badon Ghyben, W., 101
Baertschi, P., 325
Baes, C. F., Jr., 291, 292
Bailey, E. H., 125
Baily, S. M., 292
Baker, E. G., 140
Baker, E. T., Jr., 307
Baker, G. C., 29
Ballentine, R. K., 393
Bancroft, A. R., 325
Banks, R. B., 342

Banwell, C. J., 125
Baranov, K. A., 140
Barbreau, A., 394
Barnes, H. L., 290
Barnes, I., 177, 178, 192, 206, 255, 275, 276, 308
Bartholomew, C. H., 291
Bates, R. G., 290
Bates, T. R., 292
Bathurst, R. G. C., 158, 171
Baylis, J. R., 49
Beales, F. W., 141
Bear, J., 94, 158, 171, 342, 369, 393
Bechamp, A., 24
Becker, G. F., 23
Beede, J. W., 231
Belcher, D., 49
Bell, R. P., 255, 290
Bergstrom, R. E., 352
Berner, R. A., 127, 158, 171, 290
Bernstein, F., 125
Berry, F. A. F., 325
Beyerinck, M. W., 20
Bickley, F., 96, 178
Biesenberger, K., 30
Billings, G. K., 94, 326
Birman, J. H., 351
Biscardi, C. J., 141
Bischoff, K., 383
Bjorklund, L. J., 140
Black, A. P., 140
Blake, T. R., 393
Blyth, C. R., 125, 326
Boato, G., 125
Bock, E., 140
Bock, H., 232
Boersma, L., 394
Bögli, A., 140, 158, 171, 232
Boleneus, D., 171
Bond, G. W., 11
Bormann, F., 383
Bosworth, R. C. L., 110

Author Citation Index

Boussingault, J. B., 78
Brackett, R. N., 31
Bradley, E., 140
Bray, E. E., 127
Bray, U. B., 292
Bredehoeft, J. D., 125, 326, 351, 369, 393, 395
Bretz, J. H., 158, 232
Bricker, O. P., 96, 171, 178, 290, 292
Bridgman, S., 276
Brooks, R., 275
Broom, M. E., 192
Brown, G., 297
Brown, J. S., 94
Brown, R. H., 94
Brown, T. H., 383
Brown, W. E., 290
Bruevich, S. V., 125
Buchanan, J. Y., 23
Buneev, A. N., 125
Burnham, C. W., 140
Burnhan, W. L., 95
Buswell, A. M., 256
Butts, C., 255
Bykova, V. S., 127

Canbarnous, M. A., 171
Carlston, C. W., 101
Carpenter, A., 94
Carrier, G. F., 158
Carroll, D., 5, 140, 232
Cartwright, K., 342, 351, 395
Caruccio, F. T., 255
Case, L. C., 125
Cederstrom, D. J., 78
Champ, D. R., 394, 395
Charles, M. E., 327
Chase, E. B., 393
Chave, K. E., 125, 275, 276
Chebotarev, I. I., 5, 125
Cherry, J. A., 177, 394
Cherry, R. N., 206, 383
Chilingar, G. V., 125
Christ, C. L., 140, 158, 177, 255, 276, 290, 291, 383
Christensen, J. J., 291
Chughtai, A., 290
Clark, J. H., 255
Clark, T., 12
Clarke, F. E., 177, 192
Clarke, F. W., 19, 23
Clayton, R. N., 94, 326, 327
Cleaves, E. T., 177
Cloke, P. L., 177
Cloud, P. E., 275
Cobble, J. W., 255

Cole, C. R., 394
Collins, W. D., 78
Columbus, N., 140
Combarnous, M. A., 393
Cook, F. D., 327
Cooke, C. W., 78
Cooper, H. H., Jr., 94, 95, 96, 101, 110, 369
Coplen, T. B., 94, 95
Corps, E. V., 62
Corwin, G., 140
Counts, H. B., 94
Cowart, J. B., 296
Craig, H., 5, 125, 296, 308, 326, 338
Crawford, J. G., 125
Criss, C. M., 255
Croft, M. G., 11
Cropper, W. H., 191
Cushman, J., 394

Dagan, G., 94, 96
Daly, R. A., 171
Daniels, F., 275
d'Ans, J., 140
Dansgaard, W., 296, 326, 338
Darton, N. H., 78
Davies, C. W., 290
Davies, P. J., 171
Davies, W. E., 232
Davis, D. A., 97
Davis, D. E., 256
Davis, E. M., 308
Davis, G. H., 296, 338
Davis, R., 232
Davis, R., Jr., 158, 255
Davis, S. N., 276
Davis, W. M., 232
Day, P. R., 110
Deffeyes, K. S., 276
Degens, E. T., 127, 326
Deike, R. G., 95, 255
De Josselin de Jong, G., 342, 393
deLaguna, W., 393
DeMarsily, G., 394
Denbigh, K., 291
DeSitter, L. U., 125, 140
Dickson, F. W., 140
Dincer, T., 296, 338
Dittmar, W., 13
Dole, R. B., 6, 11
Dolzer, W. L., 291
Domenico, P. A., 395
Dominick, T. F., 171
Don, J., 30
Donsky, E., 94
Doyle, W. W., 97
Du Commun, J., 101

Author Citation Index

Dreyer, R. M., 178
Druga, T. R., 292
D'yachkova, I. B., 291
Dzhalilov, T. I., 142

Eardley, A. J., 140
Eatough, D., 291
Echelberger, W. F., 256
Edmunds, W. M., 177
Ellington, A. C., 276
Ellis, A. J., 125, 140, 290, 291
Elyard, C. A., 292
Elzy, E., 394
Emerson, J., 171
Emery, K. O., 125, 126, 127
Emmons, W. H., 11
Engel, A. E. J., 126
Engel, C. G., 126
Engelhardt, W. von, 126
England, A. W., 393
Epstein, S., 326
Erga, O., 158
Erickson, C. R., 191
Eriksson, E., 110
Etard, A., 19
Eugster, H. P., 94
Evans, H. T., Jr., 177
Evans, W. H., 292
Everhart, D. L., 178

Fabuss, B. M., 141
Fader, S. W., 191
Faivre, R., 275
Farlekas, G. M., 297
Farrel, M. A., 256
Farrett, A. A., 96
Fash, R. H., 125
Faust, C. R., 96
Faust, G. T., 276
Fay, R. C., 127
Feely, H. W., 140
Feldoff, A., 29
Feltz, H. R., 308
Fenner, C. N., 126
Ferguson, G. E., 96
Feth, J. H., 142, 276
Fiedler, A. G., 11
Field, R. M., 171
Finch, J. W., 232
Fisher, D. W., 11, 177
Fokeev, V. M., 142
Folk, R. L., 95, 171
Fontaine, W. M., 78
Foote, H. P., 394
Forman, R. T. T., 171
Foster, M. D., 78, 125, 140, 142, 308

Fournier, R. O., 291
Frantz, J. D., 158
Frear, G. L., 49, 140
Freeze, R. A., 6, 342, 351, 383, 394
Fried, J. J., 171, 393
Friedman, G. M., 158
Friedman, I., 94, 95, 126, 140, 296, 297, 326
Frind, E. O., 394
Fritz, P., 395
Frolov, N. M., 352
Frolova, E. V., 141
Frost, R. R., 395
Fyfe, W. S., 140

Gaida, K. H., 126
Gans, R., 29
Garaway, W. H., 140
Gardner, J. H., 232
Garg, S. K., 393
Garrels, R. M., 6, 96, 127, 140, 158, 177, 178, 191, 206, 232, 255, 276, 291, 383
Garrett, A. A., 96
Gat, J. R., 296
Gates, G. L., 140
Gear, C. W., 383
Gedroiz, K. K., 29
Gelhar, L. W., 394
George, J. H. B., 255, 290
Geraghty, J. J., 96
Giddings, M. T., 255
Gilkeson, R. H., 395
Gillham, R. W., 394
Ginter, R. L., 125
Glaister, R. P., 327
Glover, E. D., 233
Glover, R. E., 101
Goalby, B. B., 292
Godfrey, A., 177
Godwin, H., 308
Goldberg, E. D., 158
Goldshteyn, R. I., 141
Goldsmith, J. R., 255, 275, 276
Gorrell, H. A., 126, 326
Gordon, L. I., 125
Gould, S. J., 158
Grady, J. R., 127
Graf, D. L., 94, 95, 140, 255, 275, 276, 326
Gray, W. G., 393, 394
Greene, F. C., 232
Gregg, D. O., 369
Gregory, J. W., 31
Griffin, R. A., 395
Grim, R. E., 126
Gross, M. G., 158
Grove, D. B., 393
Grubbs, D. K., 128

Author Citation Index

Grund, A., 232
Gude, A. J., 291
Gulens, J., 394, 395

Hale, F. E., 49
Halevy, E., 326
Hall, C. W., 393
Hall, E. S., 395
Hall, F. R., 178
Hall, W. E., 126, 140
Halow, I., 292
Hamer, W. J., 291
Hanna, E. M., 291
Hanor, J. S., 171
Hanshaw, B. B., 94, 95, 126, 140, 177, 206, 255, 296, 308, 326, 383, 395
Harder, H., 126
Harleman, D. R. F., 343
Harned, H. S., 158, 232, 255, 291
Harrington, G. L., 11
Harris, H. B., 308
Harris, J., 326
Harris, W. H., 95, 140, 158
Harriss, R. C., 178
Hartman, R. J., 126
Hassan, A. A., 394
Hawkins, D. B., 142
Haynes, D. D., 232
Healy, H. G., 206
Hecht, F., 126
Heck, E. T., 125
Helgeson, H. C., 140, 290, 291, 383
Hem, J. D., 6, 11, 128, 142, 178, 191, 256, 291
Hemingway, B. S., 291
Hemley, J. J., 126, 291
Hendrickson, G. E., 232
Hendry, M. J., 394
Henningsen, E. R., 140
Henry, H. R., 101
Herzberg, A., 101
Hess, A., 276
Hess, P. C., 291
Hess, R. H., 192
Heston, W. M., 140
Heubner, J. S., 232
Hilgard, E. W., 24
Hill, A. E., 141
Hill, G., 95
Hill, R. A., 11, 59
Hill, V. G., 276
Himmelblau, D. M., 383
Hinish, W. W., 255
Hitchon, B., 94, 326, 327
Hodgeman, C. D., 78
Hodgson, G. W., 327
Hofman, H. O., 21

Holland, H. D., 141, 232, 255, 275
Holser, W. T., 126
Horibe, Y., 125
Horne, R. A., 171
Hostetler, P. B., 178, 232, 255, 290, 291, 292
Howard, A. D., 232
Howard, C. S., 78
Hsu, K. J., 178, 206, 276
Hubbert, M. K., 95, 101, 110, 126, 127, 141, 232, 352, 369
Huebner, J. S., 255
Hughes, G. M., 395
Hulsemann, J., 127
Hunt, J. M., 326

Iler, H. K., 140
Inch, K. J., 394
Ingersoll, A. C., 352
Ingersoll, L. R., 352
Ingerson, E., 308
Intercomp. Resource Development & Engineering, Inc., 394
International Atomic Energy Agency, 338
Irelan, B., 141
Irmay, S., 369
Irvine, R., 21
Izatt, R. M., 291

Jackson, R. E., 394, 395
Jackson, S. A., 141
Jacob, C. E., 95, 369
Jacobson, R. L., 158, 255, 256
Jaffe, I., 292
James, R. V., 394
Jamieson, J. C., 275
Jeen, M. H., 338
Jensen, F. W., 125
Johanson, S. J., 394
Johnson, J., 292
Johnson, N., 383
Johnston, J., 49, 78, 140
Johnston, O. C., 171
Johnstone, R. W., 308
Jones, B. F., 94, 95, 158, 171, 292, 395
Jones, E. V., 141
Jones, H. W., 255
Jones, P. H., 192
Jones, R. S., 232
Jones, T., 383
Jones, W. R., 126
Junk, H., 30

Kamensky, G. N., 141
Kanwisher, J., 127
Kaplan, I. R., 5, 176
Kardos, L. T., 256

Author Citation Index

Katz, B. V. E., 296
Katzef, F., 232
Kaufman, M. I., 296
Kaufman, W. J., 110, 395
Kawaguchi, H., 126
Kaye, C. A., 232
Keenan, J. H., 291
Keller, W. D., 291
Kemper, W. D., 126
Kester, D. R., 291
Keys, F., 291
Kharaka, Y., 178
Kharitschoff, K. V., 22
Khodakovskiy, I. L., 291
Kielland, J., 291
Kim, D. H., 338
King, F. H., 232
Kinsey, D. W., 171
Kinsman, D. J. J., 158
Kirkham, D., 394
Kirsipu, T. V., 232, 255
Kittrick, J. A., 291
Kleinecke, D. C., 394
Knauth, L. P., 95
Koenig, L., 191
Koenigsberger, J., 23
Kohout, F. A., 101, 141
Konikow, L. F., 393, 394
Kornfeld, G., 30
Kornicker, L. S., 127
Korzhinskii, D., 126
Kountz, R. R., 256
Kozin, A. N., 126
Kramer, J. R., 126, 383
Krasintseva, V. V., 126
Krauskopf, K. B., 178
Kreitler, C. W., 296
Krejci-Graf, K., 126
Kremát, B., 352
Krouse, H. R., 327
Krumbein, W. C., 178
Kulp, J. L., 140
Kunkle, G. R., 395

Lachenbruch, A. H., 352
Lafon, G. M., 158, 291
Lakin, H. W., 142
Land, L. S., 95, 158, 171
Landergren, S., 126
Landon, R. A., 256
Lane, A. C., 126
Lange, A. L., 62
Langelier, W. F., 59
Langmuir, D., 158, 255, 256, 276, 291
Larsen, E. S., 31
Larson, S. P., 95, 96

Larson, T. E., 11, 256
Latimer, W. M., 191, 275, 291
Lattman, L. H., 256
Lawrence, R. E., 192
Lebedinzeff, A., 20
Ledoux, E., 394
Lee, C. K., 338
Lee, I., 338
Lee, T. S., 141
LeGrand, H. E., 206
Lemberg, J., 28
Lennox, W. C., 342, 394
Leonard, A. R., 141
Lerman, A., 256
Lersch, B. M., 19
Levine, S., 292
Levinson, A. A., 327
Levorsen, A. I., 141
Lewis, J. V., 78
Lewy, B., 78
Libby, W. E., 308
Lietzke, M. H., 291
Likens, G. E., 383
Lim, Y. K., 338
Lind, C. J., 291
Lindgren, W., 95, 126
Lindsay, W. T., Jr., 292
Lindstrom, T., 394
Linford, R. G., 292
Linke, M. F., 291
Linke, W. F., 141
Linn, G. L. E., 191
Lisitsin, A. P., 126
Logan, W. N., 78
Lomtadze, V. D., 126
Lonsdale, J. T., 308
Love, S. K., 96
Lowenstam, H. A., 275
Ludwig, H. F., 59

McCarty, P. L., 256
McCrea, J. M., 327
McCrossan, R. G., 327
MacDonald, G., 96
MacDonald, G. J. F., 275
McFadden, I. M., 291
MacInnes, D. A., 49, 291
McKelvey, J. G., Jr., 126
McKelvey, V. E., 178
Mackenzie, F. T., 96, 158, 171, 178, 292, 383
Macky, W. A., 158
McLin, S. G., 394
McNeal, R. P., 141
McNulty, D. E. O., 394
Maddaus, W. O., 394
Madgin, W. M., 141

Author Citation Index

Maksareva, T. S., 142
Malinin, S. D., 141
Malmberg, C. G., 291
Malott, C. A., 232
Marot, J., 394
Marshall, C. E., 126
Marshall, R., 290
Marshall, W. L., 141, 292
Martell, A. E., 292
Maryott, A. A., 291
Marz, S., 30
Masín, J., 352
Mason, A. C., 97
Mather, J. R., 158
Matson, G. C., 232
Matthews, R. K., 140, 158, 171
Maxey, G. B., 125, 326
Mayeda, T. K., 94, 326
Mazel, M. I., 127
Medvedeva, A. P., 127
Meents, W. F., 94, 95, 140, 326
Meinzer, O. E., 126, 141, 232
Mendenhall, W. C., 6, 11
Mendieta, H. B., 141
Mercer, J. W., 96
Merritt, W. F., 394
Mesmer, R. E., 291
Meyboom, P., 352
Meyer, C. F., 158
Meyer, F. W., 206
Meyer, L., 19
Milliman, J. D., 171
Mills, R. V. A., 126
Milne, I. H., 126
Minor, H. E., 125
Moneymaker, B. C., 232
Monk, C. B., 255
Montoya, J. J., 291
Moore, C. H., 171
Moore, D., 5, 176
Moore, E. S., 255
Moore, G. W., 232
Morgan, J. J., 178
Morita, R. Y., 126
Morris, J. C., 141
Morse, J. W., 158
Mostowitsch, W., 21
Motts, W. S., 141
Mountjoy, W. T., 291
Mühlberg, M., 23
Muma, K. E., 232
Muma, M. H., 232
Munnich, K. O., 6, 296
Murray, J., 21
Murray, R. C., 141
Myers, B. M., 393

Myers, E. A., 256
Mzhachikh, K. I., 126

Nahm, G. Y., 338
Nancollas, G. H., 290
Nelson, R. W., 342, 394
Nesbitt, J. B., 256
Nigrini, A., 383
Nordstrom, D. K., 178
Nriagu, J. O., 291
Nye, S. S., 11

Ogata, A., 342, 393
Ogienko, V. S., 126
Olivier, L., 19
Olsen, H. W., 369
O'Neil, T. J., 171
Oppenheimer, C. H., 127, 128
Orlob, G. T., 110
Orr, W. L., 125, 127
Orville, P. M., 127
Oshery, H. I., 290
Osmond, J. K., 296
Ostrouvmov, E. A., 127
Owen, B. B., 291
Oxburgh, U. M., 232, 255

Palmer, C., 59
Papadopulos, I. S., 95, 351
Parizek, R. R., 256
Parker, G. G., 94, 96, 110
Parker, J. S., 62
Parker, V. B., 292
Parsons, M. L., 352
Pasler, W., 126
Pavlyuchenko, M. M., 127
Payne, B. R., 297, 326
Pearson, F. J., Jr., 308
Penman, H. L., 158
Pennick, J. M. K., 96
Perkins, T. K., 171
Perlmutter, N. M., 96
Pethybridge, A. D., 291
Phalen, W. C., 20
Philip, J. R., 395
Pickens, J. F., 342, 394
Pierchaea, C. E., 394
Pinder, G. F., 96, 369, 393, 394
Piper, A. M., 96, 232, 383
Plauchud, E., 19
Plummer, F. B., 78
Plummer, L. N., 94, 158, 171, 395
Pohl, E. R., 232
Poland, J. F., 96
Popov, A. I., 141
Potilitzin, A., 18

Pourbaix, M. J. N., 6
Power, W. H., 141
Pray, L. C., 141
Pride, R. W., 206
Prue, J. E., 291
Pucovsky, G. M., 394
Pyle, T., 94
Pytkowicz, R. M., 291

Radhakrishna, G. N., 110
Ragone, S. E., 296
Rainwater, F. H., 256
Rankama, K., 127
Rauch, H. W., 256
Raumann, E., 30
Raymahashay, B. C., 141
Reardon, E. J., 395
Reddell, D. L., 342, 369
Redfield, A. C., 326
Reed, W. E., 326
Reeder, S. W., 327
Reesman, A. L., 291
Reistle, C. E., 62
Renick, B. C., 59, 78
Reuter, J. H., 127, 326
Revelle, R., 59
Reznek, S. R., 393
Rhoades, R. F., 232, 233
Richardson, S. H., 127
Rideal, S., 30
Rifai, M. N. E., 110
Riffenburg, H. B., 78
Rima, D. R., 393
Ristvet, B. L., 158, 171
Rittenberg, S. C., 125, 126, 127
Rittenhouse, G., 127
Roberson, C. E., 128, 291
Roberts, H. H., 171
Roberts, W. H., III, 141
Robertson, G. M., 343
Robertson, J. B., 342
Robie, R. A., 178, 256, 291
Robinson, G. D., 395
Robinson, G. W., 78
Robinson, T. W., 308
Robinson, R. A., 290, 292
Robson, S. G., 393
Rodgers, J. C., 142
Roedder, E., 141
Rogers, G. S., 127
Roques, H., 233
Rose, W., 369
Rossini, F. D., 292
Roth, E., 327
Rothmund, V., 30
Round, G. F., 327

Rubey, W. W., 126, 127, 369
Rubin, J., 394
Rubin, M., 95, 140, 206, 255, 296, 297
Ruch, R. R., 395
Runnels, D. D., 158, 256
Russell, R. J., 171
Russell, W. L., 127
Rydell, H. S., 296
Ryzhenko, B. N., 292

Saffman, P. G., 342, 393
Sahama, Th. G., 127
Saltet, R. H., 20
Samedov, N. K., 142
Sanford, S., 96
Sasaki, A., 327
Sato, M., 6, 178, 292
Satterfield, C. N., 141
Sawkins, F. J., 141
Sayles, R. W., 158
Scheidegger, A. E., 233, 342, 369, 393
Schmalz, R. F., 171, 275
Schneider, R., 233, 342
Schoeller, H., 6, 127
Schoen, B., 326
Scholes, S. R., Jr., 158, 232, 255
Schulze, G., 127
Schumm, R. H., 292
Schwartz, F. W., 393, 395
Scofield, C. S., 59
Seaber, P. R., 11
Sedel'nikov, G. S., 141
Segnit, E. R., 141
Segol, G., 393, 394
Seidell, A., 141, 291
Sellards, E. H., 78
Serne, R. J., 394
Shade, J. W., 395
Shamir, U. Y., 343
Shamir, V., 96
Sharma, T., 327
Shaw, D. M., 127
Shaw, D. R., 94, 326, 327
Shawe, D. R., 141
Shimp, N. F., 94, 95, 140, 326, 395
Shishkina, O. V., 126, 127
Shternini, E. B., 141
Shuster, E. T., 256
Siebert, R. M., 290
Siegel, F. R., 275
Siever, R., 127, 275, 291
Sigvaldason, G. E., 127, 128
Sillén, L. G., 292, 383
Simpson, E. S., 343
Sinacori, M. N., 233
Sinnott, A., 96

Sipple, R. F., 233
Sisler, F. D., 128
Skibitzke, H. E., 343
Skinner, B. J., 276
Skinner, H. C. W., 275
Slichter, C. S., 393
Slusher, R., 141
Smith, R. M., 292
Smith, W. O., 158
Snavely, P. D., Jr., 125
Sopper, W. E., 256
Sotres, M. O., 394
Southwell, C. A. P., 62
Spengel, A., 30
Spezia, G., 24
Spiegel, Z., 395
Spiegler, K. S., 126
Spiro, N. S., 127
Stabler, H., 6, 11, 15
Stallman, R. W., 343, 352
Staples, B. R., 290
Starikova, N. D., 127
Staveley, L. A. K., 292
Stearns, H. T., 96
Steele, J. D., 395
Stephenson, L. W., 78
Stewart, D. B., 393
Stoessell, R. K., 171
Stokes, R. H., 292
Stoughton, R. W., 291
Stout, J. W., 256
Strahler, A. H., 158
Strahler, A. N., 158
Stringfield, V. T., 206
Stumm, W., 141, 178, 274
Stutzer, O., 20
Sullivan, E. C., 28
Summers, W. K., 395
Sunada, D. K., 342, 369
Suter, M., 192
Sveshnikova, V. N., 142
Swales, D. A., 141
Swarzenski, W. V., 142
Sweet, R., 394
Sweeting, M. M., 233
Sweeton, F. H., 291, 292
Swinnerton, A. C., 78, 231, 233

Tageeva, N. V., 127
Tamers, M. A., 308
Taylor, G., 393
Templeton, C. C., 142
Terjesen, S. G., 158
Termier, P., 31
Thatcher, L. L., 256, 297
Theis, C. V., 96, 369, 394

Theobald, P. K., 142
Thompson, D. G., 96
Thompson, H. S., 26
Thompson, M. E., 158, 232, 255, 291
Thornthwaite, C. W., 158
Thorsen, G., 158
Thorstenson, D. C., 11, 171, 178
Thrailkill, J. V., 142, 158, 233
Tickell, E. G., 62
Tillmans, J., 49
Todd, D. K., 110, 158, 233, 394
Tóth, J., 352, 383
Trask, N. J., 393
Truesdell, A. H., 158, 171, 291, 292, 395
Tsurikova, A. P., 142
Tsurikova, V. L., 142
Tunell, G., 140
Turcan, A. N., Jr., 191
Turner, F. J., 140
Turner, S. F., 142, 308
Twenhofel, W. H., 78

U.S. Bureau of Reclamation, 394
U.S. Geol. Survey, 233
Upson, J. E., 96, 142

Vacher, H. L., 158, 171
Valyashko, M. G., 127
Van Delden, A., 20
VanGenuchten, M. T., 395
Ve, A., 158
Veatch, J. O., 78
Vennard, J. K., 233
Verhoogen, J., 140
Vernon, R. O., 96, 142
Viesohn, V. W., 49
Vinogradova, E. G., 125
Vogel, J. C., 6, 296, 297
Von Buttlar, H., 297
Von Gumbel, G., 142
Voroshilov, E. A., 142
Vovk, Ts. L., 127

Wagman, D. D., 292
Wait, R. L., 95, 96, 296, 369
Waldbaum, D. R., 178, 291
Walker, A. C., 292
Walker, W. H., 352
Walton, W. C., 352
Ward, P. E., 141
Ward, W. C., 171
Waring, G. A., 78, 128, 142
Warren, M. A., 369
Waterfield, C. G., 292
Way, J. T., 26, 29
Weber, E. M., 394

Weber, W. J., Jr., 274
Weidie, A., 94
Weller, J. M., 233
Wells, R. C., 126
Wendt, I., 297
Wentworth, C. K., 97, 101, 110
Weyl, P. K., 233, 275
Whelan, T., 171
Wherry, E. T., 31
Whitaker, S., 393
White, D. E., 6, 125, 127, 142, 276, 307, 327
White, W. A., 125, 326, 395
White, W. B., 233, 256
White, W. N., 308
Whitehead, H. C., 142
Whitfield, M., 171
Wicks, P., 394
Wiegner, G., 30
Wigley, T. M. L., 158, 171, 395
Wilkins, B., 171
Williams, J. F., 31
Williams, R. E., 352
Williamson, E. D., 78
Wills, J. H., 141
Wilson, S. H., 125
Winograd, I. J., 297, 393
Winogradsky, S., 20

Winslow, A. G., 97
Witherspoon, P. A., 351, 383
Wolfe, T. E., 233
Wollast, R., 158, 292
Wood, L. A., 97
Wood, W., 97
Woodcock, A. H., 326
Woodward, H. P., 233
Wright, J. M., 292
Wright, R. C., 191
Wyllie, M. R. J., 126, 128

Yanick, N. S., 141
Yeatts, L. B., 292
Young, T. F., 291

Zaitseva, E. D., 128
Zaslavsky, D., 369
Zelenov, K. K., 142
Zelinsky, N., 21
Zen, E-an, 292, 326
Zobel, O. J., 352
ZoBell, C. E., 6, 126, 128
Zötl, J., 233
Zotov, A. V., 142
Zyka, V., 128

SUBJECT INDEX

Activity
   calcite, 238
   calculated by WATEQ, 290
Activity coefficient
   calculation with WATEQ, 284
   definition, 23
   electrolytic dissociation, 42
Activity diagrams, 176, 271
Activity product
   calculation using WATEQ, 287
   dolomite, 239
   in relation to concentration of reactants, 37
   use in groundwater, 200, 238
Activity of water, 290
Advection
   definition, 384
   importance to regional dispersion, 390–391
Anion
   and cation balance in chemical analysis, 12–17
   as part of the chemical system, 9
Anion exchange. See Ion exchange
Anion facies, 83. See also Hydrochemical facies
Amorphous silica solubility, 268
Aquifer
   chemical system in, 181–182
   definition, 180
   Floridan, 193
   flow through limestone, 211
   temperature, 344
   temperature change in shallow brine, 349
Aragonite, 143, 203, 260
Artinite, 267

Backflooding
   factor in cave formation, 224
   in phreatic zone, 228
Bacteria
   anaerobic, in brines, 117
   as sulphate reducers, 20–22, 114–115, 117
   enrichment of bromine, 117

Badon-Ghyben and Herzberg principle
   Du Commun, 98 passim
   saltwater encroachment, 91
Base exchange. See also Ion exchange
   dolomitization of limestone, 120
   in groundwater by silicates, 26–35
   minerals effect bicarbonate solubility, 76
   in soil, 26
   for water softening, 29, 31–40
Bedding plane control on cave morphology, 208
Bicarbonate
   content indicates lithology, 66
   in Floridan aquifer, 197
   in oil fields, 113
   in Pennsylvanian, 241, 243
   relation to CO_2, 64–69
   sources of, 63, 70–71
Biologic depletion of Ca in natural water, 264
Biscayne aquifer, freshwater/saltwater interface in, 102 passim
Brine
   history and study of, 92–93, 113, 138, 143
   related to ancient oceans by isotopes, 344
Brucite, 266

Calcite
   metastability of organically produced, 260
   mineralogy, 162
   precipitation of
      effect of flowrate, 222–224
      effect of mixing, 220–222
   saturation, 193, 203, 218, 244
   solubility of, 262
Calcium bicarbonate
   color of, 71
   content as function of CO_2 availability, 69
   exchange to sodium bicarbonate, 68
   in water from sand and gravel beds, 66
Calcium carbonate
   saturation index, 41, 200
   solubility in relation to CO_2 concentration, 64–65

Subject Index

Carbon 13. See also Isotopes
  comparison with carbonate content, 305
  correction of ^{14}C age, 300
  sources of carbonate ions, 295
  values in water from Carrizo, 303
Carbon 14. See also Isotopes
  age of water, 298, 303
  corrections to measured values, 306
  dating of groundwater, 199, 299-301
  half-life, 330
  hydraulic conductivity, 295
  kinetics of dissolution, 295
  rate of entropy change, 295
  saltwater encroachment, 295
  velocity of flow, 295, 298, 308
Carbon Dioxide
  degassing and carbonate precipitation, 38, 159, 168-169, 217-218, 251
  exsolution, 234
  generated by carbonaceous material, 70-71
  model sensitivity to solubility, 375
  role in solubility of calcite, 134, 244
Carbonaceous material, as source bicarbonate, 70-71, 76
Carbonate
  cements, 23-24, 159
  effect of solid phase on equilibrium, 258
  equilibria, 239, 257
  kinetic and biological factors in precipitation of, 259
  saturation, 175
  sediments-composition of recent, 258
Carbonate equilibrium
  models of, 193, 249-252
  solid phase of, 257-258
Cascade effect, 315
Cation
  as component of chemical system, 9
  facies, 86
  field analysis of, 240-241
Cation exchange. See Ion exchange
Cation facies, in Atlantic Coastal Plain, 83
Caves
  formation of, 208
  hypothesis of origin, 209-211
Chemical hydrogeology
  current goals of, 340
  description of system, 371
Chemical kinetics, effect on mass balance, 373-374
Chemical thermodynamics
  carbonate equilibrium, 175
  development of, 4
Chloride
  in oil fields, 113
  sediment membranes, 120

Clay
  as semipermeable membrane, 11, 117-119
  in ion-exchange reactions, 27
Closed-system, 163
Coefficient of dispersion, 167
Color, as indicator, 71, 74
Common ion effect
  influences on equilibrium, 376
  on diagenesis, 137-138
Concentration, effect of flowrate on, 373
Concentration gradient
  effect on zone of dispersion, 106
  in brines, 119
Conductivity, of groundwater mixed with sea water, 147
Connate water
  definition, 115-116
  effect of bacteria on, 21
Contamination of groundwater
  in Brunswick, Georgia, 356
  model of, 395
Continuity equation, 354
Convection, 384

Darcy
  flow, 212
  Law, 2
  law for forced convection, 371
Dating of groundwater. See Isotopes
Debye-Hüchel equation
  Brönstead-La Mer modification of, 43
  carbonate dissolution, 239
  explanation of, 284-286
  MacInnes assumption, 285
Density, saltwater-freshwater interface, 98-100, 102
Depth
  chemical changes in groundwater with, 67-69, 132
  chemical changes in oil field waters, 21
  increased salinity, 113
Deuterium. See also Isotopes
  variation due to mixing, 316-317
Diagenesis
  definition, 129
  due to bacterial activity, 115-116
  magnesium, 152-153
  strontium, 153-154
Diffusion, definition, 105
Diffusion coefficient, 167
Dispersion
  definition, 105, 384
  as transport mechanism, 116-118
Dispersion coefficient, factors controlling 371

Subject Index

Dissociation constants, 218, 238, 240, 278 passim
Dissolution
 of aragonite, 153
 in mixing environment, 129
 and precipitation in mixing environment, 137–138
Dolomite
 precipitation, 263
  as function of Ca/Mg ratio, 252
 saturation, 203, 234
 solubility characteristics control carbonate equilibria, 263
Domepits, 229
Du Commun, Joseph, salt/fresh water balance, 98–101

Eh-pH
 and iron equilibrium, 183
 development of relationship, 174–175
 natural systems, 188
 use of Nernst equation, 182, 289
Entholpy, 278 passim
Environmental isotopes. See Isotopes
Equilibrium, chemical
 application to iron, 180, 190
 aragonite, 193, 203
 definition, 239
 development of approach, 174–175
 dolomite, 205, 245, 253, 264
 effect of solid phase on, 258
 free energy, 183, 200
 metastable, 261
 models, 175, 193
Equilibrium constant, 238–239, 278 passim
 calcite, 202, 238
 dolomite, 240
Evaporation, factor in precipitation of Mg-CO_3 compounds, 265
Evapotranspiration, in Bermuda, 146

Facies, hydrochemical. See Hydrochemical facies
Faraday's Constant, 184, 289
Feldspar, alteration of, 270–272
Floridan aquifer
 carbonate equilibrium in, 193
 flow of saltwater in, 102
 groundwater quality model, 386
 isotope study of, 295
Flow pattern
 effect of pumping on, 181
 effect on soil temperature, 345
 models, 214–216
 types, 208, 210–211

Flow rate
 effect on concentration, 373
 measured by change in soil temperature, 344
Formation water
 definition, 115
 importance to economic geology, 309
Free energy, 183, 200, 278 passim

Ghyben-Herzberg equation, saltwater encroachment, 91
Graphic presentation
 of analytical data, 25, 50–59, 60–62
 development of methods, 9–10
 of Eh-pH, 182–183
 Stiff diagrams, 60–62
 trilinear diagrams, 50–59
Groundwater flow, types of, 210
Gypsum, 176

Hagen-Poiseville Law, 212
Huntite, mineral equilibrium, 262–263
Hydraulic flow
 in coastal aquifer, 106
 in response to potential fields, 132
Hydrocarbons
 effect on water chemistry, 18–24, 113
 as sulfate reducers, 21
Hydrochemical facies
 definition, 80
 development of concept, 10
 in the Floridan aquifer, 196
 types of, 80–83
Hydrochemical patterns, modeling of, 379
Hydrogen and oxygen isotopes. See also Isotopes
 analysis, 311
 change in ratios of, 312
 change in seawater composition of, 315
 content in rainfall, 332–333
 deuterium balance, 315
 environmental interpretation of, 328
 exchange, 232, 319
 fractionation, 310–311, 317
 hydrogen fractionation by shale membranes, 324
 oxygen in exchange with carbonates and water, 232
 sampling, 310, 329
 to study origin of formation waters, 312
 ultrafiltration, 311, 317
 values in formation water, 319, 325
Hydrogen sulfide
 effect on isotopic fractionation in formation water, 319–322
 and field brines, 20

Subject Index

Hydromagnesite, 266
Hydrostatic pressure
  calculation of, 108
  vs. lithologic pressure in saline waters of sedimentary rocks, 124
Hydrothermal solutions
  deposition of ores from, 133–134
  precipitation of calcite from, 133

Interstitial water of marine sediments, 116–117
Ion activity product, calcite and aragonite saturation, 164, 200
Ion exchange
  in clays, 120
  modeling of, 376–377
  process, 26, 63
  simulation of, 10
  in solid solution, 261
Ionic strength
  calculation of with WATEQ, 277
  definition, 43
Iron phases
  effect of bicarbonate activity on, 186
  equilibrium of, 180
  solubility, 174, 186
Isotopes. *See* Carbon 13; Carbon 14; Hydrogen and oxygen isotopes
  to define reactions, 388
  development of use in hydrology, 294
  nitrogen, $^{14}N/^{14}N$ sources of pollution, 295

Kaolinite, 270
Karst, landforms in Bermuda, 145
Kinetics, factor in mineral equilibria, 265

Laminar flow, 212
Lanfordite, 267
Law of mass action. *See also* Mass action
  development of, 4
Limestone
  aquifers, flow through, 211
  bicarbonate content in terranes of, 66
  caves, 208
  effect of pH on solubility of, 217
  factors affecting solubility of, 300
  Floridan aquifer, 193
  hydrogeology in Bermuda, 143–144
  solution in phreatic zone, 217
Limonite, 183

Magnesite, 267
Magnesium calcite, solubility of, 262
Marine organisms, role in Mg content of calcite, 261
Mass action
  equilibrium equations, 277–282
  law, 183, 200
  solution of equations with WATEQ, 286
Mass balance
  with isotopes, 316
  as mass accounting procedure, 373
  silicate dissolution, 176
Mass transfer
  factors in, 371
  significance of in solute transport, 4
  variables used in simulation of, 374
Mass transport
  combined with flow equation, 361
  of fluid, dissolved solids, and heat in aquifers, 353
  importance of equation to development of hydrochemistry, 5
Mechanical dispersion, as element in total dispersion, 105
Membrane, hydrolysis, 120
Metamorphic water, definition, 115
Metastable equilibrium, 261
Methane, as possible agent of sulfate reduction, 22
Milligram equivalents, 15
Mineral dissolution, mass transfer of, 374, 375
Mineral equilibria. *See also* Saturation index; Equilibrium, chemical
  anhydrite, 200, 204
  calcite, 259
  in Floridan aquifer, 200–206
  general, 3
  gypsum, 200, 204
Mixing
  Badon-Ghyben-Herzberg equation, 91
  definition, 166, 312
  as a diagenetic process, 129–130
  effect on calcite, 156, 220–222
  effect on carbonate equilibrium, 234
  effect on isotope composition, 313
  effect on precipitation of calcite, 163
  efficiency of in diagenesis, 135
  examples of, 130–134
  and formation or ore deposits, 134–135
  of natural waters, 130
  and precipitation if iron, 188
Models
  development of chemical, 370
  obstacles to development of geochemical, 388
Molecular diffusion, as element in dispersion, 105–106
Montmorillonite, as semipermeable membrane, 118–119

Nernst equation, 174
  applied Eh-pH, 182

Nesquehonite, 266
Neutral electrolyte, role in diagenesis, 135

Ore, formation of in mixing zone, 134–135
Osmosis, through semipermeable membrane, 111, 118
Oxidation-reduction (redox), 289
Oxygen
  dissolved, 289
  percent saturation, 242

Permutite
  development of, 29
  effect on concentration, 120
  experimental use of, 74–75
pE, 289. See also Eh-pH
pH
  changes in, due to mixing, 131, 137
  controlling factors, 271
  derivation of equation, 38–41
  effect on limestone dissolution, 217–218
  effect on siderite equilibrium, 186
  Floridan aquifer, 179
  as function of air pollution, 373
  as indicator of CO_2 loss, 165
  laboratory compared to field, 251
  in oil field waters, 122
  in relation to carbonate saturation, 38
  and solubility of iron, 186
Phreatic zone
  caves in, 208
  cementation in, 155
  definition, 211
  flow in, 225–277
  loss of head effects solution in, 216
Piezometric surface
  definition, 211
  Floridan aquifer, 194
Pipe network, flow model, 214–216
Piper diagrams. See Trilinear diagrams
Pollution, effect on carbonate saturation index, 245–246
Polymorphism, of carbonates, 258
Potential, fluid in coastal aquifers, 107–109
Potential energy flow of H_2O as of, 180
Precipitation of solid phases
  due to mixing, 131–132
  relation to pH, 137
Pressure
  hydrostatic, 108
  in WATEQ, 288–290
Protodolomite, 264
Pyrite, 183

Quartz, solubility, 268

Radicals, chemical, 7

Radiocarbon. See also Carbon 14
  Floridan aquifer, 199
Rainfall
  in Bermuda, 146
  chemistry of, 149
  isotopic composition, 328 passim
  in St. Croix, 160–161
Rate of reaction
  determination of, 374
  effect on groundwater chemistry, 153–154
  field measurement of, 373
  relation to velocity and flow rate, 386
Reaction coefficient, 8
Redox potential. See also Eh-pH.
  development of concept, 175
  effect of P_{CO_2}, 190
  in WATEQ, 289
Residence time
  calculated by isotopes, 328
  calculated from Tritium content, 336
  effect on CO_2 equilibrium, 219
  importance in mass transfer depends on kinetics, 377
Richard's equation, 372

Salinity, primary, secondary, tertiary, 17
Saltwater, seaward flow of, 103, 109
Saltwater encroachment
  as stimulus to study water chemistry, 10
  recognition of, 90–91
Saltwater/freshwater interface
  in Bermuda, 152
  equation derived by Du Commun, 98–101
  hydrodynamic quality of, 101, 102
  shape of, 103
Saltwater head, conversion to freshwater, 108
Saturation index. See also Equilibria, chemical
  definition, 239
Seawater, changes in isotope composition, 315
Seawater intrusion, as stimulus to study of water chemisty, 3, 10
Semipermeable membranes
  efficiency of shales, 313, 316
  of fine ground sedimentary rocks, 118–124
  membrane theory, 119
  as salt sieves, 114–115
  in shales, 111
Shale
  efficiency, 313, 316
  as semipermeable membrane, 111
Siderite, 183
Silicates
  development of geochemistry, 176
  in natural waters, 268
  weathering of, 268

Subject Index

SMOW, 315, 320. *See also* Isotopes
   definition, 311
Soil
   change in pH due to, 243
   in CO_2 contribution, 145-146, 149, 299
   development of in Atlantic Coastal Plain, 76
   soil temperature, factors affecting, 344
   temperature indicates flow rate, 344
   temperature in highly permeable, 347
   temperature in poorly permeable, 348
Solid solution, factors controlling, 261-264
SOLMNEQ, 176
Solubility. *See also* Saturation index
   of calcite and dolomite, 234
   of calcite, magnesium calcite, and aragonite, 262
   calculation of, 182
Solubility curve
   of calcite, aragonite, and Mg calcite, 262
   shape as a function of added salts, 139
   use of in studying mixing zone effects, 136-139
Solubility product
   calculation of with WATEQ, 287-288
   concentration of reactants ratio, 37
Solute transport equation, development of, 4
Specific capacity, definition, 188
Springs
   isotopes in, 334
   mineral equilibria in water from, 273
Stability diagrams, for silicates, 269-272
Stiff diagram
   description of, 60-62
   uses of, 61-62
Sulfate
   modifies groundwater composition, 376
   in oil field brines, 113-117
   reduction of, 18-23
   reduction of as source of CO_2, 70
Sulfide, oxidation of, 98

Temperature, annual variation in land surface temperature, 345
Temperature effects
   on depletion of stable isotopes, 331
   effect on calcite solubility, 219-220, 224, 234
   effect on rate of reaction, 45-47
   effect on saturation index, 45-47
   on isotope exchange, 317
   on solid solution, 261
   on solution and parts per thousand, 45-47
   as variable in WATEQ program, 288

Texas, Southeastern isotope study area, 251
Thermal water. *See* Hydrothermal solutions
Tide, as pumping mechanism, 166, 168
Trace elements, stabilization of calcite by, 260
Transmissivity, function of sorting and grain size, 181
Transport process, components of, 384
Travertine, 175
Trilinear diagram
   explanation of, 50-59
   flow pattern, 195
   methods of plotting, 52-54
   to show hydrochemical facies, 84
   use to distinguish water types, 54-55
   for water mixtures, 55-58
Tritium. *See also* Isotopes
   evidence of modern recharge, 331
   from nuclear bombs and testing, 330
Turbulent flow, in limestone, 212-213

Vadose Zone
   definition, 211
   flow in, 219
   P_{CO_2} in, 149
   seepage, 217
van't Hoff equation, 200, 288
Vapor phase, factors causing, 219
Velocity
   effect on importance of molecular diffusion, 385
   factor in mass transfer, 166, 167

Water density. *See* Density
   as variable in location of saltwater/freshwater interface, 98-102
Water table
   definition, 211
   fluctuation of, 159, 182
   solution of limestone caves in, 231
WATEQ, input, 290
Weathering, of silicates, 270-272

Yucatan, 176

Zeolites, in loose-exchange reactions, 27-29
Zone of dispersion. *See also* Mixing zone
   development of principle of static equilibrium in, 98-101, 102
   dispersion in, 105-106
   effect on calcite saturation, 156-157
   effects of mixing in, 131-133
   history of study, 93-94
   role of density in, 102

About The Editors

WILLIAM BACK is a research hydrogeologist with the U.S. Geological Survey, Reston, Virginia, and adjunct professor at George Washington University, Washington, D.C. He received the B.S. in geology from the University of Illinois, the M.S. in geology and hydrology from the University of California, Berkeley, the M. Pub. Admin., Conservation of Natural Resources, from Harvard University, and the Ph.D. in hydrogeology from the University of Nevada. Dr. Back has served as an advisor or consultant to the United Nations Development Programme, U.S. Agency for International Development, and Food and Agriculture Organization for the governments of Israel, Poland, Pakistan, Costa Rica, and Bolivia. He has authored or coauthored more than fifty articles, including chapters in *Advances in Hydroscience, Advances in Chemistry,* and *Advances in Groundwater Hydrology.* Dr. Back currently serves on the Karst Commission, International Association of Hydrogeologists, and on the Editorial Advisory Board for *Journal of Hydrology.* He was co-recipient with Dr. Bruce Hanshaw of the Meinzer Award by the Geological Society of America and was later selected to be that society's Birdsall Distinguished Lecturer in Hydrogeology.

R. ALLAN FREEZE is a professor in the Department of Geological Sciences and an associate dean in the Faculty of Graduate Studies at the University of British Columbia in Vancouver, Canada. He received the B.Sc. in geological engineering from Queens University in 1961 and the Ph.D. in civil engineering from the University of California, Berkeley, in 1966. Before joining UBC, he was a research scientist with the Hydrologic Sciences Division of the Canada Inland Waters Branch in Calgary, Alberta, and a research staff member at the IBM Thomas J. Watson Research Center in Yorktown Heights, New York. He is the author of over fifty technical publications in the fields of hydrology, hydrogeology, soil physics, and engineering seepage. He is coauthor with J. A. Cherry of the textbook *Groundwater* published in 1979. Dr. Freeze was awarded the Horton Award by the American Geophysical Union in 1970 and 1972 (with J. A. Banner) and the Meinzer Award by the Geological Society of America in 1974. He served as editor of the journal *Water Resources Research* during the period 1976–1980. He is a Fellow of the Royal Society of Canada.

Benchmark Papers in Geology

Series Editor: Rhodes W. Fairbridge
Columbia University

Volume
1. ENVIRONMENTAL GEOMORPHOLOGY AND LANDSCAPE CONSERVATION, Volume I: Prior to 1900 / *Donald R. Coates*
2. RIVER MORPHOLOGY / *Stanley A. Schumm*
3. SPITS AND BARS / *Maurice L. Schwartz*
4. TEKTITES / *Virgil E. Barnes and Mildred A. Barnes*
5. GEOCHRONOLOGY: Radiometric Dating of Rocks and Minerals / *C. T. Harper*
6. SLOPE MORPHOLOGY / *Stanley A. Schumm and M. Paul Mosely*
7. MARINE EVAPORITES: Origin, Diagenesis, and Geochemistry / *Douglas W. Kirkland and Robert Evans*
8. ENVIRONMENTAL GEOMORPHOLOGY AND LANDSCAPE CONSERVATION, Volume III: Non-Urban / *Donald R. Coates*
9. BARRIER ISLANDS / *Maurice L. Schwartz*
10. GLACIAL ISOSTASY / *John T. Andrews*
11. GEOCHEMISTRY OF GERMANIUM / *Jon N. Weber*
12. ENVIRONMENTAL GEOMORPHOLOGY AND LANDSCAPE CONSERVATION, Volume II: Urban Areas / *Donald R. Coates*
13. PHILOSOPHY OF GEOHISTORY: 1785-1970 / *Claude C. Albritton, Jr.*
14. GEOCHEMISTRY AND THE ORIGIN OF LIFE / *Keith A. Kvenvolden*
15. SEDIMENTARY ROCKS: Concepts and History / *Albert V. Carozzi*
16. GEOCHEMISTRY OF WATER / *Yasushi Kitano*
17. METAMORPHISM AND PLATE TECTONIC REGIMES / *W. G. Ernst*
18. GEOCHEMISTRY OF IRON / *Henry Lepp*
19. SUBDUCTION ZONE METAMORPHISM / *W. G. Ernst*
20. PLAYAS AND DRIED LAKES: Occurrence and Development / *James T. Neal*
21. GLACIAL DEPOSITS / *Richard P. Goldthwait*
22. PLANATION SURFACES: Peneplains, Pediplains, and Etchplains / *George F. Adams*
23. GEOCHEMISTRY OF BORON / *C. T. Walker*
24. SUBMARINE CANYONS AND DEEP-SEA FANS: Modern and Ancient / *J. H. McD. Whitaker*
25. ENVIRONMENTAL GEOLOGY / *Frederick Betz, Jr.*
26. LOESS: Lithology and Genesis / *Ian J. Smalley*

27 PERIGLACIAL PROCESSES / *Cuchlaine A. M. King*
28 LANDFORMS AND GEOMORPHOLOGY: Concepts and History / *Cuchlaine A. M. King*
29 METALLOGENY AND GLOBAL TECTONICS / *Wilfred Walker*
30 HOLOCENE TIDAL SEDIMENTATION / *George deVries Klein*
31 PALEOBIOGEOGRAPHY / *Charles A. Ross*
32 MECHANICS OF THRUST FAULTS AND DÉCOLLEMENT / *Barry Voight*
33 WEST INDIES ISLAND ARCS / *Peter H. Mattson*
34 CRYSTAL FORM AND STRUCTURE / *Cecil J. Schneer*
35 OCEANOGRAPHY: Concepts and History / *Margaret B. Deacon*
36 METEORITE CRATERS / *G. J. H. McCall*
37 STATISTICAL ANALYSIS IN GEOLOGY / *John M. Cubitt and Stephen Henley*
38 AIR PHOTOGRAPHY AND COASTAL PROBLEMS / *Mohamed T. El-Ashry*
39 BEACH PROCESSES AND COASTAL HYDRODYNAMICS / *John S. Fisher and Robert Dolan*
40 DIAGENESIS OF DEEP-SEA BIOGENIC SEDIMENTS / *Gerrit J. van der Lingen*
41 DRAINAGE BASIN MORPHOLOGY / *Stanley A. Schumm*
42 COASTAL SEDIMENTATION / *Donald J. P. Swift and Harold D. Palmer*
43 ANCIENT CONTINENTAL DEPOSITS / *Franklyn B. Van Houten*
44 MINERAL DEPOSITS, CONTINENTAL DRIFT AND PLATE TECTONICS / *J. B. Wright*
45 SEA WATER: Cycles of the Major Elements / *James I. Drever*
46 PALYNOLOGY, PART I: Spores and Pollen / *Marjorie D. Muir and William A. S. Sarjeant*
47 PALYNOLOGY, PART II: Dinoflagellates, Acritarchs, and Other Microfossils / *Marjorie D. Muir and William A. S. Sarjeant*
48 GEOLOGY OF THE PLANET MARS / *Vivien Gornitz*
49 GEOCHEMISTRY OF BISMUTH / *Ernest E. Angino and David T. Long*
50 ASTROBLEMES—CRYPTOEXPLOSION STRUCTURES / *G. J. H. McCall*
51 NORTH AMERICAN GEOLOGY: Early Writings / *Robert Hazen*
52 GEOCHEMISTRY OF ORGANIC MOLECULES / *Keith A. Kvenvolden*
53 TETHYS: The Ancestral Mediterranean / *Peter Sonnenfeld*
54 MAGNETIC STRATIGRAPHY OF SEDIMENTS / *James P. Kennett*
55 CATASTROPHIC FLOODING: The Origin of the Channeled Scabland / *Victor R. Baker*
56 SEAFLOOR SPREADING CENTERS: Hydrothermal Systems / *Peter A. Rona and Robert P. Lowell*
57 MEGACYCLES: Long-Term Episodicity in Earth and Planetary History / *G. E. Williams*
58 OVERWASH PROCESSES / *Stephen P. Leatherman*
59 KARST GEOMORPHOLOGY / *M. M. Sweeting*
60 RIFT VALLEYS: Afro-Arabian / *A. M. Quennell*
61 MODERN CONCEPTS OF OCEANOGRAPHY / *G. E. R. Deacon and Margaret B. Deacon*

62 OROGENY / *John G. Dennis*
63 EROSION AND SEDIMENT YIELD / *J. B. Laronne and M. P. Mosley*
64 GEOSYNCLINES: Concept and Place Within Plate Tectonics / *F. L. Schwab*
65 DOLOMITIZATION / *Donald H. Zenger and S. J. Mazzullo*
66 OPHIOLITIC AND RELATED MELANGES / *G. J. H. McCall*
67 ECONOMIC EVALUATION OF MINERAL PROPERTY / *Sam L. VanLandingham*
68 SUNSPOT CYCLES / *D. Justin Schove*
69 MINING GEOLOGY / *Willard C. Lacy*
70 MINERAL EXPLORATION / *Willard C. Lacy*
71 HUMAN IMPACT ON THE PHYSICAL ENVIRONMENT / *Frederick Betz*
72 PHYSICAL HYDROGEOLOGY / *R. Allan Freeze and William Back*
73 CHEMICAL HYDROGEOLOGY / *William Back and R. Allan Freeze*
74 MODERN CARBONATE ENVIRONMENTS / *Ajit Bhattacharyya and G. M. Friedman*
75 FABRIC OF DUCTILE STRAIN / *Mel Stauffer*
76 TERRESTRIAL TRACE-FOSSILS / *William A. S. Sarjeant*
77 GEOLOGY OF COAL / *Charles Ross and June R. P. Ross*
78 NANNOFOSSIL BIOSTRATIGRAPHY / *Bilal U. Haq*
79 CALCAREOUS NANNOPLANKTON / *Bilal U. Haq*